BASIC ELECTRONICS: theory and practice

J. A. WILSON
Technical Education Advisor · Bank Wilson Services

MILTON KAUFMAN
President, Electronic Writers and Editors

McGraw-Hill Book Company
GREGG DIVISION *New York St. Louis Dallas San Francisco Auckland Bogotá Düsseldorf Johannesburg London Madrid Mexico Montreal New Delhi Panama Paris São Paulo Singapore Sydney Tokyo Toronto*

Library of Congress Cataloging in Publication Data

Wilson, J A
Basic electronics—theory and practice.

Includes index.
1. Electronics. I. Kaufman, Milton, joint author.
II. Title
TK7816.W56 537 76-23248
ISBN 0-07-070670-0

Dedicated to Sharon Wilson for her untiring efforts in connection with this book.

By Milton Kaufman and J.A. Wilson
BASIC ELECTRICITY: THEORY AND PRACTICE

By Milton Kaufman (coauthor)
UNDERSTANDING RADIO ELECTRONICS, Fourth Edition (with Herbert Watson, Herbert Welch, and George Eby)

BASIC ELECTRONICS: THEORY AND PRACTICE

2345678910 MUMU 987

The editors for this book were Gordon Rockmaker and Alice V. Manning, the designer was Tracy A. Glasner, the art supervisor was George T. Resch, and the production supervisor was Rena Shindelman. It was set in Caledonia by Progressive Typographers.

contents

preface

This book is written with the assumption that the reader has a basic knowledge of electricity. It can be used as a self-study book or as a textbook in the classroom.

Since it is concerned with the *practical* aspects of electronics, pure theory and mathematics are held to a minimum. It is a basic training book for anyone who wants to get an overall, basic knowledge of electronics.

The circuits used for examples in the text are taken from all types of electronic equipment, such as home entertainment systems, industrial electronics, communications, and automotive electronics.

The use of long theoretical discussions is avoided. Instead, a "What can you do with it?" approach is used.

Chapter 1 gives an overall view of the electronics industry and explains how some of the basic electronic components and systems evolved. Following the first chapter there are several chapters dealing with the basic components used in electronic circuits. The circuits that use these components are discussed in the next part of the book, and then the operation of complete electronic systems is described in the concluding chapters. Helpful information on troubleshooting is included throughout the book. The last chapter deals specifically with the equipment and procedures used in analyzing and troubleshooting electronic equipment.

The following format is used for each chapter.

Introduction—Each chapter begins with an Introduction which gives readers an overview of what they will learn by reading the material. Special emphasis is placed on the practical things to be covered. At the end of each Introduction section some of the important subjects to be covered in the chapter are listed.

Instruction—This part of the chapter describes the theory and the practical application of the subject. There is a liberal use of illustrations. Frequent summaries of the important points are included so that readers get a review while these points are fresh in their minds.

Programmed Review Questions—The programmed section in each chapter permits readers to review the subject matter covered in the Instruction. If readers have a good grasp of the subject matter covered in the Instruction section, they will be able to move quickly through this part of the chapter. New material and new applications are sometimes introduced in the Programmed Review.

Experiments—The next section of each chapter describes electronic experiments. These experiments can be performed on the circuit boards described in Appendix C or on any "breadboard" type of equipment.

Most of the experiments are directly related to material covered in the chapter. In all cases they demonstrate important facts about electronic circuits and systems.

Self-Test with Answers—This is a multiple-choice test with all answers given. Most of the answers are explained for the benefit of those who may not have answered some of the questions correctly.

We are very grateful for the material and illustrations supplied by manufacturers, and to the technicians who made helpful suggestions. We also wish to thank Sharon Wilson for her dedicated work on the entire manuscript.

J. A. Wilson
Milton Kaufman

what is electronics? 1

INTRODUCTION

The word *electronics* is used so carelessly that it has almost lost its real identity. Those who say "I work in electronics" are giving only a vague idea of their job.

For example, do they design electronic equipment? Then, they are electronics *engineers*. Do they locate and repair troubles in electronic equipment? Then they are electronics *technicians*. Do they turn dials to make electronic equipment work? Then, they are *operators*.

In this chapter we will discuss a basic meaning of the word *electronics*. We will show you how electronics developed from a simple experiment to a science that has affected almost every field of manufacturing and entertainment. Some very important principles of electronics and electricity will be reviewed. These principles will serve as a foundation for your future study of electronic systems.

You will be able to answer these questions after studying this chapter:

- ☐ **What are electrons?**
- ☐ **What is electron current?**
- ☐ **How do you trace a circuit?**
- ☐ **What is required to make a system electronic?**
- ☐ **What were the first electronic components?**
- ☐ **How did radio communications develop?**
- ☐ **What is an audion?**

INSTRUCTION

What Are Electrons?

Suppose you were asked to make a list of all the different types of materials there are in the world. Even if you devoted your life to making the list, it is very doubtful that you could complete the task. You would have to include

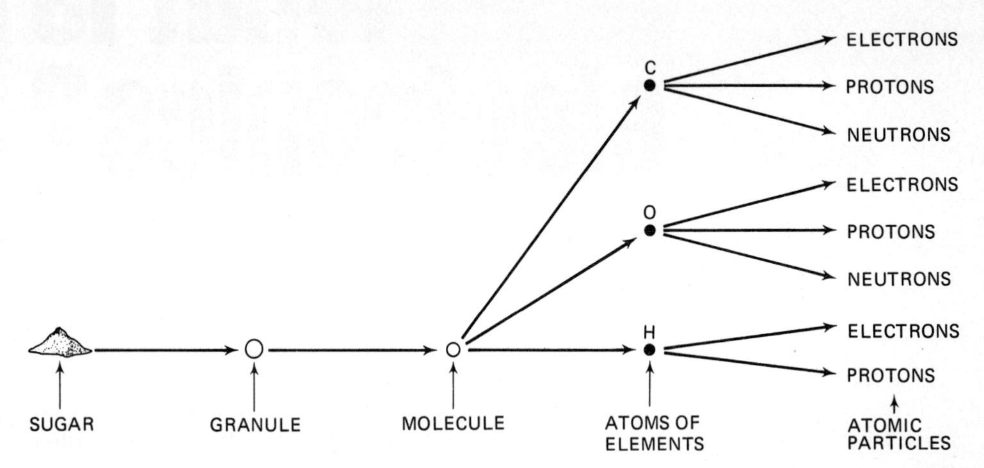

Fig. 1-1 Reduction of sugar to atomic particles.

all the different types of metals, plastics, woods, chemicals, and so on. The list would seem endless.

Early philosophers were concerned with the idea that there must be some very basic things which can be combined to make all the materials that you now know. One of the early theories was that everything in the universe is made of four basic ingredients: *water, fire, air,* and *earth.* It was thought that if you combined these ingredients in the right amounts you could make any known material. The early alchemists spent lifetimes trying to combine them in the right combination to make gold.

We know today that the basic ingredients are *not* air, earth, fire, and water. Actually, there are 92 basic ingredients—called *elements*—which, in various combinations, produce all materials that are known. Every material in the universe is either one of these elements or made by combining these elements.

Hydrogen and oxygen are both elements. By combining them in the right percentages you can make water. Salt is made with the elements sodium and chlorine, and sugar is made from the elements carbon, hydrogen, and oxygen.

You are no doubt familiar with the fact that a spoonful of sugar is made up of very tiny granules. Suppose you took a grain of sugar and began to divide it again and again. Eventually you would reach a point where the tiny particles of sugar could no longer be divided and still be called sugar. This smallest particle of sugar is called a *molecule*. Figure 1-1 shows the relationship between a sugar granule and the molecule. By definition, a molecule is the smallest particle of any material that still retains the properties of that material.

If you could divide the molecule, you would find that it is made up of combinations of elements. In the case of sugar, as shown in Fig. 1-1, dividing the molecule produces small particles of carbon, hydrogen, and oxygen. These are all elements. The smallest particle of an element that still retains the property of the element is called an *atom*.

If you could divide the parts of an atom into smaller particles, you would find out that it is made of three basic particles called *electrons, protons,* and *neutrons*. The difference between one element and another,

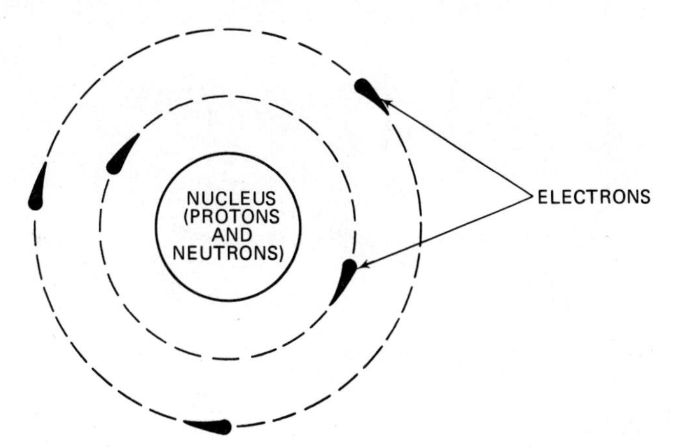

Fig. 1-2 The atom is sometimes compared with the solar system.

such as carbon and oxygen, is the number of electrons, protons, and neutrons that the elements have in their atoms.

No one has ever seen an atom, but scientists have evidence that it is constructed like a miniature solar system. It consists of a nucleus that is made up of protons and neutrons, and around this nucleus there are tiny electrons moving at great speeds. Figure 1-2 shows a drawing of one type of atom as it is believed to be. In an atom, the number of electrons moving around the nucleus is always equal to the number of protons within the nucleus.

Electrons are sometimes thought of as being negatively charged particles. A basic law of electricity is that *unlike charges attract and like charges repel.* Therefore, the negative electrons moving at great speed around the nucleus are held in their orbits by the strong attraction of the positive protons in the center.

In electronics our basic job is to remove some of the electrons from atoms and put them to work. In fact, *electronics* can be defined as being *the job of putting electrons to work.*

What Is Electron Current?

If we can get a large number of electrons to move through a wire, we have an *electric current.* The electron is so small that it takes 6,240,000,000,000,000,000 electrons moving past a point in a wire every second to make a current of one ampere (A). This number is sometimes written as 6.24×10^{18}, which means 6.24 with the decimal place moved to the right 18 places. Electrons are so tiny that it takes about a thousand, million, million, million, million electrons to make a weight of one gram (g), which is $\frac{1}{28}$ of an ounce.

At this time it will be a good idea to review the meaning of some basic terms used in electricity and electronics. Figure 1-3 shows you a comparison of the water system in a car and a simple electric circuit. This type of comparison of two systems is called an *analogy.*

In the water system of Fig. 1-3*a* the water pump is used to force water and antifreeze through the radiator and through the engine. The water picks up heat from the engine and carries it to the radiator, where it is dissipated. The

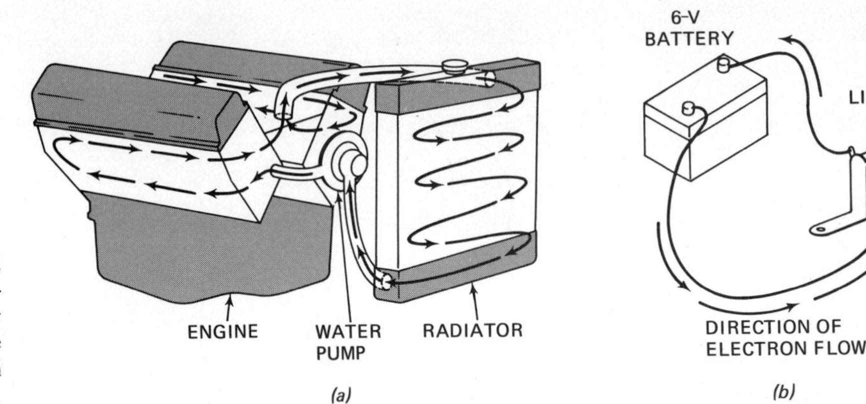

Fig. 1-3 Comparison of water and electrical systems. (*a*) Water cooling system in a car. Arrows show path of water current. (*b*) Basic electric circuit. Arrows show path of electron current.

flow of water is opposed by friction in the radiator and by friction in other parts of the system.

There are three very important features of the water system. The *pressure* for the water is supplied by the pump. The *flow* in the system consists of so many gallons per minute (gal/min) of fluid. The *opposition* is presented by the friction of the radiator and also by the friction of the system.

The electrical system of Fig. 1-3*b* can be compared with the water system. In the electrical system, the *electrical pressure* is supplied by the battery. You could use a generator instead of the battery. In either case, it is convenient to think of the source of voltage as being the electrical pressure that forces current through the circuit.

The electric current consists of a flow of electrons. The opposition to the current flow is due to the circuit resistance. This includes the resistance of the wires and connectors as well as the resistance of the lamp. Usually you can ignore the resistance of the wires and connectors because it is so small.

The relationship between the voltage [measured in volts (V)], the current (measured in amperes) and the resistance [measured in ohms (Ω)] is expressed by Ohm's law:

$$\text{Current in amperes} = \frac{\text{voltage in volts}}{\text{resistance in ohms}}$$

Using symbols,

$$I = \frac{E}{R}$$

where I = the current, in amperes
E = the voltage, in volts
R = the resistance, in ohms

How Do You Trace a Circuit?

An important characteristic of electron-current flow is that the electrons leave the source of voltage at the negative terminal and return to the source at the positive terminal. This is shown in Fig. 1-4.

You can think of the negative terminal as being a place where there is a

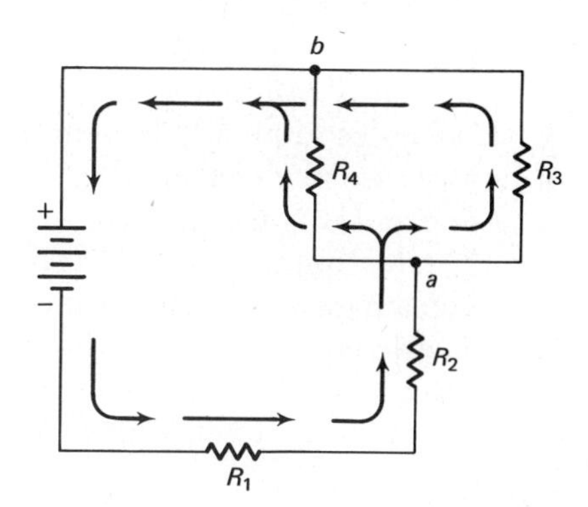

Fig. 1-4 Arrows show electron current paths.

surplus of electrons. They repel electrons in a conductor to that terminal. The positive terminal has a shortage of electrons and it attracts electrons. Therefore, the negative electrons move away from the negative voltage terminal and toward the positive terminal.

When you trace the electron-current path, always start at the negative terminal of the voltage source and end at the positive terminal. If you can trace a complete path, then you have a *closed circuit.* If you cannot get to the positive terminal, then you have an *open circuit.* The process of following the electron path is called *circuit tracing.*

You must be able to trace a circuit in order to be able to locate faulty components.

In Fig. 1-4 the arrows show that current *must* flow through resistors R_1 and R_2 in order to get back to the positive terminal. These resistors are said to be in *series.*

The current divides at point a, and part of it flows through both R_3 and R_4. Resistors R_3 and R_4 are said to be in *parallel.* At point b the electron currents join and return to the positive terminal. Since there is a complete path for current, we have a closed circuit.

What Is Required to Make a System Electronic?

The fact that electrons flow through a system does not necessarily mean that it is an electronic system. Electrons can be made to flow through a wire made of copper or aluminum, but we do not think of a wire as being an electronic device. Likewise, motors and generators operate with electron flow, but they are not considered to be electronic devices. An electronic device not only has a motion of electrons through it, but is also capable of *controlling* that motion. For a definition of an *electronic component,* then, we say that it is a device capable of controlling the number of electrons that pass through it during any given instant of time. Examples of electronic components are vacuum tubes, transistors, FETs (field-effect transistors), and cathode-ray tubes (like the picture tube of a television set).

Summary

1. All the materials in the world are made up of combinations of basic elements.

2. The smallest part of a material that still retains all the properties of the material is called a *molecule.*
3. Molecules are made by combining atoms of the 92 elements.
4. *Atoms* are made of small particles called *protons, neutrons,* and *electrons.*
5. *Electronics* is the science of putting the electron to work.
6. The electron is so tiny that it takes an enormous number of them to produce a measurable amount of current flow.
7. Electric current may be thought of as the flow of electrons, and a voltage source (such as a battery or generator) can be thought of as supplying pressure to force the current to flow through a circuit.

What Were the First Electronic Components?

When Thomas Edison was experimenting with his light bulb, he became concerned with dark deposits on the inside of the glass of the light bulb. In an effort to learn about the nature of these dark deposits, Edison performed an experiment, which is illustrated in Fig. 1-5. Figure 1-5*a* shows a picture of the experiment, and Fig. 1-5*b* shows the same components represented in schematic form. It would be impossible to draw all the complicated electronic systems that we have to work with by using pictures, so the components are generally represented by the type of schematic symbol used in Fig. 1-5*b*. It will be important for you to memorize the schematic symbols so that you will be able to "read" the schematic diagram for electronic systems.

For his experiment Edison connected a battery to the filament. Electric current flowing through the filament caused it to heat and become incandescent—that is, it began to give off light. He placed a metal plate within the glass envelope, hoping to get some of the unwanted deposit on the plate. For some reason Edison connected a microammeter between the plate and one of the filament connectors. To his very great surprise the meter showed that an electric current was flowing. This was contrary to all the basic principles that Edison knew of electricity.

Edison knew, for example, that in order for an electric current to flow, it must be able to leave the source of voltage and it must also always be able to

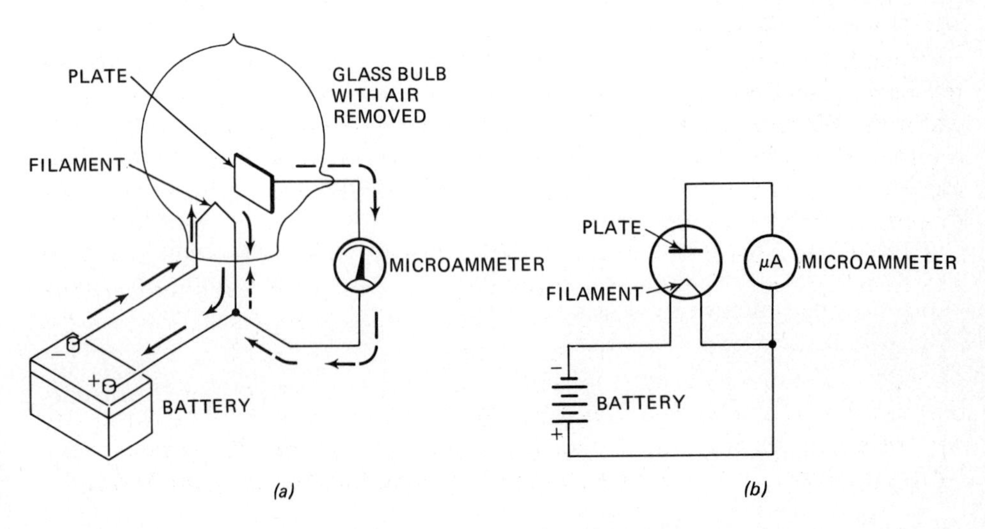

Fig. 1-5 An early Edison experiment. (*a*) Pictorial. (*b*) Schematic.

return to the source of voltage. However, in the simple circuit of Fig. 1-5 there did not seem to be a voltage source to move the current through the microammeter. Furthermore, the plate was placed in a vacuum, so that it appeared to be an open circuit. In other words, there seemed to be no complete path for current flow.

The solid arrows in Fig. 1-5*a* show the path of electric-current flow from the negative terminal of the battery back to the positive terminal. This current, as you know, comprises a flow of a great number of electrons. But remember, the current will only flow provided the circuit is complete—that is, provided there is a conducting path away from the voltage source and returning to the voltage source. This was a basic principle known to Edison, yet the meter showed that there definitely was a current flow as shown by the dotted arrows.

Edison did not pursue this experiment, but did record it. As a result, this flow of electrons and the results of his experiment are referred to as the *Edison effect.*

How Is the Edison Effect Explained?

In Edison's day the idea of electrons and electron flow had not been proposed. Today, we know that when the filament becomes red hot, it boils electrons off its surface. You will remember that the electrons are moving around the nucleus of the atom at high speed. When the atom is heated, the speed increases, and eventually an electron may achieve enough energy to escape the atom. Those electrons on the surface of the filament simply fly off into space, and some of them actually strike the plate, which is located close to the filament.

When an electron leaves an atom in the filament, the atom is short one electron. You will remember that in a normal atom the number of electrons in orbits equals the number of protons in the nucleus. If an electron leaves an atom, the atom is short one electron. It is said then to be *positively charged.* Because many electrons are leaving the heated filament, the filament becomes positively charged. The electrons that strike the plate return to the positively charged filament through the meter. This explains why the meter shows an electron-current flow.

Since Edison put his experiment away and did not try to use it for anything, the basic principle was not used for some time. Later, a man named Sir J. Ambrose Fleming of England invented a similar device and used it in a number of simple circuits. Fleming called his component a *valve* (also called a *diode vacuum tube*) because it allowed the electricity to flow in only one direction. Figure 1-6 shows this clearly.

In Fig. 1-6*a* a battery has been added to Edison's circuit between the plate and the filament. The positive terminal of the battery attracts the negative electrons within the diode (unlike charges attract). This causes a much larger current flow than in the original experiment, shown in Fig. 1-5. You see the direction of electron-current flow from the arrows in the illustration. When the diode is connected into the circuit in such a way that its plate is positive with respect to its filament, it is said to be *forward-biased.* If you reverse the battery as shown in Fig. 1-6*b*, so that the negative terminal is connected to the

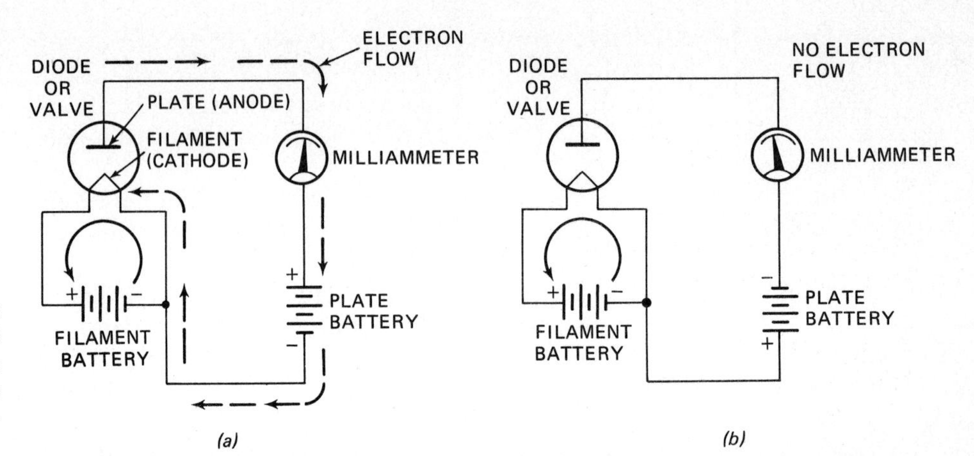

Fig. 1-6 Action of a diode as a valve. (*a*) With a positive voltage connected to the plate, electron current flows. (*b*) No current flows when a negative voltage is connected to the plate.

plate, no current flows. The negative voltage on the plate repels the negative electrons (like charges repel). The electrons are still being emitted by the filament, and they exist within the tube in the form of a cloud of electrons. This cloud of electrons is called a *space charge.* When the diode is connected into the circuit in such a way that its plate is negative with respect to its filament, it is said to be *reverse-biased.*

Fleming compared the operation of the diode with that of a valve which allows current or fluid to flow in only one direction. Electron current can flow away from the plate (toward the battery) when the plate is positive, but it cannot flow toward the plate (away from the battery) when the plate is made negative. As you will see later, this is a very important principle, and it is utilized in many electronic circuits.

In America the device is called a *vacuum-tube diode*—a term that means that it has two electrodes, a plate, and a filament suspended in a vacuum. The filament is often referred to as the *cathode,* and the plate is often referred to as the *anode.* A vacuum is needed to prevent the filament from combining with oxygen, which would cause it to burn up.

The diode was actually the first electronic component, because it not only provided a path for electron flow, but is also provided some control of electrons. In this case it controlled the direction in which the electrons were allowed to flow.

Summary

1. When a filament is heated to *incandescence*—that is, when it is made so hot that it gives off light—it emits electrons. The heated filament must be in a vacuum so it will not burn up.
2. If you put a metal plate in the vacuum, close to the heated filament, some of the emitted electrons will strike the plate.
3. When an electron leaves the filament, the filament takes on a small positive charge. In other words, the filament is missing one electron. Negative electrons reaching the plate will return through the external circuit to the positive filament to equalize this positive charge.

4. A meter connected between the plate and the heated filament will indicate a current flow.
5. If a battery is connected so that it makes the plate positive with respect to the filament, a larger amount of current flows between the plate and filament.
6. If a battery is connected so that it makes the plate negative with respect to the filament, no current flows between the plate and filament.
7. A vacuum-tube diode is made up of a plate and a filament insulated from each other and enclosed in a container from which the air has been removed. The plate is called the *anode*, and the filament is called the *cathode.*

How Did Radio Communications Develop?

Between the years 1865 and 1873 James Clerk Maxwell did a considerable amount of research on the theory of electromagnetic waves. He predicted that it would be possible to transmit these waves, which we call *radio waves,* from one point to another. However, it was not until the year 1887 that Heinrich Hertz was actually able to transmit a radio wave over a short distance. Hertz used an electric spark developed across a gap for producing the radio waves.

In 1896 Marconi first operated his wireless telegraphy system by sending Morse code over a distance of 100 yards (yd). This was the first practical use of radio waves for communication. Despite the success of the early wireless telegraphy experiments, there were many critics who claimed that long-distance communication was not possible. Critics reasoned that the radio waves would go off into space rather than follow the curvature of the earth. They also felt that the radio waves would have the same characteristics as light waves. They knew that because of the curvature of the earth, it is not possible to see beyond a distance much greater than 30 or 35 miles (mi).

Today, we do know that long-distance communication is possible. What the critics did not know is that there is a layer of ions surrounding the earth above the earth's atmosphere. This layer of ions is known as the *Kennelly-Heaviside layer,* and it is also called the *ionosphere.* Relatively low frequency radio signals are reflected back to earth by this ionized layer.

Figure 1-7 shows how long-distance communication is possible. The radio signal goes directly along the earth's surface to a receiver at point *A*. This

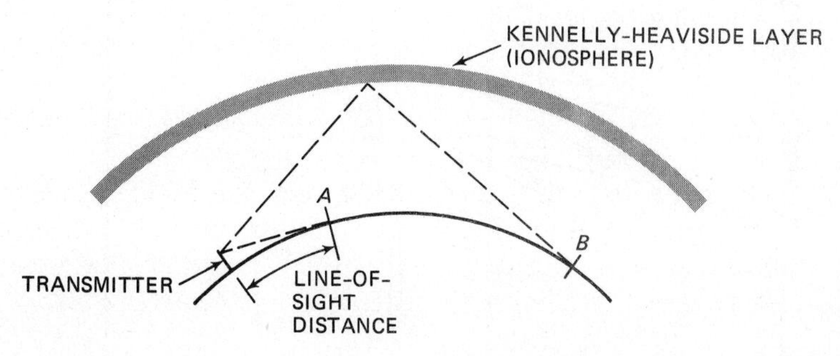

Fig. 1-7 How radio waves are transmitted beyond the line-of-sight distance.

radio signal is called the *ground wave.* Beyond point A, which is at the maximum line-of-sight distance, the wave moves into space.

The transmitter does not just send out a single sharp radio wave. Instead the waves move away from the transmitting antenna in much the same way as light waves travel away from a light bulb. Figure 1-7 shows how one part of the signal reflects from the Kennelly-Heaviside layer and strikes a receiving antenna at point *B*. This point is well beyond the line-of-sight distance from the transmitter.

Reflections from the ionosphere make it possible to receive radio signals at great distances from the transmitter. The first transatlantic broadcast of Morse code occurred in the year 1901. However, it was not until 1904 that Fleming used the diode as a detector of radio signals. In a later chapter we will discuss the operation of radio-signal detectors and you will study how the diode detector works.

What Is an Audion?

Until the year 1906 the means for transmission and reception of radio signals was very crude. The transmitted signal was produced by drawing an electric spark across a gap. The spark produced electromagnetic waves over a wide range of frequencies. Since all the spark-gap transmitters were broadcasting approximately the same range of frequencies, it was not possible to tune to one station and exclude all others. The station most likely to be heard was the one with the largest spark gap. The received signals set up very weak voltages in the receiving antennas, and there was no method of amplifying the signals.

Today each radio transmitter is assigned a specific frequency or range of frequencies, and the law is very strict in holding them to these frequencies.

In the year 1906 DeForest introduced an invention which helped to change completely the method of radio communication. He called his invention the *grid audion,* but today it is known more commonly as a *triode* (short for triode vacuum tube).

Figure 1-8 shows the basic construction of the triode as it was invented by DeForest. You will note that it has a plate and a filament just like the diode

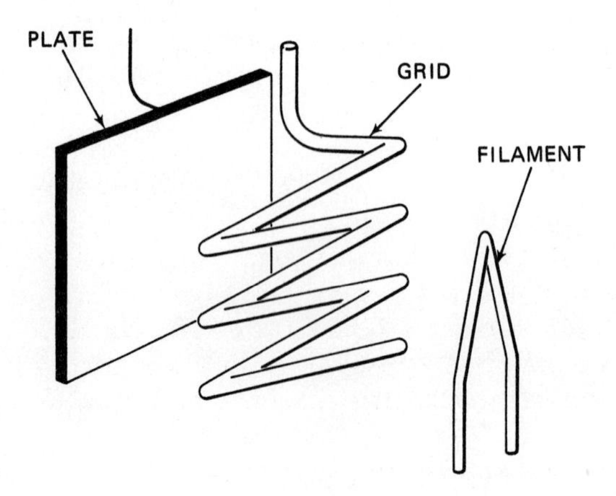

Fig. 1-8 The grid audion.

that you studied before, but an additional grid wire has been inserted between the plate and the filament. This grid wire is called the *control grid.* The filament, control grid, and plate are all housed in a glass or metal enclosure. As you saw with the diode, it is necessary to remove all air from the inside of the enclosure to prevent the hot filament from combining with oxygen. If air gets inside the enclosure, the filament will very rapidly burn up.

Electrons in the triode travel from the heated filament to the positive plate, just as they do in a diode. But they must pass through the grid wires on their way.

If you make the grid highly negative, it will repel the negative electrons back toward the filament (like charges repel). None of the electrons will reach the plate under this condition.

If there is zero voltage on the grid, a large number of electrons will reach the plate.

Thus the number of electrons that reach the plate at any instant depends upon the grid voltage. What is more important is the fact that a small change in negative grid voltage can produce a large change in the number of electrons reaching the plate. It is this important fact that makes *amplification* possible.

The triode made it possible for radio receivers to tune in weak signals from stations a great distance away.

By 1948 the triode and more complex vacuum tubes had become developed to a very high state of refinement. Besides radio communications, the tubes had by then also been used for warfare equipment (radar and sonar) and also for television. Actually the first crude television demonstration was held late in the 1920s.

Although the vacuum tube was largely responsible for advances in radio communication and other electronic devices, it has several disadvantages. Much of the power used to heat the filament is wasted. Furthermore, the large amount of heat generated by systems employing a number of tubes is undesirable. The filaments of tubes burn out periodically, and this means increased maintenance costs.

In 1948 the transistor was announced to the world. Like the vacuum tube, it amplifies weak signals. But that is where the similarity ends. Transistors do not have filaments, so they operate more efficiently, with less heat, and with lower maintenance costs.

Much of your study of electronics will deal with the operation of tubes and transistors in circuits, with more emphasis on the transistor circuits.

Summary

1. The possibility of transmitting and receiving radio waves was predicted by Maxwell, and Hertz actually accomplished it some years later in 1887.
2. Marconi was the first to use radio waves to send Morse-code signals.
3. Long-distance communication is possible because the radio waves are reflected back to earth by the Kennelly-Heaviside layer, which is also known as the ionosphere.
4. Originally radio communication used spark-gap transmitters.

5. Fleming first used the diode as a detector in 1904. This simplified the design of receivers.
6. DeForest put a control grid between the filament and plate of a diode. A small change in negative voltage on the control grid causes a large change in the number of electrons reaching the plate.
7. DeForest called his invention an audion, but later it became better known as a triode.
8. Transistors, like triodes, can amplify signals. Since transistors do not have filaments, they are more reliable than vacuum tubes and they do not generate nearly as much undesirable heat.

INSTRUCTIONS FOR ANSWERING PROGRAMMED REVIEW QUESTIONS

Start with Question 1 in block 1. Answer this question with the choice that you feel is correct. If you think that choice A is correct, proceed to block 17. If, on the other hand, you think that choice B is correct, proceed to block 9.

When you turn to the block indicated by your answer, you will learn whether your answer is right or wrong. If your answer is wrong, you will learn why. If your answer is right, you will get another question to answer.

If you have learned the material in the Instruction section, you will be able to complete this section easily.

PROGRAMMED REVIEW QUESTIONS

We will now review the important concepts of this chapter. If you have understood the material, you will progress easily through this section. Do not skip this material, because some additional theory is presented.

1. Long-distance radio communication is possible because the radio waves are reflected back to earth by a layer of ions above the earth's atmosphere. This layer of ions is called
   A. the Kennelly-Heaviside layer. (Proceed to block 17.)
   B. the ionic reflectometer. (Proceed to block 9.)

2. *Your answer to the question in block 16 is* **A**. *This answer is wrong. Ohm's law is a math formula that tells how voltage, current, and resistance are related.* Proceed to block 5.

3. *Your answer to the question in block 20 is* **B**. *This answer is wrong. Protons are very heavy compared with electrons. In fact, they are more than 1800 times heavier. Only the lighter electrons are free to move in a wire.* Proceed to block 15.

4. *The correct answer to the question in block 23 is* **B**. *Fleming was the*

man who developed the diode and used it as a detector for radio signals. Here is your next question.

A spark-gap transmitter produces a radio wave that covers a wide range of frequencies. This type of transmission has not been used for many years. Which of the following best describes the reason that spark-gap transmitters are no longer used?

A. They cost too much to produce. (Proceed to block 24.)

B. They take up so much of the radio spectrum that only a few stations could be on the air at the same time. (Proceed to block 12.)

5. *The correct answer to the question in block 16 is* **B**. *The illustration shows the experimental setup for the Edison effect.*

Ohm's law says that the amount of circuit current I (in amperes) equals the voltage E divided by the resistance R. This is shown by the equation $I = E/R$. Here is your next question.

Electron current flows

A. from the negative terminal of the voltage source to the positive terminal. (Proceed to block 18.)

B. from the positive terminal of the voltage source to the negative terminal. (Proceed to block 22.)

6. *Your answer to the question in block 12 is* **A**. *This answer is wrong. A triode amplifies a small signal into a large signal. The ability of a negative grid voltage to stop electron flow is not called amplification.* Proceed to block 25.

7. *Your answer to the question in block 17 is* **A**. *This answer is wrong. A diode has only a plate (or anode) and a filament (or cathode). It does not have an element to control the flow of electrons from cathode to anode.* Proceed to block 20.

8. *Your answer to the question in block 26 is* **A**. *This answer is wrong. Ammeters should not be thought of as being a source of electrical pressure.* Proceed to block 23.

9. *Your answer to the question in block 1 is* **B**. *This answer is wrong. The Kennelly-Heaviside layer is called the* **ionic layer** *or* **ionosphere,** *and it reflects radio waves back to earth, but it is not known as an ionic reflectometer.* Proceed to block 17.

10. *Your answer to the question in block 23 is* **A**. *This answer is wrong.* **Audion** *is the name that DeForest gave to his triode tube.* Proceed to block 4.

11. *Your answer to the question in block 21 is* **B**. *This answer is wrong. There are many different uses of the word* **electronics,** *but it is not used in reference to the combination of elements.* Proceed to block 16.

12. *The correct answer to the question in block 4 is* **B**. *There is not an unlimited number of radio and television stations that can now go on the air, even though each station is assigned a wavelength. With spark-gap transmission, only a few stations could be on the air at one time.* Here is your next question.
Which of the following best describes the characteristic of triodes that makes it possible for them to amplify?
A. The control grid can be made negative enough to stop electron flow within the tube. (Proceed to block 6.)
B. A small change in grid voltage causes a relatively large change in plate current. (Proceed to block 25.)

13. *Your answer to the question in block 25 is* **B**. *This answer is wrong. The size of the amplifying component cannot be directly related to the signal.* Proceed to block 19.

14. *Your answer to the question in block 18 is* **A**. *This answer is wrong. Start at the negative terminal of the battery in Fig. 1-11 and trace the electron-current path back to the battery.* Then proceed to block 26.

15. *The correct answer to the question in block 20 is* **A**. *One way to trace a circuit is to follow the path of the electron current from the negative terminal of the battery to the positive terminal.* Here is your next question.
Will the milliammeter of Fig. 1-9 indicate a current flow?
A. Yes. (Proceed to block 27.)
B. No. (Proceed to block 21.)

16. *The correct answer to the question in block 21 is* **A**. *The study of how elements are combined to make different materials is called* ***chemistry***. *In electronics, the object is to find ways to put electrons to work for us.* Here is your next question.
A metal plate is placed in a vacuum with a heated filament. Current flows when a wire is connected from the plate to the filament. Figure 1-10 shows the circuitry. This is known as
A. Ohm's law. (Proceed to block 2.)
B. the Edison effect. (Proceed to block 5.)

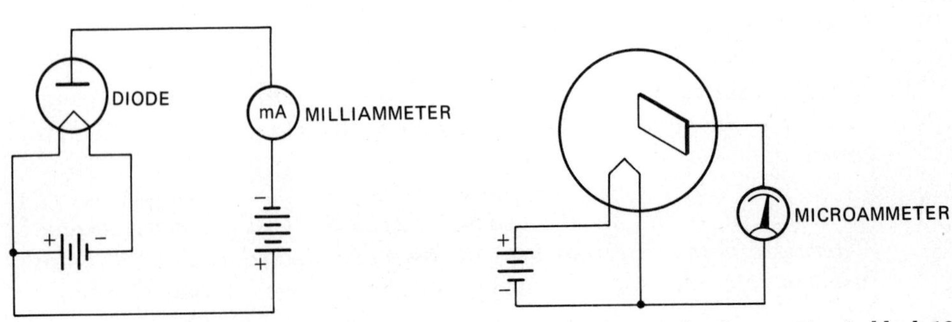

Fig. 1-9 Circuit for the question in block 15.

Fig. 1-10 Circuit for the question in block 16.

17. *The correct answer to the question in block 1 is* **A**. *The ionosphere, as it is more commonly called, is in the upper atmosphere more than 60 miles above the earth's surface. It actually consists of several individual layers which are identified by letters of the alphabet. For example, the D layer is present only during daylight hours. The F layer is divided into two layers in the daytime, designated as* F_1, *which is 140 miles above the earth's surface, and* F_2, *which is about 200 miles above the earth's surface. These layers merge into one layer at night. The E layer is about 70 miles above the earth. It also disappears after sunset.*

Each of these layers has an influence on communications. Radio technicians have learned to predict accurately the chances of carrying on long-distance communication by studying the characteristics of these ionized layers. Here is your next question.

The *grid audion*, which was invented by DeForest, is considered by many to be the start of the electronics industry. Today his invention is more commonly referred to as the

A. diode. (Proceed to block 7.)
B. triode. (Proceed to block 20.)

18. *The correct answer to the question in block 5 is* **A**. *At one time electric current was always traced through the circuit from the positive terminal of the voltage source to the negative terminal. This is called* ***conventional current flow***. *In this book we will follow the path of electrons. In other words, we will use* ***electron current flow***, *which is negative to positive.* Here is your next question.

In Fig. 1-11 the direction of electron-current flow through R_2 is

A. from point *a* toward point *b*. (Proceed to block 14.)
B. from point *b* toward point *a*. (Proceed to block 26.)

19. *The correct answer to the question in block 25 is* **A**. *The cooler operating temperature is a very important advantage of the transistor over the tube. The fact that the transistor does not have a filament means that it not only operates cooler, but also is more reliable.*

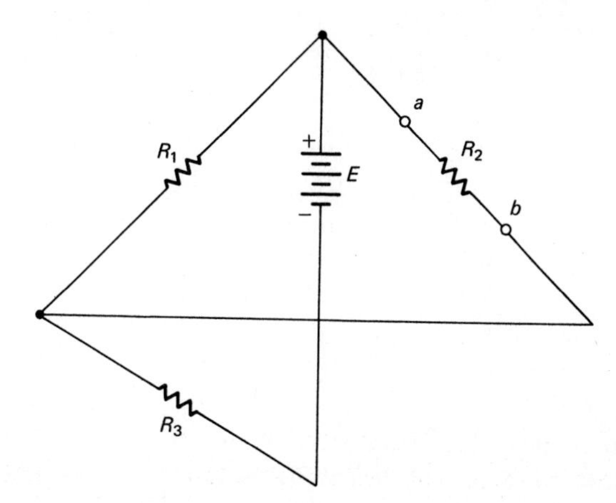

Fig. 1-11 **Circuit for the question in block 18.**

Here is your next question.
Is a dc motor an example of an electronic component? (Proceed to block 28.)

20. *The correct answer to the question in block 17 is* **B**. *Since the invention of the triode, additional grids have been added between the control grid and the anode to achieve special effects in some types of tubes.* Here is your next question.
An electric current flowing in a wire can be considered to be a flow of
A. electrons. (Proceed to block 15.)
B. protons. (Proceed to block 3.)

21. *The correct answer to the question in block 15 is* **B**. *The negative voltage on the plate of the diode repels electrons, and no electron current flows in the plate circuit.* Here is your next question.
Which of the following best describes the word *electronics?*
A. It is a study of the electron and how electrons can be put to work. (Proceed to block 16.)
B. It is a study of how the 92 basic elements can be combined to make all the materials in the world. (Proceed to block 11.)

22. *Your answer to the question in block 5 is* **B**. *This answer is wrong. In order to trace electron current, you start at the negative terminal of the voltage source and go through the circuit to the positive terminal.* Proceed to block 18.

23. *The correct answer to the question in block 26 is* **B**. *A battery or generator supplies voltage. An ammeter is an instrument used for measuring current.* Here is your next question.
Another name for the Fleming valve is
A. audion. (Proceed to block 10.)
B. vacuum-tube diode. (Proceed to block 4.)

24. *Your answer to the question in block 4 is* **A**. *This answer is wrong. Cost was not a factor in outlawing spark-gap transmission.* Proceed to block 12.

25. *The correct answer to the question in block 12 is* **B**. *With the triode, and also with the transistor, a small input signal produces a large output signal. This is called* **amplification**. Here is your next question.
Which of the following is an advantage of a transistor over a triode vacuum tube?
A. The transistor does not operate with a filament, so it runs cooler and with fewer maintenance problems. (Proceed to block 19.)
B. Since the transistor is smaller than the vacuum tube, it can be used to amplify smaller signals. (Proceed to block 13.)

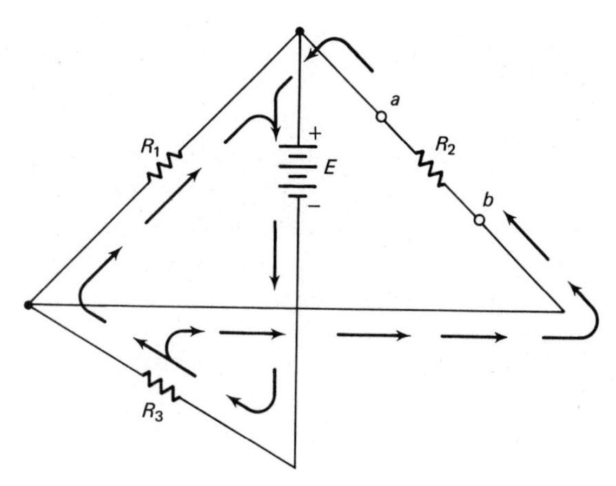

Fig. 1-12 Circuit for the answer in block 26.

26. *The correct answer to the question in block 18 is* **B**. *Figure 1-12 shows the electron current paths. Note that* R_1 *and* R_2 *are in parallel.*
Here is your next question.
The electrical pressure for forcing a current to flow in a circuit is supplied by
A. an ammeter. (Proceed to block 8.)
B. a battery or a generator. (Proceed to block 23.)

27. *Your answer to the question in block 15 is* **A**. *This answer is wrong. Note that the negative terminal of battery B is connected to the plate of the tube.* Proceed to block 21.

28. *The answer to the question in block 19 is* **no**. *An electric motor is an electric component, not an electronic component.*
You have now completed the programmed questions. The next step is to put some of these ideas to work in laboratory experiments. Proceed to the Experiment section of this chapter.

EXPERIMENT

(The experiment described in this section may be performed on the circuit board described in Appendix C or on a similar laboratory setup.)

A volt-ohm-milliammeter is needed for performing the experiments in this book.

(Note that the symbol shown on the right is used throughout this book to call your ***attention*** to an especially important explanation.)

Purpose The purpose of this experiment is to show that a diode acts as a valve to allow current flow in one direction only.

Theory In the chapter material you studied about the Edison effect and the Fleming valve. Vacuum-tube diodes were used in electronic equipment for many years, but today the *semiconductor diode* is favored. They are also called *solid-state diodes,* and they do the same basic job as the vacuum-tube types. In fact, modern solid-state diodes have almost completely replaced the vacuum-tube diodes.

There are many different companies making diodes, and thus they may be marked in different ways. The usual practice is to mark the cathode in some way. Figure 1-13 shows some of the diodes in use and also shows how you can tell which is the cathode and which is the anode. The anode of a solid-state diode is like the plate of the Fleming valve, and the cathode is like the filament. Thus electron current can flow from cathode to anode, but not from anode to cathode.

Most of the experiments in this book require a dc power supply. The one shown in Fig. 1-14 is called a full-wave supply. You will study about power-supply circuits in Chapter 6.

The diodes (X_1 and X_2) in this supply permit current to flow in one direction, so the output of the supply is a *direct current.* The input is the power-line current that flows back and forth in a wire, so it is an *alternating current.*

PART I

Test Setup The schematic diagram for the dc power supply is shown in Fig. 1-14*a*, and a pictorial diagram is shown in Fig. 1-14*b*. You should learn to wire the circuits from the schematic because that is the type of drawing that companies supply with their equipment.

Procedure

step one Wire the circuit as shown in Fig. 1-14.

step two Use the *ac voltmeter* scale to measure the voltage across the transformer secondary. This is the voltage between points *a* and *b*. This should be about 12 volts, so you must use a meter with a range greater than 12 volts.

If you are measuring a voltage and you do not know the approximate value of that voltage, you should start your measurement with the meter set to the highest range. Reduce the range one step at a time until the needle has moved up to the middle of the scale.

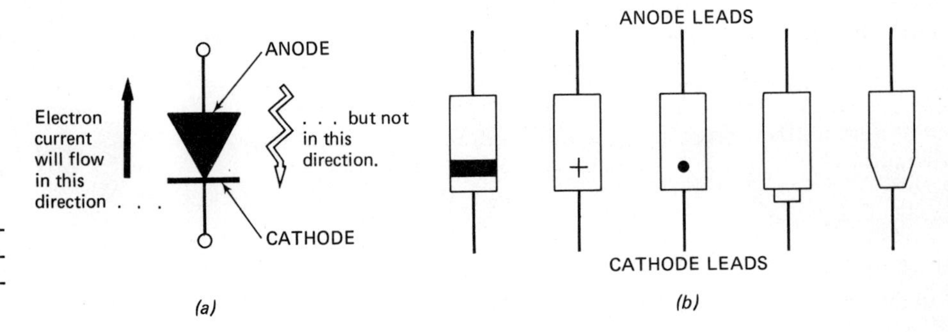

Fig. 1-13 **(*a*) Semiconductor symbol and (*b*) popular ways of identifying the cathode end of semiconductor diodes.**

step three Record the ac voltage across the secondary.

____________________ volts

step four Use a *dc voltmeter* to measure the voltage across the output of the power supply. This is the voltage between points *c* and *d*. As with ac measurements, if you do not know the approximate value of voltage being measured, you should start at the highest meter range setting. Reduce the range setting until the needle has moved up to the middle of the scale.

step five Record the power-supply output voltage. ____________________ volts

The fact that the circuit has an ac input and a dc output shows you that the diodes conduct in only one direction.

The capacitor is used as a *filter* to keep the output voltage at a steady dc value. The use of this filter capacitor will be explained to you in the chapter on power supplies (Chapter 6).

PART II

Theory Diodes are sometimes used in switching circuits. Figure 1-15 shows you one way of doing this.

In Fig. 1-15*a*, when the switch is in position *a*, the power-supply voltage delivers its power to R_{L_1}. When the switch is in position *b* the power supply delivers its power to R_{L_2}. In position *c* there is no power delivered to either load resistor.

Suppose you have a third load—in this case, a lamp. You want to have power delivered to the lamp whenever power is being delivered to either R_{L_1}

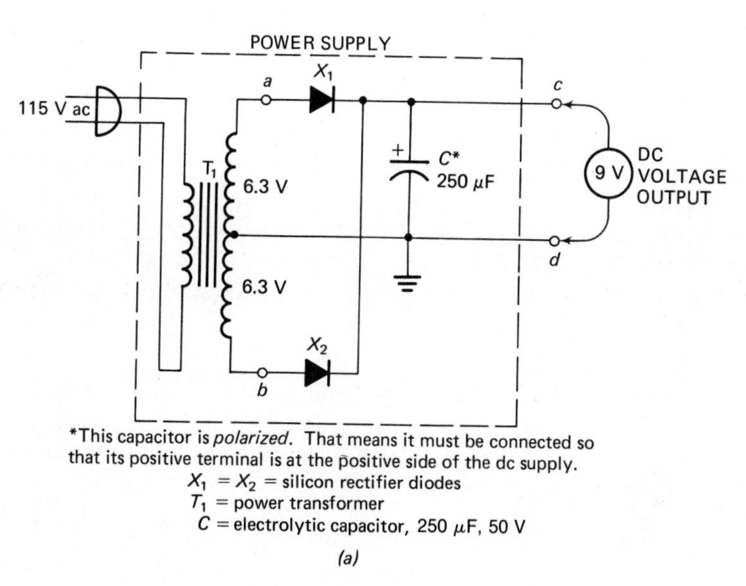

(a)

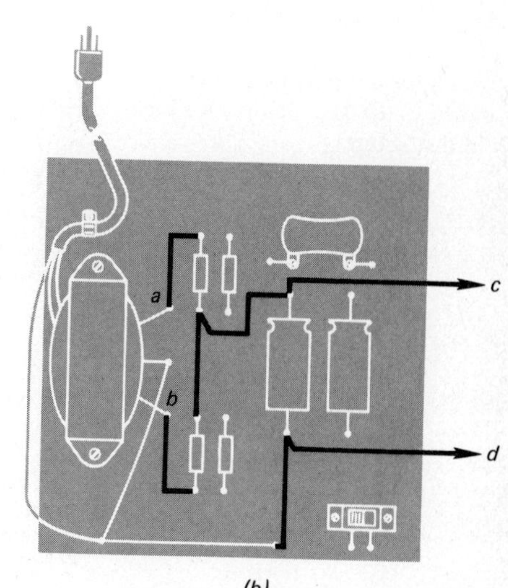

(b)

Fig. 1-14 Setup for Part I of experiment. (*a*) Schematic diagram. (*b*) Pictorial diagram.

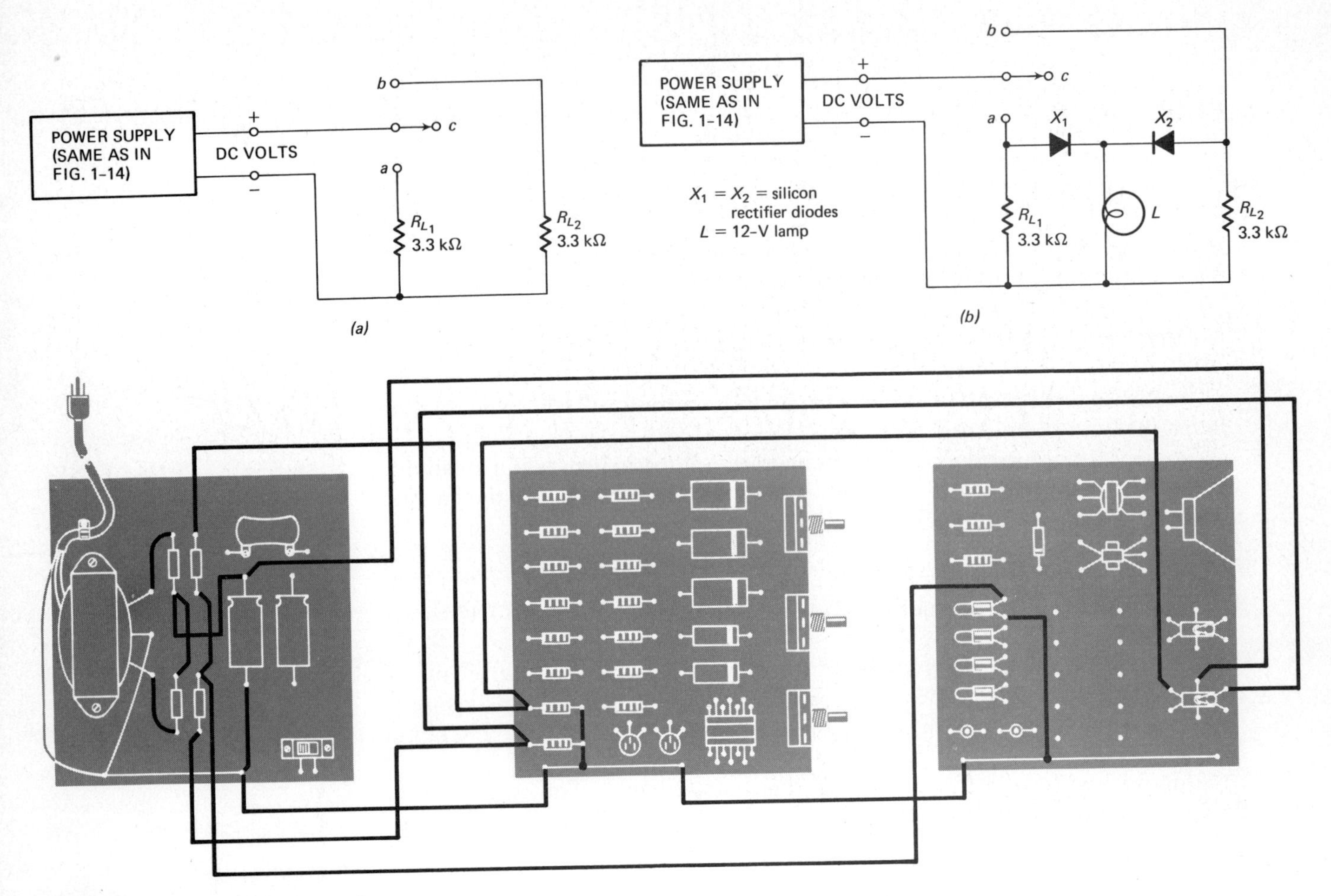

Fig. 1-15 Setup for Part II of experiment. (*a*) A three-position switch for controlling power to R_{L_1} and R_{L_2}. (*b*) Test setup. (*c*) Pictorial test setup.

or to R_{L_2}. You do not want the lamp ON when the switch is in position *c* because there is no power being delivered to a load in that case.

Figure 1-15*b* shows you how this can be done. The diodes are connected *back to back*. When the switch is in position *a*, the lamp gets current only through X_1. When the switch is in position *b*, the lamp gets current only through diode X_2. In position *c* there is no power delivered to the lamp.

You could not get the same results by connecting a wire between points *a* and *b* in place of the diodes. The wire would permit current to flow in both resistors regardless of whether the switch is in position *a* or *b*.

The diode arrangement of Fig. 1-15*b* is sometimes called a *diode* or *logic circuit*. The lamp is ON if the switch is in position *a* or in position *b*, but the lamp is OFF in position *c*.

Test Setup Wire the circuit as shown in Fig. 1-15*b*. This is the schematic drawing. Figure 1-15*c* shows a pictorial drawing of the test setup.

Procedure

step one Turn the switch to position *a* and record the following data:

Is the lamp ON? ______ Yes or No

Is there a voltage across R_{L_1}? ______ Yes or No

Is there a voltage across R_{L_2}? ______ Yes or No

If you have wired the circuit correctly, and if the components are working, then you should have a voltage across R_{L_1}, the lamp should be ON, and there should be no voltage across R_{L_2}. It is very important that the diodes be installed properly. If your circuit does not work properly you should start troubleshooting by checking the connections to the diodes.

step two Turn the switch to position b and record the following data:

Is the lamp ON? ______ Yes or No

Is there a voltage across R_{L_1}? ______ Yes or No

Is there a voltage across R_{L_2}? ______ Yes or No

You should have the lamp ON and a voltage across R_{L_2}, but no voltage across R_{L_1}.

Conclusions A diode will permit current to flow in only one direction, so it can be used to convert ac to dc. In this application the diode is called a *rectifier.*

A diode can also be used as a switching component. In this experiment you demonstrated the use of diodes in a simple logic switching circuit called an *OR circuit.* Early computers used many diodes in this application.

SELF-TEST WITH ANSWERS

(Answers with discussions are given at the end of the chapter.)

1. A vacuum-tube diode will conduct an electron current when (*a*) its plate (or anode) is positive with respect to its filament (or cathode); (*b*) its plate (or anode) is negative with respect to its filament (or cathode).
2. A radio wave can be produced by (*a*) a diode; (*b*) a grid audion; (*c*) a spark gap; (*d*) a resistor.
3. Another name for the Kennelly-Heaviside layer is (*a*) atmosphere; (*b*) ionosphere; (*c*) troposphere; (*d*) stratosphere.
4. The component that Edison used in the experiment which led to the discovery of the Edison effect was (*a*) a diode; (*b*) a triode.

5. When tracing the electron-current path in a circuit, (*a*) start at the positive terminal of the voltage source; (*b*) start at the largest resistor in the circuit; (*c*) start at the antenna; (*d*) start at the negative terminal of the voltage source.
6. Which of the following can be used to provide electrical pressure in a circuit? (*a*) A diode; (*b*) A motor; (*c*) A triode; (*d*) A battery.
7. The reason spark-produced radio waves are not used now in radio communication is (*a*) they are too hard to produce; (*b*) they are too expensive; (*c*) they cannot be amplified; (*d*) they have such a wide range of frequencies.
8. Which of the following is an advantage of a transistor over a triode? (*a*) The transistor does not require a voltage for its operation; (*b*) The transistor operates at a higher temperature; (*c*) The transistor has no filament; (*d*) The transistor has an Edison effect.
9. Fleming called his two-element tube a (*a*) valve; (*b*) diode; (*c*) audion; (*d*) tube.
10. The triode made it possible to amplify signals. It was invented (*a*) in 1926; (*b*) by Hertz; (*c*) by DeForest; (*d*) in 1922.

ANSWERS TO SELF-TEST

1. (*a*)—When the anode of a diode is positive with respect to its cathode, it is said to be *forward-biased*, and it will conduct an electron current.
2. (*c*)—A spark between two metal electrodes produces radio waves. The space between the electrodes is called a *spark gap*.
3. (*b*)—Actually, the ionosphere is made up of a number of layers that are identified by letters of the alphabet.
4. (*a*)—Edison's component consisted of a filament and a metal plate in a glass envelope. Air was removed from the envelope. This is actually a diode.
5. (*d*)—If you always start at the negative terminal and return to the battery at the positive terminal, then you will be sure to cover the complete circuit.
6. (*d*)—A battery supplies voltage which may be thought of as being electrical pressure.
7. (*d*)—The wide range of frequencies means that only a few stations will occupy an entire band. Nowadays many more stations can transmit in the same band.
8. (*c*)—The filament makes the tube operate at higher temperatures. Also the filament could burn out, so the tube is less reliable than the transistor. These are two disadvantages of tubes compared with transistors.
9. (*a*)—Fleming called it a valve because it allowed current to flow in only one direction.
10. (*c*)—DeForest announced the invention of the audion, or triode, in 1906.

what are the two-terminal components used in electronic circuits? 2

INTRODUCTION

A good way to start the study of electronic circuitry is to learn about the basic components. In this chapter you will study two-terminal components. These are components that have only two electrical connections for signal flow or current flow, or for connection to a source of voltage. Resistors, capacitors, inductors, and diodes are examples of two-terminal components that are used extensively in electronic circuits.

Appendix B shows a few schematic symbols for the components discussed in this book. Two or three different symbols may be used for some of the components. This is because the symbols used on *industrial* electronic drawings are different from those used in *consumer* electronics. Figure 2-1 shows how different symbols may be used to represent the same part. You should become familiar with all the symbols. However, in this book we will use only the consumer-electronics* symbols.

It is possible for a two-terminal component to have more than two electrical connections. An example is the directly heated vacuum-tube diode. (Look at its symbol in Fig. 2-17*a*.) This tube has two filament leads and one plate lead, making a total of three external connections. However, it is still considered to be a two-terminal component. The cathode (or filament) is considered to be one input terminal, and the plate is the other.

* *Consumer electronics* refers to such devices as TV sets, AM-FM radios, and phono amplifiers.

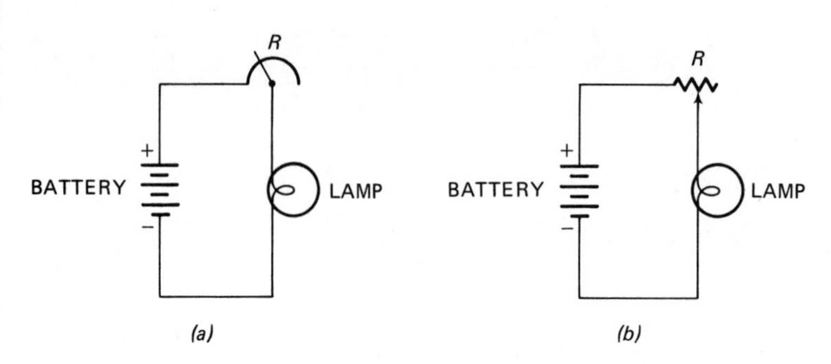

Fig. 2-1 Both circuits are identical in function, but a different symbol is used for the variable resistor (*R*). (*a*) The symbol for the variable resistor (*R*) shown in this circuit is more frequently used on *industrial* electronics diagrams. (*b*) The symbol for the variable resistor in this circuit may be used on industrial electronics schematic diagrams and is commonly used on other electronics schematic diagrams.

You will be able to answer these questions after studying this chapter.

- ☐ **What are resistors used for?**
- ☐ **What are thermistors and VDRs?**
- ☐ **What are capacitors used for?**
- ☐ **What are variable capacitors?**
- ☐ **What are inductors used for?**
- ☐ **What are diodes used for?**

INSTRUCTION

What Are Resistors Used For?

As far as their ability to conduct electric current is concerned, all materials can be classified as being insulators, semiconductors, or conductors. Insulators, such as rubber and glass, will not readily pass an electric current. Conductors, such as copper and aluminum, pass a current with almost no opposition. Semiconductors will pass an electric current but with some opposition. A resistor is an example of a semiconductor component.

Whenever current flows through a resistor there are always two effects:

- ☐ Heat is always produced.
- ☐ There is always a voltage drop.

There are three main applications of resistors. You should keep them in mind when studying electronic circuits.

- ☐ Resistors are used for limiting current.
- ☐ Resistors are used for introducing a voltage drop.
- ☐ Resistors are used for generating heat.

Figure 2-2 shows examples of these applications.

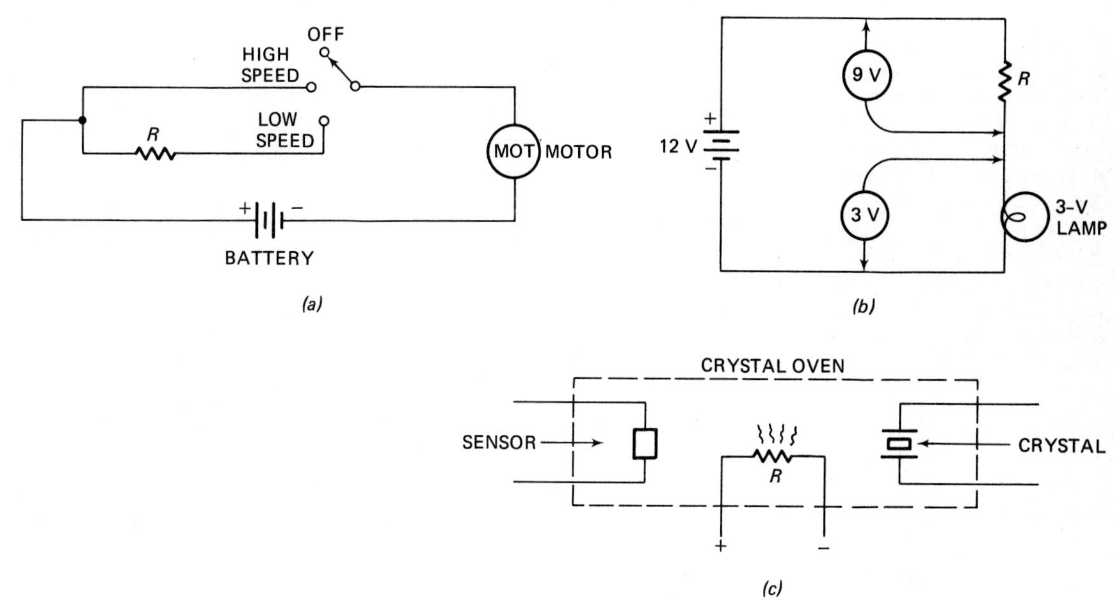

Fig. 2-2 **Fixed resistors may be used to (*a*) limit current, (*b*) divide voltage, or (*c*) generate heat.**

In Fig. 2-2*a* the speed of the dc motor is controlled by changing the amount of current through it. When the switch is in the OFF position, no current flows and the motor is stopped. When the switch is turned to the HIGH SPEED position, there is no resistance in the circuit and maximum current flows through the motor. Under this condition the motor speed is maximum. When the switch is turned to the LOW SPEED position, the motor current must flow through resistor *R*. The effect of the resistor is to reduce the current through the motor, and this reduces the motor speed. The resistor *R* is used to limit current flow.

Figure 2-2*b* shows how a resistor can be used as a voltage divider. A 3-volt lamp is to be connected to a 12-volt source. If you connect the lamp directly across the battery, it will burn out. By connecting it in series with resistor *R*, as shown in the illustration, the lamp will not burn out. The voltage drop across the resistor is 9 volts, and the remaining 3 volts is dropped across the lamp. Note that the sum of the voltage drops must equal the applied voltage, which is 12 volts in this case. This is always true.

Figure 2-2*c* shows you how a resistor can be used for generating heat. The dotted enclosure is a *crystal oven*. A crystal is a vibrating component that produces an ac voltage at a very exact frequency, but only if its temperature is held at a constant value. A crystal, a resistor, and a sensor are enclosed in the crystal oven. Current through the resistor produces heat. The sensor determines when the resistor current should be turned on and off so that the temperature in the oven does not become too high or too low.

What Is an Ohm?

Resistance is the amount of opposition to current flow in a circuit. It is measured in *ohms*. An ohm is a unit of value that tells you how much opposition a resistor offers to current flow. The larger the resistance in ohms, the greater the opposition to current.

Larger values are rated in kilohms (kΩ) or megohms (MΩ). A kilohm is 1000 ohms. A resistance of 100 kΩ means 100,000 ohms. A megohm is a million ohms. A resistance of 4.7 MΩ means 4.7 million ohms.

What Are Carbon Resistors?

Fixed resistors may be made of carbon, metal oxides, silicon, germanium, or other semiconductor materials.

Carbon resistors (or *carbon-composition resistors,* as they are more properly called) come in a wide range of resistance values and power ratings. They are the least expensive of the resistors and the most commonly used.

How Are Resistance Values Marked on Carbon Resistors?

The resistance value and percent tolerance of a carbon resistor is determined by color-coded bands. Figure 2-3 explains how the color code is used for indicating resistance values and tolerances.

The color code tells you what resistance value the resistor should have. But resistors rarely have exactly that value. They are allowed to differ from that value by an amount called the *tolerance.* For example, a tolerance of ±5 percent (read "plus or minus 5 percent") means that the resistance can be 5 percent more than the resistance given by the color code or it can be 5 percent less.

Suppose a resistor has a coded resistance value of 100 ohms and a tolerance of ±5 percent. Five percent of 100 is 5. (That is, $100 \times 0.05 = 5$.) The highest value the resistor can have and still be within its tolerance is $100 + 5 = 105$ ohms. The lowest value it can have is $100 - 5 = 95$ ohms.

Table 2-1 shows the standard values used for carbon-composition resistors. If you buy a carbon-composition resistor, it must have one of the resistance values determined in Table 2-1.

What Is the Meaning of Power Rating and How Is It Determined?

In addition to the resistance and tolerance, a resistor is given a *wattage* rating. This tells you how much heat it can handle in normal usage without burning up. As a general rule, larger resistors can handle more heat than the smaller ones.

Figure 2-4 shows the wattage ratings of carbon-composition resistors. Note that the rating is determined by its physical size.

What Are Wire-Wound Resistors?

Wire-wound resistors are made by winding a *resistance wire*—that is, a wire with a high resistance per inch—on an insulated core. The structure is

Color Code

Color	Value	Tolerance (Fourth Band)
Black	0	
Brown	1	
Red	2	
Orange	3	
Yellow	4	
Green	5	
Blue	6	
Violet	7	
Gray	8	
White	9	
Gold*	×0.1	± 5%
Silver†	×0.01	±10%
No color		±20%

* When the third band is gold, the first two digits are multiplied by 0.1.
† When the third band is silver, the first two digits are multiplied by 0.01.

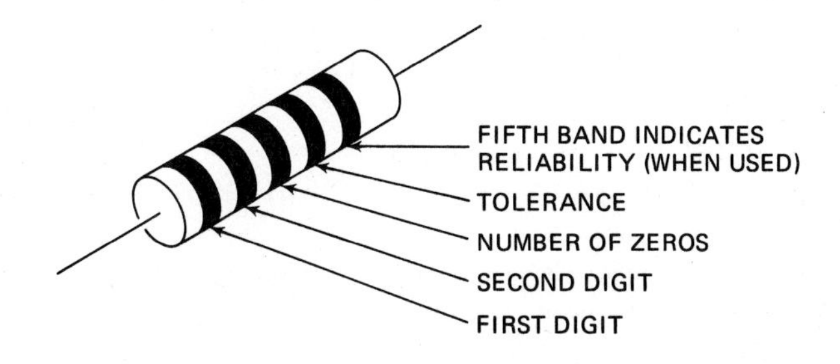

Examples

Color	Value
Red, Red, Red, Gold 2 2 00 ±5%	2200 ohms ±5% (This is 22 followed by 2 zeros.)
Yellow, Violet, Black, Silver 4 7 no zeros ±10%	47 ohms ±10% (The third band is black, meaning no zeros.)
Orange, Blue, Gold, No Color 3 6 Move decimal to left one place ±20%	3.6 ohms ±20% (The third band is gold, which means to multiply 36 by 0.1.)

Fig. 2-3 How resistance values are determined by color code.

sealed to keep out air and moisture. Wire-wound resistors are generally used when higher wattage ratings [above 2 watts (W)] are required. The resistance value and power rating of a wire-wound resistor is usually printed on its case.

Table 2-1 Standard Resistance Values of Carbon-Composition Resistors*

5% Tolerance	10% Tolerance	20% Tolerance
10	10	10
11		
12	12	
13		
15	15	15
16		
18	18	
20		
22	22	22
24		
27	27	
30		
33	33	33
36		
39	39	
43		
47	47	47
51		
56	56	
62		
68	68	68
75		
82	82	
91		
100	100	100

* Multiply values given here by any required multiple of 10.
Examples: 470 ohms ±10%, 62 kΩ ±5%, and 3.3 megohms are standard sizes, but 450 ohms, 65 kΩ, and 3.5 megohms are not standard. Nonstandard sizes can be obtained by using series or parallel combinations.

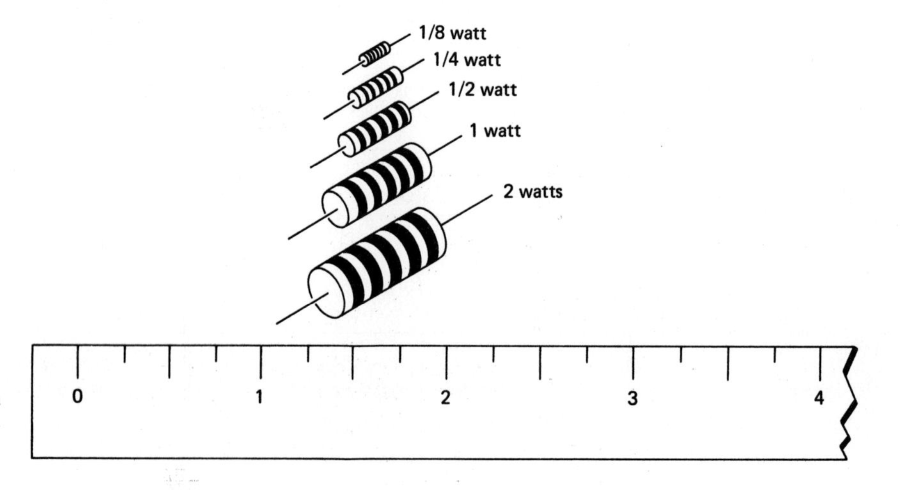

Fig. 2-4 The power rating of a carbon-composition resistor is determined by its physical size.

Summary

1. A two-terminal component has two electrical connections. Resistors, capacitors, inductors, and diodes are examples of two-terminal components.
2. When current flows through a resistor there are always two effects: heat is generated and there is a voltage drop across the resistor.
3. Resistors are used in electronic circuits for limiting current, introducing a voltage drop, or generating heat.
4. Carbon resistors have a color code. This color code tells the resistance value and tolerance.
5. The tolerance rating of a resistor tells the range of resistance values it may have.
6. The wattage rating of a resistor is determined by its size.
7. The larger the wattage rating, the more heat the resistor can handle without burning up.
8. Carbon resistors are made only in "preferred values." Other values are made by connecting them in series or parallel.

What Are Variable Resistors?

A variable resistor can be set to any desired value within its range. It can be connected into a circuit in two ways. When a variable resistor is connected into a circuit in such a way that it varies a current, it is called a *rheostat.* When a variable resistor is connected to vary a voltage, it is called a *potentiometer* or *pot.* The same general type of variable resistor can be used for both applications.

Figure 2-5 shows the two types of connections. Note that the rheostat in

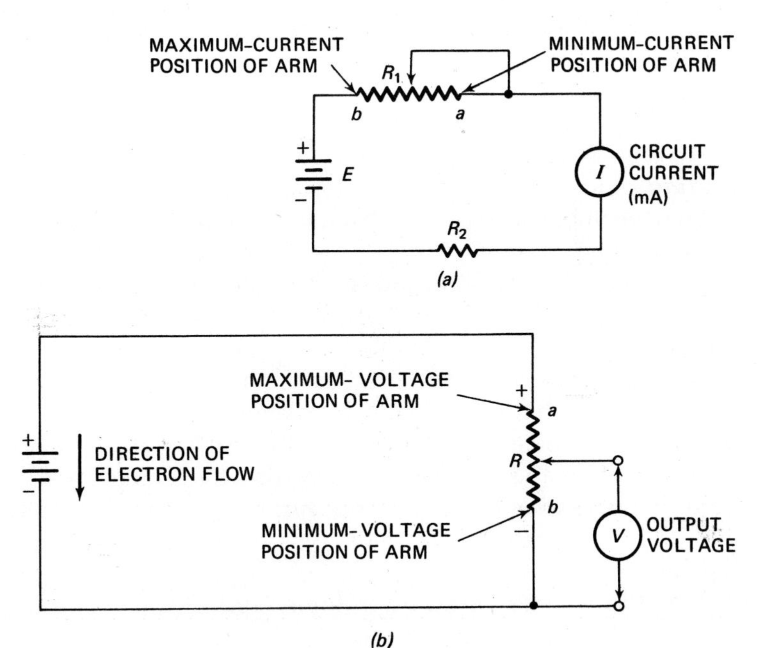

Fig. 2-5 A variable resistor can be connected as (*a*) a rheostat and (*b*) a potentiometer.

Fig. 2-5*a* has a two-terminal connection and the potentiometer in Fig. 2-5*b* has a three-terminal connection.

In Fig. 2-5*a* the variable resistor R_1 is connected in series with a meter to measure the amount of current flow. When the *arm* of the variable resistor (indicated by the arrow in the symbol) is moved to point *a*, all the resistance of R_1 is in the circuit and current is minimum. As the arm is moved toward point *b*, smaller and smaller amounts of resistance are in the circuit, so the current increases. When the arm is all the way to point *b*, there is no resistance in the circuit. The arm lead presents a short circuit across R_1 in this position, and maximum current flows.

Resistor R_2 limits the current flow in Fig. 2-5*a*. Without this resistor there might be too much current flow in the ammeter when the arm is at point *b*.

In Fig. 2-5*b* the variable resistor is connected across the voltage source and the arm is moved between points *a* and *b*. At point *a* the *maximum voltage* will occur at the output terminals. When the arm is at point *b*, there will be no *output voltage*.

How Are Variable Resistors Rated?

Variable resistors (usually called "pots") are rated by their maximum resistance. A 10 kΩ variable resistor has a range of resistance values from 0 to 10,000 ohms. Remember that "k" means *multiply by 1000*.

As with fixed resistors, pots are also rated in watts of power. This is an indication of how much heat they can dissipate without being destroyed.

Here is an important rule to remember when replacing any kind of resistor in a circuit:

Rule: **Never replace a fixed resistor or a variable resistor with one having a lower power rating in watts!**

(Note that the symbol shown on the right is used throughout this book as a ***caution*** sign. It will call your attention to things that are especially important to your personal safety and/or safe operation of equipment.)

A third method of rating variable resistors is by their *taper*. This is simply a way of saying how the resistance value changes with rotation of the shaft. (See Fig. 2-6.)

In Fig. 2-6*a* the resistance increases directly as you turn the shaft. You will note that at the halfway point the resistance equals one-half of the total resistance. This type of variable resistor is said to have a *linear* taper.

In Fig. 2-6*b* the resistance is not directly related to shaft rotation. Note here that when the shaft has been turned halfway the resistance is less than half of the total resistance. This type of resistor is said to have a *nonlinear* taper.

Variable resistors used for volume controls on radios and television sets

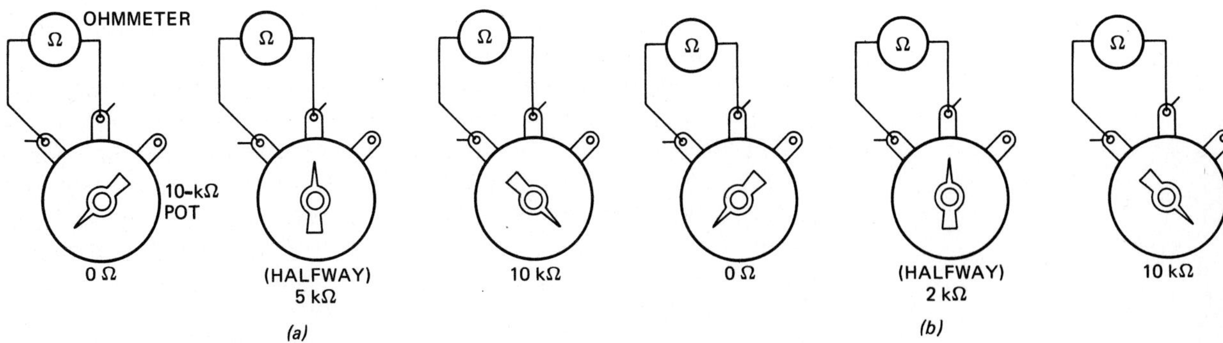

Fig. 2-6 Variable resistors have taper. (*a*) This one has a linear taper. (*b*) This one has a nonlinear taper.

have a nonlinear taper. That is because your ear does not respond the same way to small changes in volume when the sound volume is low as it does when the sound volume is high. By using a nonlinear pot, the change in volume seems to be the same when the pot is changed a given amount at low-volume and at high-volume settings.

What Are Thermistors and VDRs?

Besides fixed resistors and variable resistors, there are some special types. Two important examples are *thermistors* and *varistors*.

The resistance of a thermistor changes by a relatively great amount when its *temperature* is changed slightly. When its temperature is low, its resistance is high, and when its temperature is high, its resistance is low.

The resistance of a *voltage-dependent resistor* (*VDR*) changes by a great amount when the *voltage* across it is changed. (VDRs are also known as *varistors*.) When the voltage across a VDR is low, its resistance is high. When the voltage across a VDR is high, its resistance is low.

Thermistors are used as sensors in circuits that measure or control temperature. For example, the sensor in Fig. 2-2*c* may be a thermistor. Varistors are used to limit the voltage across components that could be damaged by large voltage values.

Summary

1. Variable resistors can be connected as *rheostats* to vary current or as *potentiometers* to vary voltage.
2. When replacing a variable resistor in a circuit, you should be careful to get the exact range of resistance values, the same power rating, and the same taper.
3. Thermistors have a large resistance change for a small change in their temperature.
4. A VDR has a large resistance change for a small change in voltage across it.

What Are Capacitors Used For?

Capacitors can be defined in either of two ways. They are components that store energy in the form of an electrostatic field. Also they are components that oppose any change in voltage across their terminals. These two definitions actually describe the main uses of capacitors in circuits—that is, to store energy and to oppose a change in voltage.

Two other important applications of capacitors are to pass a high frequency while at the same time opposing a low frequency and to divide a voltage. Figure 2-7 shows examples of capacitor applications.

Figure 2-7*a* shows a simple filter circuit. In this application the input voltage is a pulsating dc. Even though the input voltage varies from zero to maximum, the output voltage is maintained at nearly a constant dc voltage. The function of the capacitor in this application is to store energy from the input wave and then to release the energy to the output as it is needed.

Another way of looking at the circuit of Fig. 2-7*a* is to say that capacitor *C* opposes any change in voltage across its terminals. Therefore, even though the input voltage is varying, the output voltage is maintained at a fairly constant value.

Figure 2-7*b* shows you how a capacitor may be used to pass a high frequency and reject a low frequency. In this application the low frequency is a dc voltage having a frequency of 0 hertz (Hz) (the "low" frequency). The voltage applied to the circuit consists of an ac generator e_1 in series with a battery. It is desired to pass the signal from the ac generator to the output terminals, and at the same time to prevent the dc voltage from passing. Both voltages are developed across R_1. Capacitor *C* can pass the ac but cannot pass the dc voltage. Therefore the voltage developed across R_2, and hence the output-

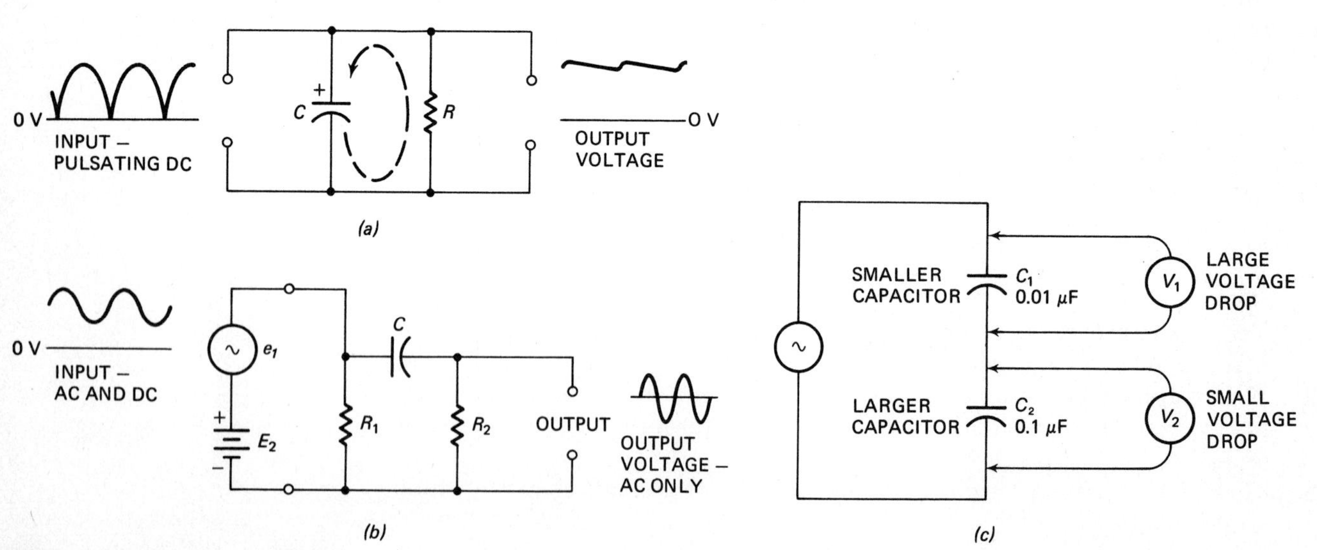

Fig. 2-7 Applications of capacitors for (*a*) storing energy, (*b*) selective coupling, and (*c*) voltage dividing.

signal voltage, consists of only the ac voltage. This type of circuit is used for coupling signals between amplifiers.

Figure 2-7*c* shows how two capacitors can be used as a voltage divider. Note that C_1 has a smaller capacitance [0.01 microfarad (μF)] than C_2 (0.1 microfarad). The two capacitors are connected in series across an ac voltage source. As you can see in the illustration, the *larger* voltage drop occurs across the *smaller* capacitor. This is an important point to remember. The larger the capacitor, the lower the voltage drop across it. This is always true regardless of whether the applied voltage is ac or dc.

What Is a Farad?

Capacitance of a capacitor is measured in *farads* (F). This is a unit of value that tells how much energy the capacitor can store. The larger the value of capacitance the greater the amount of energy the capacitor can store.

The farad unit is too large for most practical work, so *microfarads* (μF) are more often used. A microfarad is a millionth of a farad. A *picofarad* (pF) is a millionth of a millionth of a farad, or a millionth of a microfarad. You will also see the unit *nanofarad* (nF), which is one thousandth of a microfarad. Always replace a capacitor with one having the same capacitance value.

What Is Capacitive Reactance?

The opposition that a capacitor offers to the flow of alternating current is called *capacitive reactance.* It is measured in ohms. The larger the capacitance, the lower its reactance or opposition to the flow of alternating current. Also, the higher the frequency, the lower the opposition that a capacitor offers to current flow. Mathematically,

$$X_C = \frac{1}{2\pi f C}$$

where X_C = the capacitive reactance in ohms
π = 3.14
f = the frequency, in hertz
C = the capacitance, in farads

This equation simply tells you that the reactance is *inversely* related to the frequency and capacitance. In other words, if you make either the frequency or capacitance *higher,* the reactance becomes *lower.* Capacitors are sometimes called *reactive components* because they react against the flow of alternating current. Resistors are nonreactive because they oppose alternating-current and direct-current flow by the same amount.

What Is the Meaning of the Voltage Rating for a Capacitor?

There is no such thing as a perfect insulator. If you put enough voltage across any insulating material—such as glass or air—it will conduct.

This means that there is a limit to the amount of voltage you can place across a capacitor. If the voltage is too high, a spark will jump between the plates. In some types of capacitors this spark will ruin the *dielectric*—that is, the insulation between the capacitor plates.

The voltage rating of a capacitor is the value of voltage that can be safely connected across the capacitor without producing a spark.

You can replace a capacitor with one having a higher voltage rating, but never replace it with one having a lower voltage rating.

What Are Some Types of Fixed Capacitors?

A basic capacitor is made by using two metallic *plates* separated by a dielectric. The closer the spacing between the plates (the thinner the dielectric), the greater the value of capacitance. Figure 2-8 shows a simple capacitor with an air dielectric. Most fixed capacitors use a dielectric made of some type of insulating material.

Fixed capacitors have a single value of capacitance. They are usually identified by the type of material used for the dielectric between the plates. The one in Fig. 2-8 is called an *air capacitor.* We will now discuss some other examples of capacitors used in electronic circuits.

Vacuum dielectric capacitors have a high breakdown voltage rating, but come in very small capacitance values. The maximum value available is usually about 1 nanofarad (one-thousandth of a microfarad).

The paper dielectric used in *paper capacitors* may be coated with a wax or some other insulating material. When a plastic film is used instead of paper, it is called a *film capacitor.* The plates consist of a metal layer or metal foil on each side of the paper or film.

If the metal is sprayed directly on the paper or plastic when the capacitor is made, it is called a *metallized paper* or *metallized plastic capacitor.*

Figure 2-9 shows you how paper capacitors are color-coded. They usually come in values between 250 picofarads and 20 microfarads, and voltage ratings between 400 and 600 volts, with a maximum rating of about 5000 volts.

A black band around the capacitor at one end marks the lead connected to the outside foil. This lead should be connected to the ground side of the circuit whenever possible.

Mica capacitors are made by stacking small metal strips which are used for the plates and separating them by a mica dielectric.

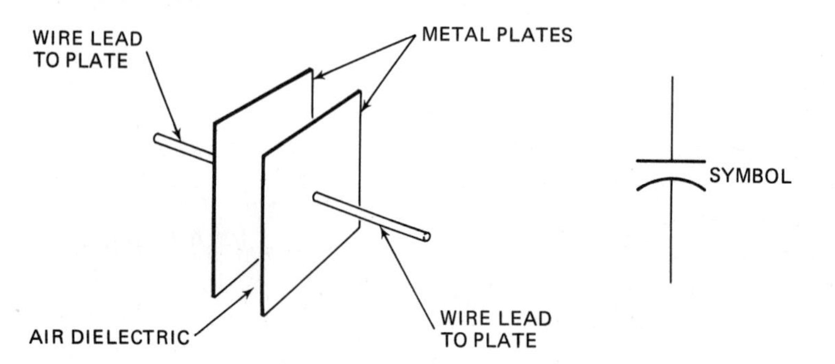

Fig. 2-8 Simple capacitor with an air dielectric.

Capacity in Picofarads

Color	Digit	Multiplier	Tolerance
Black	0	1	20%
Brown	1	10	
Red	2	100	
Orange	3	1000	
Yellow	4	10,000	
Green	5	100,000	5%
Blue	6	1,000,000	
Violet	7		
Gray	8		
White	9		10%
Gold			5%
Silver			10%
No color			20%

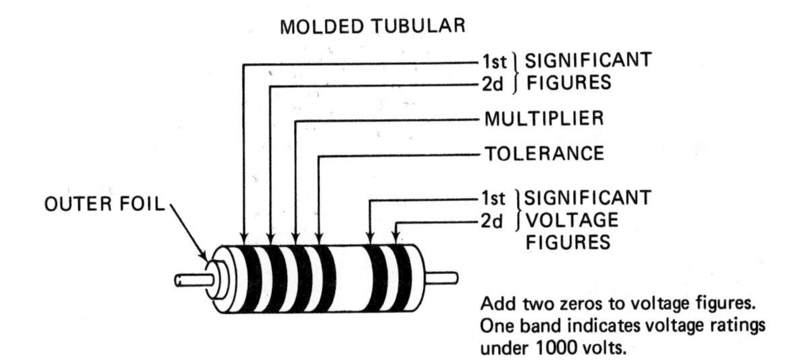

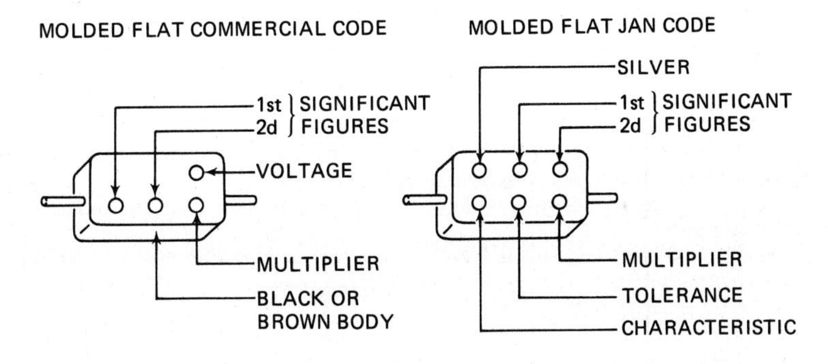

Fig. 2-9 The color code for paper capacitors.

Silver mica capacitors are made by depositing a silver coating directly onto the mica surfaces to serve as the capacitor plates. Figure 2-10 shows you how mica capacitors are color-coded. They are usually made in capacitance values between 10 picofarads and 0.01 microfarad, and with high voltage ratings of up to 5000 or more volts.

Ceramic capacitors have some type of ceramic material used as a dielectric. Figure 2-11 shows you how the color code of ceramic capacitors is determined. They provide high capacitance values in small packages. They are made in a range of capacitance values from 0.1 picofarad to 10 microfarads.

The *temperature coefficient* tells how much the capacitance changes when its temperature changes. A *positive* value means that the capacitance *increases* when the temperature goes up. A negative value means the capacitance *decreases* when the temperature goes *down.* A value called NPO (negative-positive-zero) means that the capacitance value *does not change* with temperature.

You must be sure to duplicate the temperature coefficient as well as the capacitance and voltage values when replacing ceramic capacitors. The reason is that circuits are sometimes designed with a certain temperature characteristic. Of course, it is also important to use an exact replacement for capacitance and voltage characteristics.

Glass capacitors are made by stacking layers of aluminum-foil plates and glass-ribbon dielectrics. They are made in capacitance values ranging from 0.5 picofarad to 10 nanofarads, and with voltage ratings up to 6000 volts. Their capacitance values are usually printed on the case. An important fea-

Capacity in Picofarads

Color	Digit	Multiplier	Tolerance*	Class or Characteristic†
Black	0	1	20%	A
Brown	1	10	1%	B
Red	2	100	2%	C
Orange	3	1000	3%	D
Yellow	4	10,000		E
Green	5		5% (EIA)	F (JAN)
Blue	6			G (JAN)
Violet	7			
Gray	8			I (EIA)
White	9			J (EIA)
Gold		0.1	5% (JAN)	
Silver		0.01	10%	

* Or ±1.0 picofarad, whichever is greater.
† Specifications of design involving *Q* factors, temperature coefficients, and production test requirements.
All axial-lead mica capacitors have a voltage rating of 300, 500, or 1000 volts.

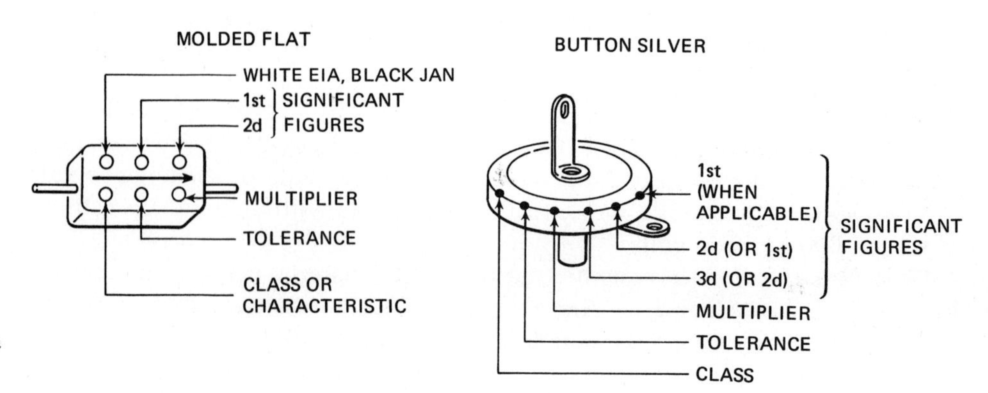

Fig. 2-10 The color code for mica capacitors.

ture of glass capacitors is the fact that they maintain their capacitance value over a long period of time.

Electrolytic capacitors have very high capacitance values because their dielectric is a *very thin layer* of metal oxide. Two types are popular. *Aluminum electrolytics* have a dielectric of aluminum oxide. Tantalum oxide is used for the dielectric in *tantalum electrolytics*. Aluminum electrolytics have been more popular because of their lower cost.

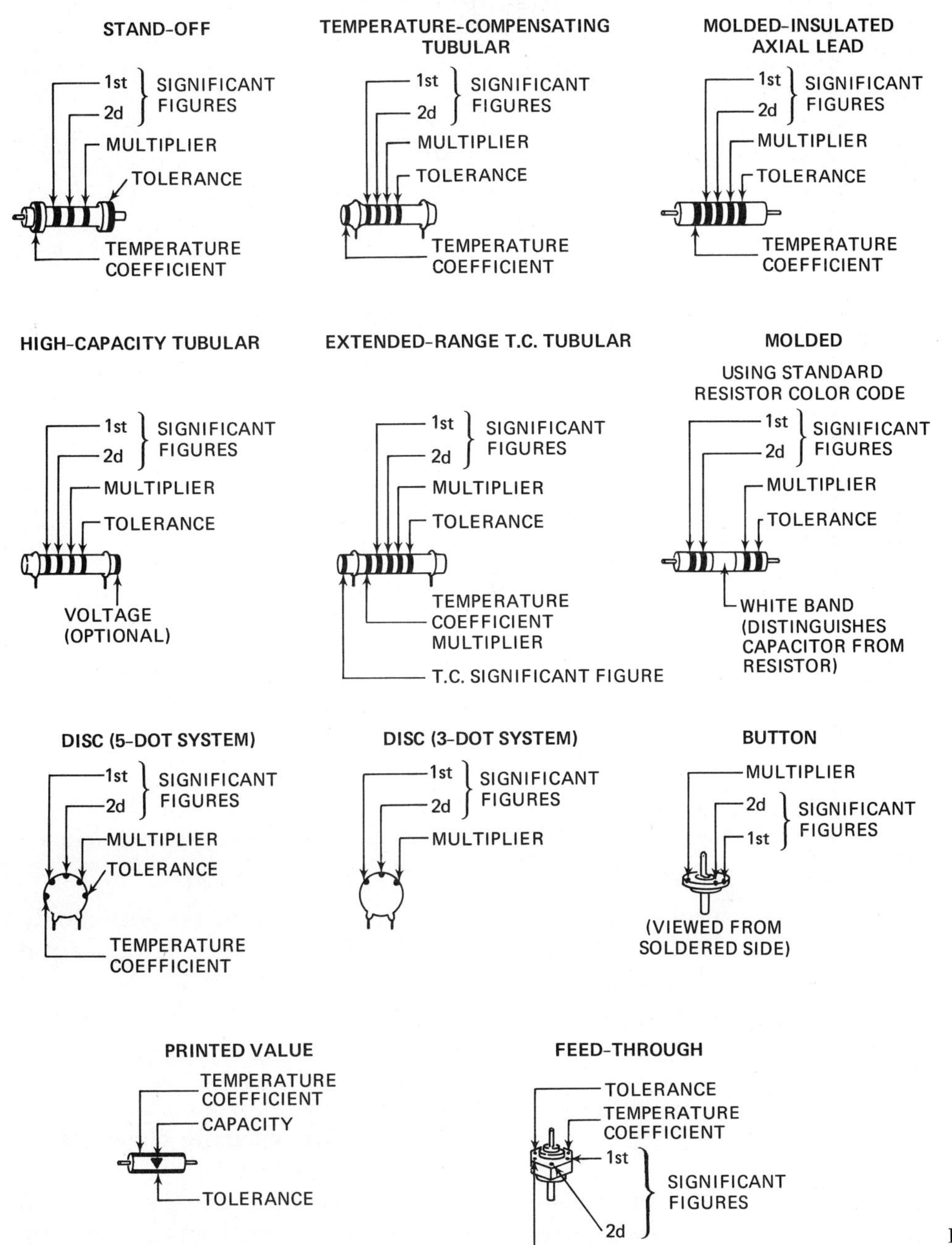

Fig. 2-11 The color code for ceramic capacitors.

Capacity in Picofarads

Color	Digit	Multiplier	Tolerance		Temperature Coefficient ppm/°C	Extended Range Temperature Coefficient	
			10 Picofarads or Less	Over 10 Picofarads		Significant Figures	Multiplier
Black	0	1	±2.0	±20%	0 (NPO)	0.0	−1
Brown	1	10	±0.1	±1%	−33 (N033)		−10
Red	2	100		±2%	−75 (N075)	1.0	−100
Orange	3	1000		±2.5%	−150 (N150)	1.5	−1000
Yellow	4	10,000			−220 (N220)	2.2	−10,000
Green	5		±0.5	±5%	−330 (N330)	3.3	+1
Blue	6				−470 (N470)	4.7	+10
Violet	7				−750 (N750)	7.5	+100
Gray	8	0.01	±0.25		+30 (P030)		+1000
White	9	0.1	±1.0	±10%	General-purpose bypass & coupling		+10,000
Silver							
Gold					+100 (P100, JAN)		

Voltage ratings are standard 500 volts for some manufacturers, 1000 volts for other manufacturers.

JAN Letter	Tolerance: 10 Picofarads or less	Tolerance: Over 10 Picofarads
C	±0.25	
D	±0.5	
F	±1.0	±1%
G	±2.0	±2%
J		±5%
K		±10%
M		±20%

Fig. 2-11 (Continued)

Electrolytic capacitors are *polarized*—that is, they can be connected only into pulsating dc circuits like the one in Fig. 2-7*a*. Capacitance values and voltage ratings for electrolytic capacitors are printed on the case. The symbol for an electrolytic capacitor shows a + sign beside one of the plates. (Check this in Appendix B.)

What Are Variable Capacitors?

Variable capacitors are usually made with an air dielectric. Figure 2-12 shows examples of variable capacitors. Capacitance is changed either by *changing the area* of the plates facing each other or by *changing the distance*

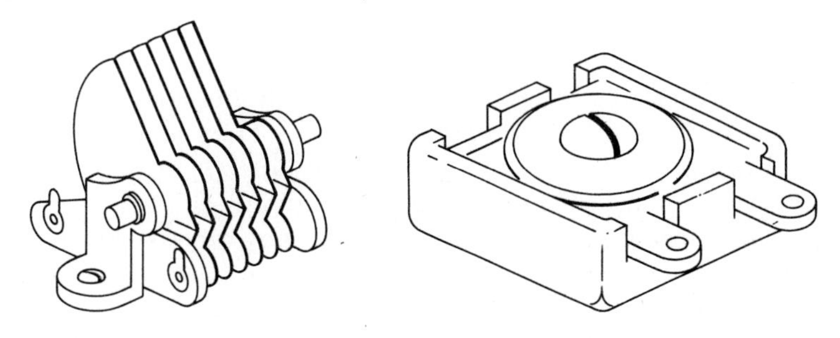

Fig. 2-12 Examples of variable capacitors.

between the plates. They are rated by the range of capacitance values through which the capacitor will vary.

The capacitance range can be adjusted by the use of *trimmer capacitors* or *padder capacitors*. Figure 2-13 shows the difference between a trimmer and a padder. In Fig. 2-13*a* the trimmer is seen to be connected in *parallel* with the variable capacitor, while the padder of Fig. 2-13*b* is seen to be connected in *series* with it. The trimmer connection is more popular. Capacitors used as trimmers are made with dielectrics of air, polystyrene, Teflon, ceramic, or mica.

Summary

1. Capacitors are components that store energy in the form of an electrostatic field.
2. Capacitors are components that oppose any change in voltage across their terminals.
3. Capacitors are used for storing energy, opposing a change in voltage, passing high frequencies and rejecting low frequencies, or dividing a voltage.
4. Capacitors are rated according to their capacitance, voltage rating, and in some cases, their temperature coefficient and percent tolerance.
5. Capacitors are identified by the type of material used for their dielectric. Examples are paper capacitors, mica capacitors, and ceramic capacitors.
6. Trimmers are small variable capacitors placed in parallel with a larger variable capacitor.
7. Padders are small variable capacitors placed in series with a larger variable capacitor.
8. Trimmers and padders are used to adjust the capacitance range of the larger variable capacitors.

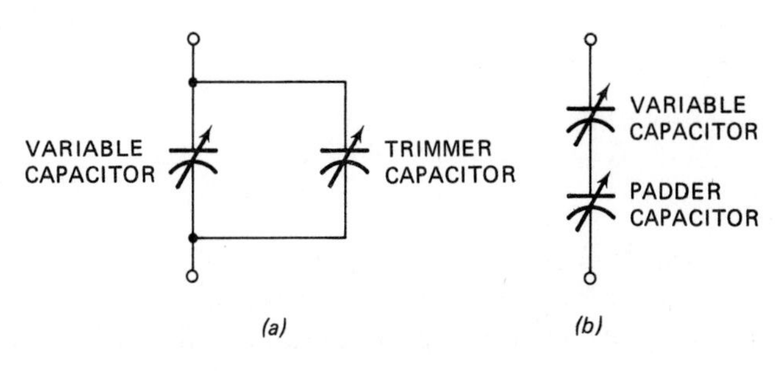

Fig. 2-13 Small adjustable capacitors may be connected as (*a*) trimmers or as (*b*) padders. Trimmers and padders have smaller capacitance values than variable capacitors.

What Are Inductors Used For?

Inductors, or *coils,* can be defined as being components that store energy in the form of an *electromagnetic* field. They may also be defined as being components that *oppose any change in current* through them. These definitions tell what inductors are used for in circuits—that is, for storing energy and for opposing a change in current. They can also be used for passing a low frequency while at the same time rejecting a high frequency.

Figure 2-14 shows how an inductor is used in a circuit for passing a low frequency and rejecting a high frequency. In this application the inductor is often called a *choke.* The complete circuit in Fig. 2-14 is called a *low-pass filter.* Low frequencies can pass through the coil, but high frequencies are rejected. At the same time, the high frequencies are shorted to ground through the capacitors, but the low frequencies see a high opposition in this path. The overall result is that only low frequencies are passed through the circuit from a to b.

What Is a Henry?

The inductance of a coil is measured in *henrys* (H). This is a unit that tells how much energy an inductor can store. The larger the inductance value in henrys, the more energy an inductor can store.

The henry unit is too large for practical electronics coils, so the units *millihenrys* (mH) and *microhenrys* (μH) are more commonly used. A millihenry is a thousandth of a henry. A microhenry is a millionth of a henry.

What Is Inductive Reactance?

Inductors, like capacitors, oppose the flow of alternating current. You will remember that the higher the frequency, the lower the opposition that the capacitor offers to alternating-current flow. *Inductive reactance,* which is the opposition that a coil offers to alternating-current flow, *increases* as the *frequency increases.* Also inductive reactance *increases* as the amount of *inductance increases.* Mathematically,

$$X_L = 2\pi f L$$

where X_L = the reactance, in ohms
π = 3.14
f = the frequency, in hertz
L = inductance, in henrys

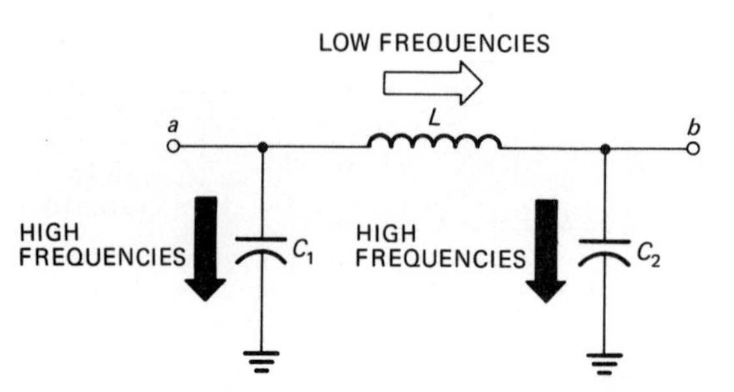

Fig. 2-14 An example of how an inductor can be used as a low-pass filter.

This equation shows you that the opposition or reactance of an inductor increases when you increase either the inductance or the frequency. Inductors are reactive components.

What Are Some Types of Inductors?

Filter chokes are used in power-supply circuits to smooth out variations in power-supply current. This application is shown in Fig. 2-15. Filter chokes are usually designed to carry relatively large currents. They are wound on *laminated* iron cores. Laminated iron is iron made in layers or sheets piled on top of each other. The inductance value of filter chokes is usually about 1.0 to 30 henrys.

RF chokes have an air core. As their name implies, they are used for opposing changes in radio-frequency currents. When they are molded in cylindrical forms they may be color-coded as shown in Fig. 2-16.

Variable inductors are designed in such a way that the core material can be moved into or out of the center. The inductance increases as the core moves into the coil. The core material may be a powdered iron or ferrite material. These materials are used because of their low loss at high frequencies. The powdered iron is molded into small cylinders under high pressure, so it appears to be made of a solid material. These inductors are used only at radio frequencies.

Ferrite beads are small beads of magnetic material that act as RF chokes. A current-carrying wire passes through the bead. It has the same effect as connecting a choke coil in series with the wire. Ferrite beads have very good inductance properties with very little loss.

Summary

1. Inductors, which are also called *coils*, are components that store energy in the form of an electromagnetic field.
2. Inductors are components that oppose any change in current through them.
3. Inductors are used for storing energy and for opposing a change in current. They are also used for passing a low frequency and rejecting a high frequency.
4. Filter chokes are used for smoothing fluctuations in power-supply currents.

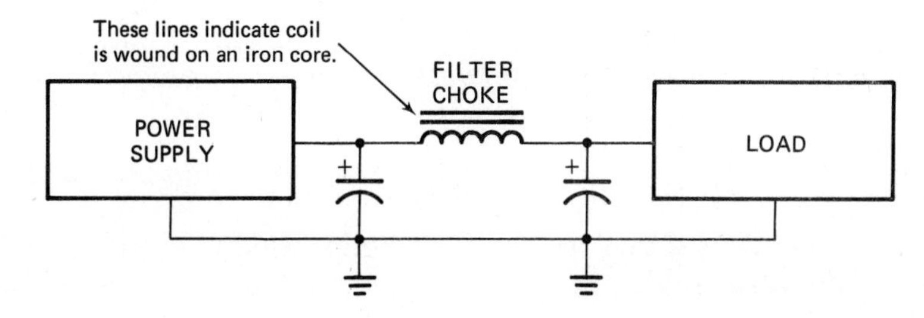

Fig. 2-15 Use of a filter choke in a power supply circuit. The + signs on the capacitor symbols mean that the capacitors are electrolytic.

Color	Significant Figure	Multiplier*	Tolerance
Black	0	1	
Brown	1	10	
Red	2	100	
Orange	3	1000	
Yellow	4		
Green	5		
Blue	6		
Violet	7		
Gray	8		
White	9		
None†			±20
Silver			±10
Gold	Decimal Point		±05

* The multiplier is the factor by which the two significant figures are multiplied to yield the nominal inductance value.
† Indicates body color.

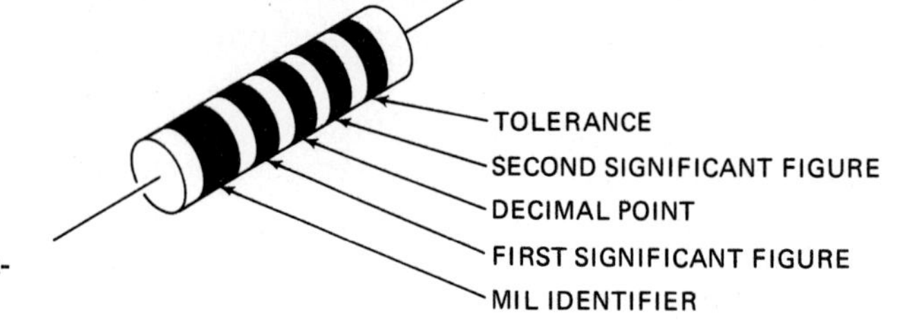

Fig. 2-16 The color code for cylindrical RF chokes.

5. Coils that are wound on laminated iron cores are used for low frequencies, while coils that have air cores are used at radio frequencies.
6. Coils that are wound on either powered iron or a ferrite material are also used at radio frequencies.
7. Ferrite beads are small beads of magnetic material that act as RF chokes.

What Are Diodes Used For?

Diodes are components that conduct electron current in *one direction* (cathode to anode) but not in the reverse direction (anode to cathode). At one time the most extensive use of diodes was in power supplies, where they are used as rectifiers. A *rectifier* is a diode used for converting the alternating current of the ac power line to a direct current. Today rectification is only one of a wide variety of functions performed by diodes.

What Are Some Examples of Rectifier Diodes?

Figure 2-17 shows the types of diodes used in rectifier circuits.

The diode shown in Fig. 2-17*a* uses a filament as the cathode. The fila-

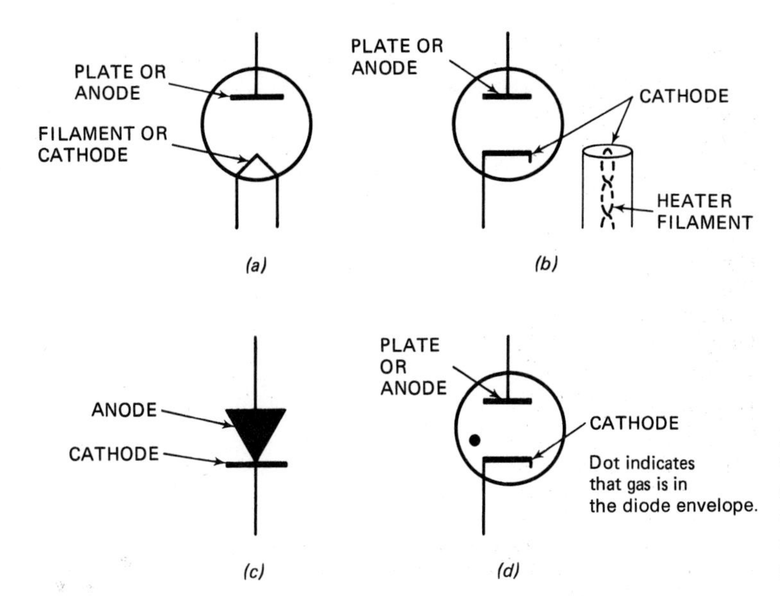

Fig. 2-17 Diodes used for rectification. (*a*) Vacuum-tube diode with filament cathode. (*b*) Vacuum-tube diode with indirectly heated cathode. (*c*) Semiconductor diode. (*d*) Gas-filled diode.

ment is heated to the point that electrons are emitted from its surface. When the plate voltage is positive, it will attract the negative emitted electrons. Thus the electron-current flow is from the *cathode to the plate.* The emitted electrons come directly from the surface of the filament, and diode tubes of this type are said to have *directly heated cathodes.*

Figure 2-17*b* shows a diode with an indirectly heated cathode. A filament, which is not shown on the symbol (see inset), heats the cathode. Electrons are emitted from the heated cathode surface. The cathode is actually shaped like a sleeve with the filament inside. This is shown in the inset of Fig. 2-17*b*. When the plate is made *positive,* it attracts the emitted electrons.

The *semiconductor* diode also conducts electron current only from its cathode to its anode. As with the tube diodes, it will not conduct from anode to cathode. The symbol for this component is shown in Fig. 2-17*c*. Semiconductor diodes are also called *solid-state* diodes. They have the advantage that a hot cathode is not required for their operation. The result is a simpler component that operates at a lower temperature and does not require filament power. The physical and chemical composition of the semiconductor diode enables it to rectify.

Figure 2-17*d* is the symbol for a gas-filled diode, sometimes called a *phanotron.* It has an indirectly heated cathode and a plate, like the vacuum-tube diode of Fig. 2-17*b*. The difference is that the inside of the tube is filled with a *gas.* This produces so-called avalanching when the tube is conducting electron current.

Figure 2-18 illustrates the condition of avalanching. An electron leaves the cathode and is attracted toward the positive plate. Before it has traveled very far it collides with a gas molecule and knocks an electron loose from the molecule. The two electrons then collide with two more gas molecules and knock two more electrons loose. The process continues, so that the total number of electrons arriving at the plate is quite large. *One important characteristic of a gas diode is its ability to conduct relatively large currents.* When this type of diode is conducting, it glows.

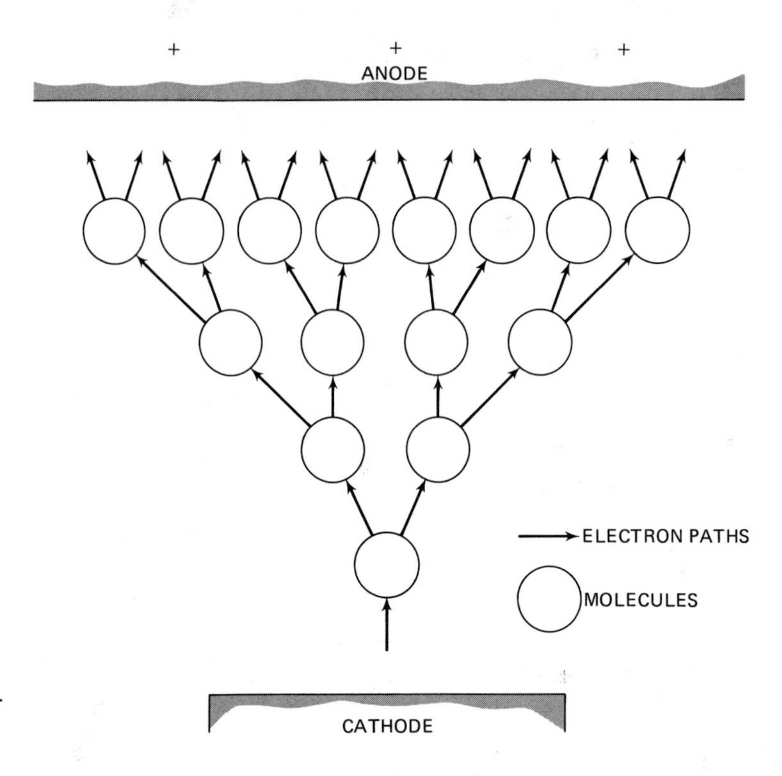

Fig. 2-18 Avalanching in a gas-filled tube.

What Are Regulating Diodes?

A *regulating diode* always has the same voltage across it, within its operating current range. Both tube and semiconductor types are shown in Fig. 2-19.

The *gas-regulator diode* of Fig. 2-19*a* does not have a heated cathode. Electrons are drawn from the cathode by *field emission.* This simply means that electrons are torn from the surface of the cathode by the positive voltage applied to the anode.

A low voltage across the gas-regulator diode will not cause any current to flow through it. As the voltage is increased, a certain value will be reached that causes the diode to *fire.* This means that it glows. Once the firing voltage has been reached, the voltage across the diode remains a constant value within the operating current rating. The current must never be allowed to exceed the value given in the data sheets for the tube.

The *neon lamp* of Fig. 2-19*b* is similar to the gas regulator. It has two electrodes which are identical in size, so electrons can pass easily in either direction through the lamp. Neon lamps are smaller than gas regulators. They are used for light sources as well as for their voltage-regulating properties.

An unusual feature of the solid-state *zener diode* in Fig. 2-19*c* is that it must be operated with *reverse bias*—that is, the anode must be negative with respect to the cathode. Normally a diode will not conduct with this polarity of voltage across it, but the zener diode will.

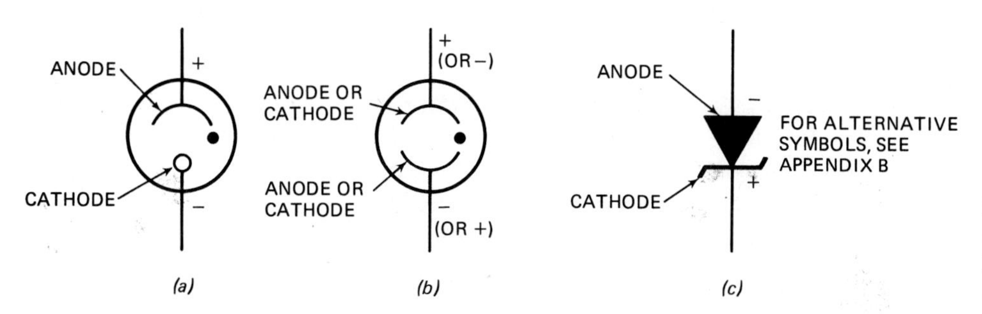

Fig. 2-19 Voltage-regulating diodes. (*a*) Gas regulator. (*b*) Neon lamp. (*c*) Zener diode. For other symbols, see Appendix B.

Regulating diodes are used in circuits where the voltage must be maintained at a *constant value* with normal changes of circuit conditions. Gas regulators are not available for regulating voltages below 50 volts, whereas zener diodes can be obtained with voltage ratings down to a few volts.

What Is an LED?

An *optoelectronic component* will have either of the following characteristics:

1. It *gives off light* when it is provided with the correct voltage. An example of this type is the LED.
2. Its characteristics change in some way when it is *exposed to light.* There are two types: photoconductive and photovoltaic.

An LED is an example of a two-terminal optoelectronic component. The initials LED stand for *L*ight-*E*mitting *D*iode.

When current flows in semiconductor material, energy is produced. This energy cannot be simply discarded; a basic law of physics says that energy cannot be created or destroyed. When a current flows here, the energy is converted to light.

Figure 2-20 shows the LED symbol. Arrows *pointing away* from a solid-state component mean that it *gives off light.*

The LED has a longer life than a filament-type lamp. Also, it runs much cooler. However, it can be destroyed more easily with an *overload* (that is, with excessive current flow).

Photoconductive diodes operate in the reverse way from LEDs. If you

Fig. 2-20 The LED symbol.

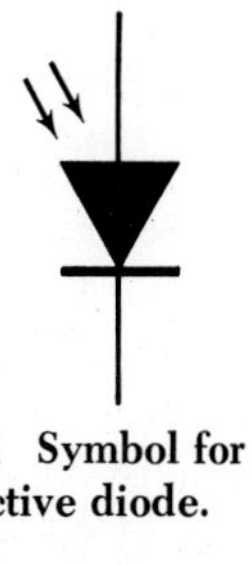

Fig. 2-21 Symbol for a photoconductive diode.

shine a light on a photoconductive diode, this causes a change in the amount of current flowing in the diode.

Figure 2-21 shows the symbol for a photoconductive diode. Note that the arrows are pointing toward the component. All semiconductor diodes would be somewhat sensitive to light, but they are packaged in light-tight cases to prevent light from affecting their operation.

Photovoltaic components generate a voltage that is dependent upon light shining on it. A *solar cell* is an example of a photovoltaic component.

The Greek letter lambda (λ) is often used with optoelectronic components. This symbol is used to mean that the component requires light for its operation or that it gives off light.

Summary

1. Rectifier diodes will conduct electron current from cathode to anode but not from anode to cathode.
2. When the anode of a diode is positive with respect to its cathode, the diode is said to be *forward-biased.* Under this condition it conducts current.
3. When the cathode of a diode is positive with respect to its anode, the diode is said to be *reverse-biased. Normally current does not flow through a reverse-biased rectifying diode.*
4. Rectifying diodes are used for converting alternating current to direct current.
5. The voltage across a regulating diode is always the same value regardless of how much current is flowing through it. It is used for providing an unvarying (regulated) voltage.
6. An optoelectronic component can have one of the following characteristics: (*a*) give off light (LED) when energized; (*b*) change its current when exposed to light (photoconductive); or (*c*) generate voltage (photovoltaic) when exposed to light.

PROGRAMMED REVIEW QUESTIONS

(Instructions for using this programmed section are given in Chapter 1.)

We will review the important concepts of this chapter. If you have understood the material, you will progress easily through this section. Do not skip this material because some additional theory is presented.

1. A bilateral circuit component is one that will conduct electron current in either direction. Which of the following is an example of a two-terminal bilateral circuit component?
   A. A phanotron. (Proceed to block 9.)
   B. A carbon-composition resistor. (Proceed to block 17.)

2. *The correct answer to the question in block 16 is* **A**. *Zener diodes are*

operated with a reverse voltage. In other words, electron current flows through the zener from anode to cathode. Even though the amount of current may change, the voltage across the zener diode remains constant. For this reason it is said to be a **voltage regulator.**
Here is your next question.
Which of the following is an application of resistors?
A. Pass low frequencies and reject high frequencies. (Proceed to block 13.)
B. Produce a voltage drop. (Proceed to block 10.)

3. *The correct answer to the question in block 17 is* **A**. *See Fig. 2-6 for examples of variable resistor taper. Slewing rate is not a term used with variable resistors. (Actually, it is the maximum rate of voltage change that can occur at power-supply terminals.)* Here is your next question.
A component that opposes any change in voltage across its terminals is the
A. rectifier diode. (Proceed to block 5.)
B. capacitor. (Proceed to block 15.)

4. *Your answer to the question in block 15 is* **A**. *This answer is wrong. The symbols for all diodes are shown in Appendix B.* Proceed to block 12.

5. *Your answer to the question in block 3 is* **A**. *This answer is wrong. A rectifier is a diode that conducts current in only one direction.* Proceed to block 15.

6. *Your answer to the question in block 11 is* **A**. *This answer is wrong. Review the symbols for diodes in Appendix B,* then proceed to block 16.

7. *Your answer to the question in block 16 is* **B**. *This answer is wrong. A choke coil offers a high opposition to any change in current, but it is not used as a voltage regulator.* Proceed to block 2.

8. *Your answer to the question in block 10 is* **A**. *This answer is wrong. A rheostat is a variable resistor connected into a circuit to control the circuit current.* Proceed to block 18.

9. *Your answer to the question in block 1 is* **A**. *This answer is wrong. A phanotron is a gas diode. Since electron current will only flow through this diode in one direction (cathode to plate), it is an example of a unilateral component.* Proceed to block 17.

10. *The correct answer to the question in block 2 is* **B**. *Figure 2-2b shows an example of a circuit where a resistor is used to produce a voltage drop.* Here is your next question.

Which of the following components may be used as a heat sensor?
A. A rheostat. (Proceed to block 8.)
B. A thermistor. (Proceed to block 18.)

11. *The correct answer to the question in block 12 is* **A**. *The letters NPO stand for Negative Positive Zero. A capacitor with this rating has no change in capacitance for small changes in temperature.* Here is your next question.
Figure 2-22 shows the symbol for
A. a neon lamp. (Proceed to block 6.)
B. a vacuum-tube diode. (Proceed to block 16.)

12. *The correct answer to the question in block 15 is* **B**. *As shown in Appendix B, there is more than one kind of symbol used for zener diodes.* Here is your next question.
A temperature coefficient rating of NPO for a ceramic capacitor means
A. that it has no change in capacitance for a small change in temperature. (Proceed to block 11.)
B. that it has a large change in capacitance for a small change in temperature. (Proceed to block 14.)

13. *Your answer to the question in block 2 is* **A**. *This answer is wrong. An inductor (not a resistor) is used for passing low frequencies and rejecting high frequencies.* Proceed to block 10.

14. *Your answer to the question in block 12 is* **B**. *This answer is wrong.* Proceed to block 11.

15. *The correct answer to the question in block 3 is* **B**. *This is one of the important characteristics of capacitors.* Here is your next question.
Figure 2-23 shows the symbol for
A. a gas diode. (Proceed to block 4.)
B. a zener diode. (Proceed to block 12.)

16. *The correct answer to the question in block 11 is* **B**. *The illustration shows the symbol for a diode with a directly heated cathode.* Here is your next question.

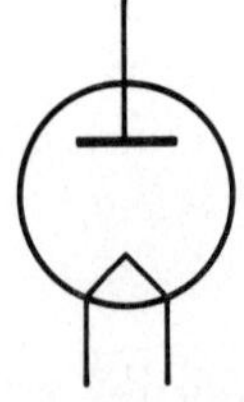

Fig. 2-22 What component does this symbol represent?

Fig. 2-23 What component does this symbol represent?

Which of the following is used for producing a constant voltage drop even though the current through it may change?
A. Zener diode. (Proceed to block 2.)
B. An RF choke coil. (Proceed to block 7.)

17. *The correct answer to the question in block 1 is* ***B****. The phanotron, or gas diode, conducts electron current from cathode to plate, but not from plate to cathode. A resistor is a bilateral component. It will conduct current in either of two directions.* Here is your next question.
The manner in which the resistance of a variable resistor changes with shaft rotation is called its
A. taper. (Proceed to block 3.)
B. slewing rate. (Proceed to block 19.)

18. *The correct answer to the question in block 10 is* ***B****. A sensor, which is also called a* ***transducer****, is used to sense the temperature in the crystal oven of Fig. 2-2c. Small changes in temperature cause large resistance changes in a thermistor.* Here is your next question.
The two-terminal component that opposes any change in current through it is the ______. (Proceed to block 20.)

19. *Your answer to the question in block 17 is* ***B****. This answer is wrong. The term* ***slewing rate*** *is not used in reference to variable resistors.* Proceed to block 3.

20. *Inductor. (A capacitor opposes changes in voltage.)*
You have now completed the programmed questions. The next step is to put some of these ideas to work in laboratory experiments. Proceed to the Experiment section of this chapter.

EXPERIMENT

(The experiment described in this section may be performed on the circuit board described in Appendix C or on a similar laboratory setup.)

PART I

Purpose The purpose of this experiment is to demonstrate the basic theory of voltage calculations and measurements.

Theory When you are using a voltmeter for troubleshooting in a circuit, you must know what the correct voltage should be for each measurement. Sometimes the manufacturer gives the voltage value on the circuit drawings. Some times you have to calculate the value from the information given.

With a little practice you can learn to calculate voltage and current values with only a small amount of mathematics. If you watch good technicians at work you will see that they usually do not need to stop and calculate voltage and current values. They can work Ohm's-law problems "in their head." We can show you *how* they do it, but you will have to practice until you can do it easily.

If you place two resistors of the same value in series across a voltage source, half of the source voltage will be dropped across each resistor. If three series resistors with the same value are used, then *one-third* of the voltage will be dropped across each resistor.

If the resistors have different values, it is still an easy matter to find the voltage drops. Figure 2-24 shows a simple series circuit with two resistors (R_1 and R_2). Resistors are connected in series when the *same* current flows through them. The arrows show the path of electron-current flow. All the electrons that pass through R_2 must also pass through R_1 in order to get back to the positive terminal of the battery. So R_1 and R_2 are in series.

Part of the battery voltage is dropped across each resistor. The resistors may be the same value or they may not. If you use V_1 for the voltage across R_1, and V_2 for the voltage across R_2, then you can easily determine how much voltage is across each resistor.

In electronics we use an equation (or formula) to make it easy to remember how to solve a problem. For the circuit of Fig. 2-24.

$$V_1 = E\left(\frac{R_1}{R_1 + R_2}\right)$$

and

$$V_2 = E\left(\frac{R_2}{R_1 + R_2}\right)$$

Another way to say this is that the voltage drop across a series resistor is found by *dividing the resistor value* by the *total resistance*, then *multiplying* the result by the *battery voltage*.

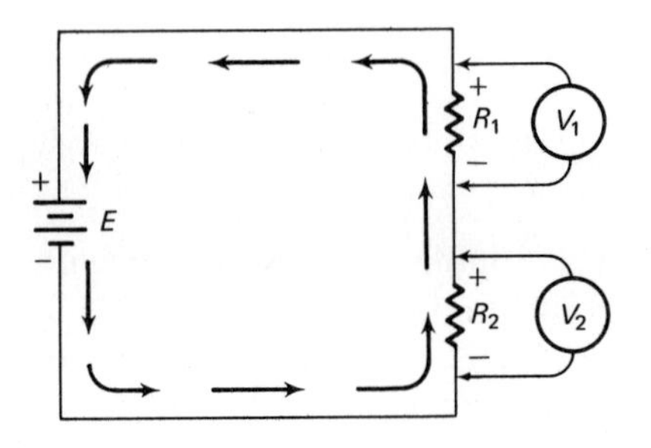

Fig. 2-24 A simple series circuit.

Let's try that for the circuit of Fig. 2-25. The battery voltage is 400 volts. The drop across R_1 is V_1.

$$V_1 = E\left(\frac{R_1}{R_1 + R_2}\right)$$

Substitute the values in the circuit and you get

$$\begin{aligned} V_1 &= 400\left(\frac{100}{100 + 300}\right) \\ &= 400\left(\frac{100}{400}\right) \\ &= 400\left(\frac{1}{4}\right) \\ &= 100 \text{ volts} \end{aligned}$$

In other words, *one-fourth* of the applied voltage is dropped across the 100-ohm resistor.

The voltage across R_2 is V_2.

$$V_2 = E\left(\frac{R_2}{R_1 + R_2}\right)$$

Substitute the values from the circuit.

$$\begin{aligned} V_2 &= 400\left(\frac{300}{400}\right) \\ &= 300 \text{ volts} \end{aligned}$$

In other words, three-fourths of the applied voltage is dropped across the 300-ohm resistor.

A very important thing to remember about voltage drops is that, when added up, they must be equal to the applied voltage. For example, in the circuit of Fig. 2-25 there are two voltage drops: 100 volts across R_1 and 300 volts across R_2. The sum of the voltage drops is 100 + 300 = 400 volts, which equals the applied voltage E.

This law is very important because it can be used in many ways in electricity and electronics. It is known as *Kirchhoff's voltage law.*

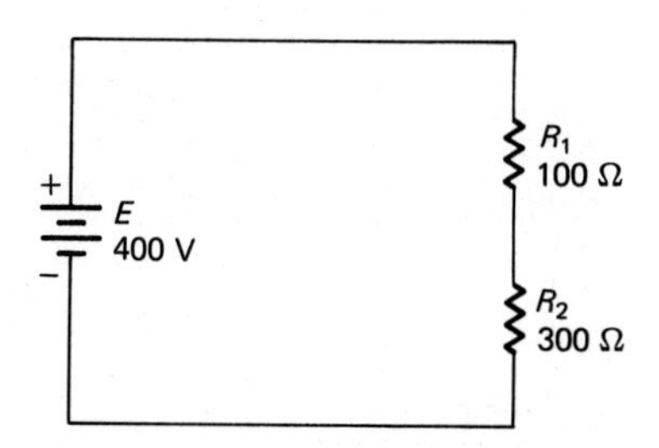

Fig. 2-25 Find the voltage across each resistor.

Test Setup Refer to the circuit in Fig. 2-26. Wire the circuit shown in Fig. 2-26*a*. The pictorial diagram of this circuit is shown in Fig. 2-26*b*. The voltage across the series resistor circuit is an ac voltage from the transformer secondary. The methods that you learned for finding voltage drops will work in this ac circuit as well as for dc circuits, provided there are no reactive components used. (Resistors are nonreactive components. Capacitors and inductors are reactive components.)

Procedure

step one Measure and record the voltage across the secondary—that is, the ac voltage between a and b. This measurement is taken with resistors R_1 and R_2 connected across the secondary winding. Record your voltage value. (You must use an ac voltmeter to make this measurement and all other ac voltage measurements.)

E = ______________________ volts ac

Use this value for E in each of the equations that follow.

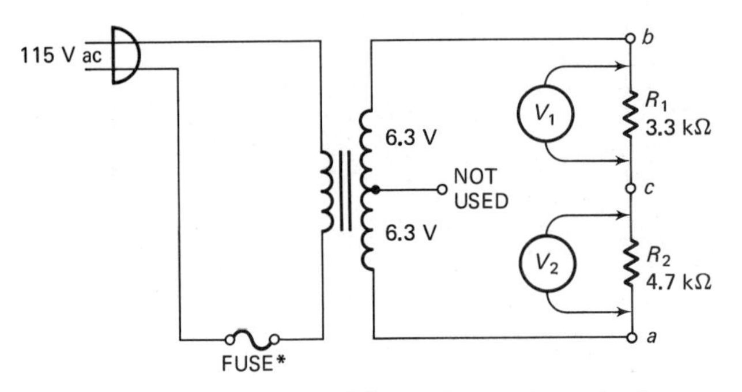

*This fuse may be an integral part of the transformer primary circuit. It will not be shown on all schematics, but it is always presumed to be part of the primary circuit.

(a)

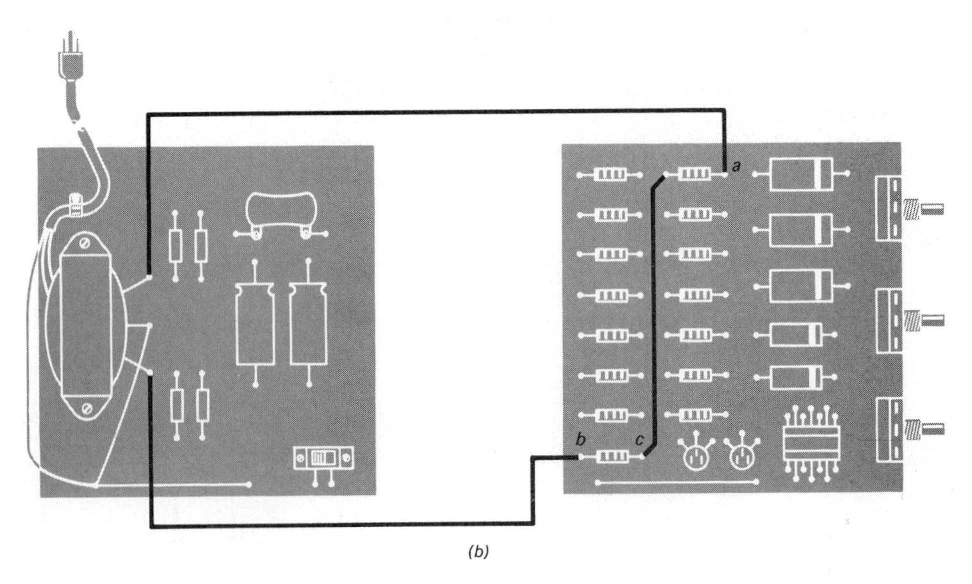

(b)

Fig. 2-26 The circuit for Part I.

You should measure about 12.6 volts across this transformer secondary. (In Chapter 4 you will study how transformers work.)

step two Calculate the voltage across R_1 using the method that was described in the theory section. For E use the value you found in Step 1. Record your calculated value.

$$V_1 = E\left(\frac{R_1}{R_1 + R_2}\right)$$

$$= E\left(\frac{3.3 \text{ kilohms}}{3.3 \text{ kilohms} + 4.7 \text{ kilohms}}\right)$$

$$= E\left(\frac{3.3 \text{ kilohms}}{8.0 \text{ kilohms}}\right)$$

The kilohms can be canceled.

$$V_1 = E(0.41)$$

V_1 = ________________ volts ac

step three Calculate the voltage across R_2. Use the value of E found in Step 1.

$$V_2 = E\left(\frac{R_2}{R_1 + R_2}\right)$$

$$= E\left(\frac{4.7 \text{ kilohms}}{8.0 \text{ kilohms}}\right)$$

$$V_2 = E(0.587)$$

V_2 = ________________ volts ac

step four Measure the ac voltage across R_1 and record the value.

V_1' = ________________ volts ac

step five Measure the ac voltage across R_2 and record the value.

V_2' = ________________ volts ac

Your measured values (V_1' and V_2') in steps 4 and 5 should agree with your calculated values (V_1 and V_2) in Steps 2 and 3.

step six Add the voltage values that you calculated in Step 2 and Step 3. Record the sum here.

$V_1 + V_2$ = ________________ volts ac

step seven Does the value that you obtained in Step 1 equal the applied voltage—that is, does it equal the voltage across the transformer secondary?

____

Yes or No

Your answer should be yes. This proves that the sum of the voltage drops equals the applied voltage.

The sum may not be *exactly* equal to the applied voltage. Remember that R_1 and R_2 are rated 3.3 kilohms and 4.7 kilohms. Their actual values will probably not be exactly equal to their rated values. Also, the voltmeter has some effect on the circuit, so it will change the voltage values slightly. You must also allow for the fact that there can be some error in the meter readings.

step eight Use Ohm's law to find the current through R_1. Use the measured voltage value from Step 4.

$$I_1 = \frac{V_1'}{R_1}$$

$$= \frac{V_1'}{3.3 \text{ kilohms}}$$

$I_1 =$ __________________ milliamperes (mA)

step nine Use Ohm's law to find the current through R_2. Use the measured voltage from Step 5.

$$I_2 = \frac{V_2'}{R_2}$$

$$= \frac{V_2'}{4.7 \text{ kilohms}}$$

$I_2 =$ __________________ milliamperes

The values in Steps 8 and 9 should be the same because the current is the same in all parts of a series circuit.

PART II

Purpose The purpose of this experiment is to demonstrate the basic theory of current calculations and measurement.

Theory You have seen that it is a simple matter to determine the voltage drops in circuits. This is a very useful thing to know when you are troubleshooting.

It is also possible to determine easily the amount of current flowing in parallel resistors. If two equal resistances are connected in parallel, then the current will divide equally. Half of the current will flow through each resistor. If three equal resistances are connected in parallel, then one-third of the total current will flow through each resistor.

If two parallel resistors have different resistance values, then there is a simple method of finding the current in each resistor. (Note: This method does not work for more than two resistors in parallel.) In Fig. 2-27 the two unequal resistors (R_1 and R_2) are connected in parallel. This means that the same current does *not* flow through each resistor. The arrows show the

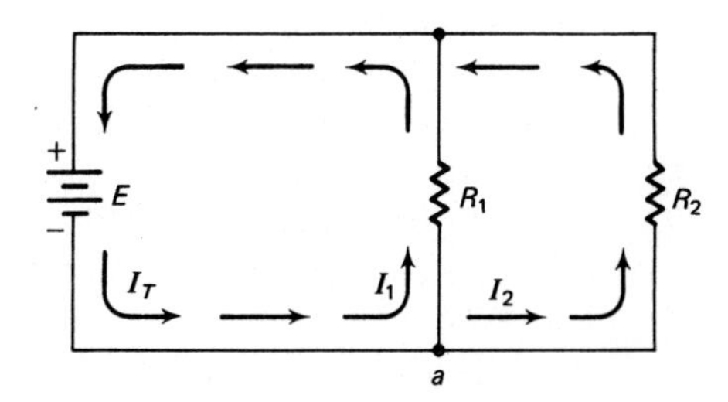

Fig. 2-27 The current divides in a parallel circuit.

electron-current paths. Note that the current divides at point *a* so that part of the total current flows through each resistor.

For the circuit of Fig. 2-27 the current values can be found by using these equations:

$$I_1 = I_T\left(\frac{R_2}{R_1 + R_2}\right)$$

and

$$I_2 = I_T\left(\frac{R_1}{R_1 + R_2}\right)$$

The two resistance values are first added (R_1 and R_2). Then the current through one resistor is found by dividing the resistance of the other resistor by the sum $R_1 + R_2$. This ratio is then multiplied by the total current I_T.

Figure 2-28 shows a circuit example. The current through resistor R_1 is

$$I_1 = I_T\left(\frac{R_2}{R_1 + R_2}\right)$$

$$I_1 = 300 \text{ milliamperes} \left(\frac{200}{300}\right)$$

$$= 200 \text{ milliamperes}$$

You can see here that two-thirds of the current flows through the smaller resistor.

To find the current through R_2,

$$I_2 = I_T\left(\frac{R_1}{R_1 + R_2}\right)$$

$$= 300 \text{ milliamperes} \left(\frac{100}{300}\right)$$

$$= 100 \text{ milliamperes}$$

Here you see that one-third of the current flows through the larger resistor.

Another very important rule in circuits is that *the sum of the currents in all parallel branches must equal the total current.* This is *Kirchhoff's current law,* and it is usually stated as follows: The sum of the currents that enter a point must equal the sum of the currents leaving that point.

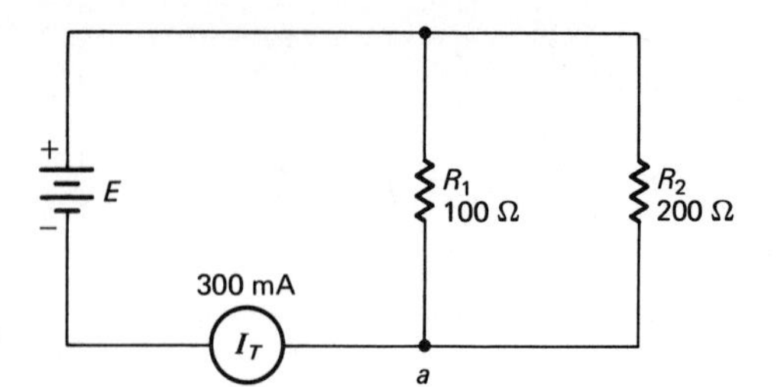

Fig. 2-28 How much current flows in each resistor?

In Fig. 2-28 the total current I_T entering point *a* is 300 milliamperes, and the total current leaving point *a* is 100 + 200 = 300 milliamperes.

Test Setup Refer to Fig. 2-29. Wire the circuit shown in Fig. 2-29*a*. The pictorial diagram for this circuit is shown in Fig. 2-29*b*. The voltage across the series-parallel resistor circuit is an ac voltage from the transformer secondary. The method described for finding current in parallel resistors can be used for both ac and dc.

Procedure

step one Measure and record V_1, the voltage across R_1.

V_1 = ______________ volts ac

step two Calculate the current through R_1 using Ohm's law. That is

$$I_T = \frac{V_1}{R_1}$$

$$= \frac{V_1}{3.3 \text{ kilohms}}$$

I_T = ______________ milliamperes

This is the total current entering point *a*.

step three Using the method described in the theory section, find current I_2 through R_2. For I_T use the value you found in Step 2.

$$I_2 = I_T\left(\frac{R_3}{R_2 + R_3}\right)$$

$$= I_T\left(\frac{4.7 \text{ kilohms}}{3.3 \text{ kilohms} + 4.7 \text{ kilohms}}\right)$$

$$= I_T\left(\frac{4.7 \text{ kilohms}}{8.0 \text{ kilohms}}\right)$$

I_2 = ______________ milliamperes

step four Using the method described in the theory section, find current I_3, which is the current through R_3. For I_T use the value you found in Step 2.

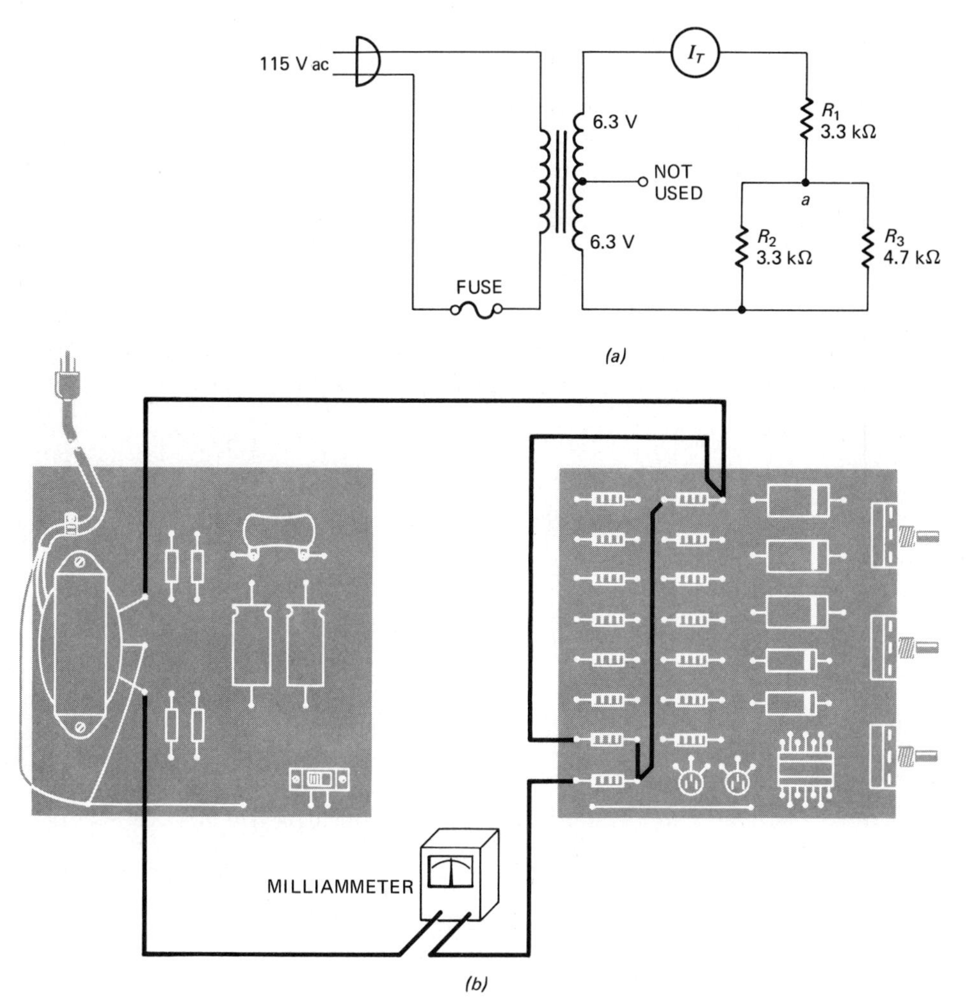

Fig. 2-29 Circuit for Part II.

$$I_3 = I_T\left(\frac{R_2}{R_2 + R_3}\right)$$

$$= I_T\left(\frac{3.3 \text{ kilohms}}{3.3 \text{ kilohms} + 4.7 \text{ kilohms}}\right)$$

$$= I_T\left(\frac{3.3 \text{ kilohms}}{8.0 \text{ kilohms}}\right)$$

I_3 = ______________ milliamperes

step five Add the current values that you obtained in Steps 3 and 4.

$I_2 + I_3$ = ______________ milliamperes

Do these currents add to equal the total current of Step 2?

____
Yes or No

Your answer should be yes. *The sum of the currents in the two resistor branches should equal the total current.*

step six Measure and record V_2, the voltage across R_2.

V_2 = ____________________ volts

step seven Calculate the current through R_2 using Ohm's law.

$$I_2 = \frac{V_2}{R_2}$$

$$= \frac{V_2}{3.3 \text{ kilohms}}$$

I_2 = ____________________ milliamperes

step eight Measure and record V_3, the voltage across R_3.

V_3 = ____________________ volts

step nine Are the voltage drops across R_2 and R_3 equal? (Compare your answers in Steps 6 and 8.)

____
Yes or No

The two voltage values should be the same because *the voltage across all parts of a parallel circuit is the same.*

step ten Calculate the current through R_3 using Ohm's law.

$$I_3 = \frac{V_3}{R_3}$$

$$= \frac{V_3}{4.7 \text{ kilohms}}$$

I_3 = ____________________ milliamperes

step eleven Do the currents in Steps 5 and 7 compare with the currents in Steps 2 and 3? Your answer should be yes. This shows that the method of calculating currents which was described in the theory section gives the correct answers.

Conclusion If there are no reactive components (inductors or capacitors) in a circuit, you can use Ohm's law for ac circuits as well as dc circuits.

Technicians use the above methods to compute the voltage or current that should be obtained in a circuit. Then they measure the actual voltage or current. If their measured value does not approximately equal their computed value, they know that there is trouble in the circuit. Before you can become good at troubleshooting you have to practice these methods of finding voltages and currents.

SELF-TEST WITH ANSWERS

(Answers with discussions are given at the end of the chapter.)

1. Which of the following semiconductor diodes is used as a voltage regulator? (*a*) Varactor diode; (*b*) Tunnel diode; (*c*) Zener diode; (*d*) Rectifier diode.
2. Which of the following is a resistor that has a resistance which depends upon its applied voltage? (*a*) VOR; (*b*) VIR; (*c*) VDR; (*d*) VAR.
3. Which of the following is not one of the typical uses of resistors in circuits? (*a*) Generate heat; (*b*) Divide voltage; (*c*) Store energy; (*d*) Limit current.
4. Which of the following statements is true? (*a*) It is easier to pass a high-frequency current through an inductor than it is to pass a low-frequency current; (*b*) It is easier to pass a low-frequency current through an inductor than it is to pass a high-frequency current.
5. A small variable capacitor is placed in parallel with a larger variable capacitor in order to change its range of capacitance values. The smaller capacitor is called (*a*) a trimmer; (*b*) a padder; (*c*) a variac; (*d*) a taper.
6. Which of the following is not one of the typical uses of capacitors in circuits? (*a*) Generate heat; (*b*) Store energy; (*c*) Pass high frequency and reject low frequency; (*d*) Introduce an ac voltage drop.
7. Which of the following statements is true? (*a*) It is easier to pass a high frequency through a capacitor than it is to pass a low frequency; (*b*) It is easier to pass a low frequency through a capacitor than it is to pass a high frequency; (*c*) Originally capacitors were used for storing energy, but this is no longer one of their uses in circuits; (*d*) Regardless of how high the dc voltage is across a capacitor, it will never permit current to flow.
8. When a variable resistor is connected so that its main function is to control circuit current, it is called a (*a*) controller resistor; (*b*) potentiometer; (*c*) varistor; (*d*) rheostat.
9. Which one of the following types of capacitors is polarized and can only be used in pulsating dc circuits? (*a*) Mica capacitor; (*b*) Paper capacitor; (*c*) Electrolytic capacitor; (*d*) Air dielectric capacitor.
10. If a capacitor has a rating of NPO it means that (*a*) it must be used in pulsating dc circuits; (*b*) it has three leads; (*c*) its capacitance decreases with a small increase in temperature; (*d*) its capacitance remains nearly constant when the temperature changes by a small amount.

ANSWERS TO SELF-TEST

1. (*c*)
2. (*c*)—A VDR is a voltage-dependent resistor.
3. (*c*)—Inductors and capacitors store energy, but resistors do not.
4. (*b*)
5. (*a*)—Variac and taper are not capacitor terms.

6. (*a*)—Resistors generate heat. Capacitors and inductors do not normally generate heat.
7. (*a*)—Compare with Question 4.
8. (*d*)—See Fig. 2-5*a*.
9. (*c*)—Electrolytic capacitors are used as filters in power-supply circuits. Electrolytic capacitors are identified on drawings by a + sign at one of the plates. (See Fig. 2-15.)
10. (*d*)—NPO stands for *N*egative *P*ositive Zero. These capacitors do not change capacitance when their temperature changes.

what are the three-terminal components used in electronic circuits? 3

INTRODUCTION

The history of electronics is usually considered to have begun when Lee DeForest invented the audion vacuum tube. Before the *audion*, or *triode*, as the modern version is called today, there was some radio communication, but there were no circuits for amplifying signals. Therefore radio communication depended primarily upon the strength of the signal being transmitted. Long-distance communication was difficult and often impossible.

Great advances in reliable communications were not possible until radio signals could be amplified. The audion tube first made this possible.

This chapter explains how the triode vacuum tube works. It also explains how different types of solid-state components work. All the amplifying components have one thing in common: They all have *three terminals*.

The material in this chapter describes only how these components operate. What happens when you put the components into circuits will be discussed in later chapters.

You will be able to answer these questions after studying this chapter.

- ☐ **How does a triode tube work?**
- ☐ **What is a tetrode?**
- ☐ **What is a pentode?**
- ☐ **What is a bipolar NPN transistor?**
- ☐ **What is a bipolar PNP transistor?**
- ☐ **What is a JFET?**

- ☐ **What is a depletion-type MOSFET?**
- ☐ **What is an enhancement-type MOSFET?**
- ☐ **What are P-type MOSFETs?**
- ☐ **What is a thyristor?**

INSTRUCTION

How Does a Triode Tube Work?

The principle of the triode is shown in Fig. 3-1. This tube has a cathode, a plate, and a grid of wire in between. A small amount of negative voltage is placed on the grid. This voltage is called the *bias*, and its purpose is to limit the number of electrons that can pass from the cathode to the plate.

The arrows show that many of the electrons pass through the spaces in the grid, but some are repelled by the negative charge on the grid. You will remember that "like charges repel," and therefore the negative grid repels some negative electrons back toward the cathode.

The more negative the grid bias voltage, the more electrons the grid repels and the fewer the number that arrive at the plate. The voltage on the grid actually controls the number of electrons that reach the plate. Current flow is measured by the number of electrons that pass a point each second, so you can see that the negative grid voltage actually controls the current at the plate. This current is called the *plate current* (I_P).

How Is a Triode Constructed?

Triodes are not made like the one shown in Fig. 3-1. Instead, they are built as shown in Fig. 3-2. In Fig. 3-2*a* you see that the plate consists of a cylinder completely around the tube's other elements. The control grid is located between the cathode and the plate. It consists of a wire grid and is used

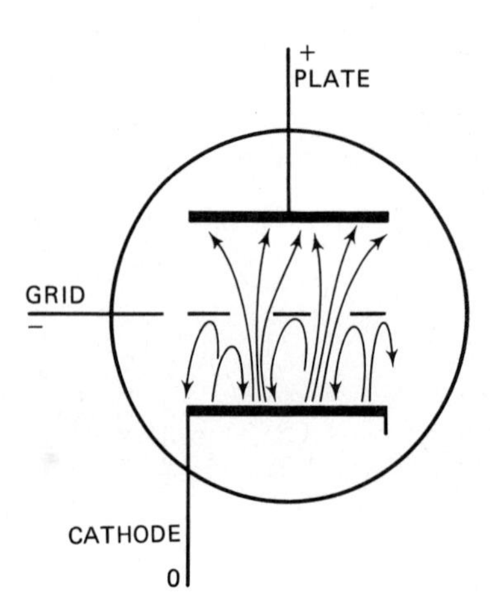

Fig. 3-1 In a triode vacuum tube the grid controls electron current.

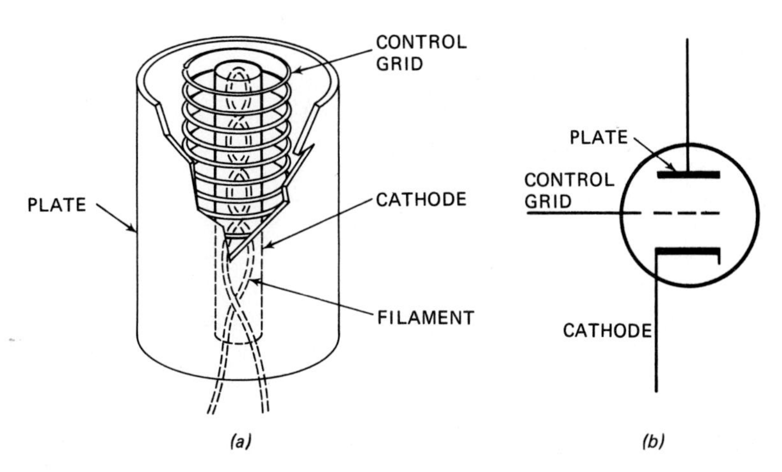

Fig. 3-2 The vacuum-tube triode. (*a*) Construction. (*b*) Symbol.

to control current flow within the tube. The cathode is heated by an internal filament. It is the cathode that actually emits the electrons. In some less common triodes, and diodes, electrons are emitted directly from the filament.

In the symbol for the triode, shown in Fig. 3-2*b*, the filament wires are not usually shown.

How Are Grid Voltage and Plate Current Related?

The relationship between the voltage on the control grid and the current flowing in the plate circuit is shown in Fig. 3-3. In Fig. 3-3*a* there is a negative voltage of 5 volts on the grid. The cathode is at 0 volts potential, and the plate is positive with respect to the cathode. In the plate circuit you see a milliammeter for measuring plate current. For a grid voltage of −5 volts the plate current is 100 milliamperes.

In Fig. 3-3*b* the grid voltage has been reduced from −5 volts to −3 volts. The grid is not as negative as it was before, and therefore it does not repel the flow of electrons from the cathode to the plate as much. More electrons per second are arriving at the plate. Since current is a measure of electrons per second, the plate current will increase when the grid voltage is made less negative. You can see in the illustration that the plate current has increased to

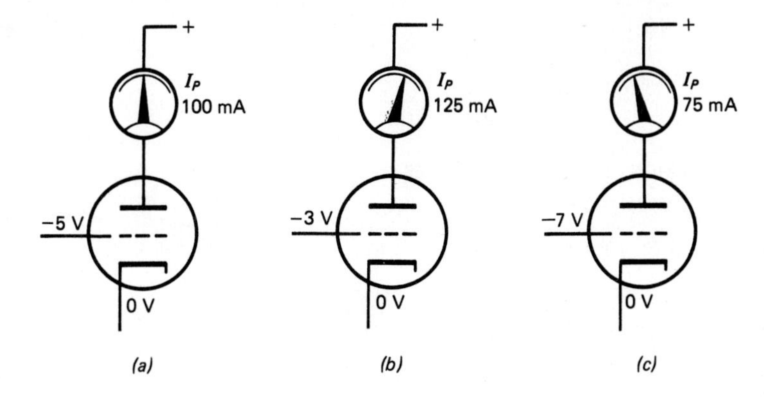

Fig. 3-3 The grid voltage controls the plate current. (*a*) With a grid voltage of −5 volts, the plate current is 100 milliamperes. (*b*) Making the grid voltage less negative (−3 volts) causes the plate current to increase. (*c*) Making the grid voltage more negative (−7 volts) causes the plate current to decrease.

125 milliamperes when the grid voltage is reduced from -5 volts to -3 volts.

As you see in Fig. 3-3*c*, making the grid bias voltage more negative than -5 volts causes the plate current to decrease. With a grid bias voltage of -7 volts the plate current is reduced to 75 milliamperes.

Summary

1. The invention of the audion made it possible to amplify weak radio signals. The modern version of the audion is called a *triode.*
2. The electrodes of a triode are called the *cathode, control grid,* and *plate.*
3. The plate of a triode is normally positive with respect to the cathode, and the control grid is normally negative with respect to the cathode.
4. Making the grid less negative increases the plate current. Making the grid more negative decreases the plate current.
5. The dc voltage on the control grid of a vacuum tube is called the *bias voltage.*

What Is a Tetrode?

You can understand one important disadvantage of triodes by reviewing the construction of any capacitor, as shown in Fig. 3-4. A capacitor is made by taking two metal plates and separating them by an insulation (or dielectric) (Fig. 3-4*a*). This situation exists within the triode tube. The grid is made of metal, and the plate is made of metal. There is a vacuum (insulator) between the grid and the plate. Therefore the plate and grid serve as two plates of a capacitor.

In Fig. 3-4*b* the grid-to-plate capacitance is represented by C_{gp}. The grid and cathode also form a capacitor. Grid-to-cathode capacitance is represented by C_{gk}. (In vacuum-tube studies you will often see the letter *k* used to represent cathode. This is so it will not be confused with capacitors, which are represented by the letter *C*.)

In triode-tube circuits, capacitance allows part of the signal to be coupled from the plate to the grid through C_{gp}. This is especially troublesome at high frequencies because it results in some loss of gain. To get around this problem, a so-called *screen grid* was placed between the control grid and the plate. The resulting tube is called a *tetrode,* and its symbol is shown in Fig. 3-5. An important advantage of a tetrode over a triode is its increased gain.

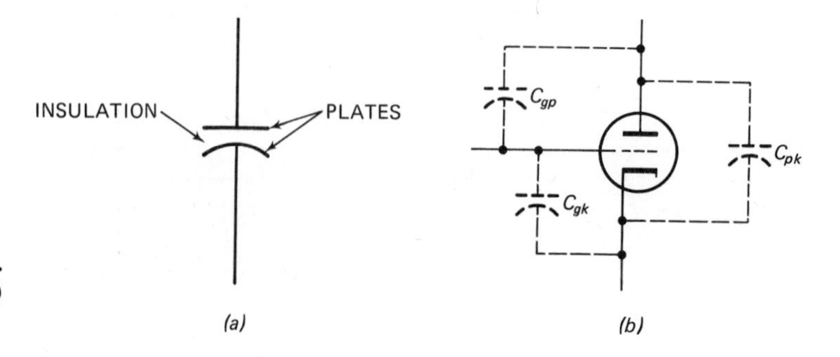

Fig. 3-4 Capacitance in a triode. (*a*) Construction of a capacitor. (*b*) Interelectrode capacitances.

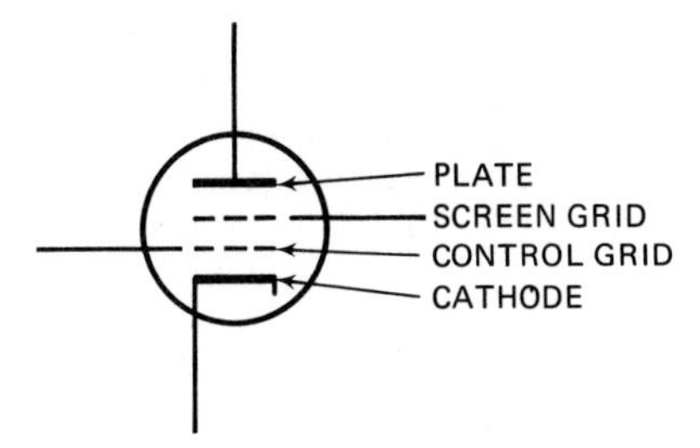

Fig. 3-5 Symbol for a tetrode vacuum tube.

What Is Negative Resistance?

While the addition of the screen grid reduces the capacitance between the control grid and the plate, it produces another problem. This problem is illustrated in Fig. 3-6. On this graph of tetrode plate voltage E_p and plate current I_p, you will note that as the plate voltage is increased from 0 to 25 volts, the plate current also increases up to point *a*.

Between points *a* and *b* the plate current decreases despite the fact that the plate voltage is still increasing. You know that when you increase the voltage across a resistor, the current through it increases. However, in Fig. 3-6 we have a case where the voltage across the tube is increased but the plate current decreases. This is sometimes called *negative resistance* because it is the opposite effect of an increase in voltage across a resistor.

Negative resistance is caused by the fact that after the voltage has reached point *a* the electron current becomes strong enough to knock electrons from the metal plate. The electrons which are jarred loose from the plate are called *secondary electrons*, and they flow to the positive screen grid. The overall result is an increase in the screen-grid current and a decrease in the plate current. In other words, more electrons are now going to the screen grid than are going to the plate.

After point *b* is reached on the curve the plate becomes so positive that it at-

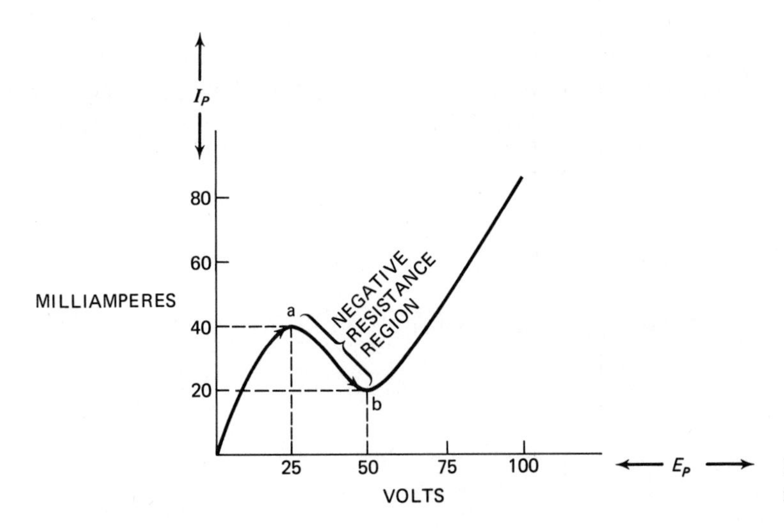

Fig. 3-6 Curve showing negative resistance of a tetrode.

tracts the secondary electrons, so they no longer go to the screen grid. Because of the negative-resistance region on the characteristic curve of Fig. 3-6, the tetrode was found to be unsuitable for some uses.

Summary

1. There are four electrodes in a tetrode. They are called the *cathode, control grid, screen grid,* and *plate.*
2. Important functions of the screen grid are to reduce the plate-to-grid capacitance C_{gp} and to increase gain.
3. Plate-to-grid capacitance in a triode limits its high-frequency performance. The capacitance causes some loss of gain.
4. The disadvantage of the tetrode is its negative resistance. However, it provides greater gain than a triode.
5. Negative resistance occurs when secondary electrons from the plate go to the screen grid. This increases screen current and reduces plate current. This effect makes tetrodes unsuitable for some uses.

What Is a Pentode?

To eliminate the effect of negative resistance a third grid, called the *suppressor grid,* was added between the screen grid and the plate. This is shown in Fig. 3-7. The suppressor grid is maintained at a negative voltage with respect to the plate and screen grid. This is done by connecting the suppressor grid to the cathode, either internally or by an external connection. Its purpose is to repel back to the plate the negative secondary electrons that are jarred loose from the plate. It prevents the secondary electrons from getting to the screen grid, and therefore the screen-grid current is never larger than the plate current.

The pentode is more valuable as an amplifier than the tetrode, but it is not without faults. Every time you add a grid inside a vacuum tube you also add to its noise production. Therefore tetrodes and pentodes are both noisier than triodes. The noise in a loudspeaker that sounds like a hissing or a waterfall is caused by tube noises.

The pentode does not have the grid-to-plate capacitance shown in Fig. 3-4 and is better for amplifying high-frequency signals than either a triode or tetrode.

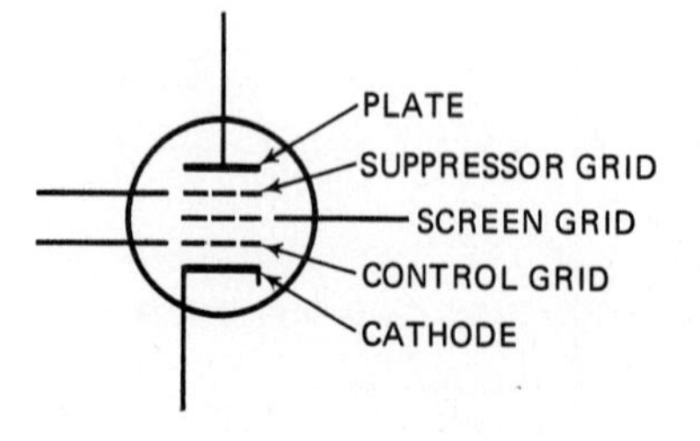

Fig. 3-7 Symbol for a pentode vacuum tube.

Summary

1. A pentode tube has five electrodes. They are called the *cathode, control grid, screen grid, suppressor grid,* and *plate.*
2. The suppressor grid eliminates the negative-resistance problem of the tetrode.
3. The suppressor grid is negative with respect to the plate. It repels secondary electrons back to the plate and prevents them from going to the screen grid.
4. Pentodes are better than triodes and tetrodes for high-frequency amplification.
5. The disadvantage of pentodes (and tetrodes) is that they add undesired noises to the desired signal.

What Is a Bipolar NPN Transistor?

Before we start the discussion of transistors it is important to review some basic principles of electric-current flow. In this book we have treated current flow as being a flow of electrons. It is presumed that the negative electrons move away from the negative terminal of the source voltage and toward the positive terminal. (Like charges repel and unlike charges attract.)

Electrons are considered to be the carriers of electric current, and they are called *negative-charge carriers.* However, there are some materials that have very few electrons available for current flow. An example is the *P-type* semiconductor material used for making transistors. In P-type material most of the current flow is due to *positive-charge carriers* which are called *holes.*

Another type of semiconductor material is called *N-type.* In this material most of the charge carriers are electrons.

All semiconductor materials (both P-type and N-type) will conduct electricity, but will not conduct as well as a conductor. When a component is made of only one kind of material (either P-type or N-type) the component is said to be *unipolar.* This means that its operation depends primarily upon only one type of charge carrier (electrons or holes).

A *bipolar* component depends upon both holes and electrons for its operation.

The relationship between electron flow and hole flow can be explained with the help of Fig. 3-8. The balls in this illustration represent electrons, and the space represents a hole. You see here that the electrons move from left to right or negative to positive. The hole moves from right to left, or positive to negative.

Whenever current flows in a material there are always two kinds of flow, electron flow and hole flow. In P-type material the current consists primarily of hole flow, but there is also some electron flow. In P-type material the holes are called *majority charge carriers,* and electrons are called *minority charge carriers.* In N-type material the electrons are the *majority* charge carriers and the holes are the *minority* charge carriers.

Figure 3-9 shows how one type of bipolar transistor is made. It consists of

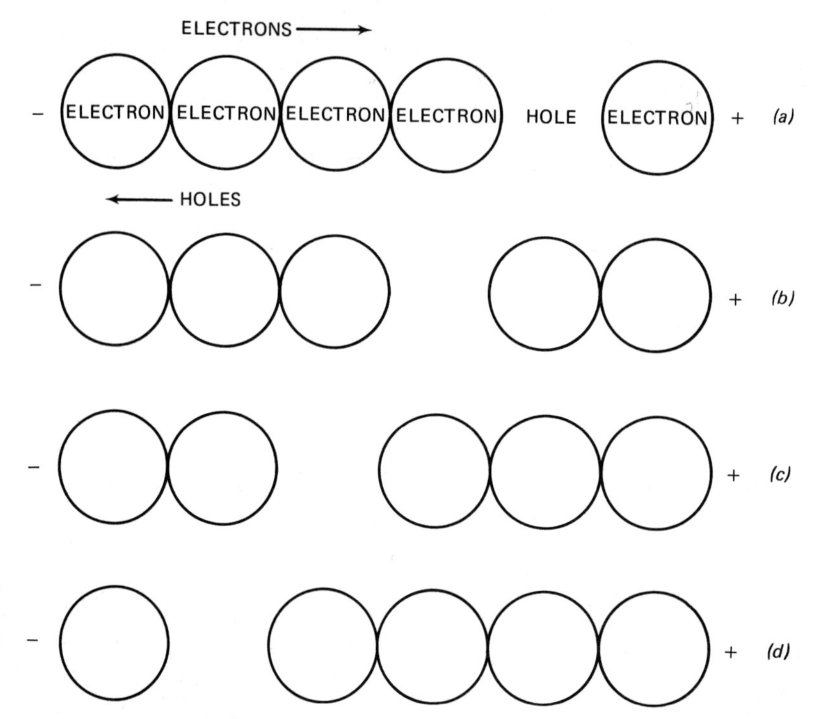

Fig. 3-8 **This illustration shows how electron flow and hole flow are related. The hole in (*a*) appears to be moving left in (*b*), (*c*), and (*d*). In the meantime, three electrons appear to have moved right one at a time until they appear as in (*d*).**

three layers of material. Two of them are made of N-type material and the other is made of P-type material. This type is called an *NPN transistor*. The charge carriers within the transistor consist of both electrons and holes. We are also interested in what kind of current flows externally to the transistor. This external current is always considered to be electrons. You see in Fig. 3-9 that the transistor has three sections, which are called the *emitter*, the *base*, and the *collector*. It will help you to remember this and also the gen-

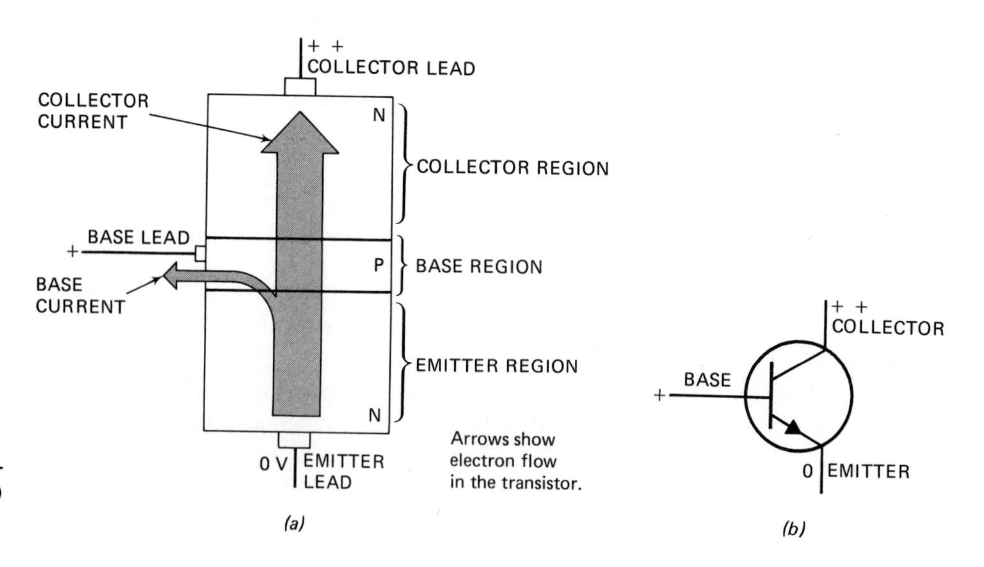

Fig. 3-9 **The NPN bipolar transistor. (*a*) Current flow. (*b*) Symbol.**

eral functions of the sections if you will simply remember that their functions relate to those of a triode vacuum tube as follows:

Emitter . . . cathode
Base . . . control grid
Collector . . . plate

The diagram in Fig. 3-9*a* shows you that a small number of electrons will pass through into the P region (base) and out the base lead of the transistor. However, most of the electrons are passed through the base to the collector.

The + sign on the base lead in Fig. 3-9*a* means that the base is positive with respect to the emitter (0 volts). The + + sign on the collector means that the collector is more positive than the base. These are the normal operating voltages. If you apply the voltages across the transistor as shown in Fig. 3-9, the current will consist primarily of electrons flowing between the emitter and the collector. However, remember that a small amount of electron flow will also go through the base.

The principle of operation of this device, and one of the most important things for you to remember, is that a small amount of current in the base will cause a large amount of current to flow in the collector. Also a small change in base current will cause a large change in collector current. The transistor is said to be a *current-operated device* because the base current controls the amount of collector current.

The symbol for the NPN transistor is shown in Fig. 3-9*b*. The emitter is represented by an arrow that is pointing out of the transistor. This arrow is *N*ot *P*ointing i*N*, which is an easy way to remember the NPN transistor symbol.

What Is a Bipolar PNP Transistor?

Figure 3-10 shows you another way that a transistor can be made. It is called a *PNP bipolar transistor*, or simply a *PNP transistor*. You can see here

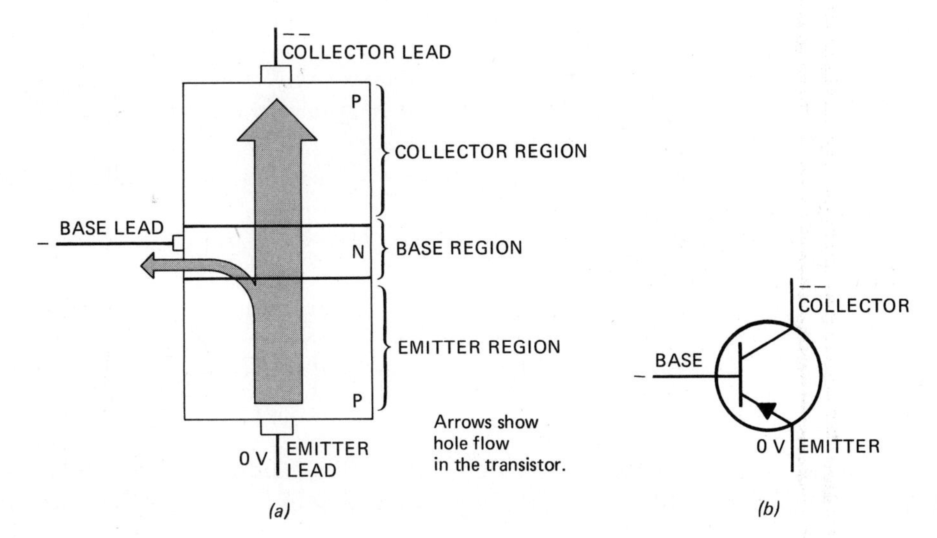

Fig. 3-10 The PNP transistor. (*a*) Hole flow. (*b*) Symbol.

that the voltage on the base and emitter are opposite to those shown in Fig. 3-9. The arrows in Fig. 3-10*a* show hole flow through the transistor. Note that the holes flow from the emitter toward the negative collector and also toward the negative base. In the emitter and collector the holes are majority charge carriers. This is because the hole is considered to be a positive charge. Remember that the electron is attracted to positive voltages and the hole is attracted to negative voltages.

As with the NPN type a small amount of base current causes a large amount of collector current. Also a small change in base current causes a large change in collector current.

The symbol for the PNP transistor is shown in Fig. 3-10*b*. Observe that the arrow in the symbol points inward. There is an important thing to remember about *all* symbols used for semiconductor or solid-state electronic components. The arrow always points *toward* an N region in the material. The vertical line in the symbol represents the base of the bipolar transistor. In the PNP type (Fig. 3-10*b*), N-type material is used for the base, and the arrow is pointing toward it. In the NPN type (Fig. 3-9) the arrow points toward the emitter, which is made of N-type material.

Summary

1. There are two kinds of charge carriers for electricity: negative electrons and positive holes.
2. In N-type material the electrons are majority charge carriers and the holes are minority charge carriers.
3. In P-type material the holes are majority charge carriers and the electrons are minority charge carriers.
4. N-type material has some of both types of charge carriers, but electrons are the primary current carriers.
5. P-type material has some of both types of charge carriers, but holes are the primary current carriers.
6. Bipolar transistors have three electrodes. They are emitter, base, and collector.
7. There are two kinds of bipolar transistors: NPN and PNP.
8. Bipolar transistors use both kinds of charge carriers—that is, they use both electrons and holes.
9. An NPN transistor normally operates with its base and its collector positive with respect to its emitter.
10. A PNP transistor normally operates with its base and its collector negative with respect to its emitter.

How Are Base Current and Collector Current Related in a Bipolar Transistor?

A very important relationship between the base current of an NPN transistor and its collector current is shown in Fig. 3-11. It shows that the *base*

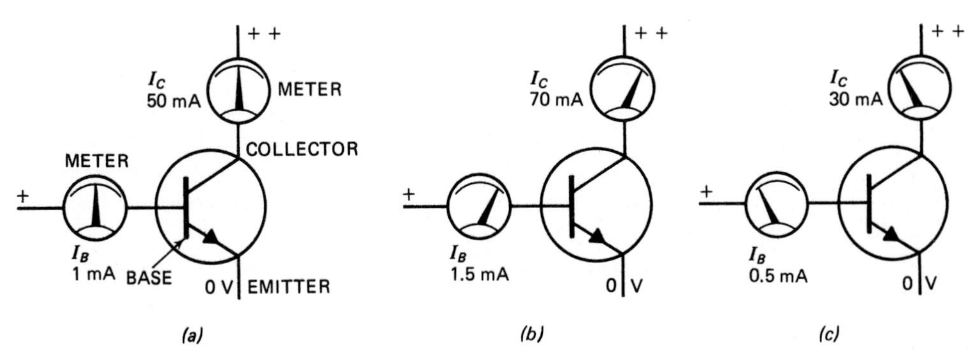

Fig. 3-11 The collector current I_C depends on the base current I_B. (*a*) 50 microamperes of collector current for a 1-milliampere base current. (*b*) Increasing the base current gives more I_C. (*c*) Decreasing the base current gives less I_C.

current controls the *collector current* in a transistor. Compare this with the vacuum tube in Fig. 3-3, where the *grid voltage* controlled the *plate current.*

In Fig. 3-11*a* you see that the base voltage is positive, and this produces a base current of 1 milliampere. When this base current I_B is present, it produces a collector current I_C of 50 milliamperes. When the base current is increased to 1.5 milliamperes, as shown in Fig. 3-11*b*, the collector current increases to 70 milliamperes. Now this is a very important point: *When the base current is increased, the collector current also increases.*

In Fig. 3-11*c* the base current is reduced to 0.5 milliampere, a lower value than it had in Fig. 3-11*b*. Decreasing the base current causes the collector current to decrease to 30 milliamperes.

To summarize, you can see that when the base current is increased the collector current increases, and when the base current is decreased the collector current decreases.

If this had been a PNP transistor (see Fig. 3-10), the action would have been the same except that the collector and the base would both be negative. *Making the base more negative in a PNP transistor increases the base current and also increases the collector current.*

Summary

1. In an NPN transistor, increasing the base current also increases the collector current.
2. In a PNP transistor, increasing the base current also increases the collector current.
3. The base current of an NPN transistor is increased by making the base more positive.
4. The base current of a PNP transistor is increased by making the base more negative.

What Is a JFET?

The letters JFET mean *J*unction *F*ield-*E*ffect *T*ransistor. It is a so-called unipolar type of transistor. This is so named because the majority charge carriers flowing through the transistor are either only electrons (as in the *N-channel JFET*) or only holes (as in the *P-channel JFET*).

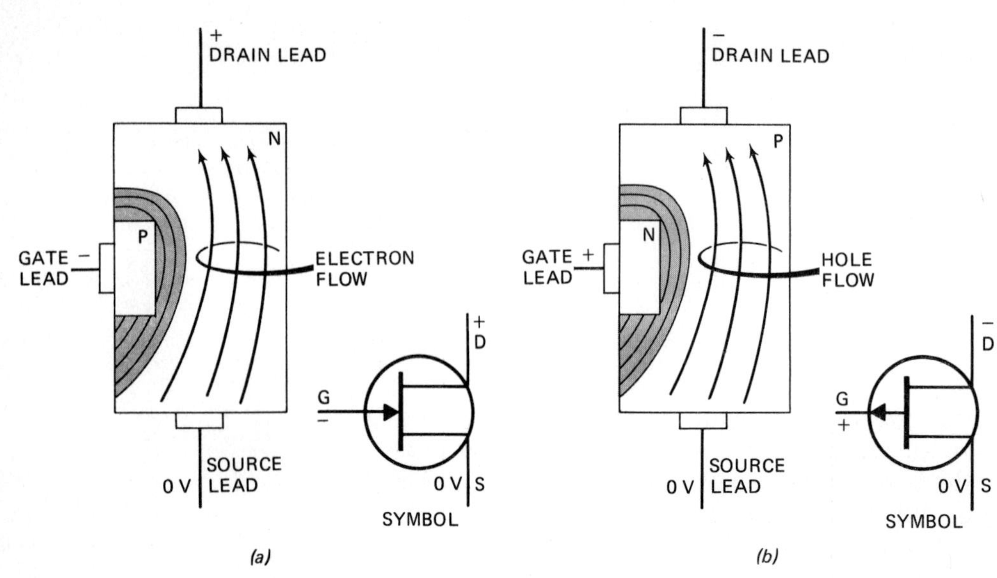

Fig. 3-12 Junction field-effect transistors. (*a*) N-channel type. (*b*) P-channel type.

The principle of operation of the JFET is shown in Fig. 3-12. In Fig. 3-12*a* you see an N-channel JFET. The P-type material is called the *gate region,* and the N-type material is called the *channel* of this transistor. The three parts of this transistor are called the *source,* the *gate,* and the *drain.* You can compare these three elements with those of a triode vacuum tube, as follows:

Source . . . cathode
Gate . . . grid
Drain . . . plate

If you compare the basic operation of the N-channel JFET of Fig. 3-12*a* with that of the triode vacuum tube of Fig. 3-3, you will find it to be very similar.

In normal operation the drain, which is the output side, is made positive. Electron current flows through the N channel. The gate lead is made slightly negative. This negative voltage on the gate causes an electric field around the gate region, which is shown as a shaded area in Fig. 3-12*a*. Electrons cannot pass through the shaded field area but can pass through the nonshaded area. The more negative the gate lead is made, the larger the field and the smaller the region through which the electrons can flow. Thus the electron current is less.

Eventually you could make the gate lead so negative that the field would reach across the transistor. This would cause electron flow to stop completely. In this JFET the voltage on the gate lead controls the number of electrons that can move through the device. In other words, the gate *voltage* controls the drain *current.* Therefore the JFET is a voltage-operated component (like a vacuum tube).

Figure 3-12*b* shows a P-channel device. The difference between the N-channel and P-channel JFETs is that the current flow through the P-channel type consists of holes rather than electrons. Also the gate is made of P-type material. The gate lead of the P-channel JFET is made positive with respect

Table 3-1 How Gate Voltage Affects Drain Current

	N-Channel	P-Channel
Polarity of drain voltage with respect to the source	Positive	Negative
Polarity of gate voltage with respect to source	Negative	Positive
To increase drain current . . .	Make gate less negative	Make gate less positive
To decrease drain current . . .	Make gate more negative	Make gate more positive

to the source, and the drain lead is made negative with respect to the source. The positive voltage on the gate repels the holes. Holes cannot pass through the field around the gate, which is shown by a shaded area. Although the charge carriers within the P-type JFET are holes, the circuit current external to the JFET consists of electrons.

You should note that the two symbols in Fig. 3-12 differ as to whether the gate-lead arrow points in or out. The vertical line in the symbol represents the channel. You will note that in Fig. 3-12*a* the channel, which is an N-type material, has the gate-lead arrow pointing toward it. In Fig. 3-12*b* the channel is P-type material and the gate-lead arrow points away from it.

Table 3-1 describes how the JFET is biased and also how the gate voltage affects the drain current. This is a very important table because it not only describes the JFET voltages and currents, but also shows the voltages and currents for the depletion-type MOSFETs that you are going to study next.

Summary

1. A JFET is a unipolar component (uses either electrons or holes).
2. There are three electrodes in a JFET. They are *source, gate,* and *drain.*
3. In a JFET, current flows from the source to the drain through a channel. The JFET is identified by the type of semiconductor material in its channel.
4. In an N-channel JFET the gate is negative with respect to the source and the drain is positive with respect to the source.
5. P-channel JFETs are normally operated with a positive gate and a negative drain. (Voltages are given with respect to the source voltage.)
6. A JFET is a voltage-operated component (like a vacuum tube).

What Is a Depletion-type MOSFET?

There is an important disadvantage of the JFET that can be understood by referring again to Fig. 3-12. We will discuss this disadvantage with relation to the N-channel JFET of Fig. 3-12*a*. You will note that the P material and N material form a junction in the gate and channel region. This junction is the same as the junction of a solid-state diode. Figure 3-13 shows how a solid-state junction diode is made.

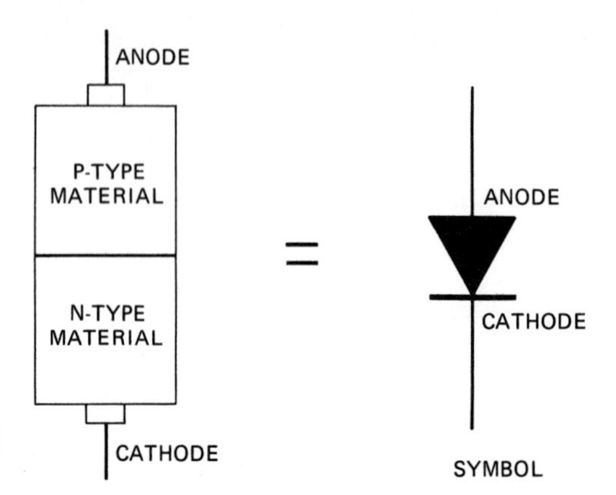

Fig. 3-13 Construction and symbol for a solid-state junction diode.

Observe that the anode is made of P-type material and the cathode is made of N-type material. You can review the action of this diode by referring to Fig. 3-14.

When the anode is positive with respect to the cathode, as shown in Fig. 3-14*a*, current flows through it. This is called *forward bias*. When the diode is *reverse-biased*, as shown in Fig. 3-14*b*, its anode is negative with respect to its cathode, and no current flows.

Refer again to the N-channel JFET of Fig. 3-12*a*. The gate and channel form a PN junction just like the one in the diode. As long as this junction is reverse-biased, there is *no current flow* in the gate lead. However, if the gate lead of the N-channel JFET is made positive with respect to the channel, the PN junction will be *forward-biased*. This will cause gate current to flow. This is highly undesirable because the JFET is supposed to operate with no gate current. It is the *voltage* that controls the drain current, and it is *always* presumed that there is no gate-current flow in the JFET circuit.

The *MOSFET* is similar to the JFET, but an *insulating* layer is placed around the gate. The insulating region around the gate is clearly shown in Fig. 3-15.

Since there is insulation around the gate it is much more difficult to cause gate-current flow even if the gate-channel junction is accidentally forward-biased. [You may still see this component referred to in older textbooks as an insulated gate FET (IGFET).]

The insulation is made of a *M*etal-*O*xide *S*emiconductor material, hence

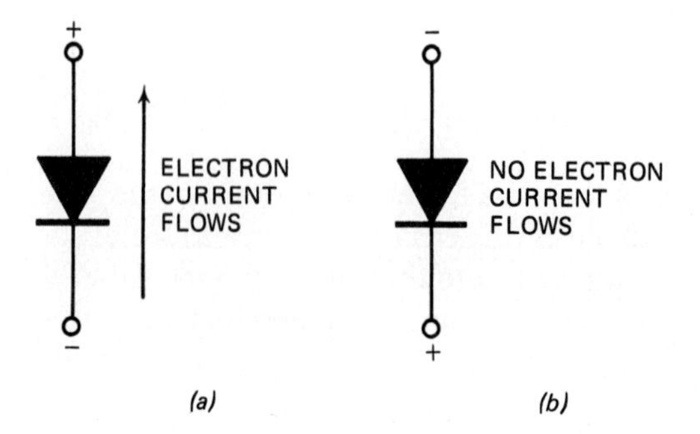

Fig. 3-14 A junction diode. (*a*) Forward-biased; (*b*) reverse-biased.

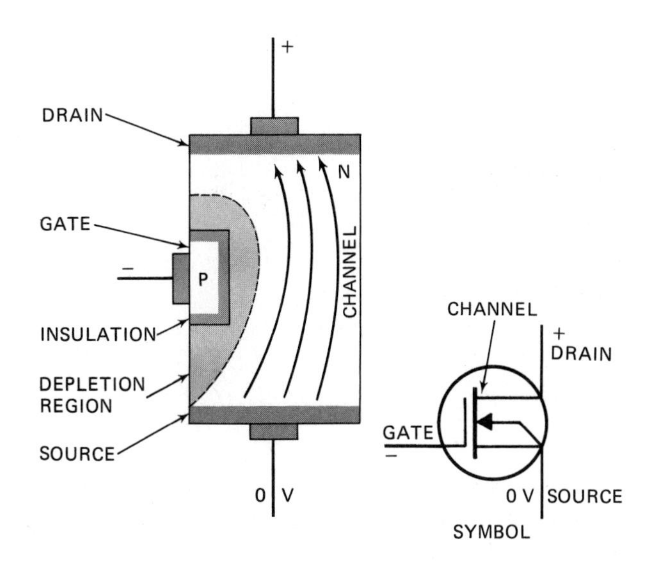

Fig. 3-15 An N-channel depletion MOSFET.

the name MOSFET. It has the same theory of operation as the N-channel JFET, and Table 3-1 can be used to determine polarities as well as the effect of gate-voltage changes.

Note the symbol for the N-channel MOSFET. The gate is shown separated from the channel, indicating that there is insulation between the gate and the channel. The arrow points toward the N channel.

As the negative bias on the gate of the N-channel MOSFET is increased, the shaded area in Fig. 3-15 increases. This means that the area for electron flow becomes smaller and that the electron current is decreased. Another way of saying this is that a more negative gate voltage causes greater depletion of the conduction path. For this reason the component in Fig. 3-15 is called an *N-channel depletion-type metal-oxide semiconductor field-effect transistor.* For the sake of simplicity it is called an *N-channel depletion MOSFET.*

What Is an Enhancement-type MOSFET?

Since there is an insulation region between the gate and the channel, it is possible to operate the MOSFET with a forward bias on the gate with respect to the channel. In fact there is a special type of MOSFET, called the *enhancement MOSFET,* that operates with forward bias. Figure 3-16 shows an enhancement-type MOSFET. In this component the nonconducting region, which is also known as the *depletion region,* is made so large that it prevents any current from flowing in the channel. The only way you can get current to flow through this component is to put a positive voltage on the gate. The positive voltage decreases the depletion region and therefore increases (or enhances) the area through which the current can flow.

Compare the symbols for the enhancement-type MOSFET in Fig. 3-16 with those of the depletion-type MOSFET in Fig. 3-15. Note that in the enhancement-type MOSFET the channel is shown as a broken line. This

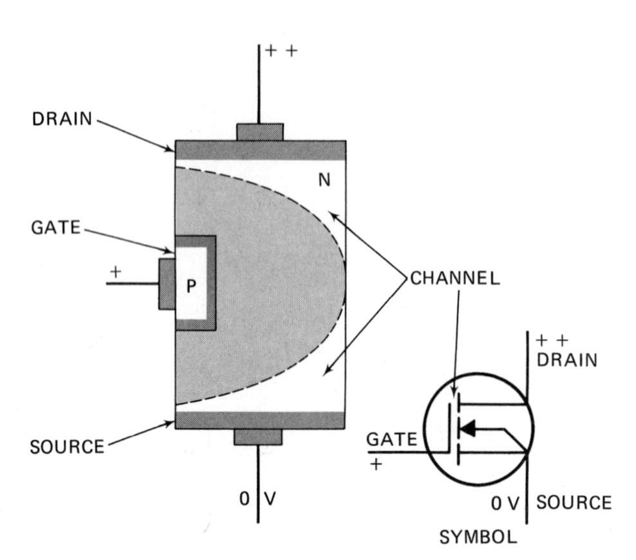

Fig. 3-16 An N-channel enhancement MOSFET.

shows that it is not a complete path for current flow unless a positive voltage is placed on the gate. (Table 3-1 does *not* apply to the enhancement MOSFET.)

Compare the voltage polarities of the N-channel enhancement MOSFET in Fig. 3-16 with the voltage polarities on the NPN bipolar transistor in Fig. 3-9, and you will see that they are operated with the same voltage polarities. You will remember that with the NPN bipolar transistor, base current must flow before collector current can flow. However, with the enhancement MOSFET a positive voltage is placed on the gate in order to get drain current to flow. But because of the insulated gate this voltage does not result in gate-current flow.

What Are P-type MOSFETs?

As you might expect, since it is possible to have N-type MOSFETs, it is also possible to have P-type MOSFETs. The P-type MOSFETs are shown in Fig. 3-17. Notice that the depletion MOSFET of the P-channel type (Fig. 3-17*a*) has voltage polarities that are opposite from the N-channel type (see Fig. 3-15). Except for this difference the operation of the N-channel and P-channel devices is quite similar. The arrows on the symbols of the P-channel depletion MOSFET (Fig. 3-17*a*) and P-channel enhancement MOSFET (Fig. 3-17*b*) both point away from the channel because the channel is made of P-type material. You will remember that such arrows normally point toward N-type material and away from P-type material.

The enhancement-type P-channel MOSFET is shown in Fig. 3-17*a*. Compare the voltage polarities with Fig. 3-16. In this component a negative voltage must be present on both the gate and the drain (more negative) to get it into operation. In other words, you must first put a negative voltage on the gate to start the hole flow of the device. It is not possible to get current flowing in an N-channel or P-channel enhancement-type MOSFET without the presence of a gate voltage. In a P-type MOSFET the gate voltage and

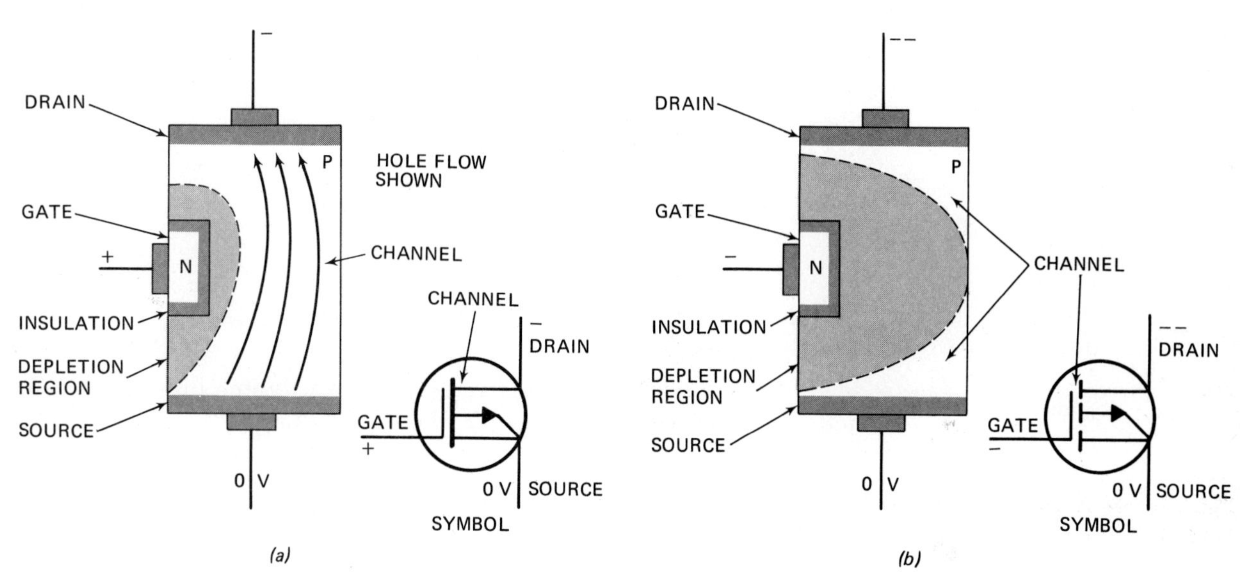

Fig. 3-17 Two types of P-channel MOSFET. (*a*) Depletion type; (*b*) enhancement type.

drain voltage are both negative during its normal operation. In an N-type MOSFET the gate and drain voltages are both negative during its normal operation.

Summary

1. MOSFETs were originally called *insulated-gate field-effect transistors* (IGFETs) because of the insulation layer between their gate and channel.
2. The insulating material between the gate and channel is made of a Metal-Oxide Semiconductor material. Hence the name MOSFET.
3. There are two types of MOSFETs: depletion and enhancement. Both types are made with either an N channel or a P channel.
4. In a depletion MOSFET the gate and drain are operated with opposite voltage polarities.
5. In an enhancement MOSFET the gate and drain are operated with the same voltage polarity.
6. Table 3-1 applies to JFETs and depletion MOSFETs, but not to enhancement MOSFETs.

What Is a Thyristor?

A *thyristor* is a two-, three-, or four-terminal semiconductor component that can be used as a fast-acting switch. The condition of a thyristor is either ON or OFF, depending on the input to the gate element.

The two most important types of thyristor components are shown in Fig. 3-18. These are the *SCR*, for Silicon-Controlled Rectifier (Fig. 3-18*a*), and the *triac* (Fig. 3-18*b*). Both of these thyristors are fast-acting switches. They are switched ON by a voltage delivered to the gate, but they cannot be

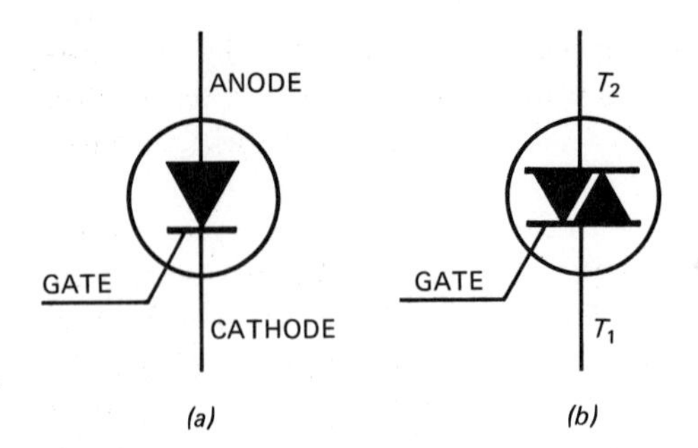

Fig. 3-18 Three-terminal thyristors. (*a*) SCR; (*b*) triac.

switched OFF with a gate voltage. In order to switch these devices OFF it is necessary to open the anode or cathode circuit.

The operation of an SCR is shown in Fig. 3-19. Here you see an SCR circuit in which the voltage to the SCR gate and the anode-current path are controlled by switches. The applied voltage E is used to supply both the anode electrode and the gate electrode, depending upon the positions of switches SW_1 and SW_2. Resistor R reduces the voltage E so that a smaller voltage will be delivered to the gate of the SCR when SW_2 is closed.

In Fig. 3-19*a* there is no current flow in the anode circuit because SW_1 is open. There is no voltage to the gate because SW_2 is open. So the lamp L is not ON.

Figure 3-19*b* shows what happens when you close both switches. The anode switch SW_1 completes the circuit to the anode through the lamp. Closing switch SW_2 allows a positive voltage to go to the SCR gate, switching ON the SCR. The lamp goes ON.

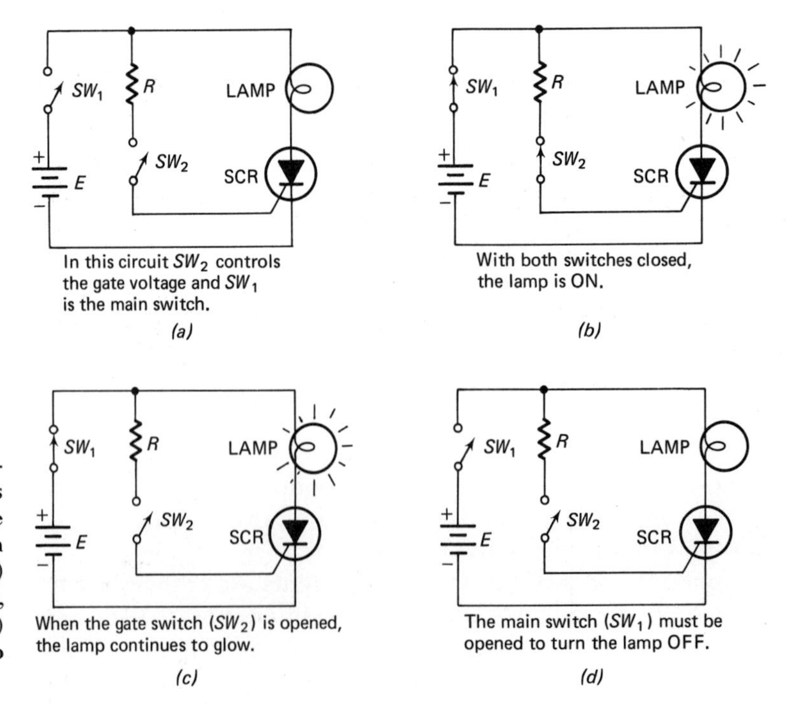

Fig. 3-19 The SCR in a lamp circuit. (*a*) In this circuit SW_2 controls the gate voltage and SW_1 is the main switch. (*b*) With both switches closed the lamp is on. (*c*) When the gate switch is opened, the lamp continues to glow. (*d*) The main switch must be opened to turn the lamp off.

In Fig. 3-19*c* the switch to the gate has been opened. Now there is no voltage applied to the gate. However, the lamp remains ON. The reason for this is that the gate has no control over conduction of the SCR once conduction has begun.

How are we going to turn the lamp OFF once the SCR has begun to conduct? If we open SW_1 as shown in Fig. 3-19*d*, the anode current stops and the lamp will go OFF. The illustrations for Fig. 3-19 only show the effect of applying voltages to the gate and anode circuits. This is not a practical circuit.

The triac in Fig. 3-18*b* does the same job as an SCR. It is a fast-acting switch. The only difference between the two components is that current can flow in either direction through a triac, whereas in the SCR it can flow in only one direction (cathode to anode). Thus triacs are used in circuits where it is desired to switch an alternating current, whereas SCRs are generally used in circuits where it is desired to switch a direct current. In both devices a very short positive pulse applied to the gate will turn the device ON.

Summary

1. A thyristor is a fast-acting semiconductor switch.
2. There are two types of thyristors that were studied in this chapter. They are the SCR and the triac.
3. Both types of thyristors can be turned ON with a positive gate voltage of short duration.
4. To turn a thyristor OFF, open the anode or cathode circuit. Once the thyristor is turned ON, the gate voltage has no control over its conduction.

PROGRAMMED REVIEW QUESTIONS

(Instructions for using this programmed section are given in Chapter 1.)

We will review the important concepts of this chapter. If you have understood the material, you will progress easily through this section. Do not skip this material because some additional theory is presented.

1. Does the transistor of Fig. 3-20 have the correct voltage polarities?
   A. Yes. (Proceed to block 9.)
   B. No. (Proceed to block 17.)

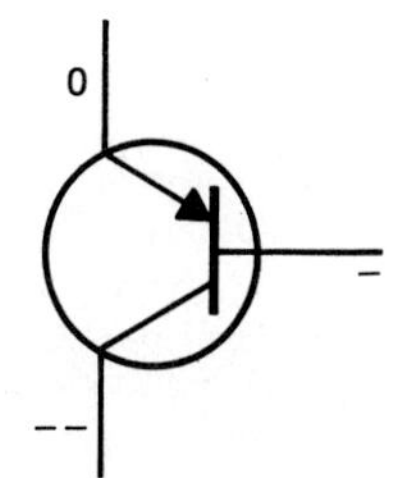

Fig. 3-20 Are the voltage polarities correct?

2. *Your answer to the question in block 9 is* **A**. *This answer is wrong. Remember that the arrow points* **toward** *the N-type material. In Fig. 3-21 the arrow is pointing away from the channel, so it could not be an N-channel MOSFET.* Proceed to block 6.

3. *Your answer to the question in block 25 is* **B**. *This answer is wrong. A MOSFET can be turned* OFF *with the proper amount and proper polarity of gate voltage.* Proceed to block 19.

4. *The correct answer to the question in block 12 is* **A**. *In a PNP transistor a change in base* **current** *causes a change in collector current. Therefore it is a current-operated component. In a triode tube a change in grid* **voltage** *causes a change in plate current. A triode tube is a voltage-operated component.* Here is your next question.
In the normal operation of a pentode-tube amplifier the screen grid is
A. positive with respect to the cathode. (Proceed to block 26.)
B. the electrode that receives the input signal. (Proceed to block 7.)

5. *The correct answer to the question in block 23 is* **B**. *A triac is called a* **bilateral component** *because it can conduct equally well in two directions.* Here is your next question.
A positive voltage on the gate of an SCR is used to start it conducting. To turn the SCR OFF
A. a negative voltage may be placed on the gate. (Proceed to block 22.)
B. reduce the anode voltage to 0 volts. (Proceed to block 12.)

6. *The correct answer to the question in block 9 is* **B**. *The symbol is for a* ***P-channel enhancement-type metal-oxide semiconductor field-effect transistor.*** *This is more commonly called a* ***P-channel enhancement MOSFET.*** *Note that the arrow points away from the channel. This is how you know the channel is made of P-type material. The broken channel line means that it is an enhancement type.* Here is your next question.
Which of the following components has "negative resistance"?
A. A tetrode. (Proceed to block 23.)
B. An SCR. (Proceed to block 15.)

7. *Your answer to the question in block 4 is* **B**. *This answer is wrong. The screen grid is normally operated with a positive voltage. The control grid receives the input signal.* Proceed to block 26.

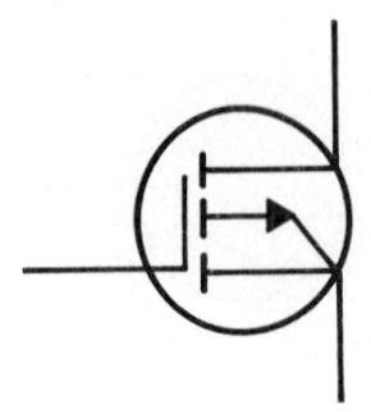

Fig. 3-21 What type of component is this?

8. *Your answer to the question in block 18 is* **B**. *This answer is wrong. There is only one type of current carrier used in a JFET, so it is a unipolar transistor, not a bipolar transistor.* Proceed to block 24.

9. *The correct answer to the question in block 1 is* **A**. *The base of a PNP transistor should be negative with respect to the emitter. The collector is more negative than the base.* Here is your next question.
Figure 3-21 shows the symbol for
A. an N-channel enhancement MOSFET. (Proceed to block 2.)
B. a P-channel enhancement MOSFET. (Proceed to block 6.)
C. an N-channel depletion MOSFET. (Proceed to block 11.)
D. a P-channel depletion MOSFET. (Proceed to block 16.)

10. *The correct answer to the question in block 24 is* **B**. *The base and the collector of an NPN transistor must both be more positive than the emitter. The collector is more positive than the base.* Here is your next question.
For the normal operation of a triode tube, if you increase the grid bias (by making the grid more negative), its plate current will
A. increase. (Proceed to block 20.)
B. decrease. (Proceed to block 25.)

11. *Your answer to the question in block 9 is* **C**. *This answer is wrong. The arrow in the symbol of Fig. 3-21 points away from the channel, so this could not be an N-channel MOSFET. Also the channel line is broken, which means that it is an enhancement MOSFET.* Proceed to block 6.

12. *The correct answer to the question in block 5 is* **B**. *Once an SCR conducts, its gate has no control over the current through it. (This is also true of triacs.) One way to stop conduction is to reduce the anode voltage to 0 volts. Then it can no longer attract electrons. Another way is to open the anode circuit.* Here is your next question.
Which of the following is a current-operated component?
A. A PNP transistor. (Proceed to block 4.)
B. A triode tube. (Proceed to block 21.)

13. *Your answer to the question in block 24 is* **A**. *This answer is wrong. You must know the correct voltage polarities for the three-terminal components. Review the NPN transistor,* then proceed to block 10.

14. *Your answer to the question in block 26 is* **B**. *This answer is wrong. The triode tube does not cause as much noise as the pentode tube.* Proceed to block 18.

15. *Your answer to the question in block 6 is* **B**. *This answer is wrong. The SCR does not have "negative resistance."* Proceed to block 23.

16. *Your answer to the question in block 9 is* **D**. *This answer is wrong. The broken channel line means that it is an enhancement MOSFET.* Proceed to block 6.

17. *Your answer to the question in block 1 is* **B**. *This answer is wrong. The transistor is a PNP type, so a negative voltage is needed on the base and on the collector.* Proceed to block 9.

18. *The correct answer to the question in block 26 is* **A**. *The basic rule is that the amount of noise is greater for tubes with more grids. Pentodes have three grids and triodes have only one grid.* Here is your next question.
Which of the following is an example of a bipolar transistor?
A. An NPN transistor. (Proceed to block 24.)
B. A junction field-effect transistor. (Proceed to block 8.)

19. *The correct answer to the question in block 25 is* **A**. *Both the SCR and the triac are examples of thyristors. A gate voltage can be used to turn them* ON, *but they cannot be turned* OFF *with a gate voltage.* Here is your next question.
Will the drain current of a normal operating P-channel JFET increase or decrease when the gate is made more positive? (Proceed to block 28.)

20. *Your answer to the question in block 10 is* **A**. *This answer is wrong. Making the grid voltage more negative will decrease the plate current.* Proceed to block 25.

21. *Your answer to the question in block 12 is* **B**. *This answer is wrong. A triode is a voltage-operated component.* Proceed to block 4.

22. *Your answer to the question in block 5 is* **A**. *This answer is wrong. An SCR cannot be shut off by making its gate negative. The same is true for a triac.* Proceed to block 12.

23. *The correct answer to the question in block 6 is* **A**. *The "negative resistance" of a tetrode is caused by secondary electrons going to the screen instead of returning to the plate.* Here is your next question.
Which of the following components conducts equally well in two directions?
A. A triode. (Proceed to block 27.)
B. A triac. (Proceed to block 5.)

24. *The correct answer to the question in block 18 is* **A**. *A bipolar transistor (NPN or PNP types) has two kinds of charge carriers flowing at the same time. They are holes and the electrons. The N-channel JFET uses only electrons as charge carriers. The P-channel JFET uses only holes as charge carriers.* Here is your next question.

In the normal operation of an NPN transistor the base is
A. negative with respect to the emitter. (Proceed to block 13.)
B. positive with respect to the emitter. (Proceed to block 10.)

25. *The correct answer to the question in block 10 is* ***B****. When the grid is made more negative, fewer electrons can pass to the plate. Therefore the plate current decreases.* Here is your next question.
Which one of the following can be turned ON with a gate voltage but cannot be turned OFF with a gate voltage?
A. Triac. (Proceed to block 19.)
B. MOSFET. (Proceed to block 3.)

26. *The correct answer to the question in block 4 is* ***A****. The input signal goes to the control grid. The screen grid has a positive voltage with respect to the cathode.* Here is your next question.
Which one of these tubes causes more noise in an amplifier circuit?
A. Pentode. (Proceed to block 18.)
B. Triode. (Proceed to block 14.)

27. *Your answer to the question in block 23 is* ***A****. This answer is wrong. A triode will conduct from cathode to plate, but not from plate to cathode.* Proceed to block 5.

28. *The drain current of a normal operating P-channel JFET will* ***decrease*** *when the gate is made more positive.*
You have now completed the programmed questions. The next step is to put some of these ideas to work in laboratory experiments. Proceed to the Experiment section of this chapter.

EXPERIMENT

(The experiment described in this section may be performed on the circuit board described in Appendix C or on a similar laboratory setup.)

Purpose The purpose of this experiment is to show you how to make voltage measurements in a typical transistor amplifier and also to show you what these measurements tell about the amplifier.

Theory All the three-terminal amplifying components that you studied in this chapter have at least one thing in common: They all require that one of the electrodes be either positive or negative with respect to another electrode. For example, the plate of a vacuum tube must be positive with respect to its cathode, the collector of a PNP transistor must be negative with respect to its emitter, and the drain of an N-channel JFET must be positive with respect to its source. Without the proper voltage across the component it cannot operate.

Another requirement that these components have in common is that there must be a *bias voltage* on the electrode that is normally used for control. The grid of a vacuum tube must be negative with respect to the cathode, the base of a PNP transistor must be negative with respect to the emitter, and the gate of an N-channel JFET must be negative with respect to the source.

Without the proper bias voltage the component may go into *saturation*. This means that the maximum possible current is flowing through it and the input signal will have little or no control over the current flow. In some cases the saturation current will be so high that the component will be destroyed by the excessive heat generated by the current.

To summarize: All the three-terminal electronic components must have the proper polarity of voltage applied to them and they must have the proper bias to limit current flow through them for normal operation.

The voltage across the component and the bias voltage are usually provided by a dc source. This is not always the way it is done. An SCR, for example, sometimes gets its operating voltage from one-half cycle of the power-line voltage. As you know, the ac power line has a positive voltage with respect to ground on one-half cycle and a negative voltage on the next half cycle. Thus an SCR can also act as a rectifier.

A dc power supply provides both the bias voltage and the operating voltage for the output circuitry. To analyze an amplifier circuit it is a good idea to start by measuring the dc voltages. If the proper dc voltages are not present, the amplifier circuit can be assumed to be not working properly. Furthermore, as you get more experience you will learn to tell what specific part of the amplifier is not operating properly simply by making the dc voltage measurements.

On the schematic diagram provided by a manufacturer, the dc voltages and polarities may be shown. But in those cases where the voltages are not provided you will need to know them. You will at least need to know the correct polarities of the voltages to be able to make the measurements properly. In each of the amplifier components discussed in this chapter the voltage polarities are given. You should review this material periodically until you have those voltage polarities memorized.

In this experiment you are going to use the transistor circuit shown in Fig. 3-22. Figure 3-22*a* shows the schematic diagram, and a pictorial diagram is shown in Fig. 3-22*b*. This is a simple transistor amplifier. Resistors R_1 and R_2 drop the positive-supply voltage E_b so that it can be used for obtaining the proper amount of base current I_B.

Resistor R_2 is made variable so the base current can be controlled over a range of values. The two-terminal connection of R_2 is called a *rheostat*. A rheostat is a variable resistor connected in such a way that it controls current.

Resistor R_3 is the collector-load resistor for the circuit. When the circuit is being used as an amplifier the amplified signal voltage is developed across R_3. The output signal for the circuit is taken from point *a*.

The emitter resistor R_4 is used to stabilize the amplifier. You will study the action of this resistor in a later chapter. (I_E stands for emitter current.)

All voltage measurements will be taken with respect to the common point (ground). This is the usual method of making such measurements in an amplifier circuit. The solid arrows show electron flow outside of the transistor.

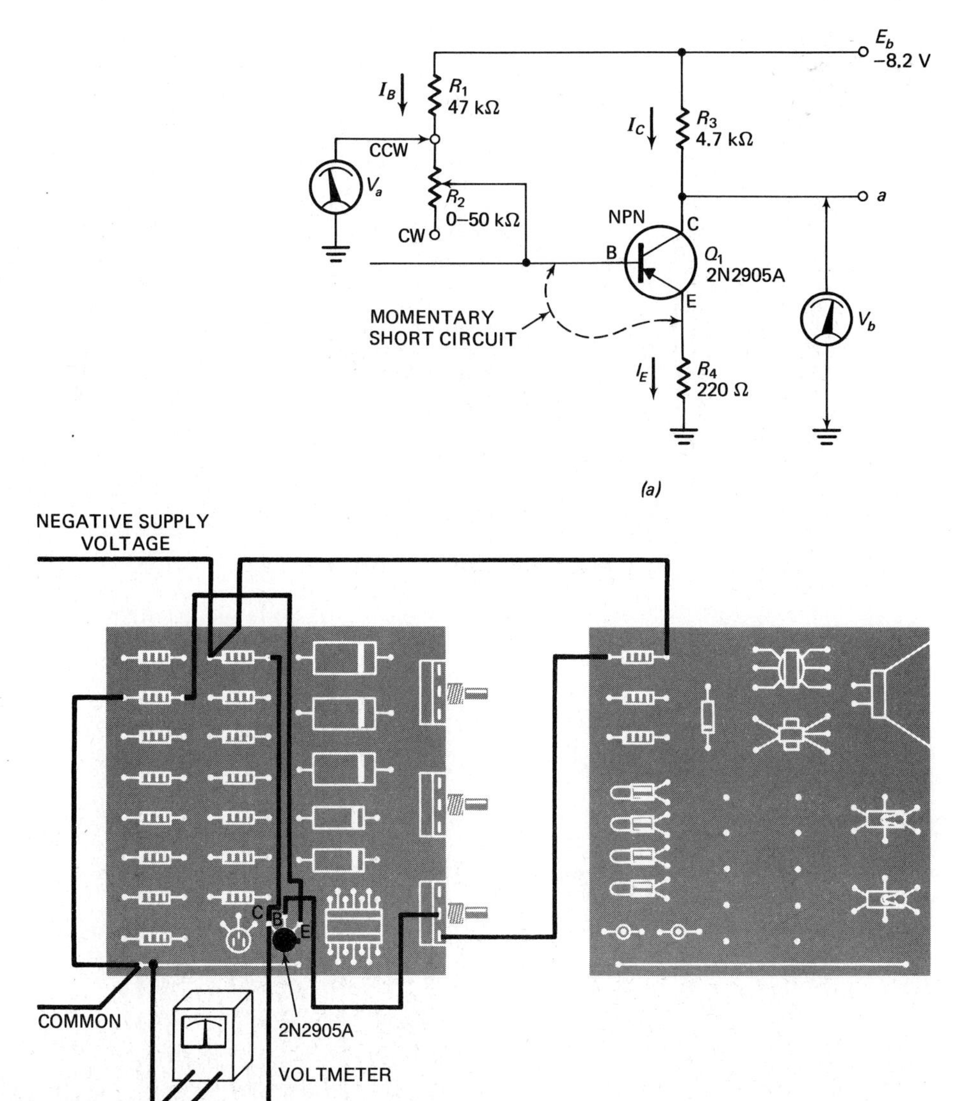

Fig. 3-22 **Transistor circuit for experiment. (*a*) Schematic diagram. (*b*) Pictorial diagram of the transistor circuit.**

Test Setup Wire the circuit as shown in Fig. 3-22. The power-supply voltage must be negative because the transistor is a PNP type.

Note that R_2 is wired so that the resistance is minimum when the variable resistor is in the counterclockwise (ccw) position. In the clockwise position (cw) the total base-circuit resistance is 50 kilohms + 47 kilohms, or 97 kilohms.

Figure 3-23*a* shows a schematic of how the power-supply voltage is obtained. This is a half-wave rectifier with a capacitor filter. Note the proper connections for the diode and the electrolytic capacitor. These components must both be wired as shown in order to make the circuit operate correctly. The pictorial diagram is shown in Fig. 3-23*b*.

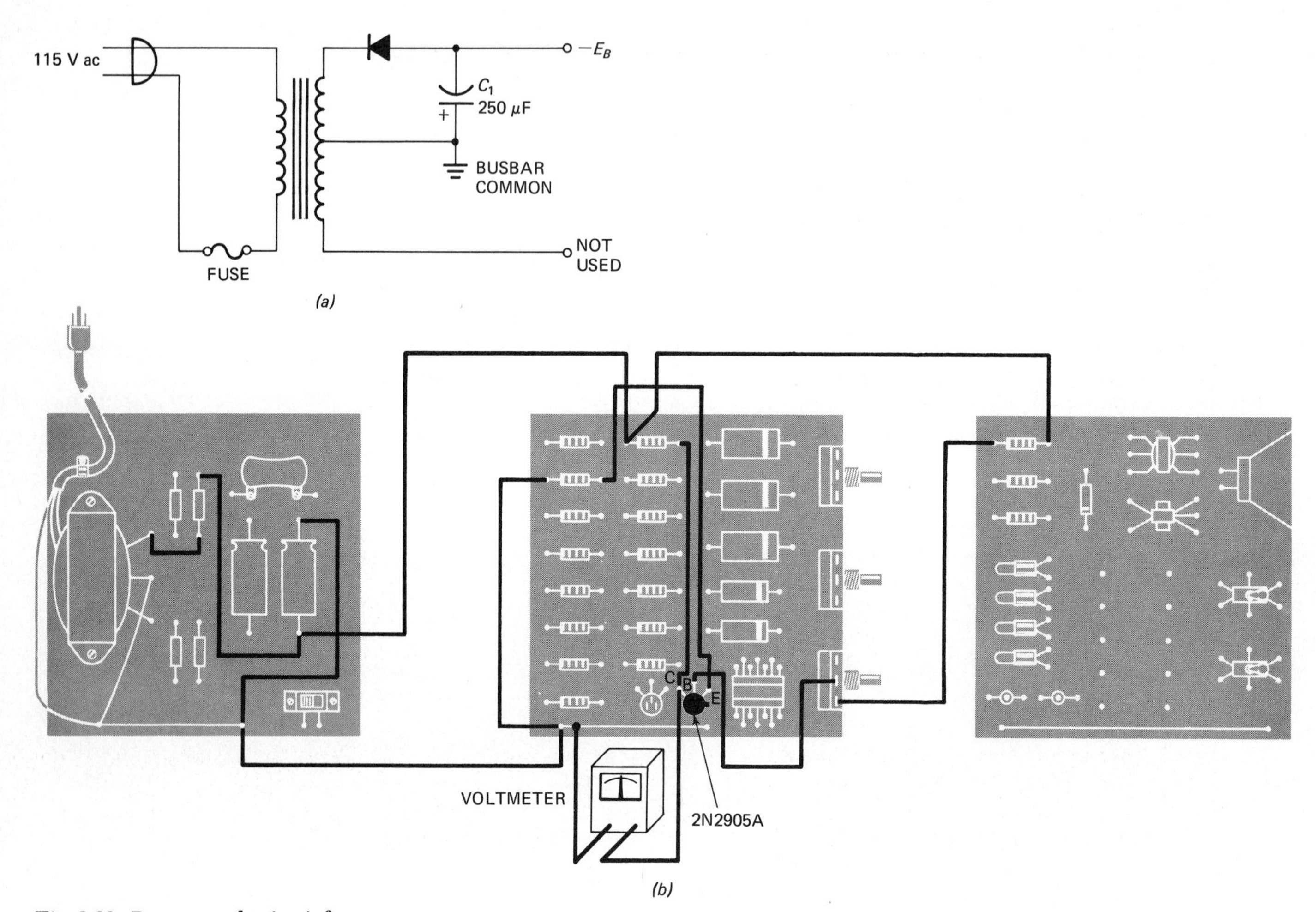

Fig. 3-23 Power-supply circuit for experiment. (*a*) Schematic diagram. (*b*) Pictorial diagram for power supply, showing connections to transistor circuit for experiment (Fig. 3-22).

In part of this experiment you will be required to short the emitter and base connections together. Do not *ever* place a short between the base and collector of a transistor! To do so would cause an excessive base-current flow and most likely would destroy the transistor. Always remember this when you are making measurements or troubleshooting in transistor circuits!!

Procedure

step one Turn R_2 to its counterclockwise (minimum-resistance) position. Measure the power-supply voltage with the transistor circuit connected to the power supply. Remember, this will be a negative voltage!

$E_b = -$ ________________ volts

step two Measure the voltage at the junction of R_1 and R_2. This voltage measurement is shown with voltmeter V_a on the schematic diagram.

$V_a = -$ ________________ volts

step three Can you find the voltage across resistor R_1 (call it V_1) with the information you now have? ______________ Yes or No

step four Your answer should be yes. You know E_b and you know V_a. The voltage across R_1 is the difference between these voltage values. Find the value of V_1.

$$V_1 = E_b - V_a = \_\_\_\_\_\_\_\_ \text{ volts}$$

(You want the *amount* of voltage drop, so you can ignore the negative signs.)

step five You know the voltage across R_1 and you know the resistance of R_1, so you can now find the current through R_1 by using Ohm's law. This is the base current I_B for the transistor.

$$I_B = \frac{V_1}{R_1} = \frac{V_1}{47 \text{ kilohms}}$$

$$I_B = \_\_\_\_\_\_\_\_ \text{ milliamperes}$$

step six Measure voltage V_b as shown in the circuit diagram.

$$V_b = - \_\_\_\_\_\_\_\_ \text{ volts}$$

step seven Is the transistor conducting? ______________ Yes or No

Your answer should be yes. If the value of V_b is different from the applied voltage, then there is a voltage drop across R_3. This means that there must be a current through R_3, so there must be a collector current flowing. This is an important measurement. Technicians will often measure the collector voltage to see if a transistor is conducting. This is part of a troubleshooting procedure.

step eight With the voltmeter connected for measuring the collector voltage, short the emitter to the base at the transistor. Be very careful *not to short anything else!* The emitter-to-base short is shown with a dotted line in Fig. 3-22. Does the emitter-to-base short cause any change in the collector voltage V_b? ______________ Yes or No

Your answer should be yes. The emitter-base short stops the transistor from conducting because it prevents charge carriers from moving across the emitter-base junction. Remember, base current must flow in the transistor in order for collector current to flow. When you short the base to the emitter there is no collector-current flow through R_3. Therefore there is no voltage drop across R_3 and V_b becomes equal to E_b.

The emitter-base short test is used to tell if the base current of a transistor can control the collector current. It is *never* used for testing tubes (grid-to-cathode short), JFETs, or MOSFETs.

step nine Find the voltage drop across collector resistor R_3. Call this voltage drop V_3. $V_3 = E_b - V_b =$ __________ volts

(Again, you can ignore the negative signs since you only want the amount of voltage drop.)

step ten You know the voltage across R_3 and you know the resistance of R_3. Use Ohm's law to find the current through R_3. This is the collector current I_C. $I_C = \frac{V_3}{R_3} =$ __________ milliamperes

step eleven Turn R_2 to its maximum clockwise (cw) position. You are now going to find the base current I_B and collector current I_C. Since the resistance in the base circuit is different the base current will also be different. If the transistor is working properly, the change in base current will cause a change in collector current.

step twelve Measure V'_a. This is the new value of V_a since the resistance of R_2 has been changed. $V'_a = -$ __________ volts

step thirteen Find the new voltage across R_1, or V'_1.

$$V'_1 = E_b - V'_a$$

$V'_1 =$ __________ volts

step fourteen Find the new value of base current that is now flowing.

$$I'_B = \frac{V'_1}{R_1} = \frac{V'_1}{47 \text{ kilohms}}$$

$I'_B =$ __________ milliamperes

step fifteen Measure the new transistor collector voltage V'_b.

$V'_b =$ __________ volts

step sixteen Find the new voltage across R_3.

$V'_3 = E_b - V'_b =$ __________ volts

step seventeen Find the collector current by Ohm's law.

$I'_C = \frac{V'_3}{R_3} =$ __________ milliamperes

step eighteen In a bipolar transistor a change in base current causes a change in collector current. The symbol Δ is used to mean "change in," so

ΔI_B means the change in I_B. Find the change in base current when the variable resistor R_2 is changed from its minimum resistance (ccw) to maximum resistance (cw).

$$\Delta I_B = I_B - I_B' = \_\_\_\_\_\_\_\_\_\_ \text{ milliamperes}$$

step nineteen Find the change in collector current when R_2 was changed from minimum resistance to maximum resistance.

$$\Delta I_C = I_C - I_C' = \_\_\_\_\_\_\_\_\_\_ \text{ milliamperes}$$

step twenty The current gain is the amount of change in collector current divided by the amount of change in base current that produced the collector-current change. Find the current gain.

$$\text{Current gain} = \frac{\Delta I_C}{\Delta I_B} = \_\_\_\_\_\_\_\_\_\_$$

The manufacturer calls the current gain β (pronounced *beta*). Actually the value should be found by taking very small changes in base current ΔI_B and then measuring the resulting change in collector current ΔI_C. The manufacturer lists the current gain for the 2N2905A as being 100. Compare this with your value. You will not get the same amount because your base current changed over a wide range and because transistors differ. Also your value of current gain here is for the entire *circuit* rather than for the transistor alone.

Conclusion In this experiment you have made dc voltage measurements to determine if a transistor circuit is operating properly. The first step in analyzing amplifier-circuit behavior is to make dc measurements. If you do not have the proper dc voltages and currents, the three-terminal amplifying component cannot be made to operate properly. For most circuits that use three-terminal amplifying components, dc voltage measurements are sufficient for determining if the circuit is working properly.

You demonstrated here that the base current in a bipolar transistor circuit controls the collector current. For a quick check of a transistor circuit some technicians short the emitter to the base and note the change in collector voltage. This method of testing should never be used for any three-terminal amplifying components *other than the bipolar* (NPN and PNP) transistors! Also it should not be used with direct-coupled amplifiers, which you will study in Chapter 8.

SELF-TEST WITH ANSWERS

1. The arrow in an electronic symbol always points toward the (*a*) base; (*b*) N material; (*c*) P material; (*d*) emitter.
2. Which of the following components can conduct current in either direction? (*a*) A rectifier diode; (*b*) A triode tube; (*c*) An SCR; (*d*) A triac.
3. In a triode-tube amplifier the input signal goes to the control grid. (*a*) True; (*b*) False.

4. A three-terminal component used for switching is called a (*a*) pentode; (*b*) triode; (*c*) thyristor; (*d*) bipolar.
5. In the normal operation of a PNP transistor, when you make the base more negative the collector current (*a*) increases; (*b*) decreases.
6. The charge carrier in P material is the (*a*) hole; (*b*) electron.
7. In an NPN transistor what is the normal voltage polarity on the collector with respect to the base? (*a*) The base is more positive than the collector; (*b*) The collector is more positive than the base; (*c*) There is no difference in the voltage on the base and collector.
8. In a junction diode the anode is made of (*a*) N-type material; (*b*) P-type material.
9. Which of these field-effect transistors will not conduct current from source to drain unless there is a bias voltage on the gate? (*a*) N-channel JFET; (*b*) Depletion MOSFET; (*c*) Enhancement MOSFET; (*d*) P-channel JFET.
10. Which of the following is true? (*a*) Two diodes can be connected together to make a transistor; (*b*) The majority charge carriers in P-type material are electrons; (*c*) All JFETs are enhancement types; (*d*) A transistor could be used as a diode because it has PN junctions.

ANSWERS TO SELF-TEST

1. (*b*)
2. (*d*)—Both the SCR and the rectifier diode will conduct electron current only from the cathode to the anode. The triode tube conducts only from the cathode to the plate.
3. (*a*)—The control grid is the signal input electrode for the triode, the tetrode, and the pentode.
4. (*c*)—Examples of thyristors are SCRs and triacs.
5. (*a*)—Making the base more negative increases the base-current flow. This increases the collector current.
6. (*a*)—You can think of the hole as being a positive charge.
7. (*b*)—This is a very important point. For some types of transistors, if you make the base and collector voltages equal even for a very short moment, the transistor will be destroyed.
8. (*b*)
9. (*c*)
10. (*d*)—Except for choice (*d*), all the statements in Question 10 are completely false.

what are the electromagnetic components used in electronic circuits? 4

INTRODUCTION

Earlier we defined *electronics* as the science of putting the electron to work. It is concerned with the use of diodes, tubes, transistors, and FETs. These are some examples of components that control the flow of electrons.

In addition to the electronic components that control electron flow, you will also find electrical components such as resistors, capacitors, inductors, transformers, and relays. The last two components are the subject of this chapter. They are called *electromagnetic* because their operation depends upon both electric currents and magnetic fields.

In this chapter you will review some of the basic principles of electromagnetism. Then you will study the application of transformers and relays in electronics.

You will be able to answer these questions after studying this chapter:

- ☐ **What is the relationship between current and magnetism?**
- ☐ **What is Faraday's law of electromagnetic induction?**
- ☐ **What is Lenz' law?**
- ☐ **What are transformers?**
- ☐ **What are some types of transformer losses?**
- ☐ **What are the uses of transformers in electronic circuits?**
- ☐ **How does a relay work?**

INSTRUCTION

What Is the Relationship between Current and Magnetism?

In the early days of experimenting with electricity it was concluded that electricity and magnetism were two completely different things. As the story goes, Hans Christian Oersted was convinced that there was no relationship between them. During one of his lectures he accidentally discovered that there is *always* a magnetic field when there is an electric current.

Oersted's experiment is illustrated in Fig. 4-1. The arrows in this illustration show the direction of electron-current flow. A compass is placed over the wire in the circuit. The needle of the compass always turns at right angles to the direction of the current flow. This is shown in the drawing. From this simple experiment you can make a rule: *When there is an electron-current flow, there is also a magnetic field. The magnetic field is at right angles to the direction of current flow.* This is always true.

The magnetic field that goes with the current is actually around the wire as shown in Fig. 4-2. Increasing the current causes the magnetic field to become stronger. Reversing the direction of the current reverses the direction of the magnetic field. A magnetic field is usually represented by lines or arrows. The arrows show the direction of the magnetic field. This is always presumed to be from the south pole of the magnet to the north pole. The word *flux* is often used instead of magnetic field or field lines. In fact, *flux* can be defined as magnetic field lines.

The direction of the magnetic field can be determined easily by what is known as the *left-hand rule.* The rule states that if you grasp the wire (mentally) with your left hand so that your thumb points in the direction of

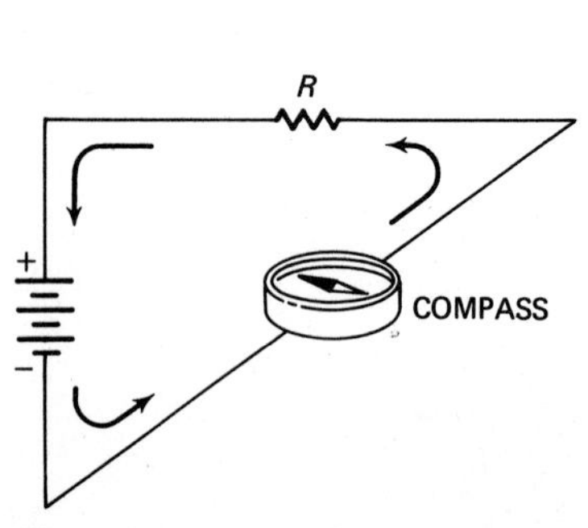

Fig. 4-1 Oersted's experiment. Arrows show electron-current flow.

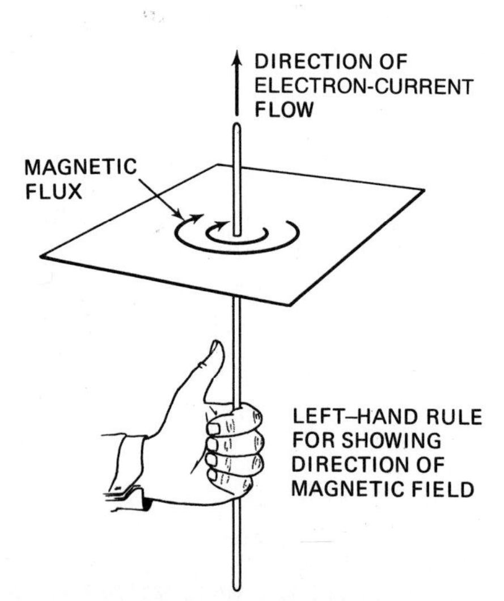

Fig. 4-2 When a current flows, a magnetic field surrounds it. The field direction can be determined by the left-hand rule.

electron-current flow, then your fingers will circle the wire in the direction of the magnetic field. This is also illustrated in Fig. 4-2. (Electron-current flow is used in this book. In other words, the direction of current flow is always assumed to be from negative to positive.)

The fact that *there is always a magnetic field with current flow* and the fact that *the strength of the magnetic field depends directly upon the amount of the current flow* are two of the most basic and important things to remember in your study of electromagnetism.

What Is Faraday's Law of Electromagnetic Induction?

An important basic law relating magnetism and electricity is *Faraday's law of electromagnetic induction.* It states that *whenever there is motion between a magnetic field and a conductor, a voltage is produced.*

The basic idea of Faraday's law is illustrated in Fig. 4-3. The meter in these illustrations reads 0 volts in the center of the scale. A positive voltage at point *a* causes the meter to deflect to the right; a negative voltage causes it to deflect to the left. In Fig. 4-3*a*, you are pulling the magnet out of the center of a coil. The magnetic field around the magnet cuts across the turns of the coil and causes a voltage to be induced. As shown in Fig. 4-3*b*, if you reverse the direction of motion of the magnet, this causes the polarity of the voltage to reverse. In Fig. 4-3*c* the magnet is stationary and the coil is being moved. Again, there is motion between the turns of the coil and the magnetic field, so a voltage is induced.

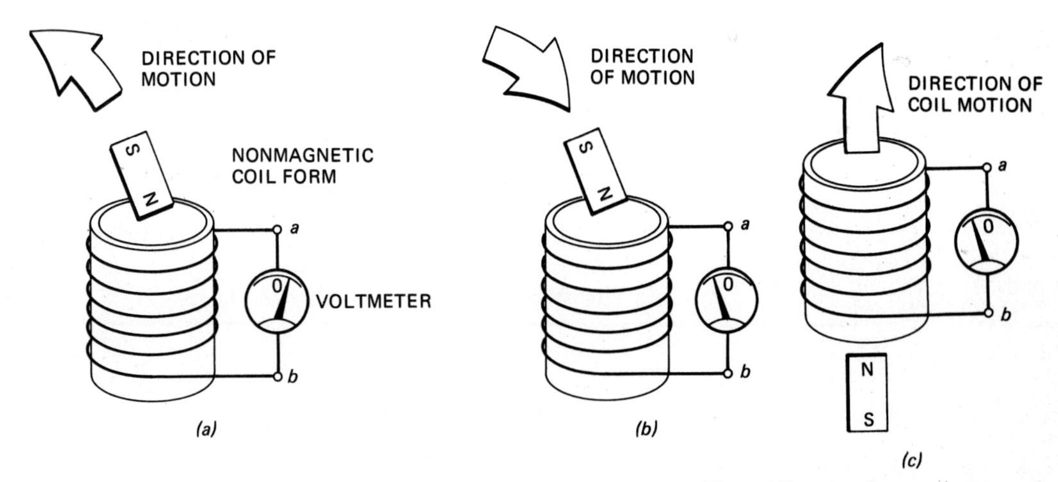

Fig. 4-3 A demonstration of Faraday's law of electromagnetic induction. (*a*) Pulling the magnet from the center of the coil causes a voltage to be induced. (*b*) Reversing the direction of motion reverses the polarity of the induced voltage. (*c*) If the magnet does not move but the coil does, a voltage is also induced.

The amount of voltage produced by the motion between a magnetic field and the conductors depends upon two things: the rate at which the conductors move through the lines of the magnetic field and the number of conductors.

Electric generators and transformers operate on the principle of Faraday's law. Electric power that is generated for use in homes and industry is produced by moving large conductors through magnetic fields. Transformers, which will be discussed later in the chapter, operate by moving magnetic fields.

What Is Lenz' Law?

Lenz' law is related to voltages generated by conductors moving in magnetic fields. (As stated before, it does not matter if you move the conductor or the magnetic field to produce the voltage.) The law states that the generated voltage produces a current with a magnetic field that opposes the motion. Figure 4-4 shows a simple experiment that demonstrates Lenz' law. In Fig. 4-4*a* the crank of a hand generator is being turned. This produces an output voltage V across the generator terminals. The only opposition to turning the crank is the friction of the parts turning within the generator.

In Fig. 4-4*b* a resistor has been placed across the terminals of the generator. The generated voltage causes the current to flow in the circuit. This current must flow in the coils of the generator. According to Lenz' law, the current in the generator coils produces a magnetic field that opposes the motion of the generator. If you were turning the crank of the generator, you would notice that it becomes hard to turn when the load resistor is connected to its terminals. This demonstrates that the induced current produces a magnetic field that opposes the motion.

Summary

1. When there is an electron-current flow, there is always a magnetic field.
2. The direction of the magnetic field around an electron current can be determined by the left-hand rule.
3. The strength of the magnetic field depends upon the amount of current.

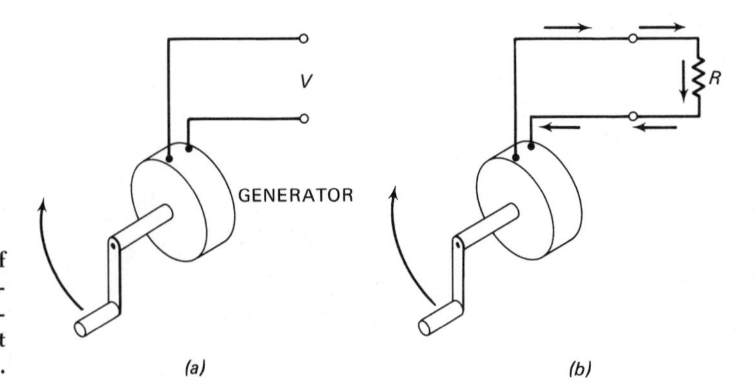

Fig. 4-4 A simple demonstration of Lenz' law. (*a*) The crank turns easily when there is no generator current. (*b*) When generator current flows, the crank is difficult to turn.

4. Faraday's law of electromagnetic induction states that whenever there is motion between a conductor and a magnetic field, there is a voltage induced.
5. Lenz' law states that an induced voltage will produce a current that has a magnetic field which will oppose the motion that produced the magnetic field.

What Are Transformers?

Figure 4-5 shows the theory of a transformer. In Fig. 4-5*a* an ac voltage is applied to a coil, called the *primary*. Since alternating current is flowing in the coil, there will be a moving magnetic field around the coil. Another coil, located close to the primary, is called the *secondary*. It is located in a position where the changing flux moves across it. Therefore the varying current (and flux) in the primary causes a voltage to be induced in the secondary. This is the principle of operation of all transformers.

The primary and secondary windings can be wound on a *soft-iron* material. Soft iron cannot be permanently magnetized. It offers very little opposition to flux. This means that flux can flow through soft iron more easily than it can flow through air. Use of a soft-iron core assures that most of the flux lines from the primary will cut across the secondary turns. This is important. If there are flux lines around the primary which do not cut across secondary turns, the amount of power delivered to the secondary circuit is decreased. To summarize: The soft-iron core makes an easy path between the primary and secondary for the flux lines. This assures that more of the flux lines cut across the secondary.

Before going any further with the discussion of transformers it will be useful to review the concept of *reluctance*. Reluctance is the opposition to flux, just as resistance is the opposition to current flow. Soft-iron materials

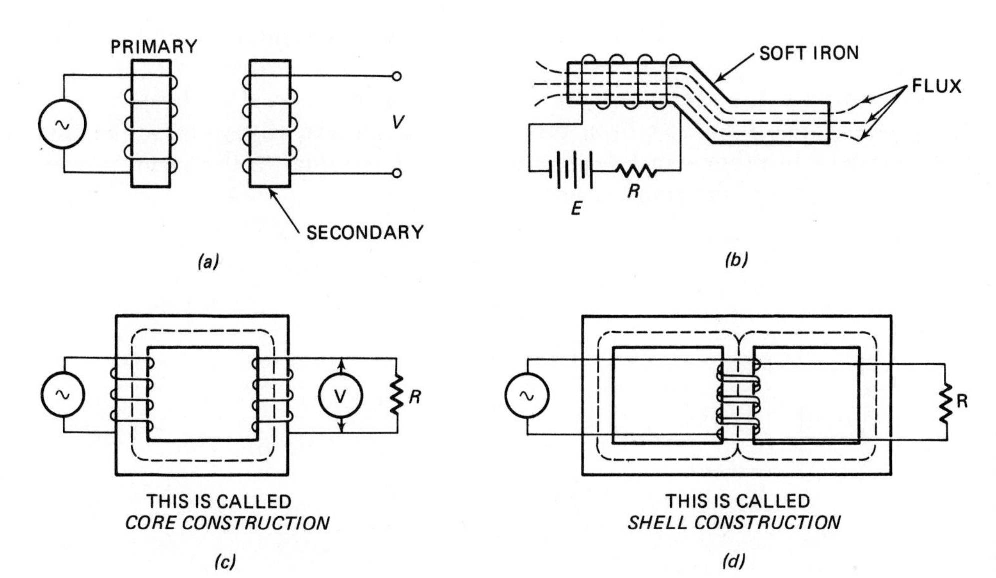

Fig. 4-5 The transformer. (*a*) A simple transformer. (*b*) The flux lines follow the iron path. (*c*) An iron-core transformer. (*d*) Another way of making an iron-core transformer.

have very low reluctance compared with air. This is just another way of saying that magnetic-flux lines will flow through soft iron more readily than through air. The illustration in Fig. 4-5*b* shows you that even though the iron has an irregular shape, the flux lines follow the iron path because its reluctance is lower than the reluctance of air.

Figure 4-5*c* shows you the primary and secondary windings of a transformer wound on an iron core. The varying current in the primary establishes a varying flux in the iron. The flux follows the path shown by a dotted line. These flux lines pass through the secondary winding, so a voltage will be induced in it.

Figure 4-5*d* shows a different way of winding the primary and secondary coils on an iron-core transformer. In this case both the primary and secondary are wound on the center leg of the transformer core.

Various symbols for transformers in electrical and electronic circuits are shown in Appendix B.

What Is an Autotransformer?

Figure 4-6 shows how an *autotransformer* is made. It consists of a winding which is tapped at some point. (See Fig. 4-6*a*.)

Assume that an ac voltage is applied across terminals *a* and *b*. An alternating current will flow in the lower portion of the winding. The alternating current establishes a flux which cuts across the complete secondary winding, producing an output voltage across terminals *c* and *d*.

It would also be possible to apply a voltage to terminals *c* and *d* and take the output voltage from terminals *a* and *b*. In either case an autotransformer always has one terminal that is common to the input (primary) and output (secondary) windings. Figure 4-6*b* shows the schematic symbol for an autotransformer.

What Are Some Types of Transformer Losses?

The power in the secondary circuit can never be as great as the power in the primary. This follows from the fact that there are always losses in the transformer. In other words, the output power is equal to the input power minus the losses of the transformer.

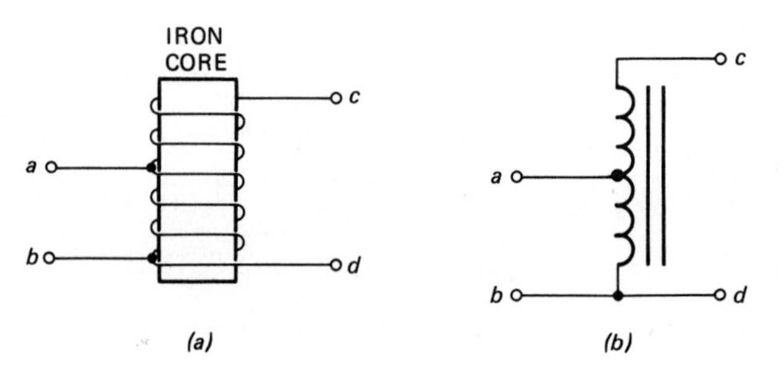

Fig. 4-6 The autotransformer. (*a*) How an autotransformer is made. (*b*) Schematic symbol.

Transformer output power = input power − transformer losses

Winding the transformer coils on iron greatly improves the efficiency at low frequencies, but there are problems caused by the iron core.

What Is an Eddy Current?

Iron does not conduct electricity as well as copper. Faraday's law says that any time there is motion between a conductor and a magnetic field, a voltage is induced. When alternating current flows in the primary winding and also in the secondary winding, there is a moving field around the coils. The varying flux from these currents cuts across the iron core. This induces currents in the core. These currents are called *eddy currents*. They are illustrated in Fig. 4-7.

Figure 4-7*a* shows an alternating-current waveform applied to a coil and the eddy currents flowing in the iron core. These eddy currents cause the iron to heat, and this heat represents a power loss. To reduce the eddy currents the iron core is built up out of thin sheet steel layers called *laminations* as shown in Fig. 4-7*b*. The path for current flow is reduced. As you know, the resistance of any conductor increases as its cross-sectional area decreases. Therefore, laminating the iron reduces the cross-sectional area and increases the resistance to eddy-current flow.

If you took a transformer apart, you would find that the laminations are insulated from each other. The insulation has no effect on the magnetic path, but it does increase the resistance to eddy-current flow.

What Is Meant by the Term Hysteresis Loss?

The word *hysteresis* is pronounced *hiss ter ēē' sis*.

Hysteresis loss occurs in transformers because the iron core becomes magnetized during each half cycle of current. This magnetism must be removed before flux can be made on the next half cycle.

Figure 4-8 shows how the magnetism in a core changes when alternating current flows in the primary winding. Figure 4-8*a* shows that when the first half cycle of current reaches its maximum value, the magnetism also reaches its maximum value. The *magnetizing force* is actually the magnetic field of

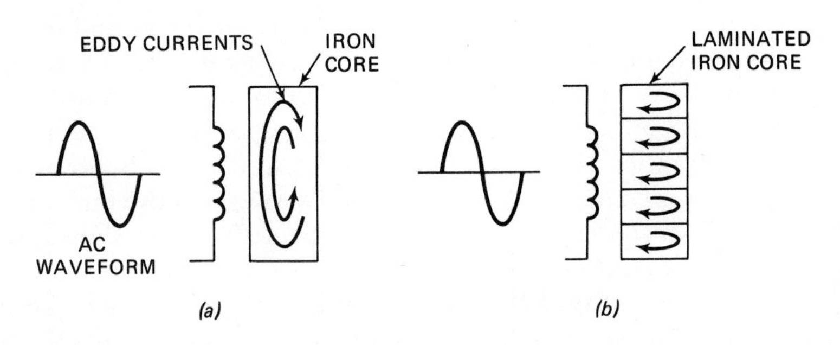

Fig. 4-7 Eddy currents cause core loss in a transformer. (*a*) Alternating current in the coil causes eddy currents in the iron. (*b*) Laminated iron reduces eddy-current loss.

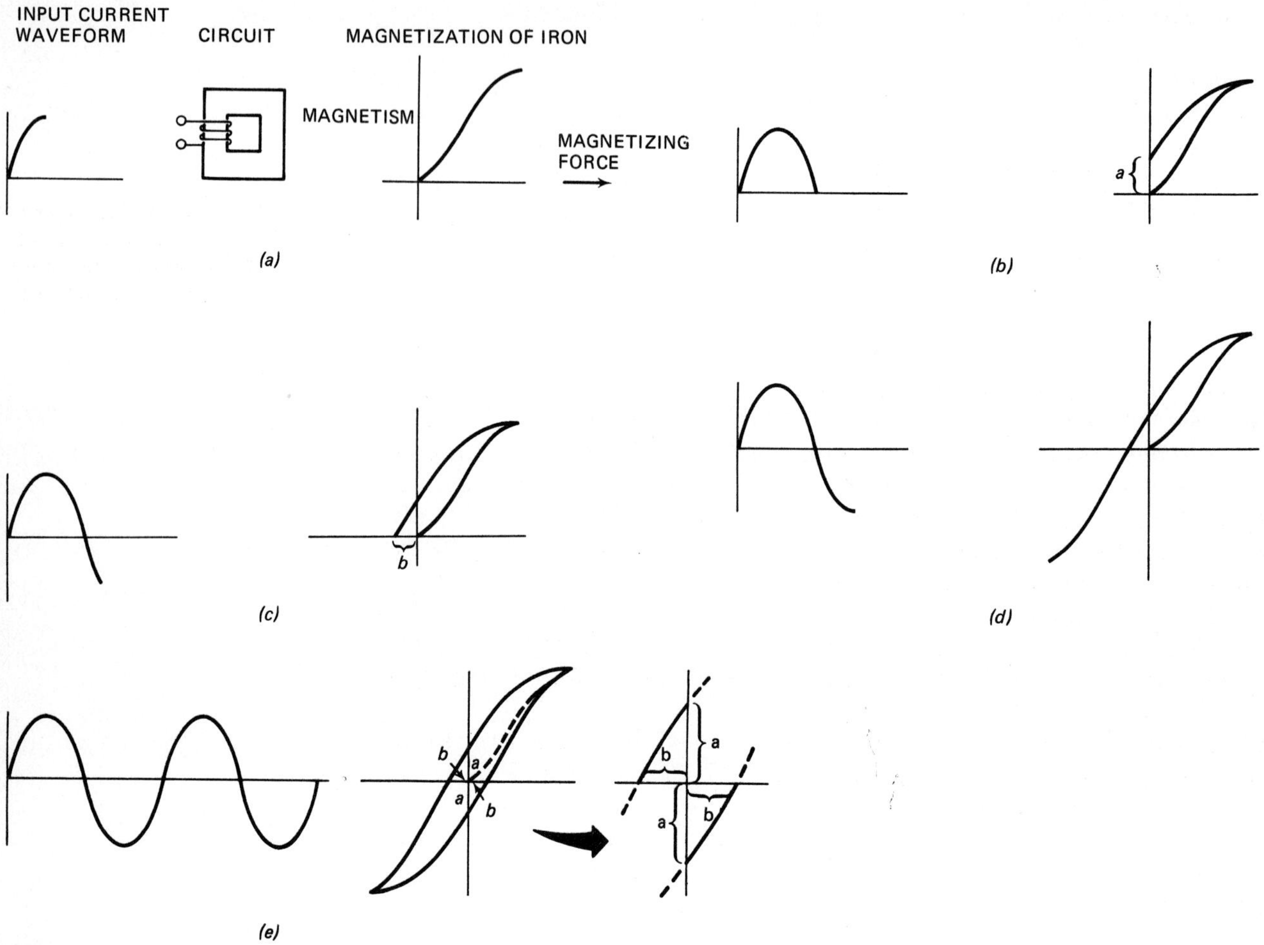

INPUT CURRENT WAVEFORM CIRCUIT MAGNETIZATION OF IRON

MAGNETISM MAGNETIZING FORCE

(a) (b) (c) (d) (e)

Fig. 4-8 Generation of a hysteresis curve. (*a*) The iron core is magnetized during the first half cycle of current. (*b*) When the current drops to zero, there is still some magnetism in the core. (*c*) The current must flow in the opposite direction to reduce the magnetism to zero. (*d*) When the current reaches its maximum value, magnetism is also maximum. (*e*) Hysteresis curve for a full cycle of current.

the current in the primary winding. The first half cycle is completed in Fig. 4-8*b*. During this period of time the current in this coil drops to zero, but the magnetism in the iron does not drop to zero. The amount of magnetism left in the iron is marked on the curve with an *a*.

The magnetism remaining in the core is called *remanence*. It would be better if the magnetism would go to zero when the current in the coil goes to zero.

Figure 4-8*c* shows that the current in the primary coil must flow in the reverse direction for a certain period of time in order to reduce the magnetic flux to zero. This means that part of the current waveform must be used to demagnetize the iron. This current causes a magnetizing force, marked with a *b* in the illustration. The magnetizing force needed to remove the magnetism is called the *coercive force*. Figure 4-8*d* shows that the flux reaches a maximum value when the current is maximum, and Fig. 4-8*e* shows the characteristic curve for several cycles of ac input.

The curve of Fig. 4-8*e* is known as a *hysteresis curve*. The amount of open space within the curve tells you how much hysteresis loss the transformer has.

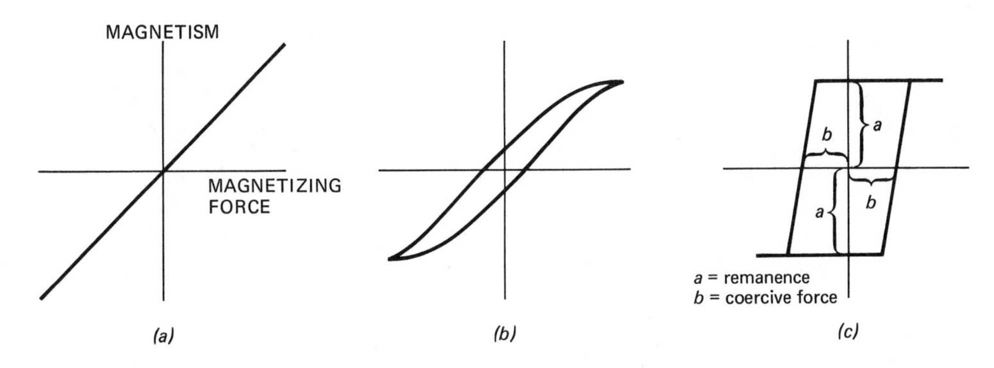

Fig. 4-9 Comparison of hysteresis loss for three types of materials. (*a*) Air. (*b*) Good transformer iron core. (*c*) Permanent magnet-type core.

The curves of Fig. 4-9 show that different materials have different hysteresis losses. The only way to hold this type of loss to a low value is to be very careful in choosing the type of core material. You will not have any control over this, so you will have to trust the manufacturer to do this for you.

Figure 4-9*a* is the hysteresis curve for air. Note that it is a straight line. There is no coercive force, and there is no remanence. Therefore there is no hysteresis loss in an air-core transformer.

Figure 4-9*b* shows the hysteresis curve for a good iron-core material. The curve is opened slightly, so there is a small amount of hysteresis loss. This cannot be avoided. Even the best core materials have a small amount of hysteresis loss.

Figure 4-9*c* shows a square-loop hysteresis curve. This is the type of hysteresis curve you would get if you used a permanent-magnet material in the core of a transformer. Note that the amount of remanence and the amount of coercive force are very large compared with the material in Fig. 4-9*a* and 4-9*b*. You would not want to use a permanent-magnet material for a transformer core. The large amount of area inside the curves means that it would have a high amount of hysteresis loss.

What Is Transformer Copper Loss?

The wires used for the primary and secondary windings of a transformer have resistance. Currents flowing in these wires produce heat loss. In transformers this is called *copper loss.* It is reduced by using wire with larger diameters.

Summary

1. A transformer is a component used for passing ac energy from one point to another.
2. Winding the primary and secondary of a transformer on an iron core greatly increases its efficiency at low frequencies.
3. An autotransformer consists of a single coil of wire with a tap. One end of the coil is common to both the primary and the secondary.
4. Eddy currents induced in the core of a transformer cause a power loss in the form of heat. Eddy currents are reduced by laminating the iron core.

5. Hysteresis loss occurs in the iron core of a transformer because energy must be used to remove magnetism during each half-cycle of primary current. It is reduced by using the right kind of iron.

What Are the Uses of Transformers in Electronic Circuits?

Transformers may be heavy, bulky, and expensive. But there are some cases where they do the best job. It is very useful for you to know how transformers are used in electronic circuits.

Will Transformers Pass DC Voltages?

In order for a voltage to be induced in the secondary of a transformer, it is necessary that the flux in the primary be expanding and contracting. If you apply a dc voltage to the primary of a transformer, the current (and the magnetic field) will rise to some value and remain there. As long as a steady direct current is flowing in the primary, there is no voltage induced in the secondary. This makes transformers useful for separating dc and ac voltages.

Figure 4-10 shows how it is done. There are two voltages on the primary winding: an ac voltage e and a dc voltage supplied by a battery E. Both voltages are applied to the primary winding at the same time. The waveform shown indicates that the primary current is varying but does not reach zero value. The varying current causes the flux from the primary to expand and contract and to cut across the secondary.

The secondary voltage is due only to the expanding and contracting flux, and is not in any way related to the steady flux of the battery current. Thus, with an ac and dc voltage applied at the primary, the secondary voltage is ac only.

What Is an Isolation Transformer?

Transformers are sometimes used for safety reasons. This can be understood with the help of Fig. 4-11. The primary of the transformer is connected

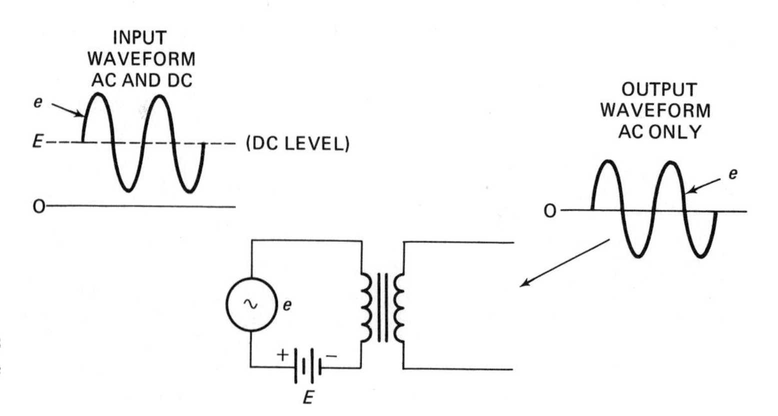

Fig. 4-10 A transformer will pass an ac voltage but will not pass a dc voltage.

to an ac generator which has one terminal grounded. You should understand that the power delivered to your home and the power delivered to industry has one line which is at ground potential, so this is a typical circuit. If you are standing on the ground or touching a ground connection, and at the same time you touch point a on the circuit, there will be a 115-volt drop across your body. This could be fatal.

On the secondary of the isolation transformer neither of the two terminals (x or y) is grounded. If you are standing on a ground point and accidentally touch point x, then point x will become grounded through your body, but you will not be shocked. Likewise, if you touch point y, point y will become grounded, but you will not be shocked. (Of course, if you touch both x and y at the same time, you may receive a fatal shock.)

What Is a Step-Up Transformer?

The amount of voltage across the secondary of a transformer depends upon the number of turns of wire in the secondary winding. If there is a greater number of turns of wire in the secondary than in the primary, the secondary voltage will be greater than the primary voltage. The symbol for a step-up transformer is shown in Fig. 4-12. Mathematically the secondary voltage E_2 is

$$E_2 = \frac{N_2}{N_1} \times E_1$$

where N_2 = the number of secondary turns
N_1 = the number of primary turns
E_1 = the primary voltage

In this equation the number of turns in the secondary N_2 divided by the number of turns in the primary N_1 is called the secondary-to-primary turns ratio.

$$\text{Secondary-to-primary turns ratio} = \frac{N_2}{N_1}$$

Suppose, for example, that 115 volts is applied to the primary of a transformer which has a turns ratio (N_2/N_1) of 6 to 1. In other words, the secondary

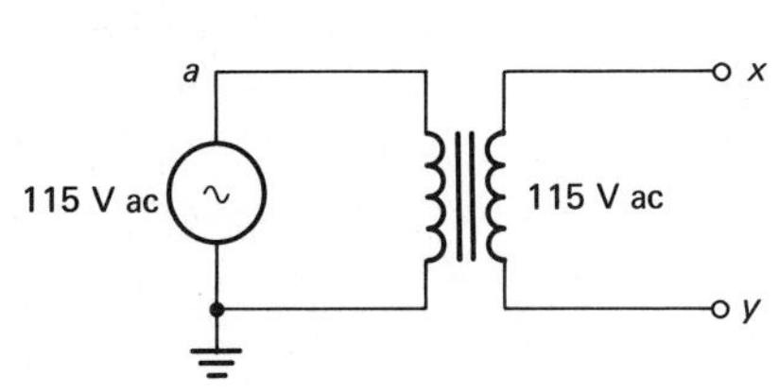

Fig. 4-11 An isolation transformer.

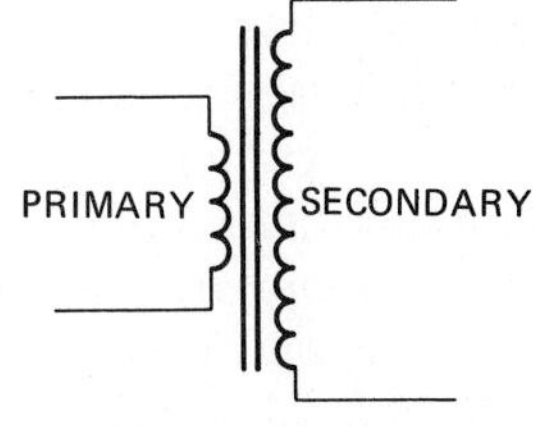

Fig. 4-12 A step-up transformer has more turns of wire in the secondary than it has in the primary.

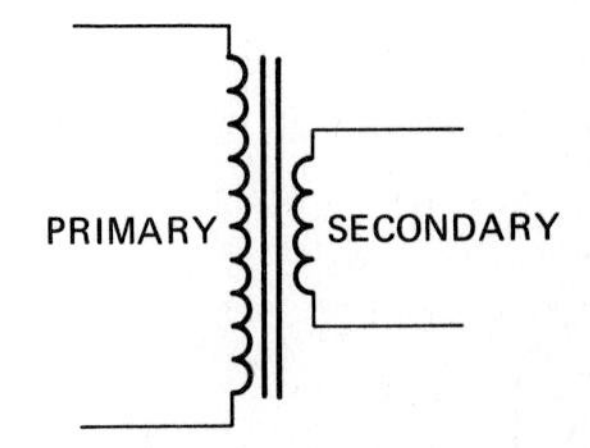

Fig. 4-13 A step-down transformer has more turns of wire in the primary than it has in the secondary.

winding has six times as many turns as the primary. The secondary voltage will then be

$$E_2 = \frac{N_2}{N_1} \times E_1$$
$$= \frac{6}{1} \times 115$$
$$= 690 \text{ volts}$$

At first it looks like you are getting something for nothing by using a step-up transformer. But if the secondary voltage is higher than the primary voltage, then the secondary current must be lower than the primary current. In other words, when the voltage is stepped up, the current is stepped down. The relationship between the turns ratio and the current in the primary and secondary is given by the equation

$$I_2 = \frac{N_1}{N_2} \times I_1$$

where I_2 = the secondary current
N_2 = the number of secondary turns
N_1 = the number of primary turns
E_1 = the primary voltage

In this equation the number of turns in the primary N_1 divided by the number of turns in the secondary N_2 is called the *primary-to-secondary turns ratio.*

$$\text{Primary-to-secondary turns ratio} = \frac{N_1}{N_2}$$

What Is a Step-Down Transformer?

Transformers can also be used to step the voltage down. If the secondary winding has one-half as many turns as the primary, then the secondary voltage will be one-half the primary voltage. The same equations for voltage, current, and turns ratio apply.

Stepping the voltage down permits us to step the secondary current up. As a general rule, step-down transformers can provide a larger secondary current but lower secondary voltage.

The symbols for step-up and step-down transformers shown in Figs. 4-12 and 4-13 are typical. You never should try to determine the turns ratio of a transformer by counting the turns in the symbol. These are just symbols. The actual turns ratio may be given by the manufacturer or it may be given on the schematic diagram of the equipment.

Remember that if the primary and secondary voltages are given, the turns ratio is also known, since the turns ratio is the same as the voltage ratio.

How Do Transformers Perform as Impedance-Matching Components?

In alternating-current circuits the opposition to current flow is called *impedance* (Z). In direct-current circuits the opposition to current flow is called *resistance.* Both impedance and resistance are measured in ohms.

An important application of transformers is to match the impedance of a generator (or electronic device) to some load. The idea of impedance matching is based on the *maximum-power-transfer theorem,* which is illustrated in Fig. 4-14.

In the circuit of Fig. 4-14*a* a battery E with an internal resistance R_i is connected to a variable resistor load R_L. As the load resistance is increased from 0 ohms, the amount of power delivered to the load resistor increases to a point and then begins to decrease. This is shown in Fig. 4-14*b*. The maximum possible power that can be delivered to the load *always* occurs when the load resistance is equal to the internal resistance of the battery. A battery is used in this example, but an ac generator could also be used.

In order to connect an ac generator to a load which has a impedance different from the internal generator impedance, an impedance-matching device is needed. Transformers can be used for this purpose. As shown in Fig. 4-14*c*, a high internal impedance in an ac generator can be used to deliver

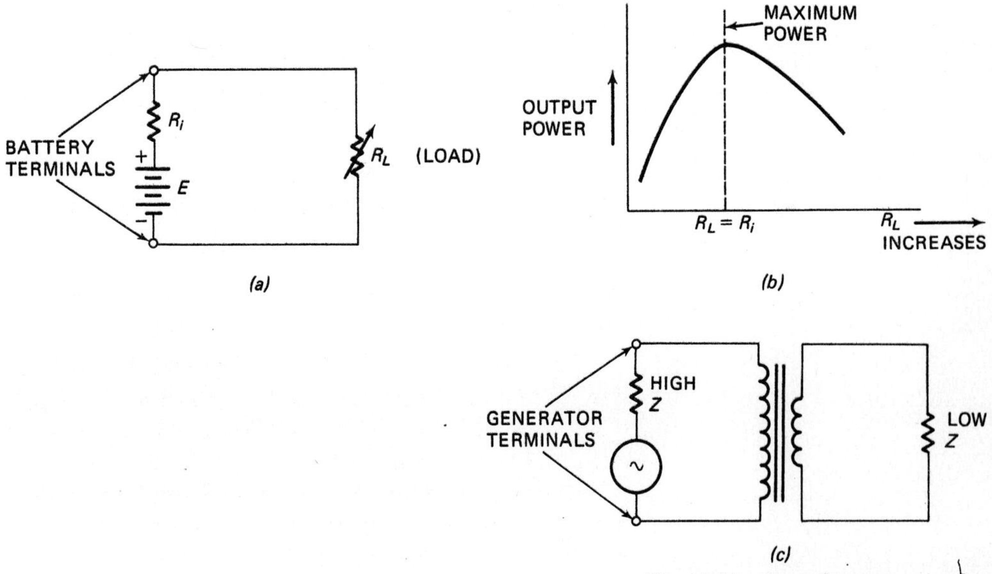

Fig. 4-14 Transformers may be used for impedance matching. (*a*) A simple circuit for demonstrating the maximum power transfer theorem. (*b*) Characteristic curve for the circuit in (*a*). (*c*) Transformer used for impedance matching.

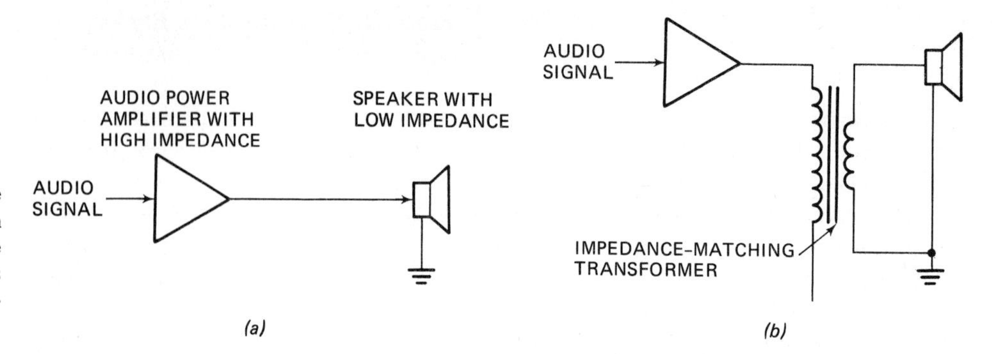

Fig. 4-15 A transformer can be used for impedance matching an amplifier to a speaker. (*a*) The amplifier and speaker impedances are not matched. (*b*) The transformer matches the impedances.

maximum power to a low impedance load by connecting the proper transformer between the two. (A transformer could also be used to match a low generator impedance to a high load impedance.)

In electronic circuits a transformer may be used to match the impedance of an amplifier to a speaker. This is shown in Fig. 4-15. The amplifier is considered to be the generator of audio power. It is supposed to deliver as much power as possible to the speaker. When they are directly connected, as shown in fig. 4-15*a*, the speaker does not receive maximum power. In Fig. 4-15*b* the transformer serves to match the amplifier and load so that maximum power can be delivered.

There are many applications in electronics where impedance matching is accomplished with a transformer.

How Is a Transformer Used for Frequency Selection?

In some circuits it is desirable to pass one frequency, or range of frequencies, from one point to another and reject all other frequencies. Transformers are ideal for this purpose because the primary and secondary of the transformers are inductances. These inductances can be tuned with capacitors. Figure 4-16 shows the basic principle. In this transformer there are three frequencies in the primary: 1000 kilohertz (kHz), 2000 kilohertz, and 3000 kilohertz. Assume that it is required to pass only the 2000-kilohertz signal to the next stage. Both the primary and the secondary of the transformer are tuned to this frequency. All frequencies other than 2000 kilohertz are rejected—that is, they are not coupled into the secondary circuit.

There are variations of the tuned-transformer circuit. In some applications only the primary or only the secondary is tuned. Sometimes the capacitors are fixed rather than variable, and the inductances of the primary and secondary (or both together) are variable. The inductance is usually varied by

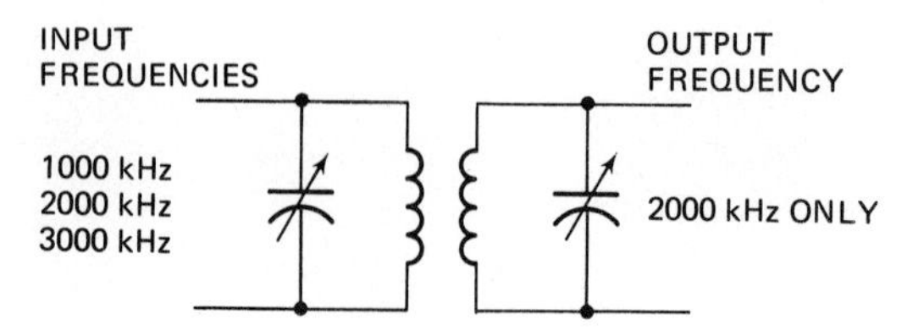

Fig. 4-16 A tuned transformer.

moving a powdered iron core or ferrite core material in or out of the transformer coil(s).

Summary

1. A transformer will pass an ac voltage from primary to secondary but will not pass a dc voltage.
2. An isolation transformer is used for protecting operators and technicians from a deadly electric shock.
3. A transformer can be used to step a voltage either up or down. When the voltage is stepped up, the secondary current is stepped down. When the voltage is stepped down, the secondary current is stepped up.
4. Transformers are used for matching a load impedance to a generator impedance. This results in the maximum possible transfer of power.
5. A transformer can be tuned so that it will pass one frequency (or a range of frequencies) and reject all others.

How Does a Relay Work?

A *relay* is an electrically operated switch. Figure 4-17 shows how a simple *clapper-type* relay works.

Figure 4-17*a* shows that the clapper-type relay has a coil wrapped around a magnetic soft-iron core material. When current flows in the coil the core becomes magnetized and attracts the *armature*. The armature moves the electrical contacts and thus switches the contacts. The contacts are located on an insulated board. When there is no current flowing in the coil a spring returns the armature to the deenergized position of Fig. 4-17*a*. The symbol for the contacts of the relay is also shown in Fig. 4-17*a*. Note that when the relay is not energized there is electrical contact between terminals *A* and *C* and there is no contact between *A* and *B*. The relay is energized by closing the switch.

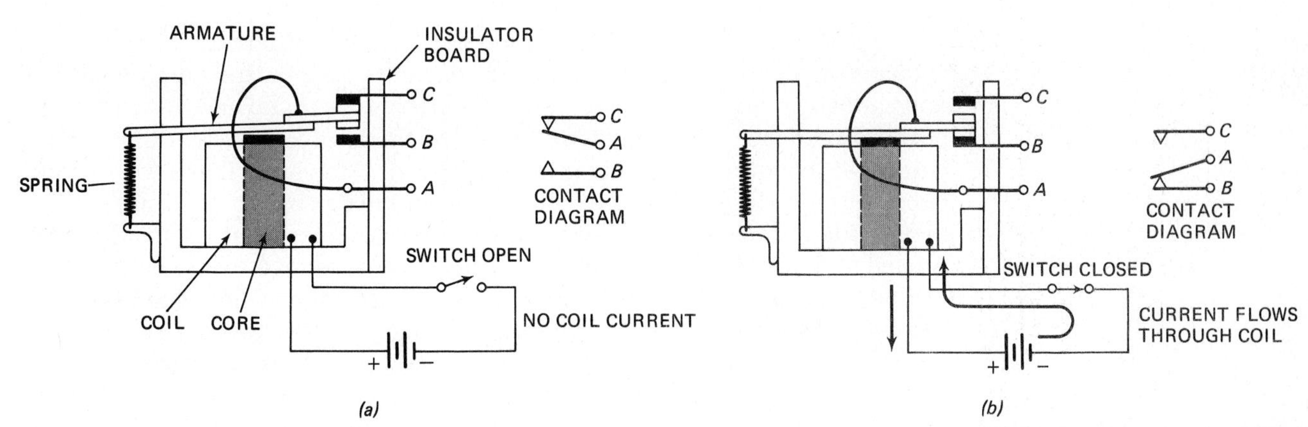

Fig. 4-17 A simple clapper-type relay. (*a*) In the deenergized position; (*b*) in the energized position.

This is shown in Fig. 4-17*b*. Current flows in the coil and magnetizes the core. The magnetized core attracts the armature. The armature moves the contacts, and the diagram shows that there is electrical contact between terminals *A* and *B*. At this time there is no contact between terminals *A* and *C*.

Contacts *A* and *C* are called *normally closed* because in the relay's normal (deenergized) position they make contact. Terminals *A* and *B* are called *normally open* because when the relay is in the deenergized position there is no contact between them. You may think that nothing has been accomplished by using the relay. You could have simply switched terminals *A*, *B*, and *C* directly rather than use a switch to control a relay. However, there are several advantages to the relay system. For one thing the coil current of the relay may be quite small, so a small switch can be used in the coil circuit. However, the contacts of the relay may be connected into high-current or high-voltage circuitry that requires large switch contacts. Thus by using a relay you can operate a small switch and close a high-power circuit.

There is another important use of relays. They are often used for remote switching. Furthermore the circuit being switched ON and OFF through terminals *A*, *B*, and *C* may be located a considerable distance from the place where you want to do the switching.

The clapper-type relay of Fig. 4-17 is used where a small amount of coil power can switch circuits that have large voltages and currents. Figure 4-18 shows some additional types of relays that you may encounter in industrial circuits.

How Do You Recognize a Relay-type Circuit?

The telephone relay of Fig. 4-18*a* is used where it is desired to switch a large number of circuits at one time.

The mechanical latching relay of Fig. 4-18*b* switches when the coil is energized and remains in the switched position even though the coil current is removed. It is necessary to reset this relay manually for the next operation.

The stepping switch of Fig. 4-18*c* is used in applications where a number of different circuits are to be switched in sequence—that is, one after another.

The *differential* relay of Fig. 4-18*d* can sense the difference between two voltages or the difference between two currents. When the voltage (or current) in two circuits is the same the coil currents are equal and flowing in opposite directions. The magnetic flux from the two coils cancel each other. Under this condition the relay is not energized. However, if one coil current becomes much larger than the other, its magnetic flux becomes strong enough to override the weaker flux and the relay becomes energized. Thus the relay is used to determine when there is a difference in voltage (or current) values.

The thermal relay of Fig. 4-18*e* is used in applications where a temperature rise is to be sensed. Suppose, for example, that a motor starts to become overheated. It is desirable to shut the motor OFF before it is destroyed. The thermal relay is located on the frame of the motor. When the temperature rises above a certain value, the relay switches the motor OFF.

The meter relay of Fig. 4-18*f* uses a meter movement. This meter movement may be connected as either a voltmeter or an ammeter. As long as the

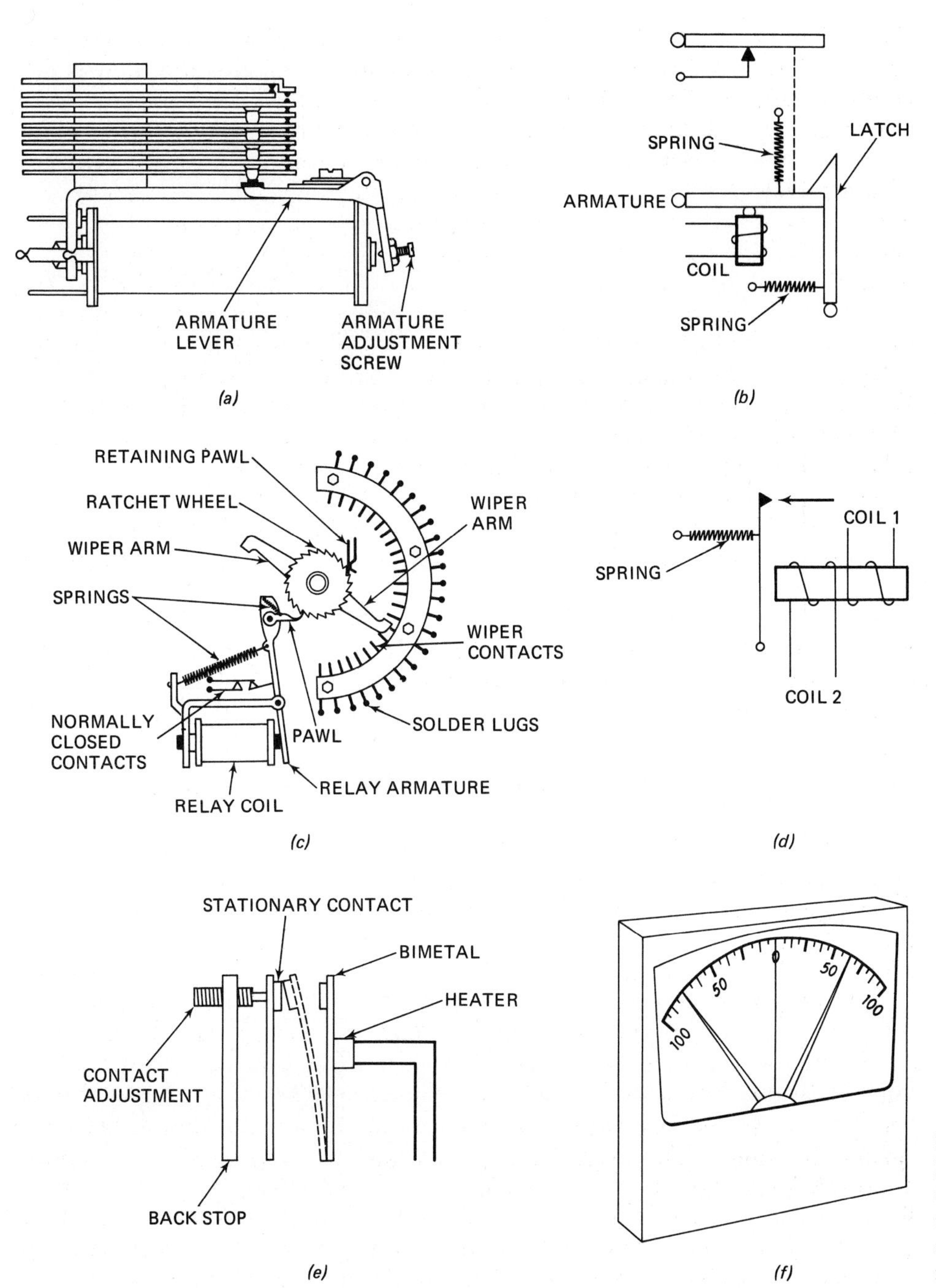

Fig. 4-18 Some of the more popular types of relays. (*a*) Telephone type; (*b*) mechanical latching relay; (*c*) stepping switch; (*d*) differential relay; (*e*) thermal relay; (*f*) meter relay.

voltage or current is within the given limits, no contact is made. However, if the voltage or current rises above or below the limit, contact is made. This contact may be used to control the operation of a machine or to shut down overloaded circuits.

To recognize a relay-type you must remember that there are two sections that must be checked. *First,* there is the relay coil. This coil must have current flowing through it when the relay is energized. *Second,* there are the relay contacts which open or close depending upon whether the relay is energized or deenergized.

The coil may be energized with a simple circuit like the one shown in Fig. 4-17. In other circuits the coil is energized by current from a tube or transistor.

What Types of Symbols Are Used for Relays?

The relay contact circuit may be drawn in either of two ways. Examples are shown in Fig. 4-19. The ANSI symbols may be slightly easier to read, but they are much more difficult to draw on large schematics. For this reason the industrial symbols have become popular.

Relay manufacturers identify the different contact layouts (which they call *pileups*) as being forms, such as Form A, B, C, and so on. Five of the most popular forms are shown in Fig. 4-19. Form A is a normally open single-pole, single-throw switch. The ANSI symbol shows the contacts as being open or in their normal deenergized position. The arrow in the ANSI symbol indicates the direction that the armature and movable contact will go when the coil is energized. In other words, if the coil is energized, the armature will push the movable contact to make connection.

The industrial symbol for Form A is not quite so easy to understand, but you should learn to identify this symbol as being a normally open switch contact. It is very important to distinguish between the industrial Form A con-

RELAY CONTACTS

ANSI SYMBOL*	INDUSTRIAL SYMBOL	TYPE	FORM
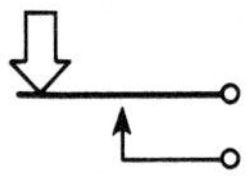	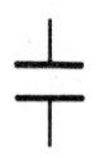	SINGLE-POLE SINGLE-THROW, NORMALLY OPEN	A
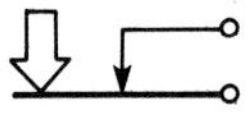		SINGLE-POLE SINGLE-THROW, NORMALLY CLOSED	B
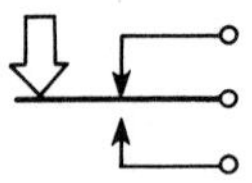	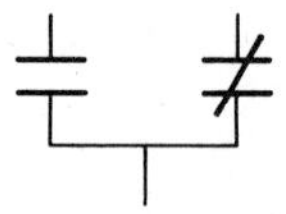	TRANSFER (ALSO CALLED BREAK-MAKE)	C
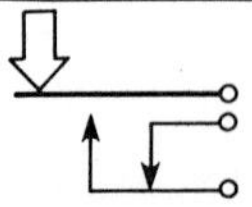	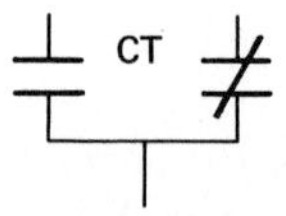 CT	MAKE BEFORE BREAK (ALSO CALLED CONTINUITY TRANSFER)	D
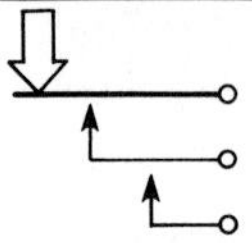	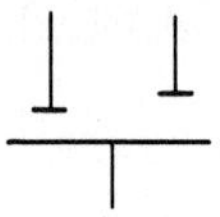	TIME SEQUENTIAL CLOSING	F

*ANSI – American National Standards Institute

Fig. 4-19 Examples of relay forms. Relay contacts are shown in their normal (deenergized) condition.

tact symbol and the symbol for a capacitor. The symbol for a capacitor has one of the two lines curved. (See capacitor symbols in Appendix B.)

Contacts in the Form B relay are normally closed. Energizing the coil will open the circuit in this case.

The Form C contact has one set of contacts closed and the other one normally open. Energizing the coil will transfer the connections from one circuit to another. The clapper-type relay illustrated in Fig. 4-17 has Form C contacts.

The Form D contact is sometimes called the *continuity transfer* (*CT*). It is designed so that when relay is energized, the normally open contact will close before the normally closed contact opens.

Form F contacts are *timed-sequence closing*. There are two normally open contacts. The relay is designed so that when the coil is energized, one set of contacts completes the circuit before the other set.

What Are Some Examples of Relay Circuits?

Figure 4-20 shows a simple relay circuit. The industrial symbols are used in Fig. 4-20*a*, and the ANSI symbols are used in Fig. 4-20*b*. Both circuits are the same. The only difference is in the type of symbols used. A dc relay is used here. Ac relays work in a similar way, but the relays are constructed in a slightly different manner. The core material and armature of ac relays are laminated to reduce the problem of eddy-current losses. Also the material is chosen for minimum hysteresis loss.

In Fig. 4-20 the relay coil *N* is in series with a switch *SW*. The letter *N* identifies this particular relay. There may be as many as thirty or forty relays in a system, and each relay coil is identified by a different letter or combination of letters. The contacts for each relay are identified by the same letter. Thus every contact marked with an *N* is operated by the coil that is marked with an *N*. When the switch is closed, current flows through the relay coil and energizes it. There are two contacts for this coil. Both are marked *N*. The contact in series with lamp L_1 is normally open. The other contact is normally closed, and it is in series with lamp L_2. Note that the two relay contacts together make a Form C contact as shown in Fig. 4-19.

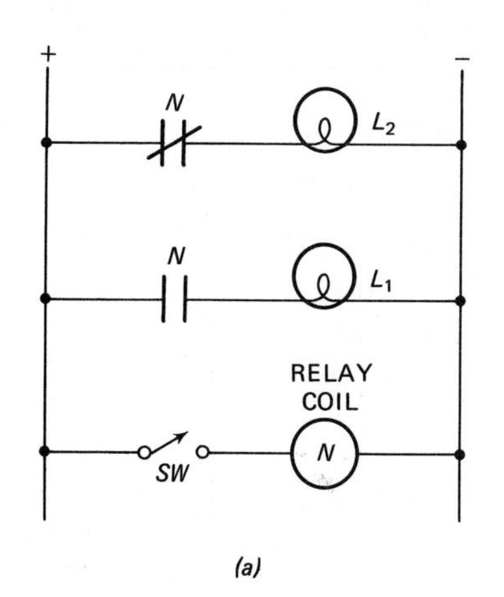

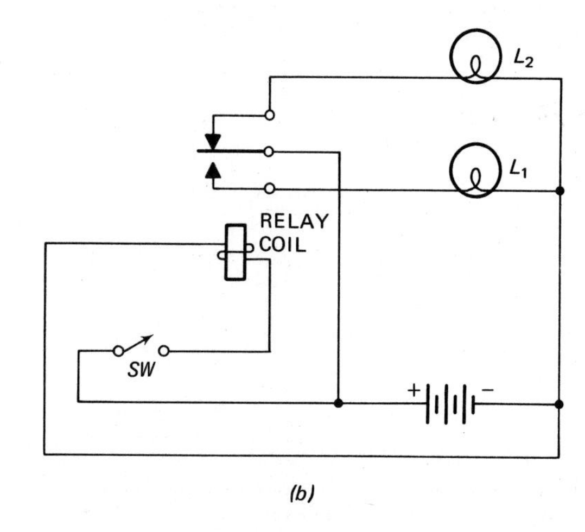

Fig. 4-20 **A simple relay circuit. (*a*) The circuit with industrial symbols. (*b*) The circuit with ANSI symbols.**

When switch *SW* is open, L_2 is ON and L_1 is not. Closing switch *SW* will cause L_1 to come ON and L_2 to go OFF.

In Fig. 4-20 the relay is shown in the deenergized condition. You should always assume that relays are in this condition when shown on schematic drawings. Note that the circuit for L_2 is complete but the circuit for L_1 is open. When the relay is energized the circuit for L_1 will be completed and that lamp will be ON. At the same time, the circuit for L_2 will be open and that lamp will be OFF.

Figure 4-21 is a simple relay circuit for starting and stopping a dc motor. Figure 4-21*a* shows the industrial symbols for the motor-control circuit. Figure 4-21*b* shows the same circuit using ANSI symbols.

The relay coil in this case is marked *A*. There are two manual "momentary contact" switches in the circuit. One is marked ON, and it is normally open. The other is marked OFF, and it is normally closed. They are both spring operated and return to the positions shown when released.

To start the motor, the ON button is pushed. This completes the circuit through the OFF switch and through relay coil *A*. The relay is energized, and both contacts marked *A* close. The relay contacts in parallel with the ON switch (now closed) allow current to flow in the circuit marked *a* and *b* through contact *A* (near *a*), through the OFF switch, and through the relay coil *A*. The relay coil is now said to be electrically *latched*. Even though the ON switch opens, the current does not stop flowing through the relay coil. The reason is that there is a path from *a* to *b* as follows: through the latching contact, through the OFF switch, and through the relay coil.

Another thing happens when the relay is energized: Contact *A*, which is in series with the motor, is closed. This causes the motor to run.

The normally closed thermal relay marked *X* is located on the motor frame. If the motor should become overheated, *X* will open and the motor circuit would be opened.

To stop the motor the OFF button is pushed. This opens the coil circuit and deenergizes the relay. Since both sets of contacts marked *A* become opened, the coil and the motor circuits are opened.

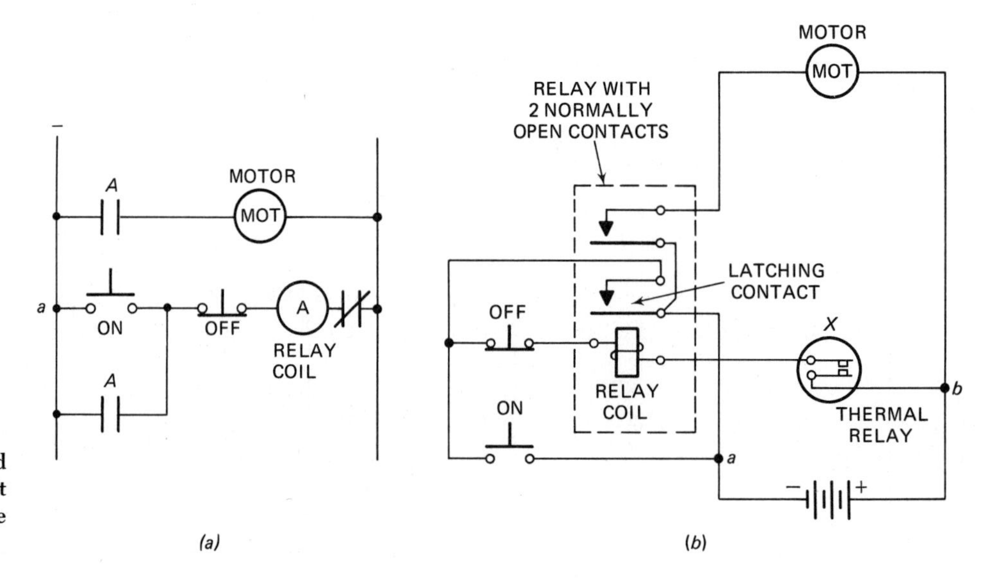

Fig. 4-21 A circuit for starting and stopping a motor. (*a*) The circuit with industrial symbols. (*b*) The circuit with ANSI symbols.

Summary

1. A relay is an electrically operated switch.
2. A relay permits a small switch to be used to control high-voltage or high-current circuits.
3. A relay is divided into two sections: the coil section and the contact section.
4. Relay-contact configurations are identified as forms, such as Form A, Form B, and so on.
5. Some relays can be electrically "latched."

PROGRAMMED REVIEW QUESTIONS

(Instructions for using this programmed section are given in Chapter 1.)
We will review the important concepts of this chapter. If you have understood the material, you will progress easily through this section. Do not skip this material because some additional theory is presented.

1. The transformer of Fig. 4-22 has a turns ratio of 1 to 1. This means that the number of turns of wire in the primary equals the number of turns of wire in the secondary. How much current is flowing in the secondary winding?
   A. 10 milliamperes. (Proceed to block 9.)
   B. 10 microamperes. (Proceed to block 17.)

2. *The correct answer to the question in block 25 is* **A**.

$$\frac{N_1}{N_2} = \frac{E_1}{E_2}$$
$$= \frac{100}{25}$$
$$= \frac{4}{1}$$

   Here is your next question.
   Which of the following is most like a switch?
   A. A transformer. (Proceed to block 14.)
   B. A relay. (Proceed to block 22.)

3. *The correct answer to the question in block 11 is* **B**. *The transformer in Fig. 4-23b has a direct connection between the primary and sec-*

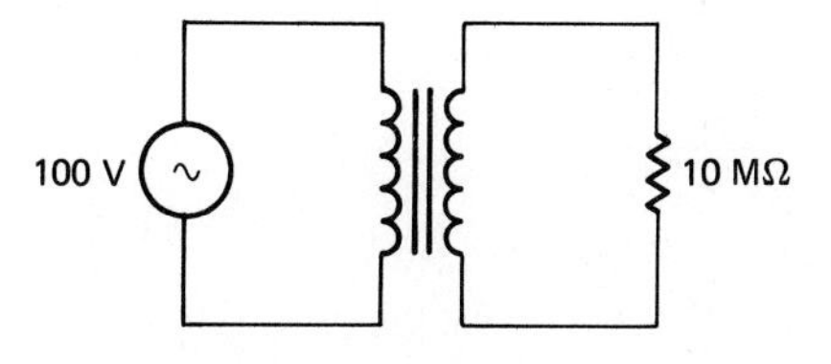

Fig. 4-22 Illustration for question 1.

ondary winding. It is an autotransformer, and it cannot be used as an isolation transformer. Here is your next question.
In the circuit of Fig. 4-24 both lamps will be ON if
A. SW_1 is closed and SW_2 is closed. (Proceed to block 15.)
B. SW_1 is closed and SW_2 is open. (Proceed to block 20.)

4. *The correct answer to the question in block 22 is* **A**. *The moving field depends upon the number of conductors and the speed at which the field cuts across the conductors.* Here is your next question.
A magnetic field is usually represented by field lines, which are also called *flux*. In which of the following materials would it be easier to get magnetic-field lines to pass through?
A. Air. (Proceed to block 16.)
B. Iron. (Proceed to block 19.)

5. *Your answer to the question in block 21 is* **A**. *This answer is wrong. It is not possible to step up the power in a transformer.* Proceed to block 11.

6. *The correct answer to the question in block 20 is* ***B***. *The magnetic field that surrounds a current flow has a direction at right angles to the current.* Here is your next question.
The direction of the magnetic field around an electron current can be determined by
A. the right-hand rule. (Proceed to block 26.)
B. the left-hand rule. (Proceed to block 13.)

7. *The correct answer to the question in block 17 is* **A**. *Laminating the core will not affect hysteresis loss, but it will reduce eddy-current losses.* Here is your next question.
Can an ac generator with a low internal impedance be matched to a high-impedance load?
A. Yes. (Proceed to block 21.)
B. No. (Proceed to block 27.)

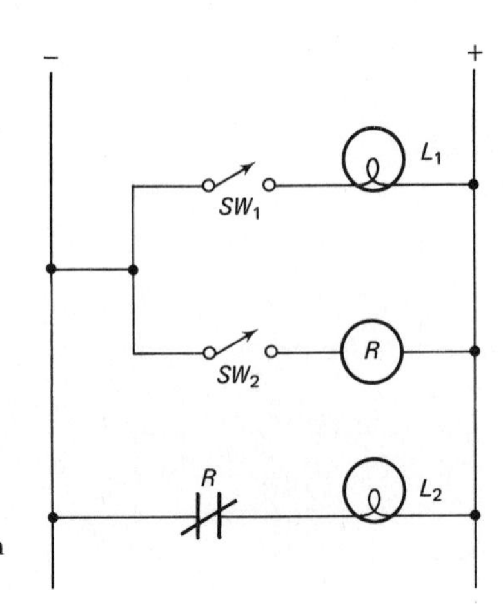

Fig. 4-24 Illustration for question 3.

8. *Your answer to the question in block 25 is* **B**. *This answer is wrong. The turns ratio is equal to the voltage ratio. In this case:*

$$\frac{N_1}{N_2} = \frac{E_1}{E_2}$$

Proceed to block 2.

9. *Your answer to the question in block 1 is* **A**. *This answer is wrong. The voltage across the secondary is equal to the voltage across the primary (100 volts). Since you know the voltage (100 volts) across the load resistor (10 megohms) you can use Ohm's law to find the amount of current. Calculate the current,* then proceed to block 17.

10. *Your answer to the question in block 22 is* **B**. *This answer is wrong. The question asks which of the choices the induced voltage does* **not** *depend upon. The induced voltage* **does** *depend upon the number of wires cut across by the moving field.* Proceed to block 4.

11. *The correct answer to the question in block 21 is* **B**. *A transformer can be used to step up current by stepping down the voltage.* Here is your next question.
Which of the transformers in Fig. 4-23 cannot be used as an isolation transformer?
A. The one in Fig. 4-23*a*. (Proceed to block 23.)
B. The one in Fig. 4-23*b*. (Proceed to block 3.)

12. *Your answer to the question in block 17 is* **B**. *This answer is wrong. Laminating the core material will not affect the hysteresis loss.* Proceed to block 7.

13. *The correct answer to the question in block 6 is* **B**. *Electron current flows from − to +, and the left-hand rule for magnetic-field direction applies.* Here is your next question.
A relay contact that is single-pole, single-throw, and normally closed is called
A. Form A. (Proceed to block 18.)
B. Form B. (Proceed to block 25.)

14. *Your answer to the question in block 2 is* **A**. *This answer is wrong. A*

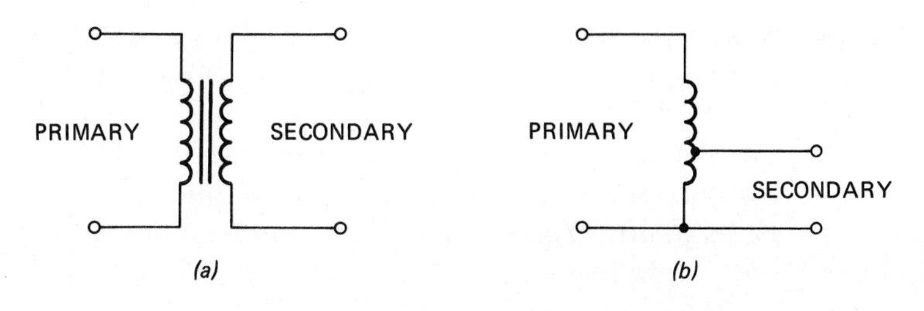

Fig. 4-23 Symbols for the question in block 11.

transformer transfers energy from one point to another, but it does not turn the circuit ON *or* OFF. Proceed to block 22.

15. *Your answer to the question in block 3 is* **A**. *This answer is wrong. If* SW_2 *is closed, relay M will be energized. This will cause the normally closed contact in series with* L_2 *to open, and the lamp will no longer be* ON. Proceed to block 20.

16. *Your answer to the question in block 4 is* **A**. *This answer is wrong. It is* ***harder*** *to get flux lines to pass through air than it is to make them pass through iron.* Proceed to block 19.

17. *The correct answer to the question in block 1 is* **B**. *The secondary voltage equals the primary voltage, which is 100 volts. The resistance is 10 million ohms, or* 10×10^6 *ohms. Without using powers of 10,*

$$I = \frac{E}{R}$$
$$= \frac{100}{10{,}000{,}000}$$
$$= 0.000\ 01 \text{ ampere}$$
$$= 10 \text{ microamperes}$$

Using powers of 10,

$$I = \frac{E}{R}$$
$$= \frac{100}{10 \times 10^6 \text{ ohms}}$$
$$= 10 \times 10^{-6} \text{ ampere}$$
$$= 10 \text{ microamperes}$$

Remember that you can consider the secondary of a transformer to be the generator of electricity as far as the secondary circuit is concerned. Here is your next question.
Hysteresis loss in a good transformer core is kept to a minimum by the manufacturer. He does this by
A. choosing the proper core material. (Proceed to block 7.)
B. laminating the core material. (Proceed to block 12.)

18. *Your answer to the question in block 13 is* **A**. *This answer is wrong. A Form A contact is normally open.* Proceed to block 25.

19. *The correct answer to the question in block 4 is* **B**. *It is easier to get the flux lines to pass through iron than it is to make them pass through air. You can say, then, that iron has a lower reluctance than air.* Here is your next question.

What type of relay can sense when there is a difference between two voltages? (Proceed to block 28.)

20. *The correct answer to the question in block 3 is* **B**. *Lamp L_2 is normally on because it is connected through normally closed contacts. As long as SW_2 is open, L_2 will be* ON. *Closing SW_1 causes L_1 to be* ON.
Here is your next question.
Oersted's experiment showed
A. that there is no relationship between electricity and magnetism. (Proceed to block 24.)
B. how electricity and magnetism are related. (Proceed to block 6.)

21. *The correct answer to the question in block 7 is* **A**. *Transformers are used to match a load impedance to a generator impedance.* Here is your next question.
Which of the following is a use for transformers in electronic circuits?
A. Step up the power. (Proceed to block 5.)
B. Step up the current. (Proceed to block 11.)

22. *The correct answer to the question in block 2 is* **B**. *A relay may be thought of as being an electrically operated switch.* Here is your next question.
Faraday's law says that when a moving magnetic field cuts across a wire a voltage is generated. The amount of voltage does *not* depend upon
A. the size of the wire. (Proceed to block 4.)
B. the number of wires that the field moves across. (Proceed to block 10.)

23. *Your answer to the question in block 11 is* **A**. *This answer is wrong. The question asks which transformer* **cannot** *be used as an isolation transformer.* Proceed to block 3.

24. *Your answer to the question in block 20 is* **A**. *This answer is wrong. You should remember that an electric-current flow always has a magnetic field around it.* Proceed to block 6.

25. *The correct answer to the question in block 13 is* **B**. *A Form B contact is normally closed, and a Form A contact is normally open.* Here is your next question.
When the voltage across the primary of a certain transformer is 100 volts, the voltage across its secondary is 25 volts. The primary-to-secondary turns ratio (N_1/N_2) is
A. 4 to 1. (Proceed to block 2.)
B. 1 to 4. (Proceed to block 8.)

26. *Your answer to the question in block 6 is* **A**. *This answer is wrong. Review the rule in Fig. 4-2,* then proceed to block 13.

27. *Your answer to the question in block 7 is* **B**. *This answer is wrong. One of the components that you studied in this chapter is sometimes used for impedance matching.* Proceed to block 21.

28. *A differential relay can sense when there is a difference in voltage between two voltages.*
You have now completed the programmed questions. The next step is to put some of these ideas to work in laboratory experiments. Proceed to the Experiment section of this chapter.

EXPERIMENT

(The experiment described in this section may be performed on the circuit board described in Appendix C or on a similar laboratory setup.)

Purpose In this experiment you will demonstrate some basic characteristics of transformer circuits and relay circuits.

PART I

Theory The secondary circuit of a transformer can be considered to be complete and separate from the primary circuit when you are tracing circuit paths. You can think of the transformer primary winding as being a load for one circuit and the transformer secondary as being a source of voltage for another circuit. However, this is only for convenience in tracing circuit paths. Remember that the primary circuit always has a definite relationship to the secondary. For example, if you increase the current in the secondary winding, *the current in the primary winding will also increase!*

This experiment will show that the primary current depends upon the secondary current.

Test Setup The schematic of the circuit is shown in Fig. 4-25*a*. Figure 4-25*b* shows a pictorial diagram of the device used to insert resistor R_1 in the primary circuit. If the primary current changes for any reason, then the voltage across R_1 will also change. This voltage is measured with an ac voltmeter.

Before you plug the circuit into the ac power line, connect the ac voltmeter across resistor R_1. Do not make any connections or touch any wires in this circuit when the plug is in the socket! There is a line voltage on the primary when the circuit is in operation. This line voltage can kill!

Procedure

step one When you have connected the circuit as shown in Fig. 4-25, check your wiring. Then insert the plug into the power line.

step two Read the value of ac voltage across R_1 with the switch OFF.

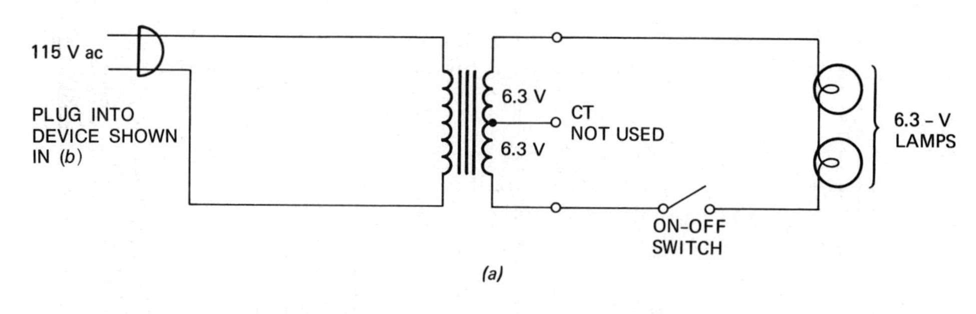

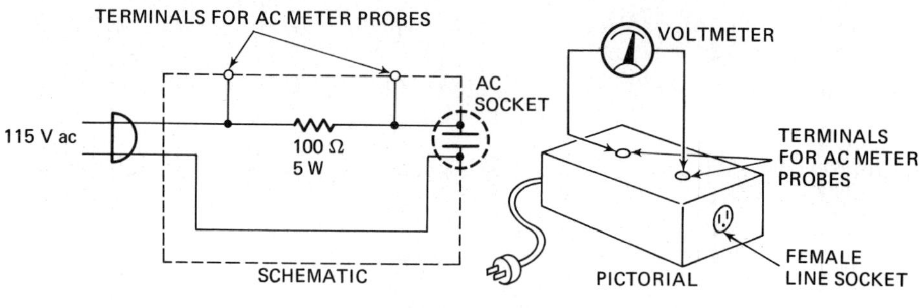

Fig. 4-25 Schematic diagram of the experiment.

(The lamps should be out.) Record the value. ________________ volts ac

step three Turn the switch ON and read the value of ac voltage across R_1.

Record the value. ________________ volts ac

step four After you have taken the readings in Step 1 and Step 2, remove the power plug from the ac line.

step five Did the voltage increase or decrease when the secondary circuit delivered power to the lamps? ________________

Conclusion Your experiment should have demonstrated that a change in the secondary current produces a change in the primary current. Furthermore, if the secondary current is increased, the primary current will also increase. The voltage across R_1 increased when the lamps were ON. This means that the current through R_1 must have increased. Remember that the voltage across a resistor is directly related to the current through that resistor.

PART II

Theory In this chapter you were shown two different ways of drawing relay circuits on schematic diagrams. It is very important for you to learn both methods. In this experiment you will wire a simple relay circuit according to its industrial wiring diagram and check its operation.

Test Setup Cover the pictorial diagram shown in Fig. 4-26*b*. Wire the relay circuit according to the schematic diagram shown in Fig. 4-26*a*.

Procedure

step one Check your circuit with the one shown in Fig. 4-26*b*.

step two With the switch SW_1 open and SW_2 closed, note that the relay is not energized and the lamp is not ON.

step three Close switch SW_1. This energizes the relay coil R. The lamp is now ON.

step four Open switch SW_1 and note that the relay remains energized. This is because the relay contact R is now closed and completes the circuit for the relay coil.

step five Open switch SW_2 and note that the relay *drops out*—that is, becomes deenergized.

Conclusion The circuit of Fig. 4-26 is known as a *latching circuit,* or *self-*

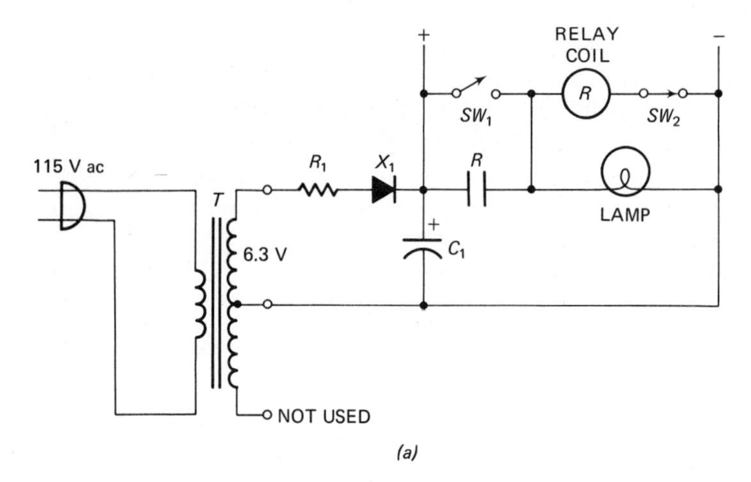

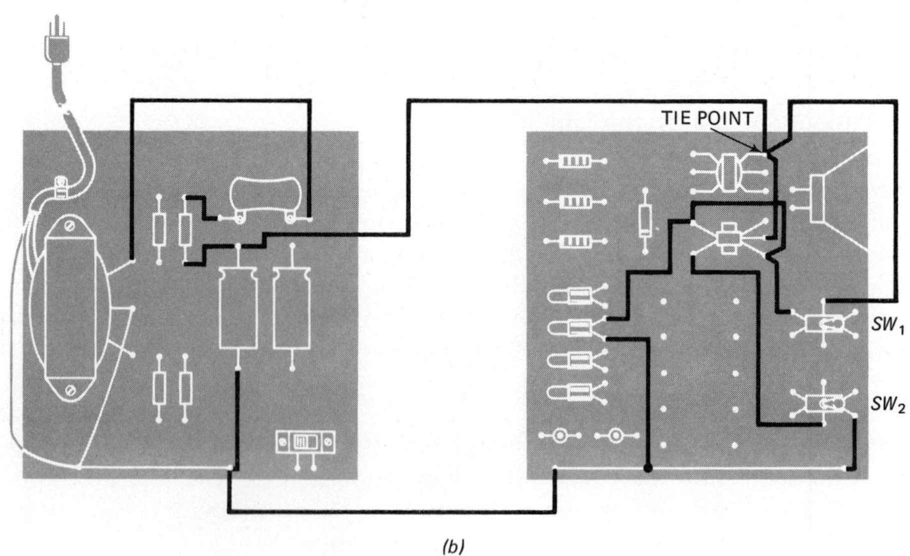

Fig. 4-26 Circuit for relay experiment. (*a*) Schematic diagram. (*b*) Pictorial diagram.

holding relay circuit. The relay is energized when SW_1 is closed and the contacts R of the relay are also closed. The coil circuit R is completed through the relay contact R, so the relay remains energized.

The semiconductor SCR circuit of Fig. 4-27 does the same thing as the relay in Fig. 4-26*a*. The SCR is turned on by applying a positive voltage to the gate through resistor R. This occurs when SW_1 is closed. Once the SCR is ON it will remain ON even though this switch is opened. To shut OFF the SCR, you must open switch SW_2.

SELF-TEST WITH ANSWERS

(Answers with discussions are given at the end of the chapter.)

1. Which one of the following types of transformers cannot be used as an isolation transformer? (*a*) An autotransformer; (*b*) An iron-core power transformer with a separate primary and secondary winding.
2. The amount of magnetic force needed to remove residual magnetism from a transformer core during each half cycle is called the (*a*) remanence force; (*b*) coercive force; (*c*) tractive force; (*d*) eliminator.
3. Eddy currents can be reduced by (*a*) using a larger diameter of wire for the windings; (*b*) reducing the primary voltage; (*c*) reducing the primary current; (*d*) laminating the iron core.
4. It is easier to establish flux lines in soft iron than it is to establish them in air. This is because the iron has a lower (*a*) resistance; (*b*) force field; (*c*) reluctance; (*d*) hold-off.
5. In the normal operation of a transformer the applied voltage goes to the (*a*) primary; (*b*) secondary.
6. When a conductor is moved through a magnetic field a voltage is always induced. The amount of voltage depends upon (*a*) the size of the conductor; (*b*) the rate at which the conductor is moved.
7. If the current in the secondary circuit of a transformer increases, the current in the primary will (*a*) decrease; (*b*) remain the same; (*c*) increase; (*d*) not be affected.
8. The turns ratio (primary-to-secondary) of a certain transformer is 2 to 3. If the voltage across the secondary is 60 volts, then the voltage across the primary must be (*a*) 40 volts; (*b*) 90 volts.
9. A certain relay is energized by a battery. An ON/OFF switch controls the

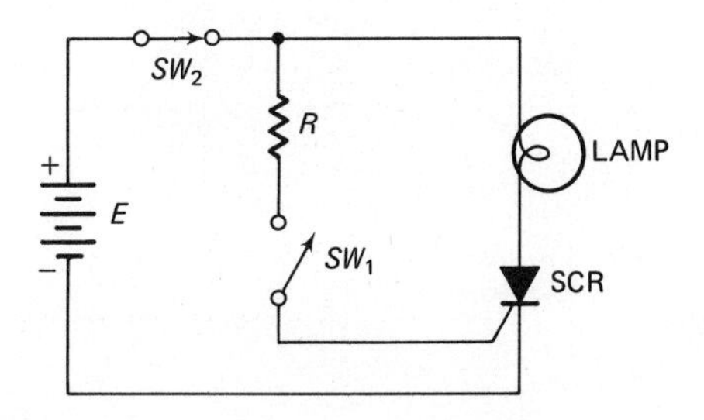

Fig. 4-27 Solid-state latching circuit.

coil circuit. If the relay is electrically latched, one of its contacts is (*a*) in series with the ON/OFF switch; (*b*) in parallel with the ON/OFF switch.

10. The part of the relay that moves the contacts is called (*a*) the rotor; (*b*) the stator; (*c*) the coil; (*d*) the armature.

ANSWERS TO SELF-TEST

1. (*a*)—There is no isolation between the primary and secondary of an autotransformer.
2. (*b*)—This is the definition of coercive force.
3. (*d*)—Laminated iron cores are used in ac relays as well as in transformers.
4. (*c*)—Reluctance is equal to the magnetic force divided by the magnetic flux. The reciprocal of reluctance is called *permeability.*
5. (*a*)
6. (*b*)—This is Faraday's law. The voltage is also dependent upon the angle at which the conductor cuts across the flux and the number (not size) of conductors. Maximum induced voltage occurs when the conductor is cutting the flux at a right angle.
7. (*c*)—This was demonstrated in the Experiment section.
8. (*a*)—There are fewer turns of wire in the primary. Therefore the voltage across the primary will be less than the voltage across the secondary.
9. (*b*)—This was demonstrated in the Experiment section.
10. (*d*)—The armature moves when the coil current starts and when it stops.

what are transducers and how are they used? 5

INTRODUCTION

In the study of transducers it is important to know the scientific meanings of the words *work* and *energy*. *Work* is defined as using a force to move an object through a distance. Mathematically,

$$\text{Work} = \text{force} \times \text{distance}$$

For example, if 1 pound (lb) of force is used to pull a sled a distance of 1 foot (ft), then the amount of work is 1 *foot-pound* (ft-lb).

If you are an office worker and you spend 8 hours (h) at your desk, you may feel that you are working. But in the scientific sense no work is actually done because you are not using a force to move objects through a distance. On the other hand, if you push a box with a force of 20 pounds over a distance of 15 feet, you have done 300 foot-pounds ($20 \times 15 = 300$) of work.

It is interesting to note that the *volt*, used in electrical measurement, is defined as the amount of work used to move a positive charge from the negative terminal of a voltage source to the positive terminal of the source. Thus the volt, like many other units of measurement in electricity and magnetism, is based on very basic laws of science.

The term *energy* also has a basic meaning when used in science. By definition, *energy* is the capacity to do work. In this definition the term *work* means force times distance. There are six well-known basic forms of energy. They are mechanical, heat, light, chemical, nuclear, and electrical. Energy in any of these forms is capable of doing work—that is, it is capable of moving an object through a distance.

In electronics we are concerned with the measurement of energy and with the use of energy. *A transducer* is a device that permits the energy of one system to control energy in another system. To simplify this you can think of a transducer as being a device that converts energy from one form to another.

Two simple examples of transducers are shown in Fig. 5-1. A microphone is an example of a transducer. It converts sound energy (a form of mechanical energy) into electrical impulses. A speaker is a transducer that converts electrical impulses into sound energy.

Transducers are used in electronic circuits for two very important purposes: for measurement and for sensing in control circuits.

Suppose, for example, that you wish to measure the temperature at some distant point. One way to do this is to convert the heat energy to electrical impulses. The electrical impulses can be sent over a wire. Then they can be converted to a meter reading. Making measurements at remote points, like the one just described, is called *telemetering*. In this example a transducer is needed to convert the heat energy into electrical impulses.

Transducers are also used as sensors. In fact the terms *transducer* and *sensor* are used to mean the same thing. Suppose you wish to control the temperature of a room with an electrical system that will turn the furnace ON when the room is too cool and turn the furnace OFF when it gets warm enough. To do this you need some type of sensor (or transducer) that will change the room temperature into an electrical signal for controlling the furnace.

In this chapter you will also study how different types of transducers work. Applications will be given so that you will have a better understanding of how the transducer works. You should keep in mind the fact that there are many uses for transducers that are not included here. Also, for most uses described in this chapter there is more than one way to accomplish the same thing. If you have a good knowledge of how the transducers work, you should have no trouble understanding electronic systems that use them.

You will be able to answer these questions after studying this chapter:

- ☐ **What is the difference between an active transducer and a passive transducer?**
- ☐ **What are some examples of the uses of transducers?**
- ☐ **What are some examples of passive transducers?**
- ☐ **What is a Wheatstone bridge?**
- ☐ **How does an ac bridge work?**
- ☐ **What are some examples of active transducers?**
- ☐ **How do microphones and speakers work?**

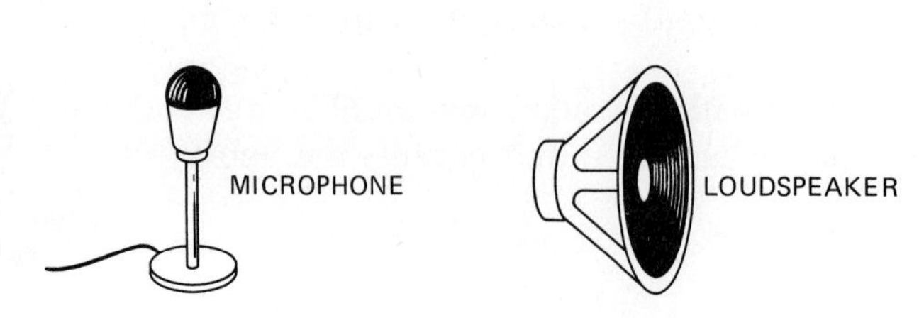

Fig. 5-1 Two examples of transducers.

INSTRUCTION

What Is the Difference between an Active Transducer and a Passive Transducer?

There are two types of transducers: *active transducers* and *passive transducers*.

If the transducer generates a voltage that is related to the amount of input energy, it is said to be an *active transducer*. An example is a photocell, which generates a voltage that is directly related to the amount of light striking it.

To understand how active transducers work you will find it helpful to review the six basic methods of generating a voltage. They are illustrated in Fig. 5-2. The method shown in Fig. 5-2*a* is by friction. This method is used to generate static electricity. Rubbing a glass rod with a silk cloth causes both the rod and the cloth to become electrified. Another example of generating static electricity is running a comb through your hair. On a dry day both the hair and the comb will become electrically charged. You can hear the sparks jumping between the comb and the hair. This method of generating a voltage is not used very often in active transducers.

Another method of generating a voltage is by chemical action. This is shown in Fig. 5-2*b*. Whenever two different kinds of metals are inserted in an acid or alkali solution a voltage is generated between them. This is the principle used in making batteries and cells. As with the friction method, you

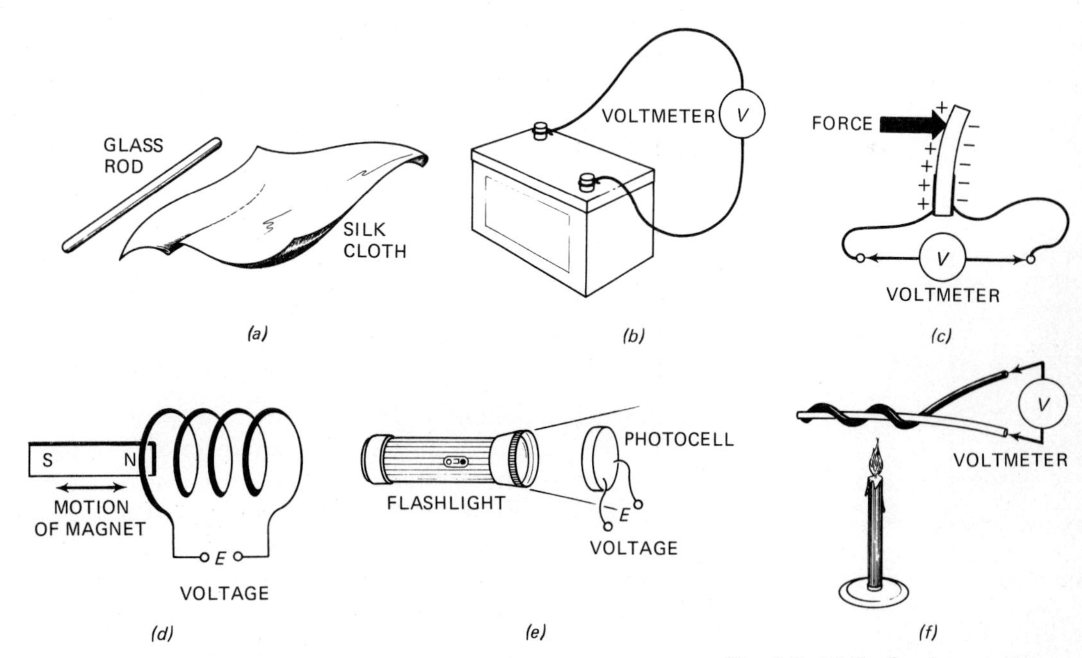

Fig. 5-2 Methods of generating a voltage. (*a*) Frictional. (*b*) Chemical. (*c*) Piezoelectric. (*d*) Moving field and conductor. (*e*) Photoelectric. (*f*) Thermal electric.

will find that this method of generating a voltage is not used very often in the manufacture of transducers.

In the third method, shown in Fig. 5-2*c*, a force or pressure is used. Certain materials (such as quartz, Rochelle salts, and barium titanate) will generate a voltage across their surface when they are under pressure or when they are under force. This is called the *piezoelectric method* of generating a voltage. The amount of voltage generated depends upon the amount of force or pressure and also upon the type of material used. The generated voltages are usually very small.

Figure 5-2*d* shows the principle of Faraday's law. This law says that when there is motion between a magnetic field and a conductor, a voltage is generated. A permanent magnet is being moved in and out of a coil in the drawing. Thus a voltage is being generated across the coil terminals. The amount of voltage depends on the number of turns of wire in the coil, the strength of the magnet, and the rate at which the magnet is moved.

Figure 5-2*e* shows a photocell which is being energized by the light of a flashlight. The photocell produces an output voltage which depends upon the amount of input light. The voltage also depends upon the type of material used for making the photocell.

Figure 5-2*f* shows the method of generating voltage by the *Seebeck* effect. Whenever the junction of two dissimilar metals is heated, a voltage is generated. The amount of voltage depends upon the amount of heat and also upon the types of metals used. The junction of the metals is called a *thermocouple.*

One of the methods of generating electricity in Fig. 5-2*c*, 5-2*d*, 5-2*e*, and 5-2*f* is usually used in active transducers.

In a *passive transducer* the energy being sensed produces changes in resistance, capacitance, or inductance. Passive transducers must always be connected into a circuit that has a source of electrical energy.

Summary

1. In the scientific meaning of the word, *work* is a force exerted through a distance. Mathematically,

$$\text{Work} = \text{force} \times \text{distance}$$

2. *Energy* is defined as the capacity to do work.
3. A transducer is a device that permits energy of one system to control energy in another system.
4. Making measurements from a remote position is called *telemetering.*
5. The terms *transducer* and *sensor* are interchangeable.
6. An active transducer produces an output voltage that is directly related to the energy being sensed.
7. There are six basic methods of generating a voltage. Four of these methods are used in making active transducers.
8. A passive transducer produces a change in resistance, capacitance, or inductance that is directly related to the energy being sensed.

What Are Some Examples of the Uses of Transducers?

Before studying the different types of transducers and how they work, it will be useful for you to study two simple examples of the uses of a transducer.

Figure 5-3 shows a simplified drawing of a temperature-controlled system. In this drawing the heated area is to be held at a constant temperature. A transducer is inserted into the heated area and connected in a network called a *bridge circuit*. Any change in temperature within the heated area will cause a change in the resistance of the transducer. This in turn will produce an output signal from the bridge circuit.

The output signal from the bridge circuit is fed to an amplifier which serves two purposes. The amplifier increases the bridge signal. It also modifies the signal so that it can be used to operate the control circuit.

The control circuit has two inputs. One input comes from the amplifier. The other is an ac power input for operating the heater in the heated area. The power to the heater is turned ON and OFF by the control circuit.

We will start with the assumption that the heated area is exactly the right temperature and the heater is turned OFF. Now suppose that the temperature begins to drop. This changes the output signal from the transducer and produces a signal from the bridge circuit. The amplifier increases the signal so that it can be used to turn ON the control circuit. The control circuit delivers ac power to the heater. The heater raises the temperature of the heated area back to the desired temperature. As soon as the desired temperature is reached the transducer output returns to normal and the amplifier no longer causes the control circuit to send power to the heater.

The power output from the transducer is much too small to be used for operating the control circuit directly. This is why an amplifier is needed in the system.

In the last section you learned about two kinds of transducers: those that produce an output voltage and those that do not. They were called *active* and *passive transducers*. The one you see in the circuit of Fig. 5-3 is passive—that is, it does not produce an output voltage. It is actually a

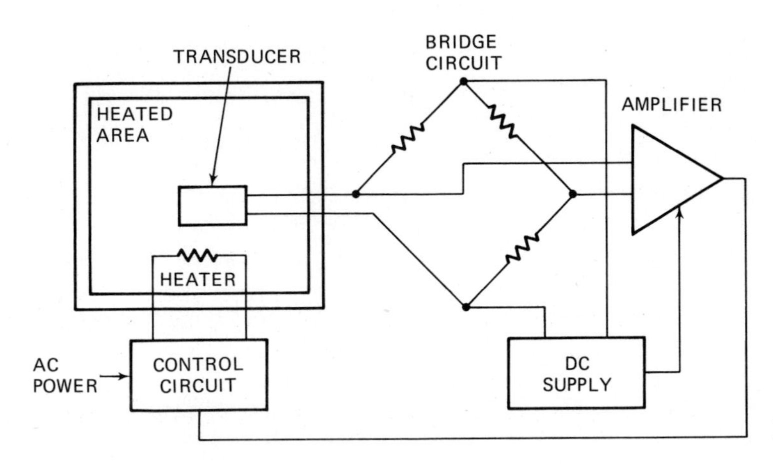

Fig. 5-3 A simple control circuit.

temperature-sensitive resistor called a *thermistor*. When the temperature in the heated area changes, the resistance of the thermistor changes.

An external dc power-supply voltage is needed for the bridge in order to convert the resistance changes of the thermistor into the voltage changes delivered to the amplifier. The dc power-supply voltage is also needed for operating the amplifier circuits.

Figure 5-4 shows another example of how a transducer may be used in a control system. The purpose of this system is to control the speed of the motor. A *tachometer* is used as the transducer. A tachometer produces a voltage that is proportional to the speed of the motor. The amplifier is needed to increase the output voltage of the tachometer.

There are two input voltages for the comparator. One is the voltage from the transducer amplifier, and the other is a reference voltage. If the two voltages are equal, there is no output from the comparator to the control circuit, and the motor speed is not changed. If the motor speed decreases, the transducer amplifier output is less than the reference voltage in the comparator. This causes a voltage to be sent to the control circuit which will raise the motor speed. On the other hand, if the motor speed is too great, the output voltage from the transducer amplifier will be greater than the reference. In this case the output of the comparator will cause the control circuit to slow the motor to the correct speed.

The circuits of Figs. 5-3 and 5-4 have two very important things in common. *First,* they both use transducers. The transducer senses heat in the control circuit of Fig. 5-3, and the transducer in the circuit of Fig. 5-4 senses the mechanical speed of the motor. A *second* thing that the two circuits have in common is that the output of the transducer is used indirectly to control the input power to the system being sensed. In Fig. 5-3 it controls the power to the heater. In Fig. 5-4 it controls power to the motor. Circuits such as this are called *closed-loop systems.*

What Are Some Examples of Passive Transducers?

With a passive (nongenerating) type of transducer the energy being sensed produces changes in resistance, capacitance, or inductance. Look again at the

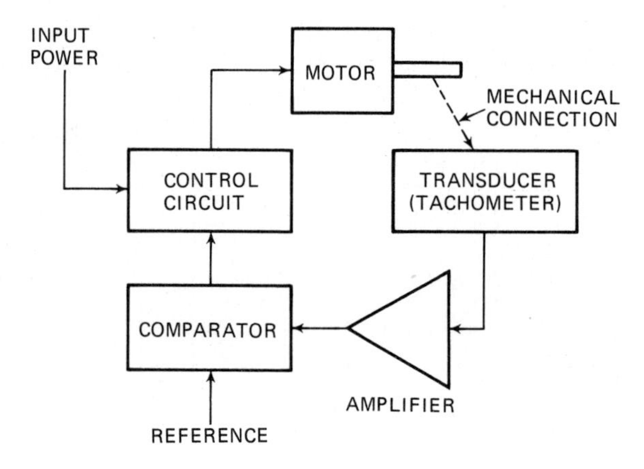

Fig. 5-4 Simple closed-loop speed control.

circuit of Fig. 5-3, and you will see that the energy being sensed is heat and that a thermistor is used as the sensor. Small temperature changes cause large changes in thermistor resistance. A thermistor is one example of a passive transducer.

How Are Resistive Transducers Made?

Figure 5-5 shows some resistive transducers. You should keep in mind the fact that passive transducers can be used with a bridge circuit in such a way that a change in resistance value produces a change in output voltage.

A thermistor (Fig. 5-5*a*) is a *therm*ally sensitive res*istor*. It was discussed in a previous section.

Figure 5-5*b* shows you how a variable resistor can be used as a transducer by causing a mechanical force to directly control the resistor arm. As the mechanical force moves from left to right the resistance between terminals *x* and *y* increases.

A photosensitive resistor (Fig. 5-5*c*) has a resistance that depends upon the amount of light falling on it.

The Hall generator of Fig. 5-5*d* has an amount of resistance that is directly related to the amount of magnetic field present.

The moisture sensor of Fig. 5-5*e* has a resistance that is related to the amount of moisture present in the atmosphere. Changes in the amount of resistance are directly related to changes in humidity.

The resistor strain gauge in Fig. 5-5*f* has a very fine wire running through a flexible material. Usually it is cemented to the material being placed under stress. As the stress is applied the material deforms, causing the wire to stretch. When the wire is stretched, its diameter changes, and hence its resistance changes.

It is important to know the scientific meaning of the terms *stress* and *strain*. *Stress* is a force that tends to cause a body to change shape, and *strain* is the amount of change produced in a body as a result of stress being applied. Thus a strain gauge measures change of shape in a body.

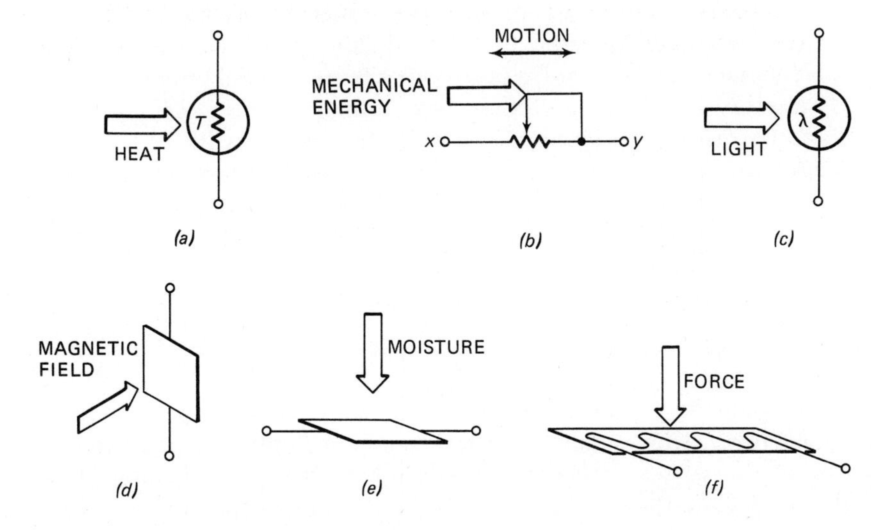

Fig. 5-5 Examples of resistive transducers. (*a*) Thermistor. (*b*) Variable resistor. (*c*) Photosensitive resistor. (*d*) Hall generator. (*e*) Moisture sensor. (*f*) Strain gauge.

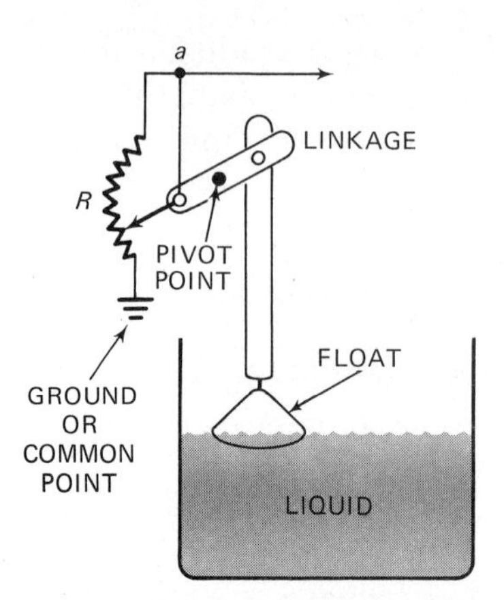

Fig. 5-6 Variable resistor *R* used as a fluid-level indicator.

In Fig. 5-6 you see how a variable resistor can be used to determine the level of fluid in a tank. The float moves up and down with the fluid level. As the float moves, it changes the amount of resistance of the variable resistor R. Resistor R is connected as a rheostat in such a way that when the tank is full, the resistance is 0 ohms. When the tank is empty, there is maximum resistance between point a and ground. As with other resistor transducers, the output may be connected into a bridge circuit.

There are other variations of the circuit of Fig. 5-6 which are also important. Instead of the linkage being connected to a float, as shown in the illustration, it may be connected to any object that moves over a limited distance. By this arrangement it is possible to obtain an output resistance that varies with position.

Summary

1. Transducers that sense motor speed are called *tachometers.*
2. When a transducer that is connected to the output of a system is used to control the input to that system, it is called a *closed-loop system.*
3. A resistive transducer changes resistance value when there is a change in what is being sensed.
4. You have seen examples of resistive transducers being used to sense heat, motion, light, magnetic-field strength, moisture, force, and fluid level.
5. Other uses of resistive transducers are also possible.
6. A thermistor is a thermally sensitive resistor. Its resistance changes over a wide range for small changes in its temperature.
7. Stress is a force that tends to change the shape of a body, and strain is the amount of shape change that takes place when a stress is applied.

How Are Capacitive Transducers Made?

In a capacitive transducer the energy being sensed controls the amount of capacitance. The capacitance of a capacitor depends upon three things: the area of the plates directly opposite each other, the distance between the plates, and the type of material used for a dielectric.

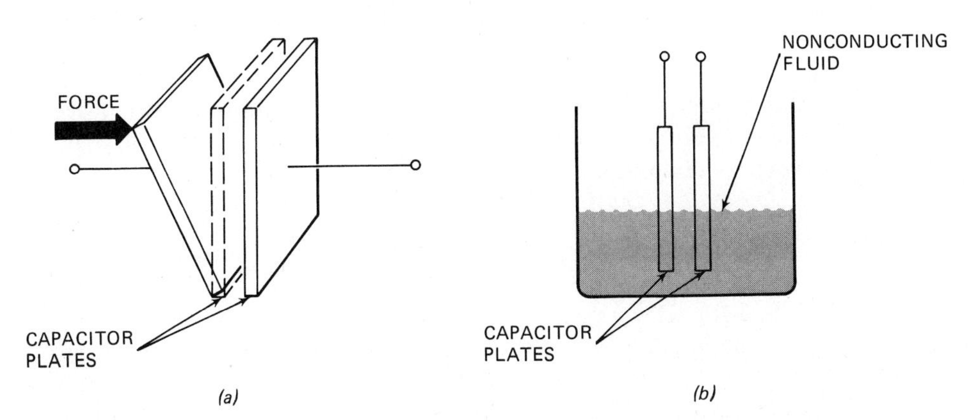

Fig. 5-7 Capacitive transducers. (*a*) Changing spacing between plates. (*b*) Changing type of dielectric (see text).

Figure 5-7 shows two examples of capacitive transducers. In Fig. 5-7*a* the amount of capacitance is varied by changing the distance between the plates. This distance is varied by the application of an external force, represented by an arrow. As the force pushes the plates closer together the capacitance increases.

Figure 5-7*b* shows another example of a capacitive transducer. In this case the level of a nonconducting fluid is being sensed by changing the dielectric of the capacitor. When the tank is full, the dielectric consists of the nonconducting fluid. As the fluid is emptied from the tank, more and more air dielectric is added between the plates, and this reduces the capacitance. Thus the amount of capacitance is directly related to the level of the fluid.

In another version of the capacitive transducer an external force inserts the dielectric into or out of the area between the plates.

When the capacitance of a capacitor transducer is changed, it causes a change in a circuit. For example, you may use the capacitance change to change a meter reading. There are also special bridge circuits that convert capacitance changes into voltage changes.

How Are Inductive Transducers Made?

With an inductive transducer the energy being sensed causes a change in the inductance of a coil. Inductive transducers are also known as *variable reluctance transducers*. In some cases an inductive transducer may be active, but the discussion here is only for passive transducers.

The inductance of a coil depends upon the number of turns of wire on the coil, the coil's shape, and the type of magnetic circuit around which the coil is wound. Inserting a soft-iron slug into the center of a coil causes its inductance to increase. An inductive transducer is shown in Fig. 5-8. Here the ap-

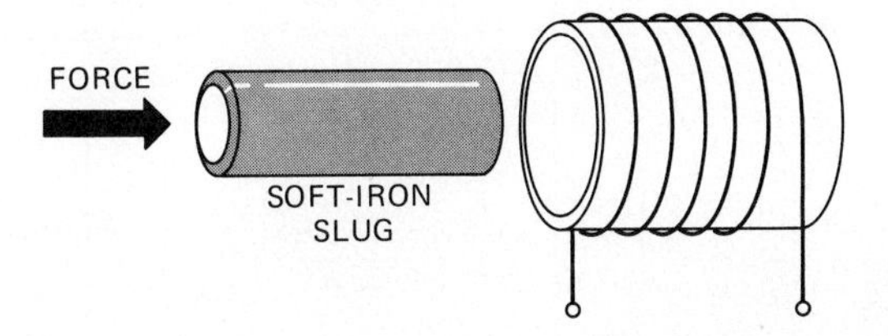

Fig. 5-8 Passive inductive transducer.

plication of an external force causes a soft-iron slug to be moved into or out of the center of a coil.

As with capacitive transducers, you may use the change in inductance to produce a voltage change in a special bridge circuit. It may also be used in other ways, such as changing a meter reading.

What Is a Wheatstone Bridge?

(Refer to Fig. 5-9.) It is a common practice to connect passive transducers into one leg of a bridge circuit like the one shown in Fig. 5-9*a*. This circuit is called a *Wheatstone bridge.* One of its most important applications is in measuring an unknown resistance value. When there is zero current between *a* and *b*, the Wheatstone bridge is said to be *balanced.* Under this condition the following equation applies:

$$R_x = \frac{R_1 \times R_3}{R_2}$$

where R_1, R_2, R_3, and R_x = resistance values in the same units, such as ohms, kilohms, or megohms

R_x = an unknown resistance value

The procedure for using the Wheatstone bridge to measure resistance is simple. The unknown resistor R_x is connected into the circuit. Then you adjust R_3 until there is no current measured by the meter. This means that the bridge is balanced. The value of resistance for R_3 is read on the dial. The values of R_1 and R_2 are also known. Thus when the bridge is balanced, the above equation can be used to find the resistance of R_x.

The Wheatstone bridge is sometimes used instead of an ohmmeter for measuring resistance values because of its greater accuracy.

Changing the value of any resistor in the bridge causes a change in current from *a* to *b* in the bridge. This means that when the bridge is out of balance,

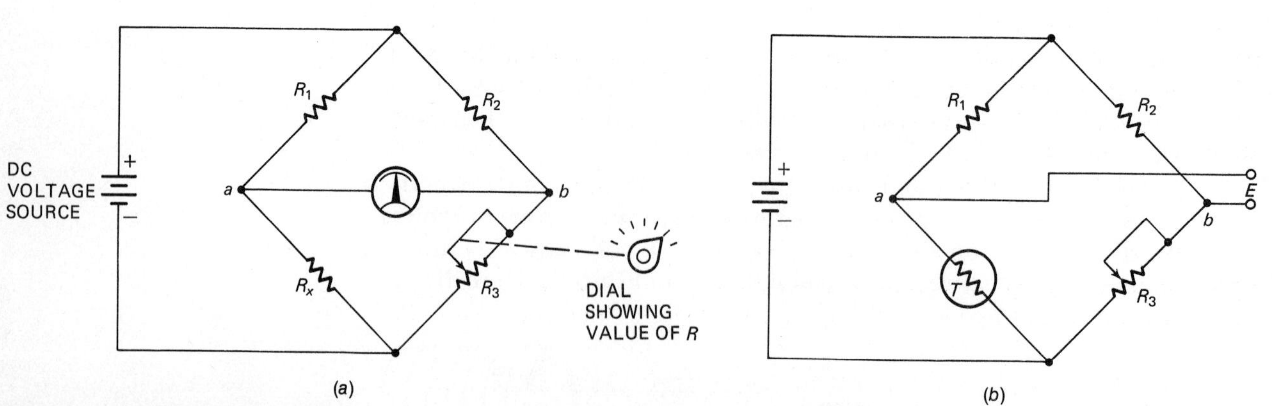

Fig. 5-9 The Wheatstone bridge circuit. (*a*) For measuring the resistance of an unknown resistor R_x. (*b*) As a transducer output circuit.

there is a voltage between these two points. Remember that a voltage is needed to cause a current flow.

Figure 5-9*b* shows how a Wheatstone bridge can produce a change in output voltage when there is a change in temperature. The change in temperature causes a change in thermistor resistance. When the thermistor resistance changes, there is a change in the output voltage E because the bridge is unbalanced.

The thermistor and bridge in the control system of Fig. 5-3 work on this principle. The amount of voltage is directly related to the amount by which the bridge is out of balance.

What Is the Advantage of Using a Wheatstone Bridge?

At first you might think that not much has been gained by using the thermistor and the bridge circuit. Suppose instead that it is connected into the simple circuit shown in Fig. 5-10. Again, the thermistor is located at some point where it is needed to sense temperature changes. When the temperature changes, the resistance of the thermistor also changes. This causes a different value of voltage E across it. The reason for this change in voltage is based on simple Ohm's law, which says that the voltage across a resistance is dependent upon the current through it and also upon the resistance. Changing the resistance of the thermistor will change the voltage across it. The question is: Why not use this simple circuit instead of the more complex one shown in Fig. 5-9?

To answer this question you must first consider the applied voltage, which in the circuits of Figs. 5-9 and 5-10 is a battery. (In practice this may be either an ac or a dc voltage.) Suppose the applied voltage in Fig. 5-10 drops in voltage value. This will reduce the amount of current flowing through R_1 and through the thermistor, causing a lower voltage E across the thermistor. Thus there will be a change in output voltage regardless of the fact that there is no change in temperature being sensed. This means that you will get a false output reading.

In the circuit of Fig. 5-9 this does not happen. When the applied voltage decreases, the voltage at point a decreases. However, the voltage at point b also decreases by the same amount. The difference between the voltages at points a and b remains the same. Therefore E does not change when the applied voltage changes!

How Does an AC Bridge Work?

Instead of using a dc voltage for operating the bridge circuit, you can use an ac voltage, as shown in Fig. 5-11. When an ac voltage is used the arms of the

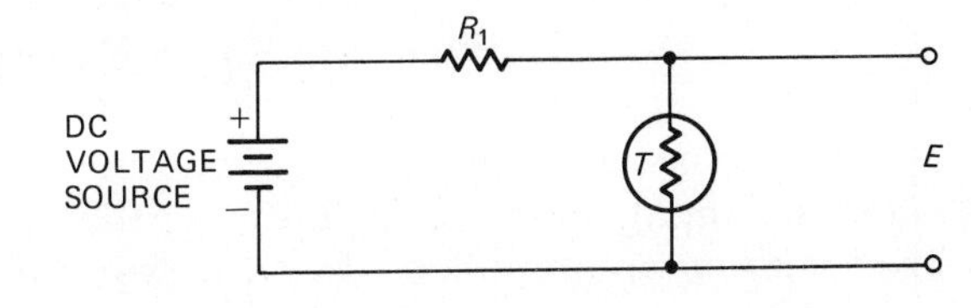

Fig. 5-10 A simple thermistor circuit.

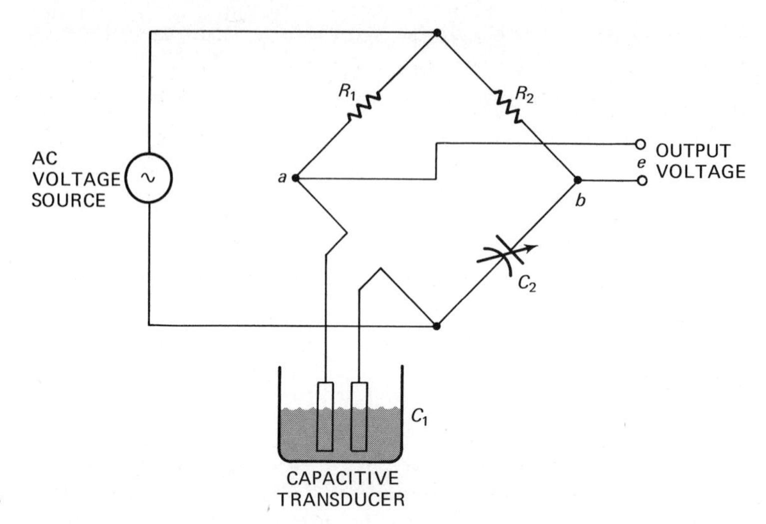

Fig. 5-11 An ac bridge circuit.

bridges may be resistances, reactances, or impedances. In Fig. 5-11 a capacitive transducer is used for measuring liquid level. This transducer is like the one shown in Fig. 5-7*b*. It is connected into one of the arms of the ac bridge. Variable capacitor C_2 is connected into another arm. The basic theory of operation is still the same. If the bridge were balanced, there would be no output ac voltage e. Suppose this is the condition when the tank is filled. As the tank begins to empty, the capacitance of C_1 changes. This causes the bridge to become unbalanced and produces a change in output voltage e.

The output voltage could be used in a control circuit for refilling the tank, or you could use it with a calibrated meter to show how much liquid is in the tank.

Summary

1. A Wheatstone bridge can be used for measuring an unknown resistance value.
2. The Wheatstone bridge circuit can also be used as a transducer output circuit.
3. An advantage of using a Wheatstone bridge for a transducer is that the output voltage is not affected by changes in power-supply voltage.
4. Either an ac or dc supply voltage can be used with a Wheatstone bridge circuit.
5. If capacitive or inductive transducers are used as one leg of the bridge, then the supply voltage must be ac.

What Are Some Examples of Active Transducers?

Active transducers produce an output voltage which depends upon the amount of input energy. In a few cases the voltage may be large enough to be used without an amplifier. In most cases, however, amplifiers are used to

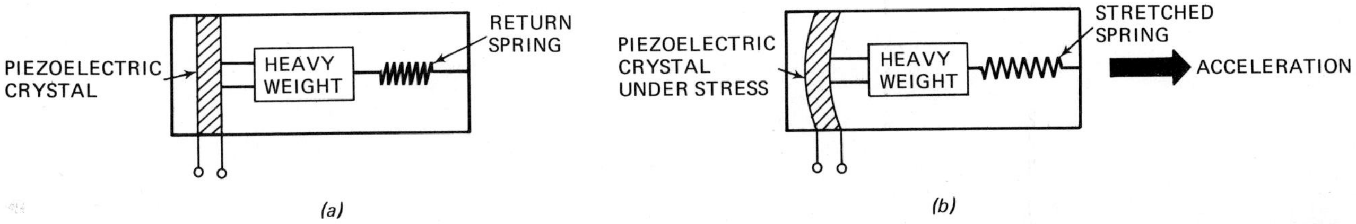

(a) (b)

Fig. 5-12 An accelerometer. (*a*) In a position at rest. (*b*) Under acceleration.

convert the weak signal voltages from the active transducer into large power variations to operate some device (such as a relay or a motor).

In this section you will learn some ways of using active transducers. It is not possible to show all the ways in one book, so you should keep in mind the fact that these are only examples.

How Is a Piezoelectric Transducer Used?

A *piezoelectric* transducer generates a voltage that is directly related to the amount of input force or pressure on a crystal. Figure 5-12 shows you how such a transducer can be used as a sensor for acceleration.

An *accelerometer* is a transducer that is designed to sense acceleration. In the scientific meaning of the word, *acceleration* is the rate of change of speed.

In the accelerometer of Fig. 5-12 a piezoelectric crystal is used as a sensing element. A heavy weight rests against the crystal. As you can see in Fig. 5-12*a*, there is no output voltage because there is no stress on the crystal. When there is acceleration, as shown in Fig. 5-12*b*, the crystal is under pressure. This occurs because the heavy weight cannot accelerate instantly. The voltage produced by the crystal depends upon the amount of pressure exerted by the weight. This pressure, in turn, depends upon the amount of acceleration.

How Is a Photoelectric Transducer Used?

A photoelectric transducer may be either active or passive. With the passive type, changes in light energy produce changes in resistance. This is the principle of the *photoresistive transducer*, shown in Fig. 5-5*c*.

An active photoelectric transducer produces an output voltage which depends upon the amount of light. *Photocells* are examples of active photoelectric transducers. For the applications described in this section, the phototransducer could be either active or passive. The designer makes this choice.

Figure 5-13 shows you how photoelectric transducers can be used for reading punched tapes. Tapes like this are used for controlling machinery. Slots are punched into the tape. There is a special code used for making these slots. As the tape passes in front of the photocells, light passes through the slots. Voltages are produced in the photocells. These voltages are related to the slots, so they are also related to the tape code. In the position shown light is passing through four slots and striking four of the photocells. When these four photocells are generating a voltage, the code may mean that the machine should stop, or it may mean it should change speed. In other words, the code instructs the machine to do something.

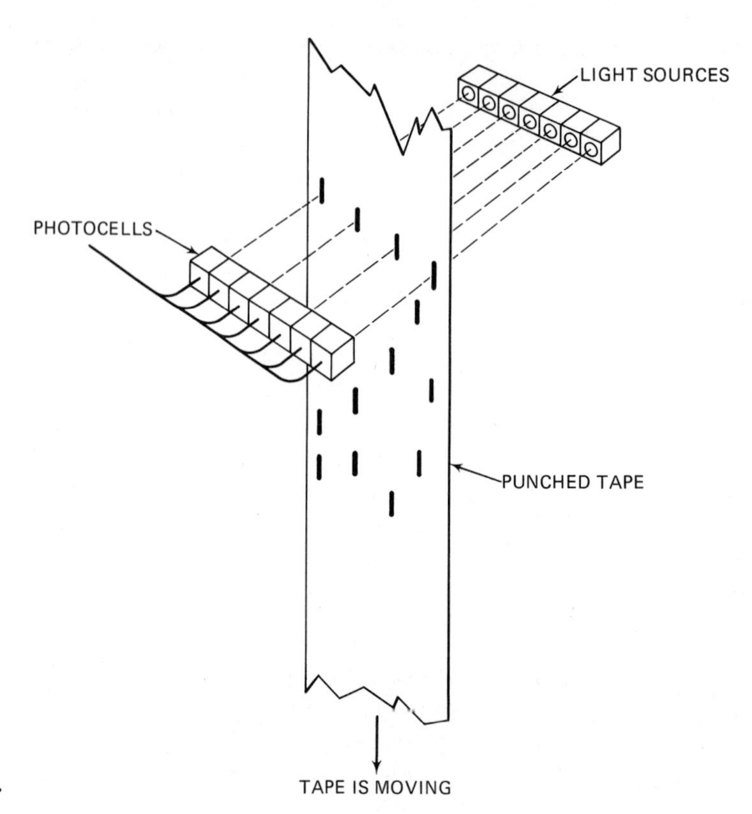

Fig. 5-13 A punched-tape reader.

Figure 5-14 shows how photocells can be used on an assembly line for counting boxes or other objects. As the box passes between the light and the photocell, the output voltage from the photocell drops to zero. The electric counter is wired in such a way that when this happens, one digit is added to the total.

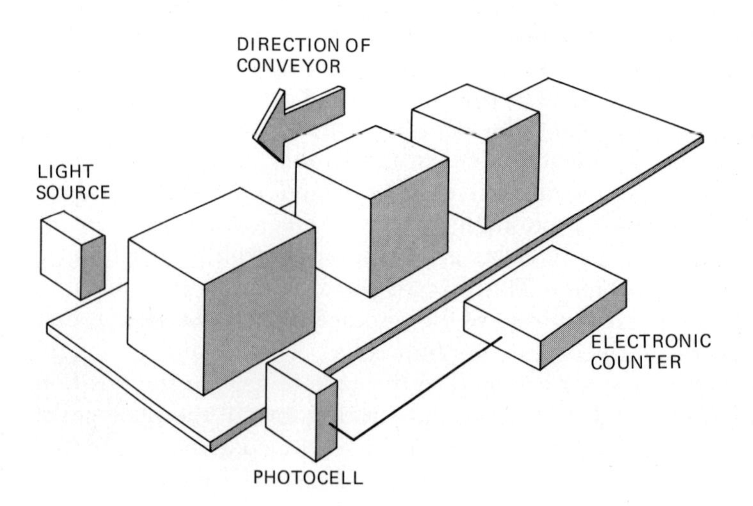

Fig. 5-14 An assembly-line counting system.

How Is an Electromagnetic Transducer Used?

An electromagnetic transducer operates by moving a magnetic field across a conductor. It does not matter if the magnetic field is moved through the conductor or if the conductor is moved through the magnetic field. In either case a voltage will always be generated.

The amount of voltage depends upon the rate at which the magnetic field moves across the conductor. It also depends upon the number of conductors. In other words, moving two conductors through a magnetic field will produce more voltage than moving one conductor.

You will recall that a tachometer is a transducer used for measuring motor speed. Figure 5-15 shows an example of an electromagnetic tachometer. Two permanent magnets are located on an aluminum wheel. As each magnet moves past the coil, a voltage is induced. This is because the magnetic field is moving across the conductor. The number of pulses per minute can be counted electronically and divided by two (because in this case there are two pulses per revolution), and the speed in revolutions per minute (r/min) can be displayed or recorded. Two magnets are used in order to keep the wheel in balance.

Instead of using a punched tape, which you saw in Fig. 5-13, some electronic control systems use a magnetic tape. Figure 5-16 shows you how to get the information from a magnetic tape. Although there is only one head shown, it would be common practice to use eight heads. This would give eight inputs. Note also that there are eight inputs in the punched tape of Fig. 5-13.

The surface of the magnetic tape is magnetized at points along the length of the tape. The magnetic fields on this tape do the same type of job as the holes in the punched tape. The tape is moved against the head, and when a magnetized point passes the air gap, a magnetic flux travels through the soft-iron core. This magnetic flux cuts through the coil around the head and generates a voltage in it.

The tape of Fig. 5-16 is similar to the system used with your tape recorders in the home. In that case the magnetic fields on the tape are related to noise or voice instead of a code. It is important that you compare the two tape systems of Figs. 5-13 and 5-16. There are only two possible output voltages in each system. There may be *0 volts,* corresponding to no hole in the punched tape or no magnetic field on the magnetic tape, or there may be a

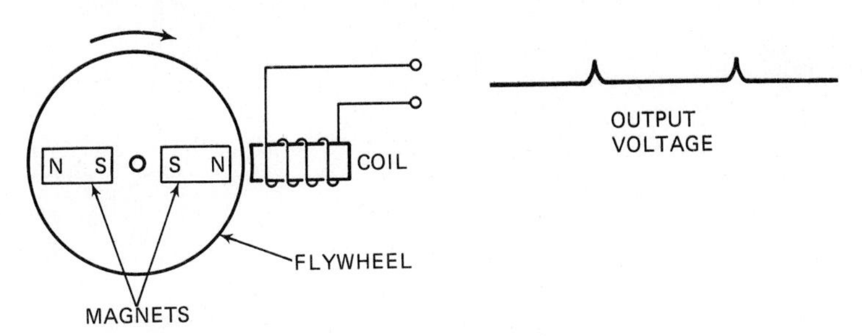

Fig. 5-15 An example of a tachometer.

maximum output voltage, corresponding to a hole in the punched tape or a magnetized spot on the magnetic tape. Therefore the output can have only two possible conditions: maximum voltage or 0 volts. There are no different levels of voltage, as you would have in a tape recording of music.

How Is a Thermoelectric Transducer Used?

The transducer you saw in the system of Fig. 5-3 feeds a bridge circuit. The bridge circuit is needed for a passive transducer. In place of this arrangement a thermoelectric generator (such as a thermocouple) could be used for a sensor. This type of sensor is shown in Fig. 5-2*f*. If it were used in the system of Fig. 5-3, the output would be a voltage. You would still need an amplifier to increase the strength of the output voltage. However, you would not need the bridge circuit. The rest of the system would be unchanged.

Figure 5-17 shows you another use of a thermoelectric transducer. In this case a pilot flame is used for igniting a gas heating system. The flame also heats the thermocouple junction. The output voltage of the thermocouple holds the control valve open. If the flame should die out for some reason, the voltage of the thermocouple will be reduced to 0 volts and the control valve will shut off the gas. This prevents gas from escaping into the room when the pilot flame is out.

How Do Microphones and Speakers Work?

A microphone is a transducer that converts sound waves into electrical waves. The electrical waves can be amplified in electronic systems and heard by using a loudspeaker.

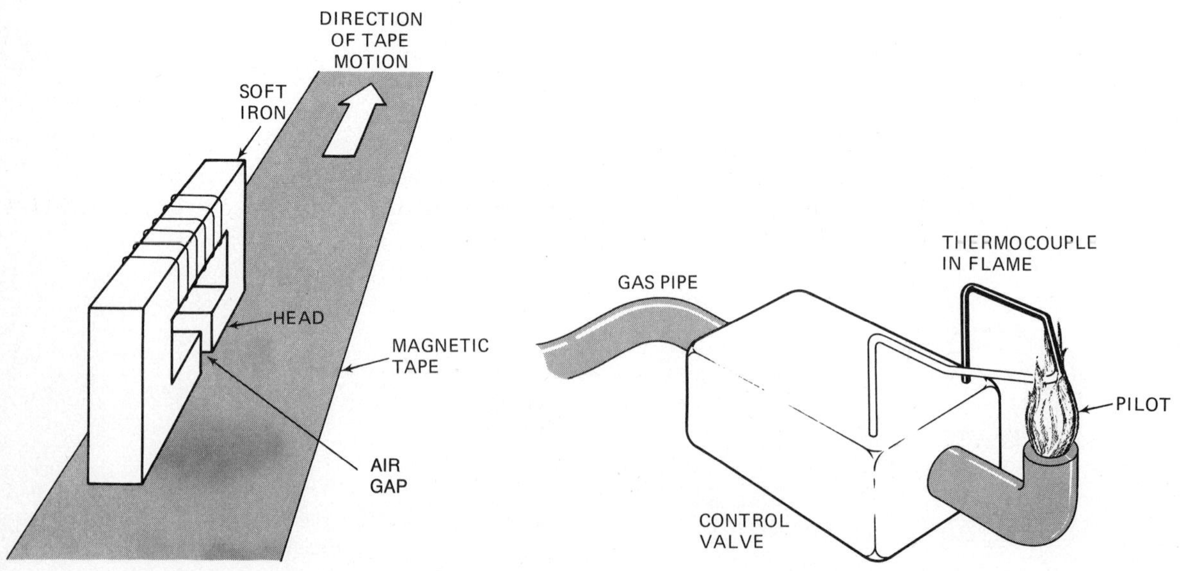

Fig. 5-16 A magnetic-tape reader.

Fig. 5-17 Application of a thermoelectric transducer.

A speaker is a transducer that converts electrical impulses into sound waves. Since a microphone produces a voltage that is related to a sound, it is possible for the microphone to be either an active or a passive transducer. Speakers must be passive transducers, since their output is intended to be sound waves. Remember that an active transducer must produce an output *voltage* related to some input.

How Does a Microphone Work?

Figure 5-18 shows you how three types of microphones work. The carbon microphone of Fig. 5-18*a* has a large number of carbon granules in a small case. When these granules are pressed together, their resistance is low. When they are loosely packed, their resistance is high. The resistance varies between these values when sound waves reach the diaphragm.

Sound waves cause the microphone diaphragm to move back and forth. This motion causes the granules to be packed tightly, then loosely. At any instant the resistance of the carbon granules depends upon the position of the diaphragm. This causes the current in the circuit to change because the current depends upon the resistance. Since the current changes with the sound waves, you can call it an *audio current.*

When the audio current flows through resistor R, there is an audio voltage across R. This audio voltage can be amplified and used as needed.

The carbon microphone is an example of a passive transducer. By itself it cannot produce an output voltage. It is necessary to have a dc voltage source to get the current through R. This current is necessary to produce the audio voltage.

Figure 5-18*b* shows the principle of the crystal microphone. The crystal is

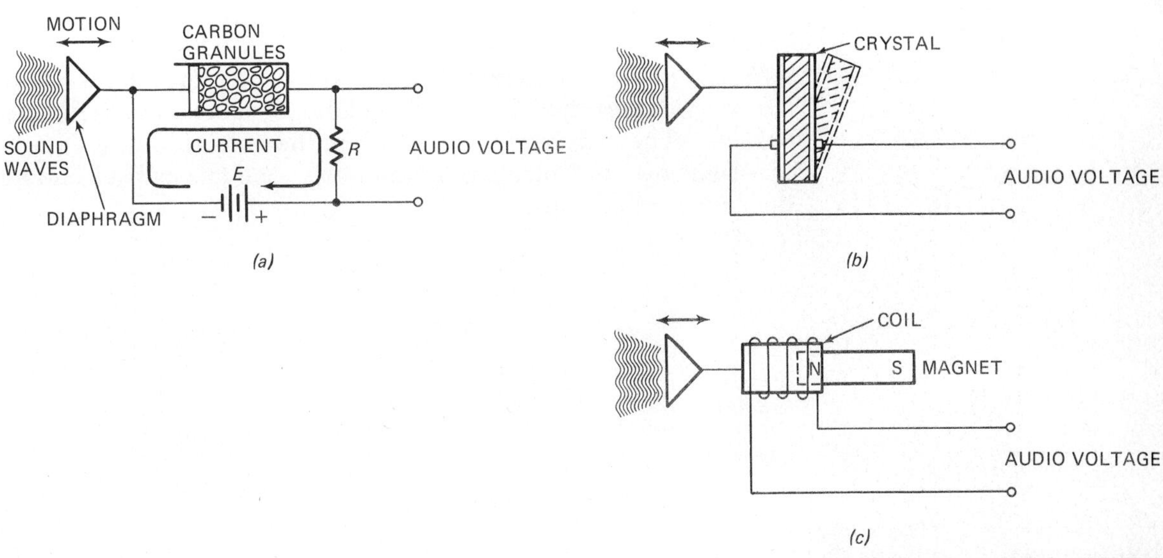

Fig. 5-18 Microphones. (*a*) Carbon. (*b*) Crystal. (*c*) Dynamic.

a piezoelectric material that generates a voltage when it is under pressure. The pressure of the sound waves moving the diaphragm back and forth causes the crystal to vibrate, and this produces an output voltage. The output voltage is directly related to the sound wave striking the microphone. The crystal microphone is an active transducer, since it can produce an output voltage by itself.

Figure 5-18*c* shows you how a dynamic microphone works. In this case the diaphragm is attached to a coil wound on a very light insulated form. The coil form is round, and it slips over a permanent magnet. As the sound waves strike the diaphragm, the diaphragm moves the coil back and forth in the magnetic field. This causes an audio voltage to be induced in the coil, since a voltage is always generated when conductors are moved in a magnetic field (Faraday's law). The audio voltage is directly related to the sound waves which move the diaphragm. This is also an active transducer.

The three types of microphones shown in Fig. 5-18 are used in many electronic systems.

How Do Speakers Work?

Figure 5-19 shows two types of speakers used in electronic systems.

Figure 5-19*a* shows you how a *dynamic* speaker works. Its operation is based on the fact that when a current is flowing through a conductor or coil, a magnetic field is produced. If an audio current is caused to flow through the coil, the varying magnetic field of the coil will react with the field of the permanent magnet. This causes the speaker cone to move back and forth to produce changes in air pressure. The resulting changes in air pressure are the sound waves which produce the sound in your ear.

Figure 5-19*b* shows you the parts of an *electrostatic* speaker. It is made with two plates similar to the plates of a capacitor. One of the plates is fixed, and the other is free to move. When an audio voltage is applied to the speaker terminals the two plates are charged with opposite polarities. The movable plate is attracted to the fixed plate, since unlike charges attract. The dotted lines show the position of the movable plate during peak values of voltage. The cone springs back when the voltage is at the minimum value.

When an audio voltage is applied, the cone moves back and forth and produces the sound waves.

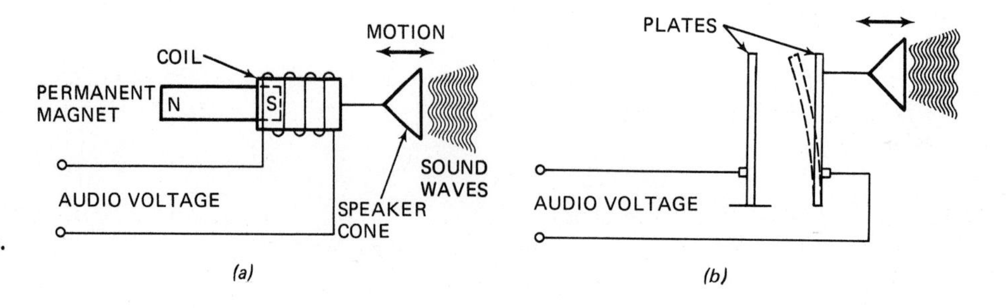

Fig. 5-19 Speakers. (*a*) Dynamic. (*b*) Electrostatic.

Summary

1. Piezoelectric transducers convert a force or pressure into a voltage value.
2. Photoelectric transducers convert light intensity into a voltage value.
3. Electromagnetic transducers convert a motion into a voltage value.
4. Thermoelectric transducers convert heat into a voltage value.
5. A microphone may be a passive transducer or it may be an active transducer.
6. A speaker is a passive transducer.

PROGRAMMED REVIEW QUESTIONS

(Instructions for using this programmed section are given in Chapter 1.)

We will review the important concepts of this chapter. If you have understood the material, you will progress easily through this section. Do not skip this material because some additional theory is presented.

1. Which of the following is an example of a closed-loop control system?
   A. The thermostatic control in a house in which the temperature of a room is maintained at a constant value. (Proceed to block 9.)
   B. A thermistor probe connected into a bridge circuit. (Proceed to block 17.)

2. *The correct answer to the question in block 18 is* **A.** *When the bridge is balanced, there is no voltage across the center leg.* Here is your next question.
   The capacity to do work is defined as
   A. transduction. (Proceed to block 14.)
   B. energy. (Proceed to block 20.)

3. *The correct answer to the question in block 22 is* **A.** *A photocell is an example of a transducer made by using the photoelectric method of generating a voltage. The chemical method is not commonly used.* Here is your next question.
   Which of the following may be an example of an active transducer?
   A. A thermistor. (Proceed to block 21.)
   B. A tachometer. (Proceed to block 27.)

4. *Your answer to the question in block 19 is* **A.** *This answer is wrong. Review the description of the LED in Chapter 2, then* proceed to block 10.

5. *Your answer to the question in block 20 is* **B.** *This answer is wrong. An accelerometer does not measure changes in acceleration. Instead, it measures a change in speed.* Proceed to block 26.

6. *The correct answer to the question in block 27 is* **A.** *A strain device produces a change in resistance that is related to the amount of twist or distortion produced. This is called* **strain. Stress** *is the force that produces strain.* Here is your next question.
Which of the following is *not* true?
A. Bridge circuits must be supplied with a dc voltage. (Proceed to block 18.)
B. Bridge circuits may be supplied with either an ac or a dc voltage. (Proceed to block 24.)

7. *The correct answer to the question in block 9 is* **A.** *An active transducer generates a voltage that depends upon input energy. Speakers do not generate voltage, so they are not active transducers.* Here is your next question.
Another name for a transducer is
A. a sensor. (Proceed to block 19.)
B. a passive circuit element. (Proceed to block 25.)

8. *Your answer to the question in block 18 is* **B.** *This answer is wrong. Review the chapter material on the Wheatstone bridge, then* proceed to block 2.

9. *The correct answer to the question in block 1 is* **A.** *The thermostatic control in a home is very much like the temperature control system shown in Fig. 5-3. It is a closed-loop control system.* Here is your next question.
A speaker is an example of
A. a passive transducer. (Proceed to block 7.)
B. an active transducer. (Proceed to block 13.)

10. *The correct answer to the question in block 19 is* **B.** *The LED (light-emitting diode) was discussed in Chapter 2. It is a transducer because it produces light energy as a result of input electric energy. It is passive because it does not generate a voltage.* Here is your next question.
Which of the following types of transducers is normally used with a bridge circuit?
A. An active transducer. (Proceed to block 16.)
B. A passive transducer. (Proceed to block 22.)

11. *Your answer to the question in block 26 is* **A.** *This answer is wrong. Piezoelectric crystals generate a voltage when a pressure is exerted on them.* Proceed to block 23.

12. *Your answer to the question in block 27 is* **B.** *This answer is wrong. Study Fig. 5-5f, then* proceed to block 6.

13. *Your answer to the question in block 9 is* **B.** *This answer is wrong. Speakers do not generate a voltage.* Proceed to block 7.

14. *Your answer to the question in block 2 is* **A.** *This answer is wrong.* ***Transduction*** *is a word that means "the act of conveying over." It is not a word commonly used to mean energy.* Proceed to block 20.

15. *Your answer to the question in block 22 is* **B.** *This answer is wrong. In the chemical method of generating a voltage, two dissimilar metals are immersed in an acid or alkali solution and a voltage is generated. This is not a common way of making an active transducer.* Proceed to block 3.

16. *Your answer to the question in block 10 is* **A.** *This answer is wrong. A bridge circuit is used with passive transducers.* Proceed to block 22.

17. *Your answer to the question in block 1 is* **B.** *This answer is wrong. A thermistor is a passive device that is normally connected into a bridge circuit. This is not an example of a closed-loop system.* Proceed to block 9.

18. *The correct answer to the question in block 6 is* **A.** *A Wheatstone bridge usually has a dc voltage applied, while capacitance and inductance bridges are used with an ac voltage source.* Here is your next question.
 When a bridge circuit is balanced the voltage across the center leg is
 A. 0 volts. (Proceed to block 2.)
 B. maximum. (Proceed to block 8.)

19. *The correct answer to the question in block 7 is* **A.** *Transducers are called* **sensors** *because they are used to sense energy changes in some system.* Here is your next question.
 An LED is an example of
 A. an active transducer. (Proceed to block 4.)
 B. a passive transducer. (Proceed to block 10.)

20. *The correct answer to the question in block 2 is* **B.** *In the introduction of this chapter* **energy** *was defined as the capacity to do work.* Here is your next question.
 An accelerometer is a transducer used for sensing a change in
 A. speed. (Proceed to block 26.)
 B. acceleration. (Proceed to block 5.)

21. *Your answer to the question in block 3 is* **A.** *This answer is wrong. A thermistor is a* ***thermally-sensitive resistor.*** *Resistors are passive circuit elements.* Proceed to block 27.

22. *The correct answer to the question in block 10 is* **B.** *An active transducer is sometimes called a "self-generating" transducer. Passive (nongenerating) transducers are frequently connected into bridge circuits.* Here is your next question.
Which of the following methods of generating a voltage is normally used for making transducers?
A. Photoelectric. (Proceed to block 3.)
B. Chemical. (Proceed to block 15.)

23. *The correct answer to the question in block 26 is* **B.** *The method of generating a voltage by motion between a conductor and a magnetic field is sometimes called the* **mechanical method.** *This method is used in producing industrial power and for other large-power systems. The generator or alternator used in automotive electrical systems also uses this method of generating a voltage.*

Quartz, tourmaline, Rochelle salt, barium titanate, and lead zirconate are all examples of materials that exhibit the piezoelectric effect. Here is your next question.
In the bridge circuit of Fig. 5-20 variable resistor R_3 is adjusted for a balance. The bridge is balanced when $R_3 = 10$ kilohms. What is the value of resistor R_x? (When you have calculated the value of R_x, proceed to block 28.)

24. *Your answer to the question in block 6 is* **B.** *This answer is wrong. The question asks which of the two statements is* **not** *true. Reread the question, then* proceed to block 18.

25. *Your answer to the question in block 7 is* **B.** *This answer is wrong. A transducer may be a passive circuit element. (A passive circuit element is a component that does not generates a voltage.) This is not another name for a transducer.* Proceed to block 19.

26. *The correct answer to the question in block 20 is* **A.** ***Acceleration** is defined as the rate of change in speed. This is what an accelerometer measures.* Here is your next question.

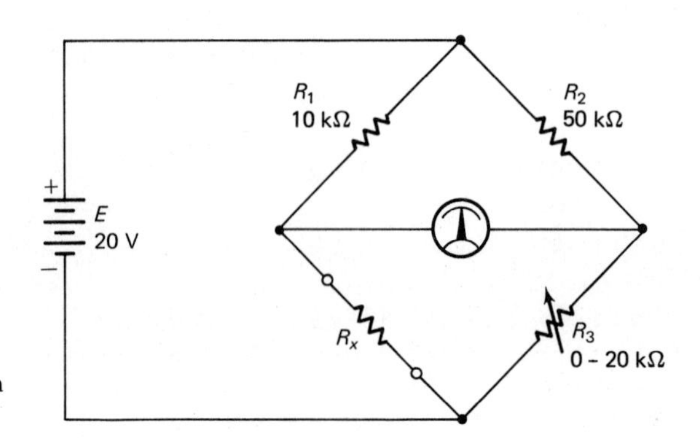

Fig. 5-20 Circuit for the question in block 23.

When a pressure is exerted on certain types of crystal materials a voltage is generated. This is called
A. the mechanical method of generating a voltage. (Proceed to block 11.)
B. the piezoelectric method of generating a voltage. (Proceed to block 23.)

27. *The correct answer to the question in block 3 is* **B.** *A tachometer measures the speed of a motor. One way of doing this is to connect a small ac or dc generator to the motor output so that the voltage generated is related to the motor speed.*

There are other ways of making tachometers, and so the question in block 3 says "which of the following **may** *be an example of an active transducer." Thermistors are always passive.* Here is your next question.
A certain component is made in such a way that an external force changes the resistance of that component. This is a
A. strain gauge. (Proceed to block 6.)
B. stress gauge. (Proceed to block 12.)

28. *The unknown resistor has a value of 2 kilohms. The value is determined as follows:*

$$R_x = \frac{R_1 \times R_3}{R_2}$$

$$= \frac{10 \text{ kilohms} \times 10 \text{ kilohms}}{50 \text{ kilohms}}$$

$$= \frac{100 \text{ kilohms}}{50 \text{ kilohms}}$$

$$= 2 \text{ kilohms} \qquad \text{Answer}$$

You have now completed the programmed questions. The next step is to put some of these ideas to work in laboratory experiments. Proceed to the Experiment section of this chapter.

EXPERIMENT

(The experiment described in this section may be performed on the circuit board described in Appendix C or on a similar laboratory setup.)

Purpose In Part I of this experiment you will show that there is an output voltage from a bridge circuit when it is out of balance. In Part II you will show how a light beam can be used to control a load such as a lamp or a motor.

PART I

Theory Figure 5-21*a* shows the schematic of a bridge circuit with an adjustable resistor in one arm. There will be no output voltage when the variable resistor is adjusted so that the total resistance of arm *ac* is equal to 10 kilohms. The pictorial diagram is shown in Fig. 5-21*b*.

A voltmeter can be used to measure the output voltage of the bridge. However, a diode must be placed in series with the meter because you can get a reverse of polarity for the output voltage. In other words, the voltage at point *a* could be either positive or negative with respect to the voltage at point *b*, depending upon whether arm *ac* is greater than 10 kilohms or less than 10 kilohms. The diode prevents a reverse voltage from damaging the voltmeter.

The resistor in series with the variable resistor prevents the resistance in arm *ac* from being adjusted to a point where the unbalance is so great that it produces an excessive output voltage. This helps to protect the voltmeter.

Test Setup Connect the circuit as shown in Fig. 5-21. Figure 5-21*a* shows a schematic drawing of the circuit, and Fig. 5-21*b* shows the pictorial diagram.

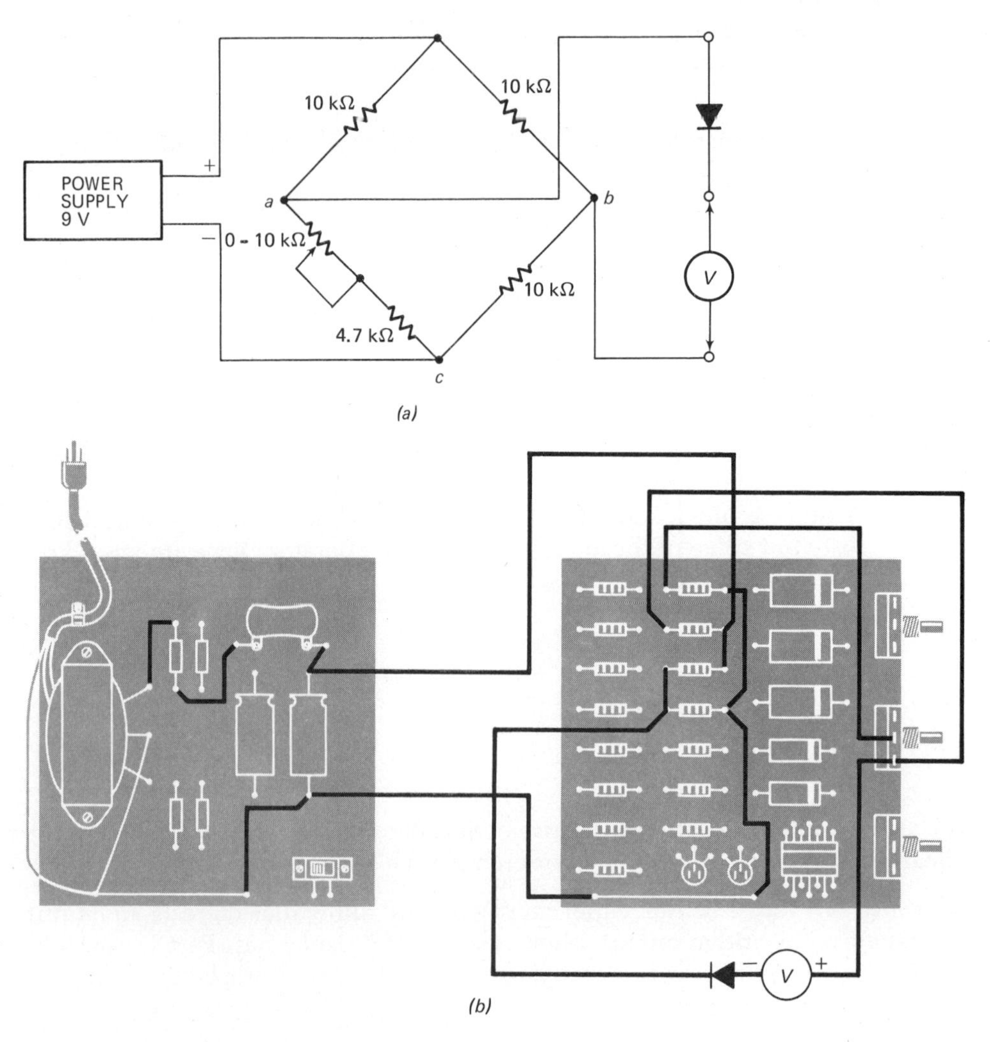

Fig. 5-21 Test setup for Part I of the experiment. (*a*) Schematic diagram. (*b*) Pictorial diagram.

Procedure

step one Adjust the variable resistor until a voltage reading is obtained on the voltmeter. Then adjust the resistor very carefully until the meter reads 0 volts. What value of resistance should the variable resistor have to get a zero output voltage? ________________

After you have adjusted the variable resistor to obtain 0 output volts, remove it from the bridge and measure its resistance with an ohmmeter. Record its value. ________________

(The value should be approximately 5 kilohms. When the variable resistor is adjusted to a value of 5 kilohms, then leg *ac* will have a total resistance of 10 kilohms. This will cause the bridge to be in balance.)

Conclusion Changing the resistance of a bridge leg causes a change in output voltage. If a resistor transducer is used in place of the variable resistor, the output voltage will then vary with changes in the transducer resistance.

PART II

Theory A photoconductive resistor is a passive transducer. Its resistance is high when it is not exposed to light, and its resistance is low when exposed to light.

The amount of resistance change is relatively small, so an amplifier can be used to make it a more sensitive control transducer. Figure 5-22 shows the circuit.

In Chapter 3 you learned that a PNP transistor will conduct when its collector and its base are both negative with respect to its emitter. The arrows in Fig. 5-22 show the current paths for the collector current (solid arrows) and the base current (dashed arrows) when switch *SW* is closed. Note that the collector current flows through the relay coil.

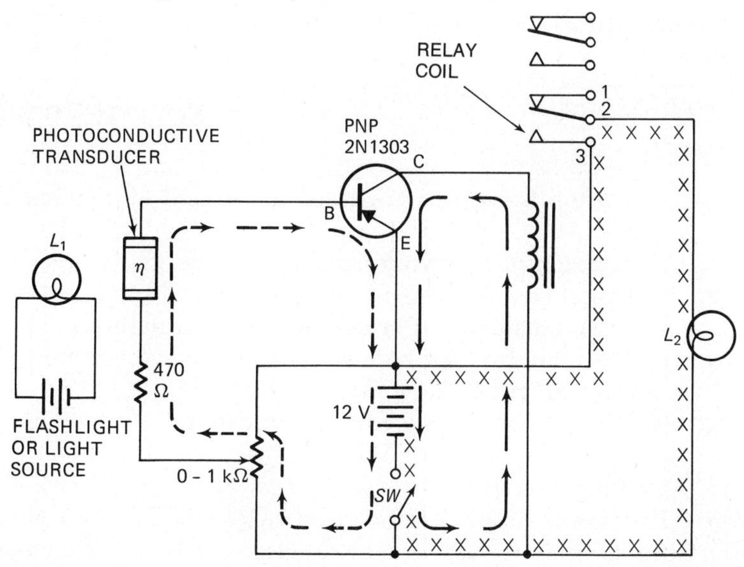

Fig. 5-22 Test setup for Part II of the experiment.

When there is no light on the photoconductive resistor the resistance of the base circuit is high. This means that the base current is low. With a low base current the collector current is also low, and the relay cannot be energized.

When light shines on the photoconductive resistor its resistance decreases. This increases the base current and causes a large increase in collector current. The collector current is now large enough to energize the relay.

When the relay is energized the 12-volt supply is connected to the lamp L_2 through the relay contacts. Follow the circuit by starting at L_2, moving through the battery and switch, and also through the relay contacts (1, 3) and back to L_2. This path is shown with x's. Note that when both *SW* and the relay contacts are closed, lamp L_2 is ON.

Although the circuit is used to light a lamp, a motor could be turned ON by the output. Shining a light on the photoconductive resistor would cause the motor to run.

Test Setup Wire the circuit as shown in Fig. 5-22. Use a dc power supply instead of a battery. Be very careful to make the power-supply connections with the switch OFF and the positive and negative terminals connected to the correct points.

Procedure

step one Close *SW*. Adjust the 0–1-kilohm resistor so that L_2 is ON when L_1 is within a few inches of the photoconductive resistor.

step two Move lamp L_1 until the relay drops out and L_2 goes OFF.

step three Rewire the relay contacts so that lamp L_2 is ON when the photoconductive resistor is not exposed to light. You can do this by moving the wire from relay terminal 3 to relay terminal 1. Show that when L_1 is moved close to the photoconductive resistor, L_2 goes OFF.

Conclusion The photoconductive resistor is a passive transducer. Its resistance depends upon the amount of light on it. (The circuit of Fig. 5-22 would be improved if the transducer was connected into a bridge circuit.)

The amplifier makes the system more sensitive to light changes, and this permits a small change in light to turn a load ON or OFF.

SELF-TEST WITH ANSWERS

(Answers with discussions are given at the end of the chapter.)

1. The capacity to do work is called (*a*) energy; (*b*) power.
2. Making measurements from a remote position is called (*a*) remote control; (*b*) translation; (*c*) reflexing; (*d*) telemetering.
3. A transducer that does not generate a voltage is called (*a*) active; (*b*) passive; (*c*) inverted; (*d*) a bridge.
4. Which of the following methods of generating a voltage is not used in active transducers? (*a*) Piezoelectric; (*b*) Thermoelectric; (*c*) Friction; (*d*) Photoelectric.
5. The power supply used with a bridge in a passive-transducer output circuit (*a*) must be ac; (*b*) must be dc; (*c*) may be either ac or dc; (*d*) cannot have an output voltage greater than 10 volts.

6. A certain capacitive transducer works in such a way that an external force moves the plates farther apart or closer together. As the plates are moved apart, the capacitance (*a*) increases; (*b*) decreases.
7. An advantage of using a bridge circuit for the output of a passive transducer is (*a*) lower cost; (*b*) lower power demand on the power supply; (*c*) amplification of signal; (*d*) an output that is not sensitive to power-supply voltage changes.
8. In a control system the commands for operation of machinery may be on magnetic tape or a (*a*) punched tape; (*b*) variable resistor.
9. Which of the following is not a type of microphone discussed in this chapter? (*a*) Electrostatic; (*b*) Carbon; (*c*) Dynamic; (*d*) Crystal.
10. A tachometer is used for sensing (*a*) temperature; (*b*) light energy; (*c*) motor speed; (*d*) pressure.

ANSWERS TO SELF-TEST

1. (*a*)—Power is the rate at which energy is expended or the rate at which work is done.
2. (*d*)—This is a definition of telemetering.
3. (*b*)—The term *passive* is also used for all components that do not generate a voltage.
4. (*c*)—Two methods of generating a transducer voltage are seldom used: the friction method and the chemical method.
5. (*c*)—If the bridge has reactive components (inductors or capacitors), then the power supply must be ac. If the bridge has only pure resistances in the legs, the power supply can be either ac or dc.
6. (*b*)—Remember this important point about capacitors: the closer the plates, the higher the capacitance.
7. (*d*)—The bridge is not sensitive to changes in power-supply voltage. Also it is not sensitive to changes in bridge temperature.
8. (*a*)—See Fig. 5-13.
9. (*a*)—See Fig. 5-18.
10. (*c*)—See Fig. 5-15.

what circuits are used in electronic power supplies?

6

INTRODUCTION

From its name you might presume that a *power supply* supplies power. But this is not true. If the equipment is operated from the ac power line, it is the power company that supplies the power. The power supply converts the ac power to dc power for operating the electronic components.

Power is usually in one of the four forms shown in Fig. 6-1. The dc power represented in Fig. 6-1*a* can be obtained from a battery. In some industrial applications dc power may be available from generators.

Figure 6-1*b* shows you a graph of *single-phase ac* power. This is the type of power delivered to most homes, and it is used in many industrial applications. When power is delivered over long distances, a voltage drop occurs along the line due to wire resistance. Transformers are placed along the line to raise the voltage back to the desired value. You can see these large transformers mounted on power poles. This is the advantage of using ac power, rather than dc power, for the homes. It would not be possible to use a transformer to step up dc voltage, so there would be no way to make up the voltage loss along the line.

A disadvantage of single-phase ac power is that the applied voltage drops to zero once each half-cycle. Note that the voltage is 0 volts at 0°, 180°, and 360°. At the instant the voltage is zero, there is no power delivered to the load.

Some types of ac motors require two voltages that are out of phase. Two voltages are out of phase if they do not reach their maximum at the same time. The two voltages can be delivered to the same load. This is called *two-phase ac* power, and it is shown in Fig. 6-1*c*. The two phases of voltage are 90° out of phase. In this drawing waveform *a* leads waveform *b*—that is, it is ahead of waveform *b*—by 90°. This means that when *a* is at its maximum value *b* is at its zero value, and when *b* is at its maximum value *a* is at its zero value. Since

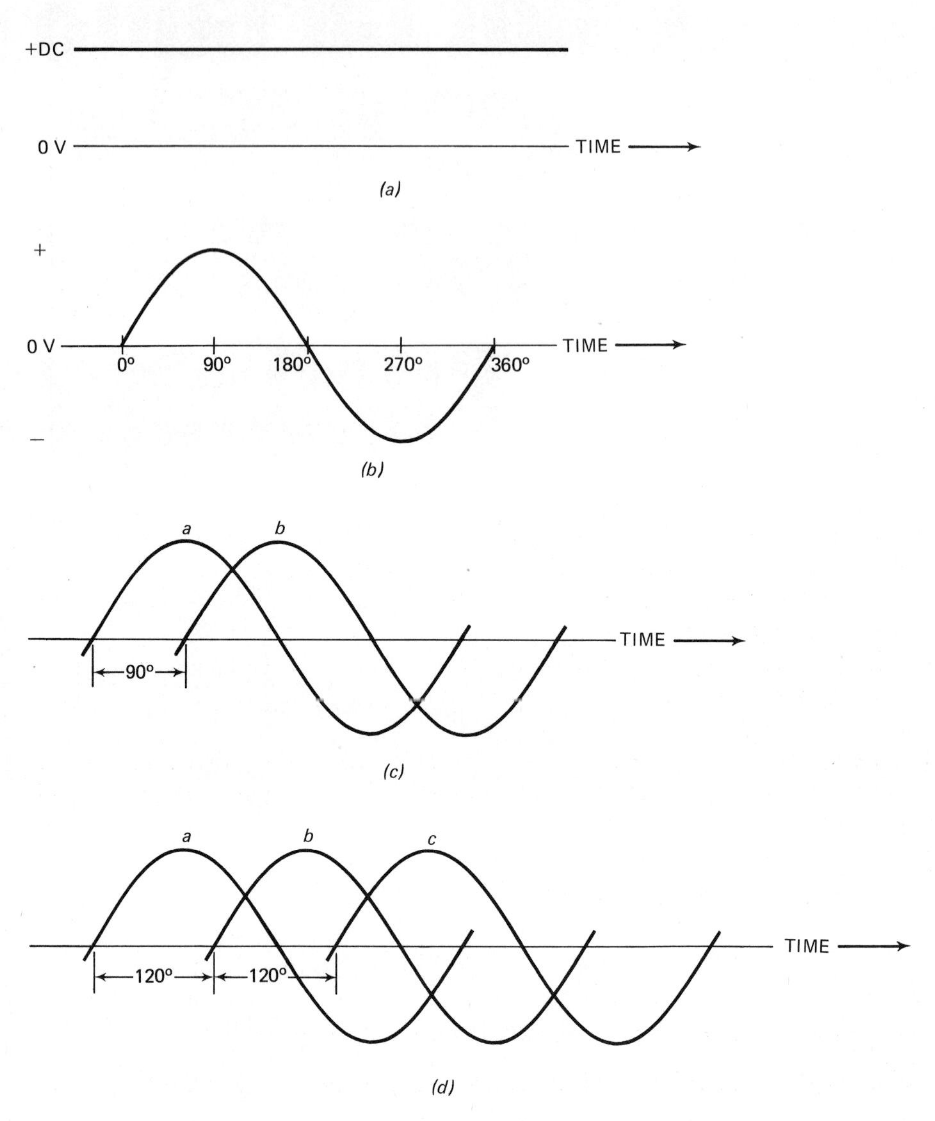

Fig. 6-1 Types of power used in industry. **(*a*) Dc. (*b*) Single-phase ac. (*c*) Two-phase ac. (*d*) Three-phase ac.**

the two voltages never go to zero at the same time, there is always power being delivered to the load. This is an advantage over single-phase ac power.

With three-phase power three different voltages are generated at the same time. You can see this in Fig. 6-1*d*. The voltages are 120° out of phase with each other. The three-phase voltage never drops to zero. As waveform *a* drops to zero waveform *b* is nearing its peak, and as waveform *b* drops to zero waveform *c* is nearing its peak. Therefore it has the same advantage as two-phase power.

The power supply in an electronic system usually *converts* the ac power to dc power, so it could be called a *power-altering system*. There are only four possible kinds of power-altering systems. You can see these in Fig. 6-2. A transformer is used for changing ac at some voltage (or current) into ac of another voltage (or current). A rectifier system is used to change ac into dc.

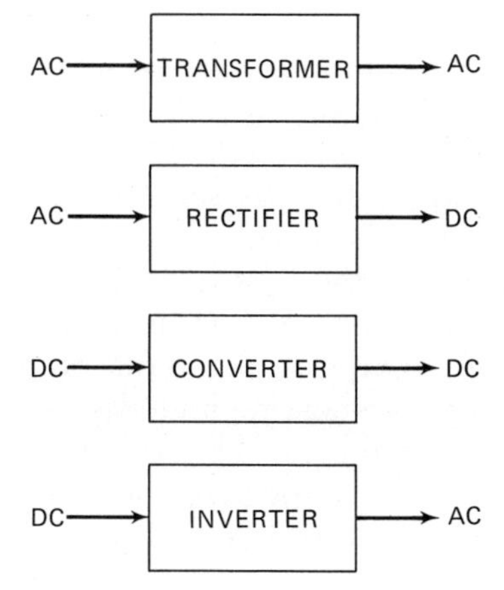

Fig. 6-2 Four basic power converters.

The rectifier is a very important part of an electronic power supply. To change dc voltage (or current) to another value of dc voltage (or current) a *converter* is used. If dc is available and you want to change the dc to ac, a circuit called an *inverter* is used.

Of the four basic systems for changing power shown in Fig. 6-2, the first two (transformers and rectifiers) are used extensively in electronic power supplies.

In this chapter we will discuss the type of power system that takes the ac power from the line and changes it to dc power. This type of system is called a *power supply*. Every electronic system that is designed to operate from an ac power line has such a power supply.

You will be able to answer these questions after studying this chapter.

- ☐ **What is the difference between unregulated and regulated power supplies?**
- ☐ **What are the components of a regulated power supply?**
- ☐ **What types of circuits are used in unregulated power supplies?**
- ☐ **How are rectifier diodes connected in series and parallel?**
- ☐ **What are the causes of poor regulation?**
- ☐ **What are the types of circuits used in power-supply regulators?**
- ☐ **How does an electronic regulated power supply work?**

INSTRUCTION

What Is the Difference between Unregulated and Regulated Power Supplies?

An *unregulated power supply* delivers a voltage and load current that depends upon the amount of load resistance. If you change the value of load resistance across the terminals of an unregulated supply, then there will be a change in the voltage across the load. Also there will be a change in current through the load.

There are two kinds of *regulated power supplies.* A *voltage-regulated* supply has the same output voltage regardless of the amount of load resistance. A *current-regulated* supply produces the same output current regardless of the load resistance.

Of course, there is a limit to how much you can change the load resistance. For example, if you placed a short circuit (0 ohms) across the output terminals of a voltage-regulated supply, then the voltage across the terminals would drop to zero. (There can be no voltage drop across 0 ohms resistance.)

The output voltage of a regulated supply will be constant over a given range of load current values. The manufacturer must give you this range of values.

The output current of a regulated supply will be constant over a given range of load resistance values. As with the voltage-regulated supply, the manufacturer must give you the range of values.

What Are the Components of a Regulated Power Supply?

We will start by getting an overall view of a typical regulated power supply. Figure 6-3 shows the system in block diagram form.

The input power from the power company is ac. It is applied to the primary of a power transformer. The transformer serves two purposes: (1) to isolate the power supply circuitry from the ac line and (2) to raise or lower the line voltage to the required value.

The transformer secondary is connected to a *rectifier circuit* which changes the ac voltage to a dc voltage. The components in rectifier circuits

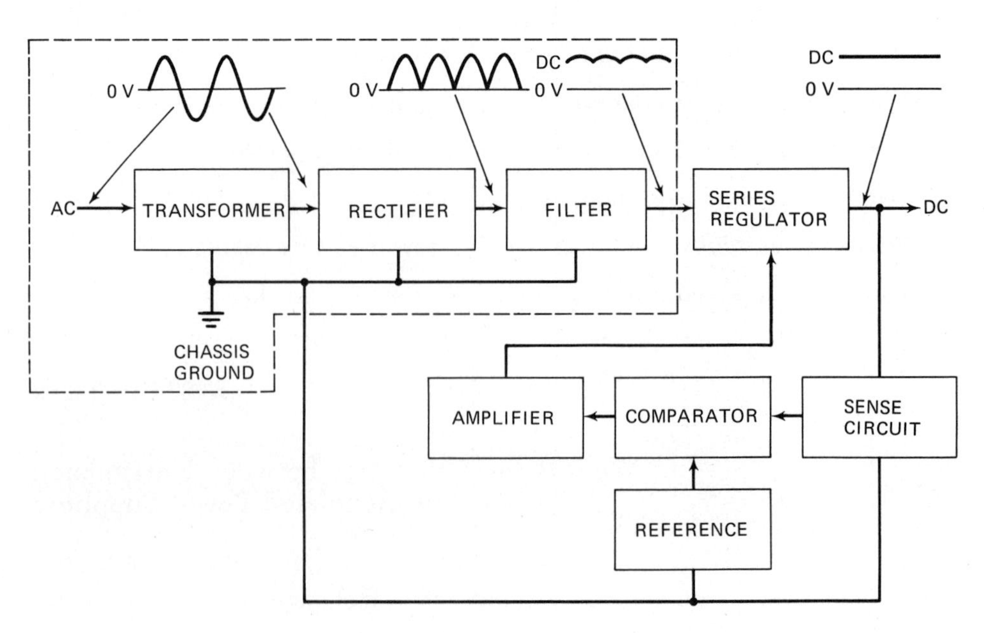

Fig. 6-3 Block diagram of a regulated electronic power supply.

may be vacuum-tube diodes, semiconductor diodes, SCRs, or gaseous diodes. Semiconductor diode rectifier circuits are the most popular.

The output of the rectifier circuit is not a pure, smooth dc voltage. Instead, it is a form of pulsating dc. You see a typical waveform on the block diagram. This voltage could not be used for supplying dc to amplifier systems, so a *filter circuit* is employed. The only purpose of the filter is to smooth out the pulsations to produce a pure (or nearly pure) dc output voltage.

The three blocks of Fig. 6-3 that have been discussed so far compose what is called an *unregulated power supply*. The output voltage from the filter may be delivered directly to a load requiring a dc voltage. However, if the load current in an unregulated power supply varies from moment to moment, there will also be changes in output voltage. In some circuits it is important to have a very steady dc output voltage regardless of whether the load current changes or not. In such applications a regulator circuit is needed. This is the part of the block diagram of Fig. 6-3 you see outside the dotted lines.

The *sense circuit* senses any change that may occur in the output voltage. (Again, these changes in voltage occur as a result of changes in load current.)

The output of the sense circuit is fed to a *comparator* which also has an input from a reference voltage. If the sense voltage and the reference voltage are the same, then you have no change in the output voltage from the comparator. If the output voltage of the power supply changes, there will be a corresponding change in output from the sense circuit. This causes the comparator to supply a different voltage to the amplifier, depending upon the amount of change sensed.

The *amplifier* amplifies the output from the comparator and supplies a control voltage to the *series regulator*. Its purpose is to control the dc output voltage of the circuit. If the output voltage of the supply tries to increase or decrease, the series regulator automatically adjusts the output to its correct value.

Summary

1. Electric power supplied to electronic systems is in one of the following forms: dc, single-phase ac, two-phase ac, or three-phase ac.
2. A power-supply circuit converts power from one form to another form. It does not actually generate the power.
3. Transformers convert ac from one value to another value and isolate the primary circuit from the secondary circuit.
4. Rectifiers convert ac to dc, while inverters convert dc to ac.
5. A converter converts dc from one value to another value.
6. A regulated power supply maintains its output voltage at a constant value even though the load resistance may change.

What Types of Circuits Are Used in Unregulated Power Supplies?

In the block diagram of Fig. 6-3 the transformer, rectifier circuit, and filter circuit make up what is known as an unregulated power supply. The power transformer was discussed in Chapter 4. The circuits used for rectifiers will be discussed here. In each case the circuit is shown with semiconductor rectifiers, but you should remember that vacuum-tube diodes or gaseous diodes are also used.

How Does a Half-Wave Rectifier Work?

Figure 6-4 shows the circuit for a half-wave rectifier. A power transformer is shown in the drawing, but this type of rectifier circuit can also work without a transformer.

The ac voltage appears across the transformer secondary at points *a* and *b*. On one-half cycle, point *a* is negative with respect to *b*. The negative voltage on the anode of the diode prevents it from conducting. On the next half cycle, point *a* is positive with respect to *b*. The positive voltage on the anode of the diode causes it to conduct electron flow in the path shown by the arrows.

The input and output waveforms for the half-wave rectifier circuit are shown in the drawing. An important disadvantage of this circuit is that the output voltage is 0 volts for one-half cycle of input. Therefore this power supply delivers no dc power for one-half cycle of input.

How Does a Full-Wave Rectifier Work?

Figure 6-5 shows a circuit for a full-wave rectifier. This type of rectifier circuit *must* be used with a transformer. The transformer is needed to split the phase so that each diode (D_1 and D_2) will conduct for one-half cycle.

On one-half cycle point *a* is positive and point *b* is negative. The center tap of the secondary (point *c*) must be 0 volts, since it is halfway between the equal positive and negative voltages. Assume that point *a* is positive and point *b* is negative. The solid arrows show the path of electron-current flow. Diode D_2 is not conducting during this period because of the negative voltage on its anode.

On the next half-cycle point *b* is positive and point *a* is negative. Diode D_1 is cut off with the negative voltage on its anode. The conduction path during this half-cycle is shown by the dotted arrows. The arrows show that electron current flows in the same direction through R_L when either D_1 or D_2 is conducting.

The input- and output-voltage waveforms of the full-wave rectifier circuit are shown in the drawing. The shaded areas are for the waveforms when D_1 conducts. Note that the output waveform drops to zero only for an instant

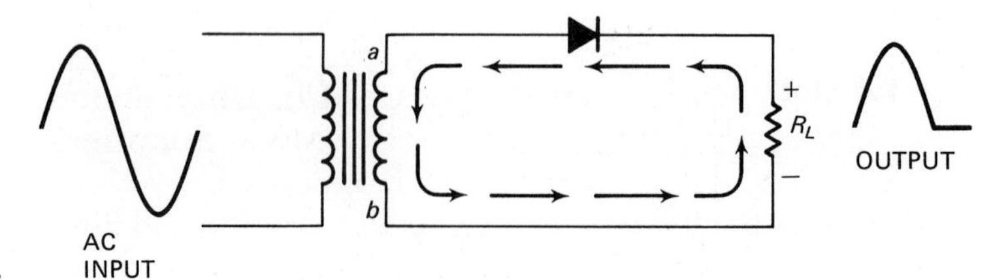

Fig. 6-4 A half-wave rectifier.

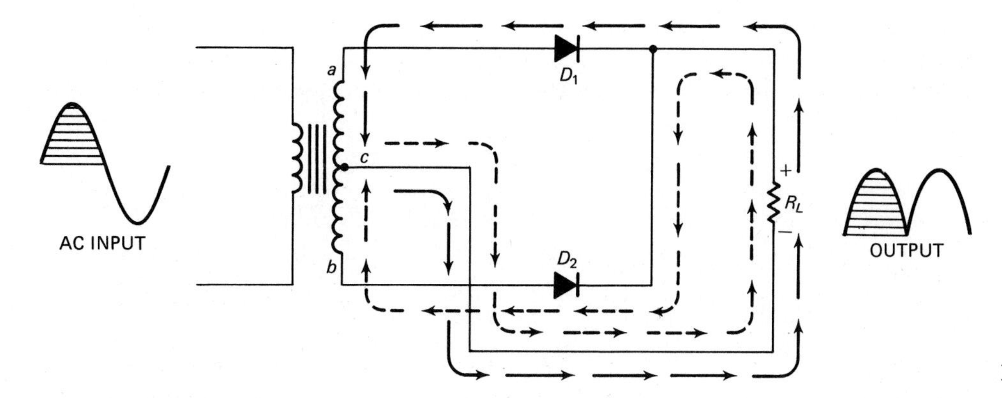

Fig. 6-5 A full-wave rectifier.

during a full cycle of input power. Also there is an output voltage for both half cycles of input. In many circuits this is an advantage over the half-wave supply, shown in Fig. 6-4.

What Is a Bridge Rectifier?

Figure 6-6 shows a bridge rectifier circuit. This is another kind of full-wave rectifier. The output waveform shows that current flows through the load resistor R_L for both half cycles of the input. A transformer is not required for this kind of rectifier even though one is shown. A bridge rectifier has the advantage of always using the entire secondary transformer winding. Thus, if you use the same transformer, the output dc voltage will be twice the voltage you get from Fig. 6-5.

The voltage across the secondary is ac. On the first half-cycle, point *a* is positive with respect to *b*. The conduction path during this half cycle is shown by dotted arrows. It is from point *b* through diode D_3, through the load R_L, through D_4, and to positive point *a* on the transformer secondary. Diode D_1 is cut off during this half cycle because of the negative voltage on its anode

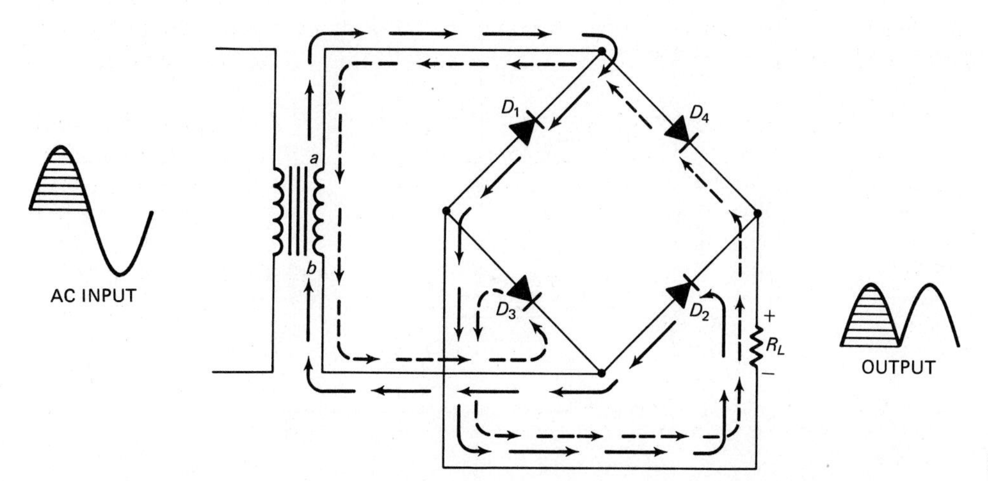

Fig. 6-6 A bridge rectifier.

at point a, and diode D_2 is cut off because of the positive voltage on its cathode at point b.

On the next half cycle point a is negative with respect to b. The current during this half cycle is shown with solid arrows. It flows through D_1, through the load R_L, through D_2, and to positive point b.

Current flows in the same direction through R_L during both half cycles. Therefore it is a full-wave rectifier. An important advantage of the bridge rectifier over the full-wave rectifier of Fig. 6-5 is that a transformer is not required. However, if the ac voltage of the power line must be stepped up or stepped down, a transformer will be used.

How Do Voltage Doublers Work?

Figure 6-7 shows two rectifier circuits which provide an output voltage which is approximately twice the peak input voltage across points a and b.

Figure 6-7a shows a *half-wave voltage doubler* circuit. On the first half cycle point a is positive with respect to point b. The negative voltage at point b cuts diode D_1 OFF and turns diode D_2 ON. Electron flow for this half cycle is shown by the dotted arrows. This current charges capacitor C_2 to the peak value of the voltage across points a and b.

On the next half cycle of input power point a is negative with respect to point b. During this half cycle the positive voltage at b turns ON diode D_1 and cuts OFF diode D_2. Electron flow is shown by the solid arrows. This electron-current flow charges capacitor C_1 to the peak value of the voltage

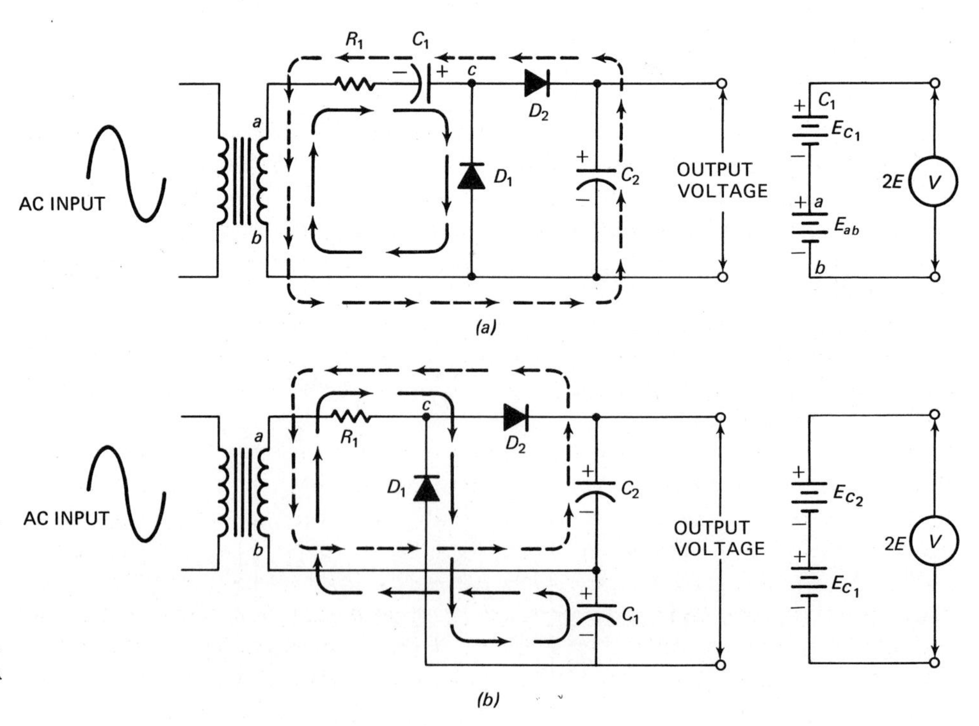

Fig. 6-7 Voltage-doubler circuits. (a) A half-wave doubler. (b) A full-wave doubler.

across points a and b. Resistor R_1 limits the current so that the current rating of the diodes will not be exceeded.

When D_1 conducts you should remember that capacitor C_1 is charged to the peak value of line voltage. This capacitor holds the peak voltage during the half cycle that D_1 is not conducting. The polarity of the voltage across C_1 is marked. You will note that this voltage is in series with the voltage across the secondary when point a is positive with respect to the voltage at point b. The insert for Fig. 6-7a shows how the two voltages add. The total voltage (E_{C_1} + E_{ab}) is applied when D_2 conducts, so C_2 charges to *twice* the peak value of the voltage across the transformer secondary.

If you measured the voltage across the transformer secondary with an ac voltmeter and found the secondary voltage to be 100 volts, you might expect that the voltage across capacitor C_2 would be 200 volts. This is not true. Remember that when you measure an ac voltage with a voltmeter, you are measuring the rms value! The capacitors charge to the peak value. The peak value is 1.414 times this rms value. Therefore, it you measure 100 volts across the transformer secondary with a voltmeter, the peak value is 1.414×100, or 141 volts. The output voltage across C_2 will be twice 141 volts, or 282 volts.

The circuit of Fig. 6-7a is said to be a *half-wave doubler* because the voltage across the output (which is also across C_2) is produced during only one-half of the input cycle.

The circuit of Fig. 6-7b is a *full-wave voltage doubler* circuit. On the first half cycle point a becomes positive with respect to point b. Diode D_1 is turned OFF by the positive voltage, while D_2 is turned ON. The conduction path is shown by the dotted arrows. Capacitor C_2 is charged to the peak voltage across a and b during this half cycle.

On the next half cycle point a becomes negative with respect to point b. The negative voltage at point a shuts diode D_2 OFF and causes diode D_1 to conduct. The conduction path is shown by the solid arrows. Capacitor C_1 charges to the peak voltage across a and b during this half cycle.

The voltage across the power-supply output is the sum of the voltages across the capacitors. This voltage is obtained by adding the two capacitor voltages in series in much the same way as adding the battery voltages you see in the inset.

If the input voltage across points a and b in the full-wave doubler of Fig. 6-7b is measured and found to be 100 volts, the output voltage will be 282 volts. Remember, a voltmeter measures the rms value, but the capacitors charge to the peak value.

An advantage of the full-wave doubler over the half-wave doubler is that the output voltage changes less with varying amounts of current load. Note that in Fig. 6-7a a low resistance connected across the output terminals will cause capacitor C_2 to discharge to a low value. This will occur during each half cycle when C_2 is not being charged. With the full-wave doubler one or the other of the two capacitors (C_1 and C_2) is being charged during each of the half cycles of input power. Since C_1 and C_2 are connected in series and the load is connected across both, the load cannot reduce the output voltage greatly at any time.

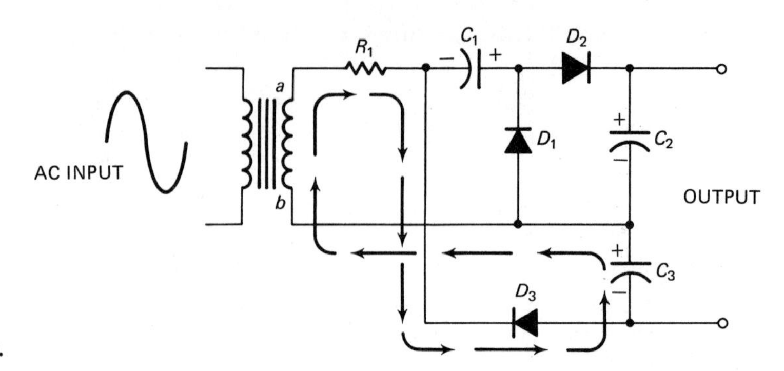

Fig. 6-8 A voltage tripler circuit.

The voltage-doubler circuits have the advantage that they can produce a large output dc voltage for a given input ac voltage. However, these types of circuits have poor regulation, which means that the output voltage varies considerably with large changes in load current.

How Does a Voltage Tripler Work?

It is possible to get even more than double the input voltage in rectifier and capacitor circuits. You can see an example of this in Fig. 6-8. This circuit consists of a half-wave voltage doubler made up of capacitors C_1 and C_2 and diodes D_1 and D_2. The only thing that is different from the half-wave doubler of Fig. 6-7*a* is that diode D_3 and capacitor C_3 have been added. The voltage across capacitor C_2 will be twice the line voltage, as explained before. Capacitor C_3 charges to the peak value of voltage on the half cycles when point *a* is negative with respect to point *b*. The arrows show the path of charging current for C_3. The peak ac voltage across C_3 plus twice the peak ac voltage across C_2 adds to make three times the peak ac voltage across the output terminals.

The voltage tripler does not give you the maximum output voltage that can be obtained in this manner. A voltage quadrupler, for example, can be made by combining the outputs of two voltage doublers in series. There is a tradeoff, however. The higher the output voltage obtained, the poorer the regulation.

How Are Rectifier Diodes Connected in Series and Parallel?

One of the things you need to know about diodes is their Peak Inverse Voltage (PIV) rating. Manufacturers give the PIV rating of the diodes they make. This is the maximum reverse voltage that can be applied across a diode before a breakdown occurs. In other words, it is the maximum voltage with the anode negative and the cathode positive.

When a breakdown occurs, the diode conducts electron current from the anode to the cathode. Its rectifying ability is usually destroyed when this happens.

It is a good idea to use a diode with a higher PIV rating than you expect will

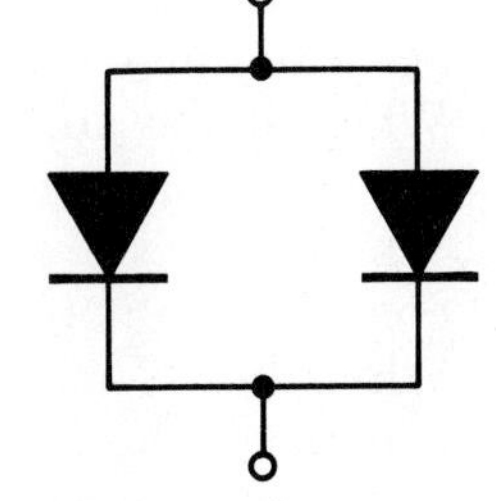

Fig. 6-9 Diodes connected in series.

Fig. 6-10 Diodes connected in parallel.

be placed across it. If the diode will have a reverse voltage across it of 100 volts when it is connected in the circuit, then you should use a diode with a PIV rating of at least 200 volts.

You might need a larger peak inverse voltage rating than you can get with any of the diodes you have on hand. You can get a higher PIV rating by connecting diodes in series.

Figure 6-9 shows how diodes are connected in series. If one diode in this circuit has a PIV rating of 100 volts and the other also has a PIV rating of 100 volts, then the peak inverse voltage rating of the combination is the sum, or 200 volts.

In addition to the PIV rating, manufacturers also rate diodes by the *maximum forward current* they can safely conduct. The maximum forward current is the most current you can allow when the diode is forward-biased (anode positive and cathode negative). You may need a larger forward-current rating than you can get with any of the diodes on hand. This can be done by connecting the diodes in parallel as shown in Fig. 6-10. Here one diode in the parallel combination has a forward current rating of $\frac{1}{2}$ ampere and the other also has a forward-current rating of $\frac{1}{2}$ ampere. When the two are connected in series, the maximum allowable forward current for the combination is 1 ampere.

When the diodes are connected in parallel, as shown in Fig. 6-10, their peak inverse voltage rating is equal to the lowest PIV rating of the combination. For example, suppose one diode has a PIV rating of 50 volts and the other has a PIV rating of 100 volts. When the diodes are connected in parallel, the PIV rating of the two is 50 volts.

Summary

1. A half-wave rectifier is the simplest, but no power is delivered to the load for one-half cycle of each input cycle of voltage.
2. A full-wave rectifier usually requires a transformer for its operation. This makes it more complex and bulkier than a half-wave rectifier. But its output voltage drops to zero only for an instant at the end of each half cycle.
3. A bridge rectifier is a full-wave rectifier that can provide twice the output dc voltage of an ordinary full-wave rectifier.
4. A voltage doubler produces a large output voltage, but it has the disadvan-

tage that large, changing load currents cause large changes in output voltage. In other words, it has poor regulation.

5. When diodes are connected in series, their PIV ratings are added. When they are connected in parallel, their forward-current ratings are added.
6. When diodes are connected in series, the forward-current rating of the combination is equal to the lowest current rating of the group.
7. When diodes are connected in parallel, the PIV rating of the combination is equal to the lowest PIV rating of the group.

What Is a Filter Circuit?

The rectifier circuits discussed in this chapter have an output dc voltage that is *pulsating*. In other words, the output is a varying dc value. As shown in the block diagram of Fig. 6-3, the filter circuit usually follows the rectifier circuit. Its purpose is to smooth out the voltage variations.

The simplest filter is shown in Fig. 6-11. It is an electrolytic capacitor C connected directly across the load on the power supply R_L. The input voltage to the power supply is the pulsating dc coming from the rectifier circuit. You will remember that one of the functions of a capacitor in a circuit is to store energy. That is the purpose of C in this circuit.

On one-half cycle the voltage across the input terminals (that is, the pulsating voltage from the rectifiers) goes positive. This is shown in Fig. 6-11*a*. During this period electron current flows through R_L and also flows into ca-

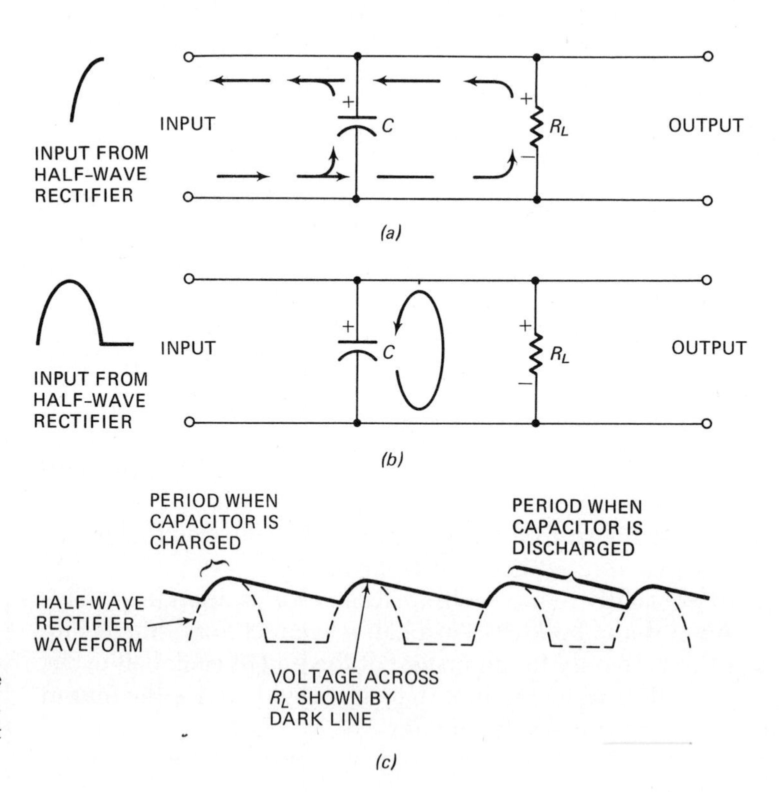

Fig. 6-11 Operation of a capacitive filter. (*a*) During capacitor charge. (*b*) During discharge. (*c*) Output waveform.

pacitor C to charge it to the peak value of the input wave. The arrows show the paths for electron flow.

It is presumed in this discussion that the input to the filter is from a half-wave rectifier circuit. Therefore, after the positive peak is passed, the input voltage drops to 0 volts for the next half cycle. When the input voltage drops to zero, the capacitor begins to discharge through R_L. This is shown in Fig. 6-11*b*. Instead of the output voltage (across R_L) dropping to zero, as it would if there were no capacitor, the discharge current from C maintains a voltage across R_L.

If the circuit is properly designed, the filter capacitor will discharge only a small amount of its energy before the next positive pulse comes from the rectifier circuit. The capacitor charges each time the input voltage is positive.

You can see the output waveform of the power supply, which is the voltage across R_L, in Fig. 6-11*c*. The dotted line indicates the half-wave rectifier input. This would be the output voltage across R_L if the capacitor were removed. The solid line represents the filtered output voltage.

Inductors are also components that store energy, so they can be used as filters as well as capacitors.

Figure 6-12 shows you four common filter circuits used in electronic power supplies. Figure 6-12*a* is a simple *capacitive filter* of the type just discussed. In Fig. 6-12*b* an inductor (choke) is used to filter the pulsations. Since an inductor opposes any change in current through it, the pulsating current flowing through R_L will be smoothed by the inductor action.

Figure 6-12*c* shows a *pi filter,* which gets its name from its similarity to the Greek letter pi. This circuit contains both capacitors and inductors for smoothing the pulsations. It is also called a *capacitive-input* filter because a capacitor C_1 is the first component in the filter circuit as seen from the rectifier circuit.

Figure 6-12*d* shows a *choke-input* circuit. This is also known as an *L filter.*

In comparing the capacitive-input filter with the choke-input filter there

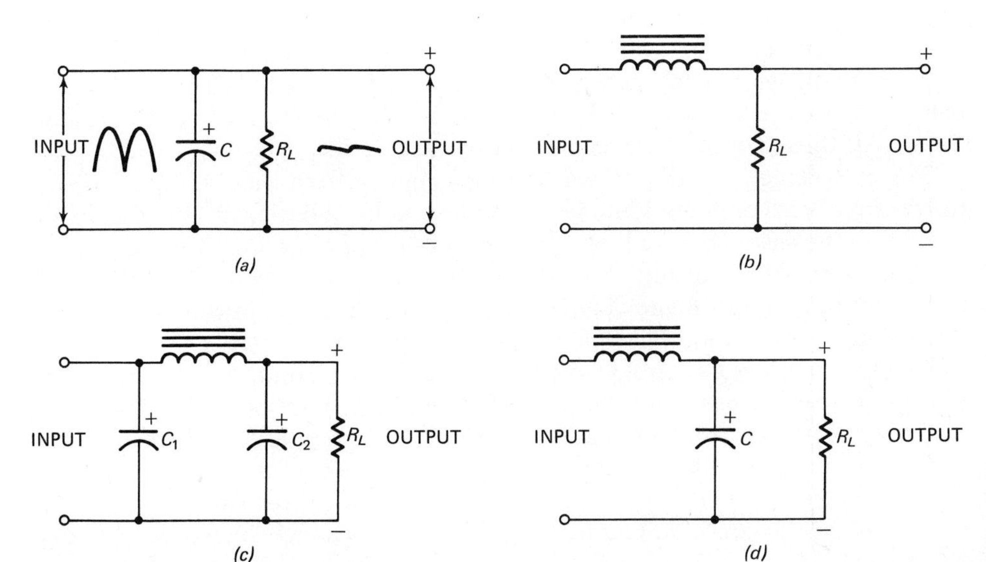

Fig. 6-12 Filter circuits for power supplies. (*a*) Capacitive filter. (*b*) Choke filter. (*c*) π filter. (*d*) L filter.

are two very important things for you to remember. With the capacitive-input filter the output voltage across the load resistance is greater than for a choke-input filter. However, the *regulation* of the capacitive-input filter is not as good as the regulation of a choke-input filter. Regulation is a measure of how well the power-supply output voltage remains constant under varying load current values. The choke-input filter has better regulation, but the output voltage is not as great.

The greater output voltage of the capacitive-input filter comes as a result of the fact that the input capacitor C_1 charges to the peak value of the input pulsating voltage, and this serves as the output voltage of the supply.

When the power supply is electronically regulated, it is common practice to use a simple capacitive filter like the one shown in Fig. 6-12*a*. This simple filter does not give the smooth output that can be obtained with a more elaborate filter system, but the regulating circuit takes out any remaining ripple. In fact, some regulating circuits are called *electronic filters* because of their ability to remove ripple from the power-supply voltage. A regulated power supply will be discussed later in this chapter.

What Are the Causes of Poor Regulation?

When it is necessary to have a constant voltage or current even though the power-supply load is changing, electronic regulator circuits may be used. Remember that the load of a power supply is the total current that it must supply. Anything that causes a large change in the output voltage during normal operation will result in poor regulation. The purpose of the regulator is to prevent the output voltage of the power supply from changing with changes in load current.

If the regulator is designed in such a way that the load current does not flow through it, it is called a *shunt regulator*. If the load current flows through the regulator, it is called a *series regulator*.

Remember that regulation is a measure of how well the power supply holds its output voltage constant when there is a changing load. The unregulated part of the power supply in Fig. 6-3 includes the transformer, rectifier, and filter. All these components can affect the regulation.

The ac voltage from the power line may change from time to time. If the power supply is connected into power lines that have large variations in load, as is often the case when industrial machinery is operated from the same line, then the ac input to the supply can vary from moment to moment. If the amplitude of the input voltage changes, it follows that the amplitude of the output voltage of an unregulated power supply will also change.

Changes in power-supply load current will cause changes in the amount of current flowing through the transformer windings. These windings have resistance, so the current changes cause voltage changes. The result is a varying secondary voltage when the power-supply current is varying.

The diodes used in rectifier circuits have a forward resistance. In other words, when the diode is conducting, there is resistance in the conduction path between the cathode and the anode. As the load current of the power supply changes, the voltage drop across the diode may change, causing a changing output voltage. (This change in voltage drop does not occur by the same amount for all types of diodes used.) Thus the type of rectifier may affect the regulation of the supply.

If the inductors in the filter circuit had no resistance, they would not have any effect on power-supply regulation. However, the inductors do have resistance. There is a voltage drop across the filter inductor, and this voltage drop varies with changes in power-supply load current.

With the filter circuits of Fig. 6-12 a resistor is sometimes used in place of the inductor. The resistor limits the discharge time for the filter capacitors and thus helps to maintain the output voltage at a constant value. However, the dc voltage drop across the resistance also varies with the load current. This will cause variations in the power-supply output voltage that are greater than if an inductor had been used. Thus this type (resistor – capacitor) of filter is used only where the current drain on the power supply is light.

To summarize, all the components in the low-voltage unregulated power supply can affect the regulation. Also changes in input ac voltage can affect the output voltage, another factor that may lead to poor regulation.

Summary

1. The purpose of a filter is to smooth out the dc variations from the rectifier circuit.
2. A full-wave rectifier output is easier to filter than a half-wave rectifier output.
3. Some common types of power-supply filters are the simple capacitive filter, the pi filter, and the *L* filter.
4. The choke-input filter gives better regulation than the capacitive-input filter.
5. The purpose of a regulator is to keep the output voltage the same if the load current changes.
6. A shunt regulator does not have the load-current flow through it. A series regulator does.
7. Some of the factors that may affect regulation are ac voltage changes, load-current changes through the resistance of the transformer windings and rectifiers, and load-current changes through the resistance of the filter inductors and/or resistors.

What Are the Types of Circuits Used in Power-Supply Regulators?

It will be helpful to study a few basic circuits that are used in regulated power supplies before analyzing a complete regulated supply. The circuits described in this section will be combined into a complete voltage regulator in the next section.

What Is a Sense Circuit?

The principle of the sense circuit is shown in Fig. 6-13. Here you see three resistors—R_3, R_4, and R_5—connected between a +30-volt source and ground. There is a voltage drop across each resistor. When added, the voltage drops must equal 30 volts.

The drop across R_5 is 10 volts, the drop across R_4 is 2 volts, and the drop across R_3 is 18 volts. The voltage at point *a* is 12 volts positive *with respect to ground.* This is the power-supply voltage (30 volts) minus the voltage drop across R_3 (18 volts).

The voltage drop across R_4 is equal to the difference between the +12 volts at point *a* and the +10 volts at point *b*. In other words, it is equal to 2 volts.

Note that R_4 is a variable resistor. The arm (represented by the arrow) of this resistor can be moved between point *a* and point *b*, and so the output voltage at the arm can be adjusted between +10 volts and +12 volts. There is a 2-volt range of adjustment. Suppose that you set the voltage at +11 volts with respect to ground. Remember that this voltage is dependent upon the setting of R_4 and the supply voltage of 30 volts.

If the supply voltage goes up to 31 volts, then the voltage at the arm of R_4 will also go up. Likewise, if the applied voltage drops to 29 volts, the voltage at the arm of R_4 will drop (or become less positive). You can assume that the voltage at the arm will change whenever the power-supply voltage changes. You can also say that the voltage at the arm of R_4 can be used to sense any change in the power-supply voltage.

There are variations of a sense circuit. Not all of them have a variable resistor, and some sense circuits contain only two instead of three resistors. Then the sense voltage is taken at the junction of the two resistors.

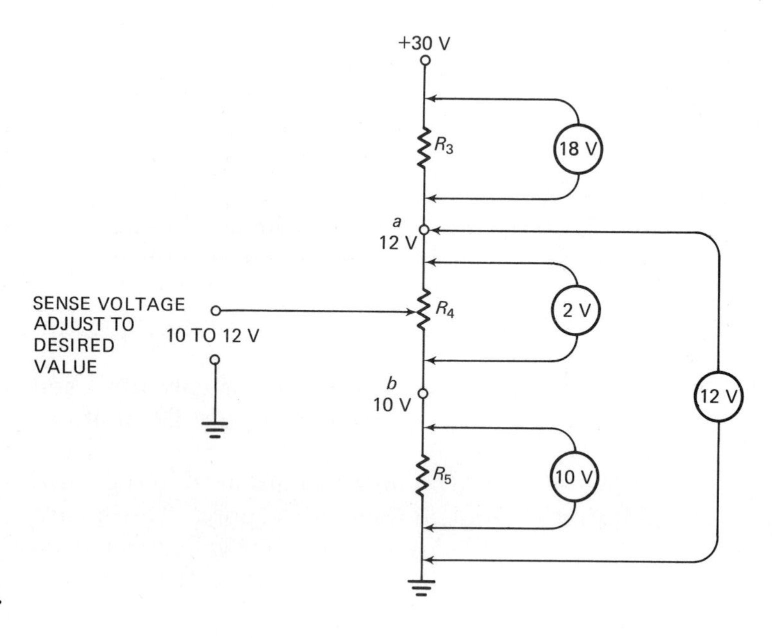

Fig. 6-13 Sense circuit.

How Is a Reference Voltage Obtained?

The reference voltage in most solid-state regulated power supplies is obtained from the constant-voltage drop across a zener diode. The operation of this reference circuit can be understood by referring to Fig. 6-14.

In Fig. 6-14*a* the power-supply voltage (10 volts) is applied across the resistor *R* and the zener diode *X* connected in series. The voltage across the zener diode is 3 volts. This is a rated zener voltage and will not change regardless of whether the power-supply voltage changes or not.

In Fig. 6-14*b* the power-supply voltage is raised to 11 volts. According to Kirchhoff's voltage law the sum of the voltage drops must equal the applied voltage. The voltage across the zener diode is still 3 volts, so the voltage across the resistor must have been raised to 8 volts. This makes a total voltage drop of

$$8 \text{ volts} + 3 \text{ volts} = 11 \text{ volts}$$

which is equal to the applied voltage.

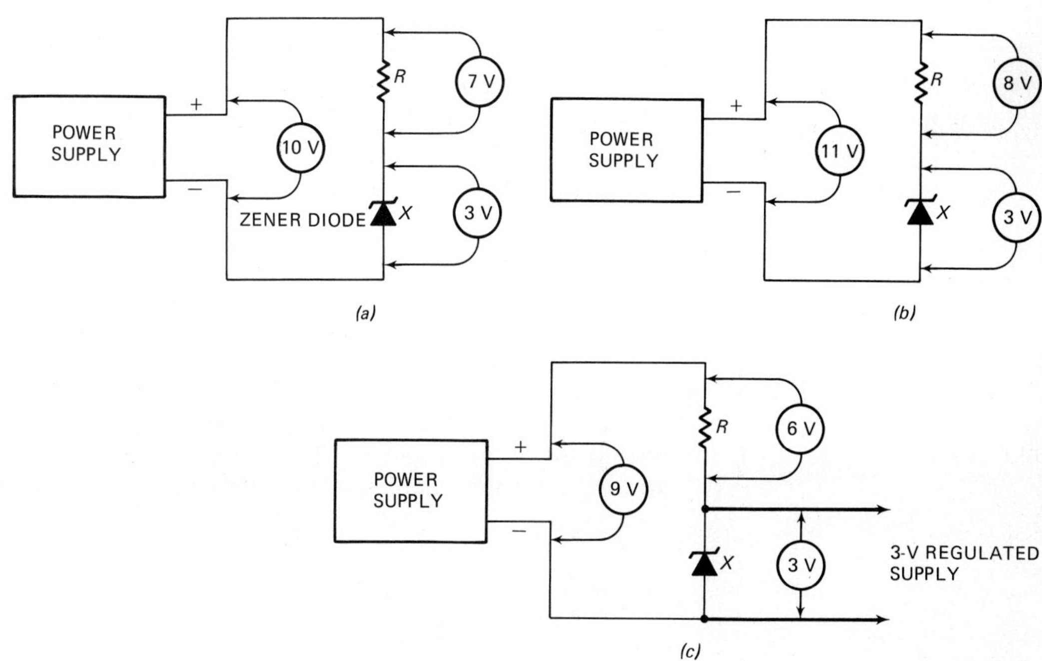

Fig. 6-14 The zener diode connected as a voltage regulator. (*a*) Part of the voltage drops across *R* and adds to the zener voltage, to equal the supply voltage. (*b*) Increasing the supply voltage will increase the voltage across *R*. (*c*) Decreasing the supply voltage decreases the voltage across *R*.

In Fig. 6-14*c* the power-supply voltage has dropped to 9 volts. The zener voltage drop is still 3 volts, so the voltage across the resistor must drop to 6 volts. This is necessary in order to make the total voltage drop equal to the applied voltage.

The circuit of Fig. 6-14 can be used as a 3-volt regulated power supply for certain low-power applications. In electronic regulated supplies the voltage across a zener diode may be used as the reference. Refer again to Fig. 6-3 and note that the reference voltage is one of the two voltages applied to the comparator. The other voltage comes from the sense circuit, which you studied in Fig. 6-13.

What Is a Comparator Circuit?

The comparator produces an output voltage that is related to the difference between a reference voltage and a voltage from the sense circuit. It is also known as a *sense amplifier* and a *voltage-comparison amplifier.*

Refer to Fig. 6-3 and note again that there are two inputs to the comparator. One is from the sense circuit, and the other is from the reference. There is no output from the comparator when the sense voltage equals the reference voltage. This is the condition when the power-supply output voltage is correct.

If the output voltage is not correct, the sense voltage changes. Since the sense voltage no longer equals the reference voltage, there will be an output from the comparator.

How Do Series Regulators Work?

The block diagram of Fig. 6-3 shows that a *series regulator* is inserted between the unregulated input voltage and the regulated output voltage of the supply. Since the power-supply current may be large, the series regulator must be a power amplifier. Figure 6-15 shows you how large current ratings can be obtained with power amplifiers.

Power amplifiers may be connected in parallel, as you see in Fig. 6-15*a*. The base voltages on the two transistors are identical, since they are connected to the same point. The collector current divides as shown by the arrows. Any number of power amplifiers can be connected in parallel to get a

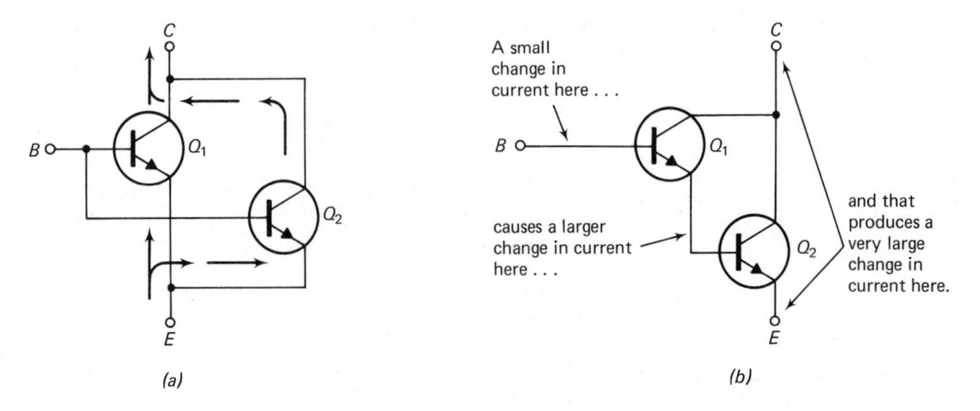

Fig. 6-15 Power transistors connected (*a*) in parallel and (*b*) in a Darlington configuration.

desired current rating. Despite the fact that there are two transistors in the parallel circuit, they may be considered as a single large power amplifier, with three terminals marked *C*, *B*, and *E*.

Figure 6-15*b* shows another way of getting a larger power rating for power transistors. This is a scheme known as a *Darlington amplifier.* Note that the base current for Q_2 is determined by the amount of current flowing through Q_1. A small change on the base of Q_1 will cause a larger collector-current change for Q_1, and this produces a very large change in the collector current through Q_2.

The advantage of the Darlington circuit is that it makes the total current through the system between terminals *E* and *C* more directly related to the current in the base *B* at all times.

How Does an Electronic Regulated Power Supply Work?

Figure 6-16 shows you an electronic regulated power supply. This supply has the same circuitry as the one shown in Fig. 6-3. Looking at the individual circuits, they are as follows:

- ☐ The sense circuit: R_3, R_4, and R_5
- ☐ The reference circuit: R_1 and X_1
- ☐ The comparator and amplifier are combined into one circuit: Q_2 with R_2 as the collector-load resistor
- ☐ Series regulator transistor: Q_1

The operation of this power supply is quite simple, but it is important for you to take it a step at a time. Notice that the unregulated input to this circuit is 50 volts. This means that there is a voltage drop across Q_1 of 20 volts.

The amount of voltage drop across Q_1 depends directly upon its base voltage. As you make the base voltage more positive, the drop across Q_1 de-

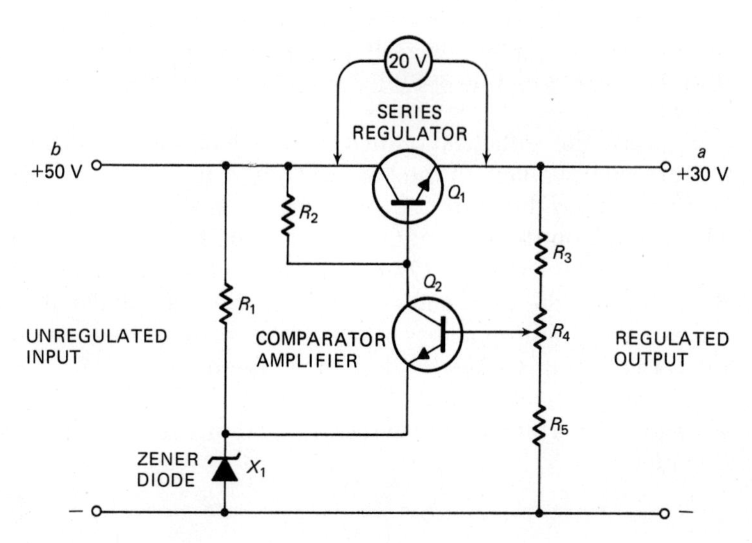

Fig. 6-16 An electronic regulator circuit.

creases. As you make the base of Q_1 less positive, the drop across Q_1 increases.

The whole idea of this regulated supply is to adjust the voltage drop across Q_1 to make up for any change in output voltage. For example, if the voltage at point *a* tries to change from 30 to 31 volts, then the drop across Q_1 will be increased to 21 volts. After the drop across Q_1 increases, the output voltage will again be 30 volts. To summarize, a rise of 1 volt at point *a* is taken up by an increase in drop across Q_1, so the output voltage stays at a fixed value.

If the output voltage at point *a* decreases, then the voltage drop across Q_1 will decrease. This will cause the voltage at point *a* to rise again to its fixed value.

Here is how the circuit controls the voltage drop across Q_1. The sense circuit senses any change in the output voltage that may occur. If the output voltage changes, the sense circuit delivers a voltage change to the base of Q_2.

Transistor Q_2 is an amplifier with its emitter tied to the voltage reference (zener diode) and its base circuit tied to the sense output R_4. It does two jobs. *First,* it compares the sense and reference voltages. *Second,* it amplifies the difference voltage whenever the two are not equal. If there is any change in output voltage at point *a*, there will also be a (lesser) change in the base voltage of the transistor. This will cause a change in conduction in Q_2. The collector current of Q_2 flows through R_2, so any change in conduction of Q_2 will cause a change in voltage drop across R_2.

The voltage at the base of the series-regulator transistor Q_1 is equal to the unregulated input of 50 volts at point *b* minus the voltage drop across R_2. When the output voltage at point *a* is correct, the base voltage of Q_2 is 40 volts.

Suppose the voltage at point *a* becomes more positive. This causes the base voltage of Q_2 also to become more positive, so transistor Q_2 will conduct harder. The collector current of Q_2 will increase, and increase the voltage drop across R_2. The greater voltage drop across R_2 makes the voltage at the base less positive. Remember that whenever you make the base of Q_1 less positive, the voltage drop across Q_1 will increase. There will be a voltage increase across Q_1 exactly equal to the amount that the voltage started to rise at point *a*.

So when the voltage at point *a* started to rise, the circuit adjusted the voltage across Q_1 to bring the voltage at *a* back down to 30 volts. It takes a little time to tell about this action, but in actual circuits it all happens instantaneously.

Suppose the voltage at point *a* tries to decrease to 29 volts. This will cause a decrease in voltage in the base of Q_2 and reduce its collector current. That in turn reduces the current through R_2 and reduces the voltage drop across R_2. The voltage on the base of Q_1 will then be more positive than it was before, and this causes transistor Q_1 to conduct harder. Remember that when Q_1 conducts harder, the drop across it is less. So the voltage at point *a* tries to decrease, and the drop across Q_1 decreases also. The lower drop across Q_1 returns the output voltage to its 30-volt value.

The overall result is that any change in voltage at point *a* will cause a change in the voltage drop across Q_1, and this returns the voltage at point *a* to 30 volts.

Summary

1. A sense circuit "senses" any changes in output voltage. This information is used to help regulate the output voltage. It is fed to the comparator.
2. The reference voltage is a fixed voltage which is applied to the comparator in addition to the sense voltage.
3. A zener diode may be used to obtain the reference voltage.
4. A comparator produces an output voltage, which is proportional to the difference between the sense and the reference voltages. This output is used as the control voltage for the regulator.
5. Series and shunt regulators may be either tubes or transistors. They work because their resistance may be varied by a control voltage.
6. Power amplifiers may be used in pairs in order to increase their total power-handling capacity.
7. The resistance of a series or shunt tube may be changed by varying its grid voltage. The more positive the grid voltage, the lower the tube resistance.
8. The resistance of a series or shunt transistor may be varied by changing its base voltage. For an NPN transistor, making the base more positive lowers its resistance. For a PNP transistor, making the base more negative lowers its resistance.
9. With a series regulator the voltage drop across the series regulator transistor (or tube) varies as needed to maintain a constant output voltage.
10. In Fig. 6-16 transistor Q_2 functions as a combination comparator and amplifier. The emitter voltage is fixed by the zener diode. Any changes of base voltage (caused by output-voltage variations) are amplified by Q_2 and fed to the series regulator Q_1.

PROGRAMMED REVIEW QUESTIONS

(Instructions for using this programmed section are given in Chapter 1.)
We will review the important concepts of this chapter. If you have understood the material, you will progress easily through this section. Do not skip this material because some additional theory is presented.

1. In the circuit of Fig. 6-17 the input voltage is 115 volts ac. The dc output voltage E will be about
   A. 115 volts dc. (Proceed to block 9.)
   B. 162 volts dc. (Proceed to block 17.)

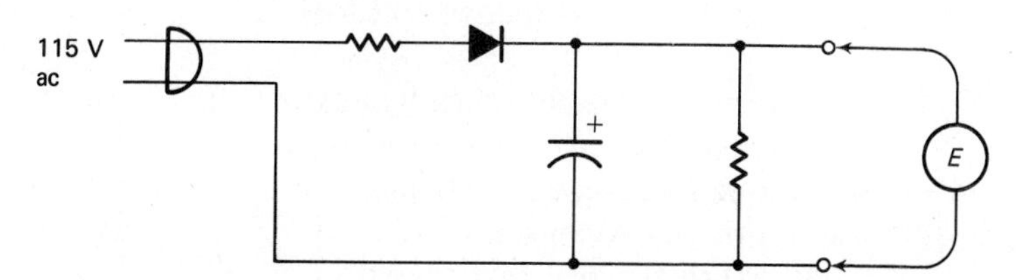

Fig. 6-17 This is the circuit for the question in block 1.

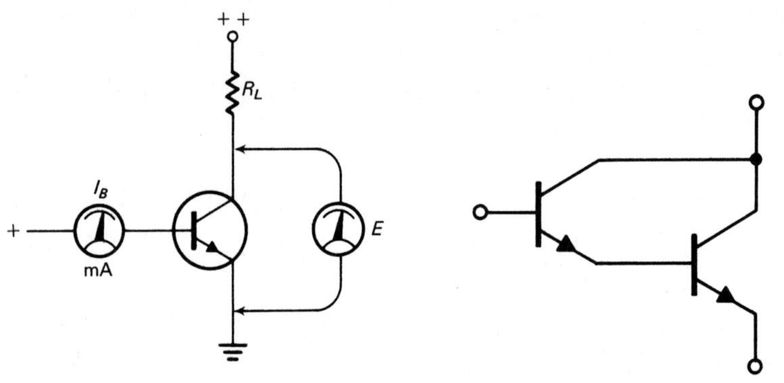

Fig. 6-18 This is the circuit for the question in block 2.

Fig. 6-19 This is the circuit for the question in block 8.

2. *The correct answer to the question in block 17 is* **A**. *This is a review question. Your answer is based on what you learned in Chapter 3. The base current of a bipolar transistor directly controls the collector current.* Here is your next question.
In the circuit of Fig. 6-18, increasing the base current I_B will
A. decrease the voltage E across the transistor. (Proceed to block 7.)
B. increase the voltage E across the transistor. (Proceed to block 10.)

3. *Your answer to the question in block 7 is* **B**. *This answer is wrong. Review the subject of filters, then* proceed to block 19.

4. *Your answer to the question in block 19 is* **B**. *This answer is wrong. Series and parallel diode connections are shown in Figs. 6-9 and 6-10. When you connect diodes in parallel, they cannot withstand a greater amount of reverse voltage.* Proceed to block 27.

5. *Your answer to the question in block 18 is* **A**. *This answer is wrong. The collector current will increase, and the voltage drop across R_L will also increase.* Proceed to block 23.

6. *The correct answer to the question in block 20 is* **B**. *A rectifier circuit has an ac input and a dc output.* Here is your next question.
Which of the following could be used as the series regulator in a power supply?
A. A Darlington amplifier. (Proceed to block 24.)
B. A thyristor. (Proceed to block 25.)

7. *The correct answer to the question in block 2 is* **A**. *Remember, increasing the base current also increases the collector current and decreases the voltage between the collector and emitter. You must know this in order to understand the regulated-power-supply operation.* Here is your next question.

In a filter circuit the capacitors
A. store energy. (Proceed to block 19.)
B. reduce the output voltage to a desired value. (Proceed to block 3.)

8. *The correct answer to the question in block 27 is **B**. A bridge rectifier is a type of full-wave rectifier that does not require a transformer for its operation.* Here is your next question.
Figure 6-19 shows
A. parallel transistors. (Proceed to block 13.)
B. a Darlington amplifier. (Proceed to block 18.)

9. *Your answer to the question in block 1 is **A**. This answer is wrong. Always remember that power-line voltages are given in rms rather than peak values. The filter capacitor in the circuit of Fig. 6-17 charges to the peak value, not the rms value.* Proceed to block 17.

10. *Your answer to the question in block 2 is **B**. This answer is wrong. Increasing the base current lowers the opposition to current flow from emitter to collector. It also reduces the voltage drop across the transistor.* Proceed to block 7.

11. *Your answer to the question in block 20 is **A**. This answer is wrong. An inverter is a circuit that changes a dc voltage to an ac voltage.* Proceed to block 6.

12. *Your answer to the question in block 23 is **A**. This answer is wrong. A choke-input filter results in better regulation. Regulation is a measure of how well the power supply holds a steady output voltage when you change the load.* Proceed to block 16.

13. *Your answer to the question in block 8 is **A**. This answer is wrong. When transistors are connected in parallel, their emitters are connected to the same point. Compare the circuit of Fig. 6-19 with the one shown in Fig. 6-15, then* proceed to block 18.

14. *Your answer to the question in block 17 is **B**. This answer is wrong. Increasing the base current moves more charge carriers into the base region, so the collector current increases.* Proceed to block 2.

15. *Your answer to the question in block 26 is **B**. This answer is wrong. A choke coil is an inductor. It opposes any change in the current flowing through it.* Proceed to block 20.

16. *The correct answer to the question in block 23 is **B**. A capacitive-input filter has a higher output voltage, but its regulation is not as good as a choke-input filter.* Here is your next question.

The components used in a sense circuit are
A. capacitors. (Proceed to block 22.)
B. resistors. (Proceed to block 26.)

17. *The correct answer to the question in block 1 is* **B**. *The rms voltage is given. The peak voltage is obtained as follows:*

$$1.414 \times E \text{ (rms)} = E \text{ (peak)}$$
$$1.414 \times 115 = 162 \text{ volts (peak voltage)}$$

The filter capacitor charges to the peak voltage, and this is the approximate output voltage of the supply. Here is your next question.
When you increase the base current of a bipolar transistor, the collector current
A. increases. (Proceed to block 2.)
B. decreases. (Proceed to block 14.)

18. *The correct answer to the question in block 8 is* **B**. *You can connect two NPN transistors or two PNP transistors to make a Darlington amplifier. Also you can purchase a Darlington amplifier with both transistors in the same case.* Here is your next question.
In the circuit of Fig. 6-20, increasing the base current will cause the voltage at point *a* to become
A. more positive. (Proceed to block 5.)
B. less positive. (Proceed to block 23.)

19. *The correct answer to the question in block 7 is* **A**. *The filter capacitors store energy, then return the energy to the circuit. The subject of capacitors was discussed in Chapter 2.* Here is your next question.
To increase the peak inverse rating of rectifier diodes,
A. connect them in series. (Proceed to block 27.)
B. connect them in parallel. (Proceed to block 4.)

20. *The correct answer to the question in block 26 is* **A**. *Figure 6-14*

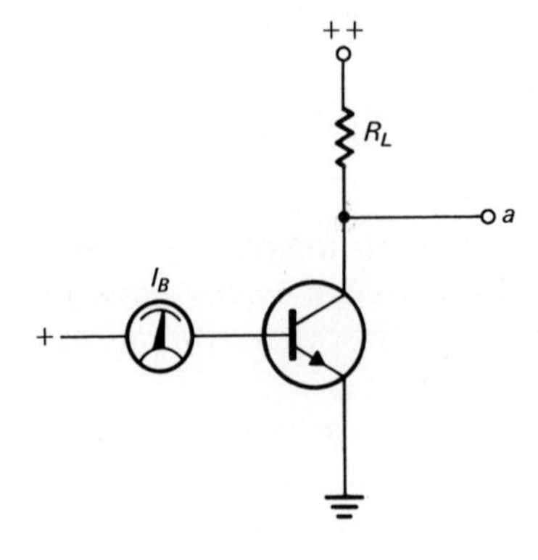

Fig. 6-20 **This is the circuit for the question in block 18.**

shows how a zener diode is connected into a circuit to get a fixed value of reference voltage. A neon tube can be used to obtain higher values of reference voltage. Here is your next question.
A circuit that converts ac to dc is called
A. an inverter. (Proceed to block 11.)
B. a rectifier. (Proceed to block 6.)

21. *Your answer to the question in block 27 is* **A**. *This answer is wrong. Study the action of the bridge rectifier as shown in Fig. 6-6, then* proceed to block 8.

22. *Your answer to the question in block 16 is* **A**. *This answer is wrong. Figure 6-13 shows a sense circuit. Study this circuit, then* proceed to block 26.

23. *The correct answer to the question in block 18 is* **B**. *The collector current increases when the base voltage increases. This makes a greater voltage drop across R_L, and R_L becomes less positive.* Here is your next question.
Which of these filters would result in a higher output voltage?
A. Choke-input filter. (Proceed to block 12.)
B. Capacitive-input filter. (Proceed to block 16.)

24. *The correct answer to the question in block 6 is* **A**. *A Darlington amplifier is a type of power amplifier. The series regulator circuit normally uses a power amplifier.* Here is your next question.
Could the regulator circuit of a power supply be made with vacuum tubes? (Proceed to block 28.)

25. *Your answer to the question in block 6 is* **B**. *This answer is wrong. A thyristor is used as a very fast acting switch, but it cannot be used as a series regulator.* Proceed to block 24.

26. *The correct answer to the question in block 16 is* **B**. *A sense circuit uses two or more resistors in series across the power-supply output. Figure 6-16 shows an electronic regulator circuit. The sense circuit is made up of R_3, R_4, and R_5.* Here is your next question.
Which of the following components may be used to produce a fixed value of reference voltage?
A. A zener diode. (Proceed to block 20.)
B. A choke coil. (Proceed to block 15.)

27. *The correct answer to the question in block 19 is* **A**. *When diodes are connected in series, their peak inverse voltage ratings add. When they are connected in parallel, their forward-current ratings add.* Here is your next question.
A bridge rectifier circuit is
A. a half-wave rectifier. (Proceed to block 21.)
B. a full-wave rectifier. (Proceed to block 8.)

28. *The correct answer to the question in block 24 is yes. A power amplifier tube can function as a regulator.*
You have now completed the programmed questions. The next step is to put some of these ideas to work in laboratory experiments. Proceed to the Experiment section of this chapter.

EXPERIMENT

(The experiment described in this section may be performed on the circuit board described in Appendix C or on a similar laboratory setup.)

Purpose This experiment demonstrates the operation of a voltage doubler and a bridge rectifier. Also it shows how to measure power-supply regulation.

Theory An important rating of a supply is its *percent regulation*. Mathematically,

$$\text{Percent regulation} = \frac{\text{no-load voltage} - \text{full-load voltage}}{\text{no-load voltage}} \times 100$$

where *no-load voltage* = the power supply voltage when it is not delivering current to an external circuit
full-load voltage = the power supply voltage when it is delivering its maximum rated current to an external circuit

If the output voltage of a supply does not change when it is delivering a full-load current, then it has 0 percent regulation. A drop in power-supply voltage when the supply is delivering its full-load current means that its percent regulation is greater than 0. Thus a low percent regulation is desirable.

Test Setups Figures 6-21 and 6-22 show the circuits to be used in this experiment.

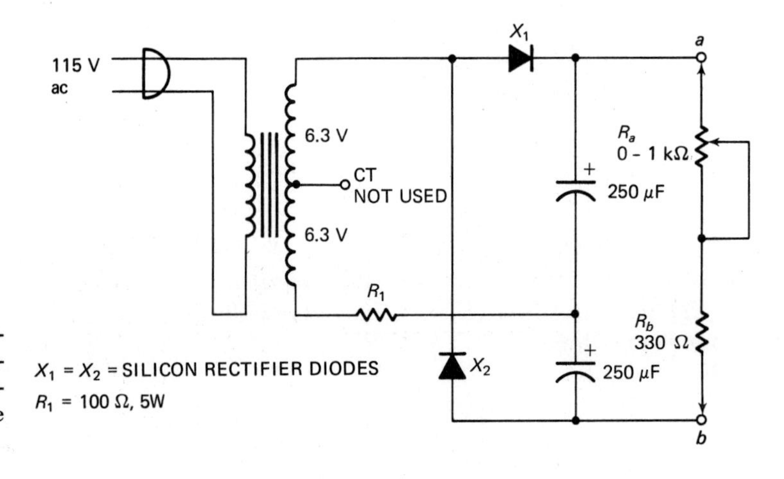

Fig. 6-21 Full-wave voltage doubler schematic for Part I of the experiment. Resistor R_1 is a surge-limiting resistor to protect the diodes.

Procedure

PART I

step one Wire the voltage-doubler circuit as shown in Fig. 6-21.

step two Measure the rms value of ac voltage E_s across the transformer secondary and record the value.

E_s = ________________ volts

step three Determine the peak value of the voltage E_{pk} measured in Step 2.

E_{pk} = peak voltage = $1.414 \times E_s$ = ________________ volts

step four Double the value of peak voltage E_{pk} obtained in Step 3.

$2 \times E_{pk}$ = ________________ volts

This is the value of dc voltage that should be obtained at the output of the voltage doubler circuit.

step five With no-load resistance connected, measure the dc output voltage E_0 of the supply. This is the voltage between points *a* and *b*. Record the value.

E_0 = ________________ volts

Does this equal the value determined in Step 4? Can you explain the reasons for the difference? (Note: This will be discussed in the conclusion of this experiment.)

step six Place the 1-kilohm rheostat R_a and 330-ohm fixed resistor R_b across the output. Adjust the load current to 150 milliamperes. If a milliammeter is not available, set the current by measuring the voltage across the 330-ohm resistor. The current will be 150 milliamperes (0.150 amperes) when the voltage across R_b equals 45 volts dc. This is determined by Ohm's law:

$$\begin{aligned} E &= I \times R \\ &= 0.150 \times 330 \\ &= 49.5 \text{ volts} \end{aligned}$$

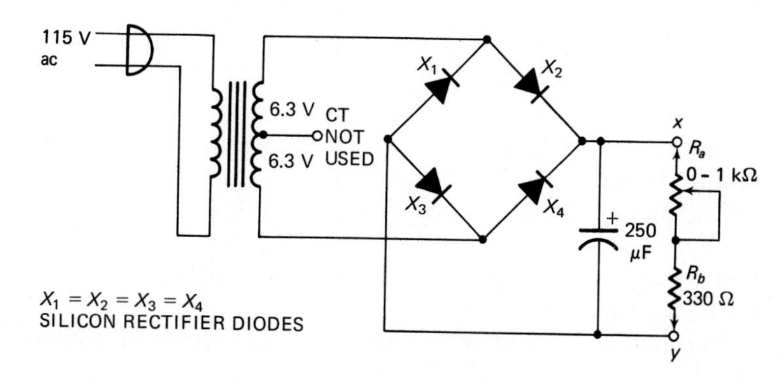

Fig. 6-22 Bridge rectifier schematic for Part II of the experiment.

Measure the output voltage with the 150-milliampere current flowing. Call this voltage E_0'.

$E_0' =$ ________________ volts

step seven Using the values determined in Steps 5 and 6, determine the power-supply regulation.

$$\text{Percent regulation} = \frac{E_0 - E_0'}{E_0} \times 100 = \text{\_\_\_\_\_\_\_\_\_\_ percent}$$

This calculation is based on the assumption that 150 milliamperes is the maximum full load for the supply. This value may be too small for your supply, but it is used as an example. In actual practice the manufacturer or designer may define the full-load rating of the supply.

PART II

step one Wire the bridge rectifier circuit as shown in the schematic, Fig. 6-22.

step two With no-load resistance connected, measure the output voltage E_0 of the supply and record the value.

$E_0 =$ ________________ volts

step three Connect R_a and R_b between points x and y across the output. Adjust the load current to 150 milliamperes as you did in Part I. Measure and record the output voltage E_0' with a 150-milliampere load current flowing.

$E_0' =$ ________________ volts

step four Using the values measured in Steps 2 and 3, determine the power-supply regulation:

$$\text{Percent regulation} = \frac{E_0 - E_0'}{E_0} \times 100 = \text{\_\_\_\_\_\_\_\_\_\_ percent}$$

As before, this calculation is based on the assumption that the full-load current of the power supply is 150 milliamperes.

step five Which of the two supplies had the better regulation? (This will be discussed in the conclusion.)

Conclusion The regulation of a voltage-doubler rectifier circuit is not as good as that of a bridge rectifier. In fact, the voltage doubler has the poorest regulation of any of the rectifier circuits you studied in this chapter.

The actual output voltage of a voltage doubler is often less than the calculated value. The surge-limiting resistor prevents the capacitor from charging to the true value. (Also a meter with a low ohms-per-volt rating may lower the terminal voltage.)

Because of its poor regulation, the voltage-doubler circuit is used only in circuits where the load current is fairly steady. It is not often used as a rectifier in a supply that has an electronic voltage regulator. If an electronic regulator is needed, it is assumed that a steady output voltage is needed, and a nondoubling full-wave rectifier circuit is generally used.

SELF-TEST WITH ANSWERS

(Answers with discussions are given at the end of the chapter.)

1. Conversion of dc power to ac power is accomplished with (*a*) a transformer; (*b*) a rectifier; (*c*) a converter; (*d*) an inverter.
2. Three diodes are connected in series. The diodes have the following ratings:

Diode	PIV	Forward Current
1	1000 volts	0.25 amperes
2	1000 volts	0.1 ampere
3	500 volts	0.5 ampere

   What is the maximum current that can be passed through the series combination? (*a*) 0.25 ampere; (*b*) 0.1 ampere; (*c*) 0.5 ampere; (*d*) 0.85 ampere.
3. Regarding the three diodes in Question 2, what is the PIV rating of the series combination? (*a*) 1000 volts; (*b*) 500 volts; (*c*) 833 volts; (*d*) 2500 volts.
4. In which of the following power-supply rectifier circuits is a voltage-regulation circuit least likely to be used? (*a*) Bridge; (*b*) Full-wave; (*c*) Voltage-doubler.
5. Which of the following components is used to obtain a reference voltage in a regulated supply? (*a*) Zener diode; (*b*) Tunnel diode; (*c*) Darlington amplifier; (*d*) Variable resistor.
6. Which of the following components is used in sense circuits for power-supply regulators? (*a*) Zener diode; (*b*) Resistor; (*c*) Power transistor; (*d*) Inductor.
7. Which of the following is the correct equation for finding percent load regulation of a power supply?

(*a*) $$\text{Percent regulation} = \frac{\text{no-load voltage} - \text{full-load voltage}}{\text{no-load voltage}} \times 100$$

(*b*) $$\text{Percent regulation} = \frac{\text{full-load voltage} - \text{no-load voltage}}{\text{full-load voltage}}$$

8. Which of the following components is used as a series regulator? (*a*) Zener diode; (*b*) Resistor; (*c*) Power transistor; (*d*) Inductor.
9. Which of the following components is used as a voltage reference? (*a*) Zener diode; (*b*) Resistor; (*c*) Power transistor; (*d*) Inductor.
10. Which of the following is normally considered to be better? (*a*) A power supply with a 100 percent regulation; (*b*) A power supply with a 0 percent regulation.

ANSWERS TO SELF-TEST

1. (*d*)—Do not confuse the inverter with the rectifier, which converts ac to dc.
2. (*b*)—If the current is larger than 0.1 ampere, diode 2 will burn out.
3. (*d*)—The PIV rating is the sum of the ratings of each diode.
4. (*c*)—Voltage doublers have relatively poor regulation. If regulation is needed, it is best to start with a rectifier circuit that has good regulation. This does not mean that regulator circuits are *never* used with voltage-doubler circuits.
5. (*a*)—Described in Chapter 2.
6. (*b*)—A variable resistor may be used to set the sense voltage to a desired value.
7. (*a*)—Remember that a low percent regulation is desirable. Also the equation in (*b*) would give a negative number, since the no-load voltage is always greater than the full-load voltage.
8. (*c*)—Power transistors may be connected in parallel to get a larger current rating, or a Darlington amplifier may be used.
9. (*a*)—Zener diodes are used to obtain a fixed value of voltage in many circuits.
10. (*b*)—The best possible regulation would occur if the percent regulation were 0 percent.

what is bias and how is it obtained? 7

INTRODUCTION

Suppose you are driving a car at a speed of 30 miles per hour. For safety reasons you should be able to go much faster so that you can pass another car or get out of the way of trouble. Likewise, you should be able to stop the car. Therefore 30 miles per hour is your cruising speed.

In special cases it is desirable to run a car at almost full speed at all times. An example is a race car. However, this is the exception.

Just as you do not expect to be operating an automobile at its full speed at all times, you should not expect to constantly operate a vacuum tube or transistor at full maximum-output current. If you did, you would not be able to increase the output current when the input signal increases. For this reason amplifiers are usually operated in such a way that their no-signal output current is less than maximum. To accomplish this the control electrode of the amplifier is operated with a dc voltage called the *bias*. The bias voltage sets the dc output current of the amplifying component when there is no ac signal input.

In this chapter you will study how the bias voltage is obtained.

You will be able to answer these questions after studying this chapter.

- ☐ **What are the polarities of bias voltages?**
- ☐ **How is the output current controlled?**
- ☐ **How is bias obtained in tube circuits?**
- ☐ **What is contact bias?**
- ☐ **What is automatic bias?**
- ☐ **How is bias obtained in bipolar transistor circuits?**
- ☐ **How are field-effect transistors biased?**

INSTRUCTION

What Are the Polarities of Bias Voltages?

Some of the amplifying devices require a positive bias voltage on the control electrode, while others require a negative bias voltage. When you are looking for troubles in an electronic circuit, one of the important methods that you will use is to measure the bias voltage. It will be necessary for you to know if this voltage should be positive or negative before you can take its measurement.

You should make it a point to memorize the polarities of the voltages on the various electrodes of the amplifying devices. To make this easier we have summarized the typical voltages in Fig. 7-1. All the amplifying devices that require a positive voltage on the control electrode are shown in the first column, and the ones that require a negative voltage on the control electrode are shown in the second column. In the same illustration we have shown the polarity of the voltage on the control electrode, the output electrode, and the input (cathode, emitter, source) electrode.

The voltage on the input electrode is always considered to be zero with reference to the other voltages on the electrodes. For some devices, such as the NPN bipolar transistor, you will note a positive sign on the base and a double positive sign on the collector. This means that the collector voltage and the base voltage are both positive, but the collector is more positive than the base. The same relationship exists for the enhancement-type field-effect transistors.

Summary

1. Tubes and transistors are not operated with continuous maximum dc output current.
2. The dc bias voltage on the grid of a vacuum tube sets the value of dc plate current.
3. The dc plate current will be less than the maximum possible value when there is a dc grid bias.
4. The dc bias current flowing in the base circuit of a bipolar transistor sets the value of the dc collector current.
5. The dc collector current will be less than the maximum possible value when there is a normal dc base bias current.
6. The dc bias voltage on the gate of an FET sets the value of dc drain current.
7. The dc drain current of an FET will be less than the maximum possible value when there is a normal dc gate bias voltage.
8. Since the dc output current is less than the maximum possible value, it can be increased and decreased by a signal.

How Is the Output Current Controlled?

You have learned that the bias voltage or current sets the no-signal value of dc output current. The output current that flows when the control electrode

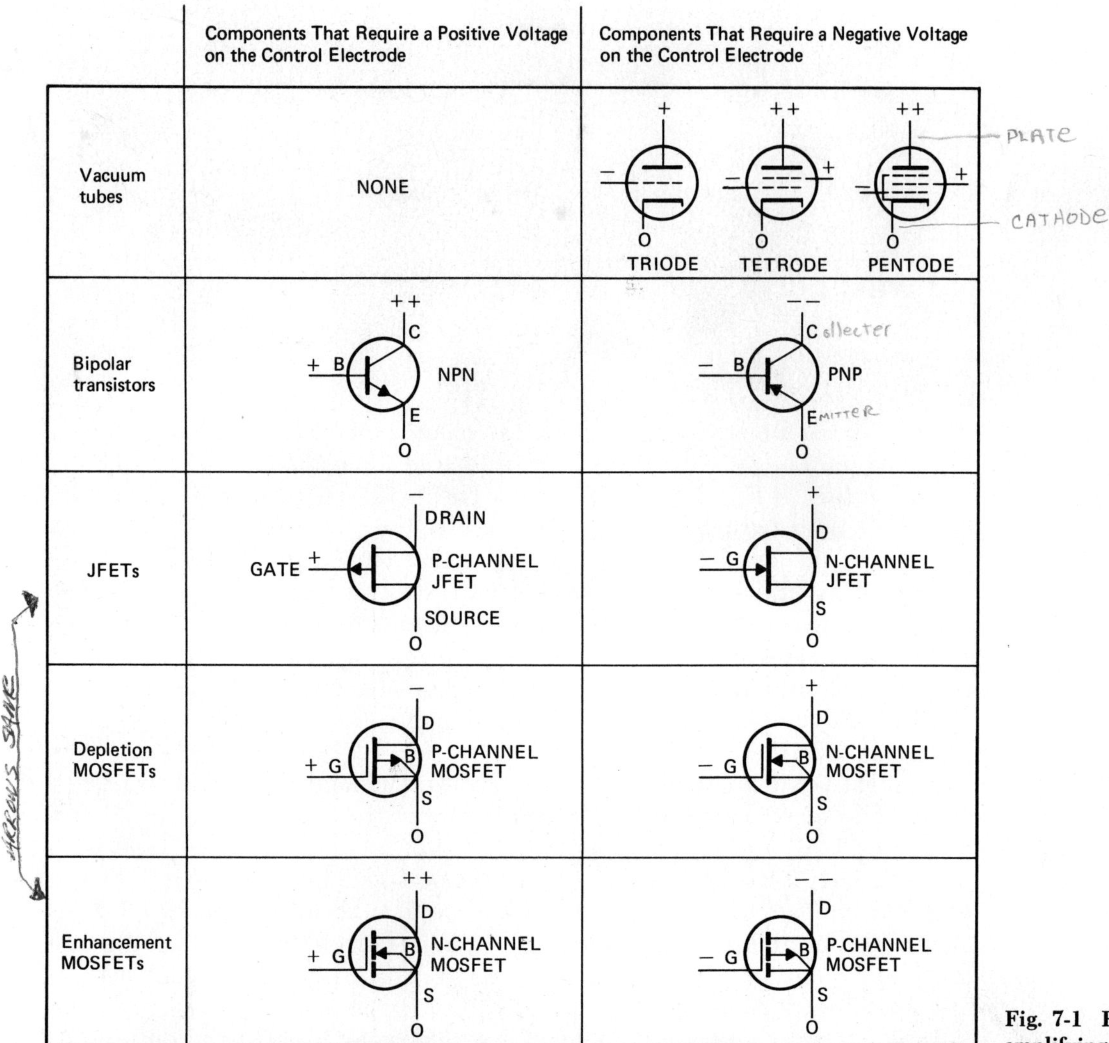

Fig. 7-1 Polarities of voltages on amplifying components.

is at the no-signal bias voltage or current is called the *idling current.* The effect of changing the voltage on the control electrode will now be discussed.

How Does Grid Voltage Control Plate Current?

Figure 7-2 shows how the bias voltage on a vacuum-tube grid determines the plate current and how changing this bias voltage will change the plate current. In Fig. 7-2*a* there is a negative grid bias voltage set by resistor *R*. In this circuit resistor *R* is connected as a *potentiometer.* A potentiometer is a variable resistor connected in such a way that it controls voltage. In this circuit resistor *R* controls the grid voltage.

Since *R* is connected directly across the battery, current flows through it at all times. The voltage drop across *R* is marked with a minus or plus sign. Notice that the plus sign is at the ground point, so it is at the cathode side of the

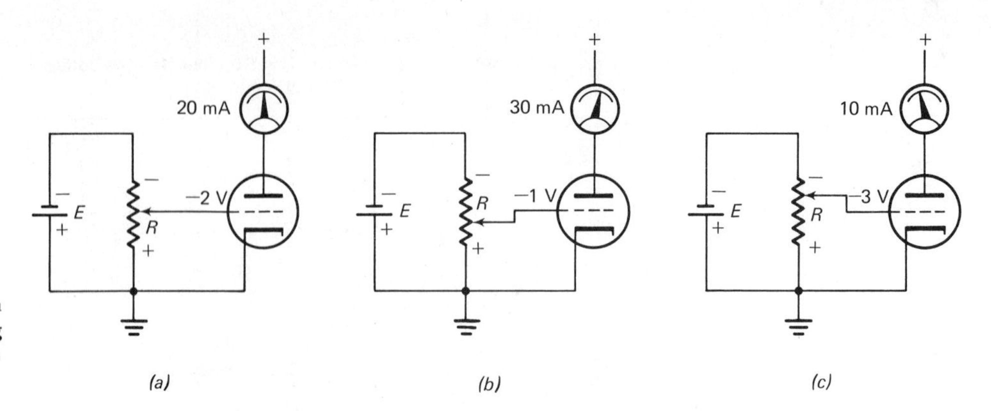

Fig. 7-2 Plate current varies when grid voltage changes. (*a*) Idling current; (*b*) less negative grid; (*c*) more negative grid.

tube. Moving the arm of the resistor can change the grid voltage from 0 volts to a maximum negative voltage with respect to the cathode.

The setting in Fig. 7-2*a* is −2 volts. This is the no-signal bias voltage for the tube. With a grid bias voltage of −2 volts, the plate current is 20 milliamperes. (In the circuit of Fig. 7-2*a* the idling current is 20 milliamperes.)

In Fig. 7-2*b* the arm of the resistor is moved toward a less negative point of the voltage across *R*. At this setting the grid voltage is − 1 volt. Making the grid voltage less negative causes the plate current to increase. Now the plate current is 30 milliamperes. Note that a change of 1 volt on the grid produced a 10-milliampere change of current on the plate.

In Fig. 7-2*c* the arm of the variable resistor has been moved toward the more negative side of the voltage across *R*. The voltage on the grid is now −3 volts. The more negative grid voltage reduces the plate current to 10 milliamperes.

In this circuit we have presumed that the no-signal bias voltage is −2 volts. This bias voltage sets the idling plate current at 20 milliamperes. This is neither the maximum nor the minimum value of current that can flow through the tube. Increasing and decreasing the grid voltage permits the plate current to increase or decrease as desired.

In practice the voltage on the grid is varied by an input signal rather than with a variable resistor. However, the principle of operation is the same.

How Does Bias Current Affect Collector Current?

Figure 7-3 shows the relationship between the base current of a bipolar NPN transistor and its collector current. The base current flows through variable resistor *R* to a positive supply voltage. The resistor is connected as a rheostat, so its purpose is to control the amount of current that flows in the base circuit.

In Fig. 7-3*a* resistor *R* is adjusted so that the base current is 2 milliamperes. This is the no-signal bias current, and it sets the collector current at 20 milliamperes.

In Fig. 7-3*b* the arm of resistor *R* has been moved so that it offers less resistance to base-current flow. The base current has now increased to 3 milliamperes, and the collector current has now increased to 30 milliamperes. It

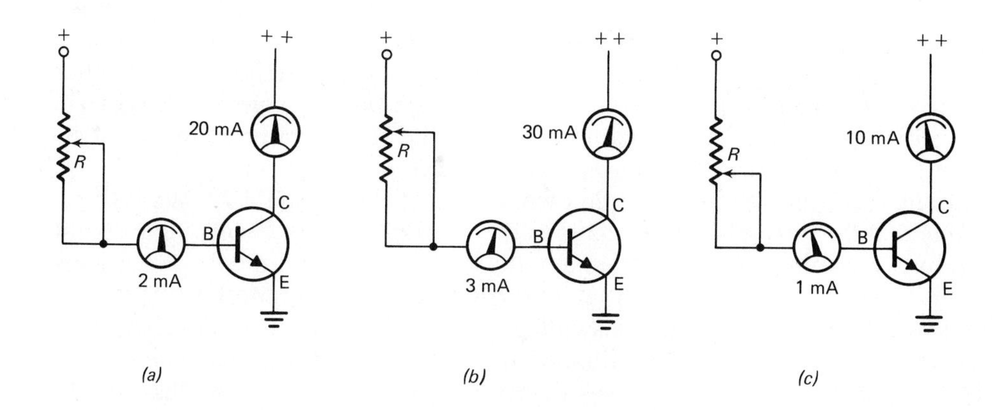

Fig. 7-3 Collector current varies when base current varies. (*a*) Idling current; (*b*) increased base current; (*c*) reduced base current.

takes a change of 1 milliampere in base current to produce a 10-milliampere change in the collector current. The transistor is said to *amplify* the current. When used this way the word *amplify* means to produce a large change in collector current by making a small change in base current.

In Fig. 7-3*c* the variable resistor has been adjusted so that it increases the resistance of the base circuit. This reduces the base current to 1 milliampere. Now the collector current has dropped to 10 milliamperes.

Let us summarize the action in Fig. 7-3. We have assumed that the no-signal base bias current is 2 milliamperes. This bias current is permitted to increase or decrease above and below 2 milliamperes. As the base current is changed, the collector current is also changed.

When the no-signal base bias current is 2 milliamperes the collector current is 20 milliamperes. This is neither the maximum nor the minimum of current that can flow in the transistor. This is important because changing the base current will permit the collector current to either increase or decrease in step with the changes in the base current.

Instead of using a variable resistor to change the base current, we could apply an input signal to the base. The signal voltage will either add to the no-signal bias or subtract from the no-signal bias. Thus with no input signal the base current will be 2 milliamperes, and the signal would cause the base current to increase and decrease much the same way as changing the resistance of *R* varies the base current in Fig. 7-3.

How Does Bias Voltage Affect Drain Current?

The operation of JFETs and depletion MOSFETs is very similar to the operation of vacuum tubes.

When we use the term *FET* we are referring to both JFETs and MOSFETs. We will use the terms *JFET* or *MOSFET* when we want to refer to a specific device. Both vacuum tubes and FETs use a voltage (on their control electrodes) to control the amount of current that flows through them.

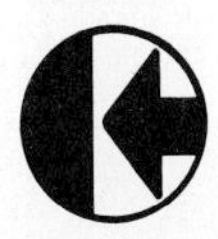

However, remember that all vacuum tubes require a negative voltage on their grid, but some types of FETs require a positive voltage on the gate. In other words, the polarities of the bias voltage may not be the same for all types of amplifying devices. Refer again to Fig. 7-1 and refresh your memory on polarities of FET bias voltages.

Figure 7-4 shows one example of bias on a P-channel JFET. We will assume that the no-signal bias voltage in this application is +3 volts. This is set up by a variable resistor *R* across a battery. In this case the voltage across *R* becomes more positive as you move further away from ground.

In Fig. 7-4*a* the no-signal bias voltage for the JFET is set at +3 volts. This sets the drain current at 20 milliamperes. In Fig. 7-4*b* the bias voltage has been reduced, and this increases the drain current. Remember that in a P-channel JFET the more positive the gate voltage, the lower the drain current.

In Fig. 7-4*c* the bias voltage has been increased to 4 volts. This reduces the drain current to 10 milliamperes. Thus, changing the gate voltage causes changes in the drain current. Like the tube, a small change in control voltage causes a relatively large change in current through the device.

In all the devices we have shown, the bias current or voltage sets the amount of current that flows through the device. The output current has been always set in such a way that changing the bias voltage can either increase or decrease the output current.

Summary

1. In a vacuum tube a small change in grid voltage causes a large change in plate current.
2. In a bipolar transistor a small change in base current causes a large change in collector current.
3. In an FET a small change in gate voltage causes a large change in drain current.
4. Vacuum tubes and FETs are voltage-operated devices. Bipolar transistors are current-operated devices.

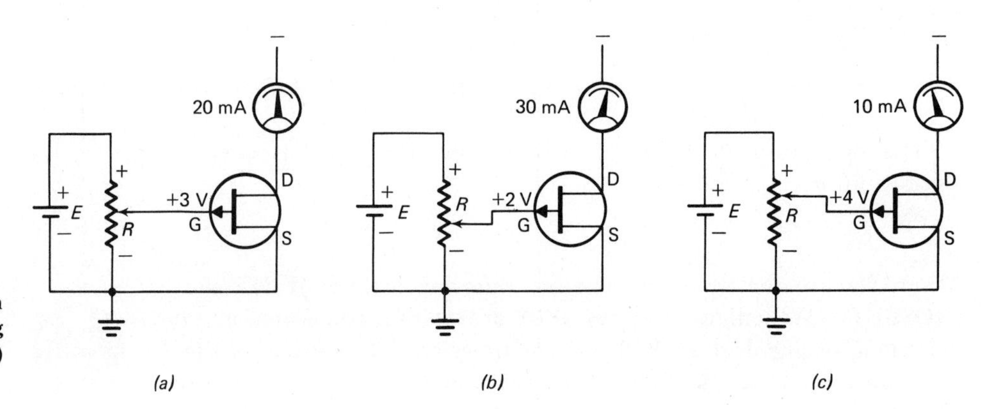

Fig. 7-4 Drain current varies when gate voltage changes. (*a*) Idling current; (*b*) less positive gate; (*c*) more positive gate.

How Is Bias Obtained in Tube Circuits?

There are a limited number of methods used to obtain the dc bias voltage for tubes and transistors. It is very important to know these methods because they are often the key to knowing how a circuit works.

When you are working on a circuit for the first time a good place to start is the dc operating voltages. Trace the dc paths in the circuit to see how the amplifying device is connected to the supply. Also, check the method of getting the bias voltages. This will aid in knowing exactly what the circuit does.

In this section you will learn the types of bias circuits. Do not skip over the material on tube bias. Since the FET is similar in many ways to the tube, you will find that their bias circuits are similar. (Of course, the values of components and voltages will be different, but the principle is the same.)

What Is C Battery Bias?

In the early days of electronics, batteries were used for operating the vacuum tubes. A separate battery was used to operate the filaments, and it was called an *A battery.* Another battery was used to operate the plate circuit, and it was called the *B battery.* We still call the main power-supply voltage the *B supply* because of this.

A third battery was used to obtain the grid bias voltage, and this was called the *C battery.* Later, when the power supply was obtained from the commercial power lines, the use of batteries was discontinued. The grid bias voltage was obtained by other means. However, in portable equipment the batteries are still used.

Figure 7-5 shows a battery bias circuit. The input signal to the stage is delivered through coupling capacitor *C* and developed across grid resistor *R*. The *C* battery, with its negative side toward the grid, provides the grid bias voltage.

Since there is normally no current flow in the grid circuit of a tube, the C battery did not have to be able to supply current. For this reason the battery could be quite small and simple. In practice it would last a very long time.

In some electronic devices, such as instruments, C batteries are used even though the filament and plate supplies are obtained from a power supply that operates from the power lines. The reason for this is that the C battery can be

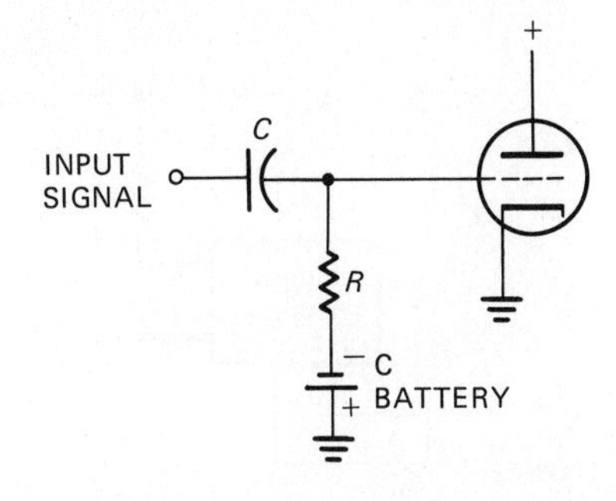

Fig. 7-5 Battery bias.

used to supply a very pure dc operating voltage (with no ripple), and this is important for certain applications. Other types of bias, as you will see presently, often require filtering. Even with the filtering the bias voltage is not a pure dc.

How Is Power-Supply Bias Obtained?

A separate power supply can be used to obtain the grid bias. This is illustrated in Fig. 7-6. Again the input signal passes through the coupling capacitor C, and the signal is developed across resistor R. The power supply operates from a commercial power line, and it supplies a negative voltage for grid bias.

This type of bias has been used extensively in large transmitters for broadcast, and it is also used to some extent in high-fidelity audio equipment. If the power supply is well designed, the amount of undesirable ac ripple is almost nonexistent. However, the disadvantage of the system is that a separate power supply is expensive compared with other methods of obtaining grid bias.

What Is Contact Bias?

Figure 7-7 shows one of the simplest methods of obtaining bias for operating a vacuum tube. To understand its operation you must remember that the control grid of the tube is actually a grid of fine wires between the cathode and the plate. Most of the electrons that go from the cathode to the plate pass through the holes in the grid mesh. However, a few of the electrons will actually strike the grid. There must be some way for these electrons to get off the grid and get back to the cathode.

As shown in Fig. 7-7, resistor R provides a path for the electrons to pass from the grid to ground and eventually back to the cathode. In this case the resistance is 4.7 megohms. The number of electrons that flow through this resistor is extremely small. As a matter of fact, the current may be less than 0.5 microampere.

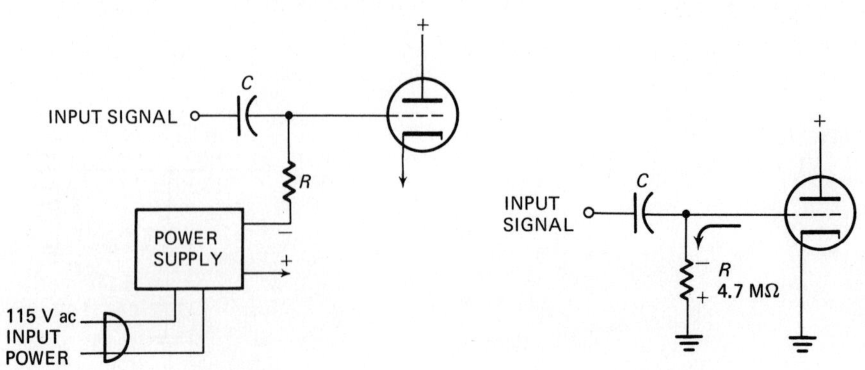

Fig. 7-6 Power-supply bias.

Fig. 7-7 Contact bias.

Assume for a moment that the actual grid current in the circuit of Fig. 7-7 is 0.25 microampere. You can calculate the grid bias voltage with a simple Ohm's law equation.

First, you must convert the current in microamperes to a value of current in amperes. To do this you move the decimal place to the left six places.

$$\tfrac{1}{4} \text{ microampere} = 0.25 \text{ microampere} = 0.00000025 \text{ ampere}$$

Next, convert the resistance value in megohms to a value of resistance in ohms. To do this you move the decimal point to the right six places.

$$4.7 \text{ megohms} = 4{,}700{,}000 \text{ ohms}$$

Then, by Ohm's law,

$$\begin{aligned}\text{Voltage in volts} &= \text{current in amperes} \times \text{resistance in ohms}\\ &= 0.000\,000\,25 \times 4{,}700{,}000\\ &= 1.175 \text{ volts}\end{aligned}$$

The grid bias voltage in our example is 1.175 volts.

Contact bias cannot be used for every type of vacuum tube. It is required that the tube have an extremely high *gain*. This is another way of saying that a very small amount of change in grid voltage will produce a large change in plate current. For a tube to have a large gain it is necessary for the grid to be located very close to the cathode. At that point the grid will intercept a larger number of electrons.

What Is Automatic Bias?

One of the most popular methods of biasing a tube is shown in Fig. 7-8. It is called *cathode bias* or *automatic bias*. The plate current of a tube must flow through the cathode. The arrow at R_K shows that the current is flowing toward the cathode, making the positive side of the voltage drop across R_K at the cathode. The grid is actually at dc ground potential. (Grid resistor R_G is 75 kilohms, which is relatively small, and so the presence of contact bias can be completely ignored.)

With the grid at ground potential and the cathode at a positive potential,

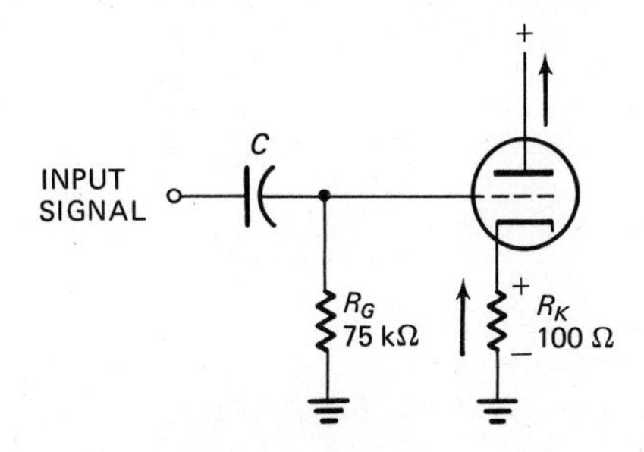

Fig. 7-8 Cathode bias.

you can say that the cathode is positive with respect to the grid. Another way of saying this is that the grid is negative with respect to the cathode. This means that the proper bias has been obtained for the grid of the tube.

If you were working in a circuit of this type and wanted to know if the tube was operating properly, the best method would be to measure the voltage across the cathode resistor. If this voltage is the proper value, then you know that the tube has the correct bias voltage.

You also know that plate current must be flowing in the circuit.

What Is Grid-Leak Bias?

The type of bias shown in Fig. 7-9 is called *grid-leak bias*. It is obtained from a part of the input signal, and therefore it is also called *signal bias*. We will discuss this bias with reference to two half cycles of input signal.

In Fig. 7-9*a* the input signal is on the positive half-cycle. This places a positive voltage on the left side of coupling capacitor C. Since unlike charges attract, the negative electrons in the tube will flow from the cathode to the grid (grid current) to the capacitor, as shown by the arrow. These electrons gather on the right side of coupling capacitor C, charging the side negative, as shown.

In Fig. 7-9*b* the input signal is going negative. At this time no current can flow from cathode to grid or from the grid to any other element. Therefore the electrons flow through the grid resistor in a downward direction, as shown by the arrow. The voltage across the resistor R_G is negative at the grid side. This makes the grid negative with respect to the cathode, as needed for proper operation of the tube.

Return for a moment to Fig. 7-9*a*. You should understand that grid current (cathode to grid) does not flow during the complete positive half cycle of input signal. Once capacitor C has become charged, it is necessary for the grid current to flow for only a short part of each positive half cycle to recharge C.

Figure 7-10 shows a serious disadvantage of grid-leak bias. Remember that the bias voltage depends upon the input signal. If the input signal is lost for some reason (due to a trouble in the previous stage), then there will be no bias on the tube. If you operate many types of tubes without a grid bias, the

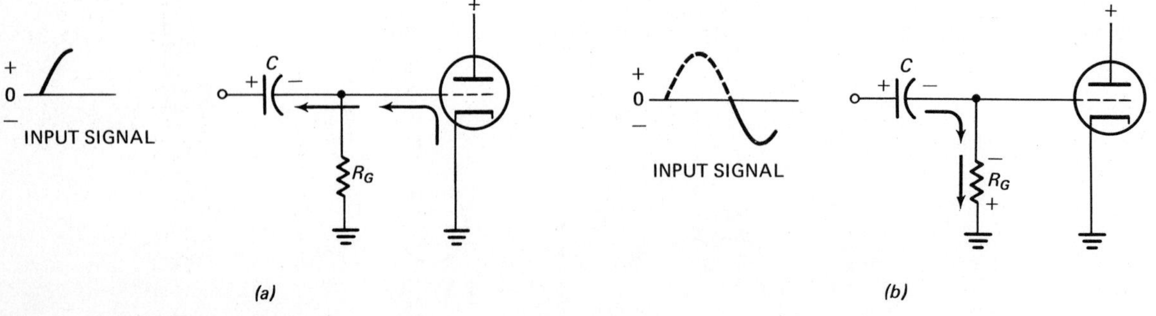

Fig. 7-9 Grid-leak bias. (*a*) Positive half cycle of input signal; (*b*) negative half cycle of input signal.

plate current may become excessively high and may damage the tube. This is illustrated in Fig. 7-10*a*. In some tubes with glass envelopes you may actually see the plate of the tube turning red hot if grid bias is lost.

To prevent this from happening many grid-leak bias circuits have some provision for protecting the tube. One simple way of doing it is to put a C battery in series with the grid resistor. Then the grid bias voltage becomes the voltage across R_G plus the voltage of the battery. If the signal is lost, the battery will keep a bias voltage on the tube to protect it.

Another way to protect the tube is to connect a small cathode resistor into the circuit. This is shown in Fig. 7-10*b*. If the plate current starts to increase to an excessive value because the signal is lost, the voltage drop across the cathode resistor makes the cathode positive with respect to the grid. As you learned when you studied automatic bias, this is the same as making the grid negative with respect to the cathode. The tube is thus protected by the automatic cathode-bias resistor.

Compare the circuit of Fig. 7-10*b* with the one shown in Fig. 7-8. Note that the connections and component layouts of the circuits are identical. This is a very important point to remember: *You cannot learn electronics simply by memorizing circuit connections.* You must also understand how the circuit works. In the above two circuits R_K in Fig. 7-10*b* would normally be much smaller than that of Fig. 7-8.

What Is AVC Bias?

The last type of grid bias to be discussed is shown in Fig. 7-11. In order to explain this type of bias it is necessary to use the block diagram of a simple radio.

The antenna picks up the signal and delivers it to a signal amplifier. The amplifier signal from this stage goes to another signal amplifier, and then to another signal amplifier. The three signal amplifiers all serve the same purpose: to increase the signal voltage. This is necessary because the amount of

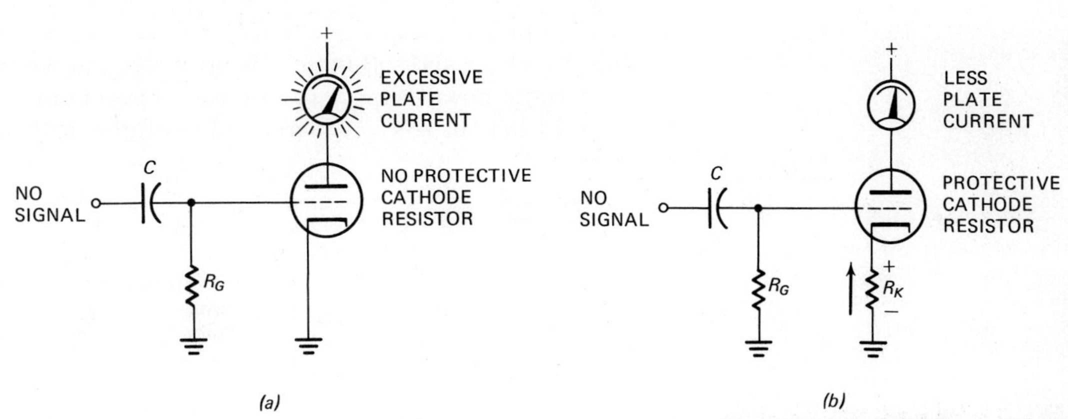

Fig. 7-10 Grid-leak-bias circuit with no signal. (*a*) With no protection; (*b*) with protection.

signal received by the antenna is so very small. A typical value of signal amplitude would be 50 microvolts (μV)—that is, 50 millionths of a volt. Such an extremely small signal cannot be used for loudspeaker operation.

An annoying problem with radio reception is that different stations may be received at radically different signal strengths. This would make it necessary to frequently readjust the volume control. In addition, as in an auto radio, the volume of any one station may vary at different locations. Also a condition of fading may occur on a particular station, causing periodic changes in listening volume.

To help eliminate the problem of the changing degrees of volume, an Automatic-Volume-Control (AVC) circuit is used in radios (and TV sets). Refer again to Fig. 7-11. Some of the output signal from the last signal amplifier is rectified and filtered, much as you rectify and filter the ac voltage from the power line to obtain a dc voltage. The dc voltage in this case is called the *AVC voltage* and is negative. The AVC voltage is fed to the grid of the first signal amplifier and used to control the gain of that amplifier. The more negative the grid voltage, the lower the gain. Conversely, the less negative the grid voltage, the higher the gain.

The AVC voltage is always such that when the signal is weak, the negative AVC voltage is low, and so the gain of the first signal amplifier is high; and when the signal is strong, the negative AVC voltage is high, and so the gain of the first amplifier is low. The gain of the first signal amplifier controls the volume of the sound in the loudspeaker.

Almost all receivers have a circuit similar to the one shown in Fig. 7-11. In television receivers it is not called AVC. Instead, it is called *AGC*, which stands for Automatic Gain Control, since the action controls the strength of the picture as well as the sound.

Summary

1. Knowing the method used to obtain bias for an amplifying device is a clue to understanding the circuit operation.
2. Some of the bias circuits used for tubes will also work for FETs.
3. Battery bias is obtained with a C battery. An A battery is used for filament current, and a B battery is used for plate voltage.
4. A separate power supply can be used for obtaining the dc bias voltage.
5. Contact bias uses a very large value of grid resistor to obtain the bias

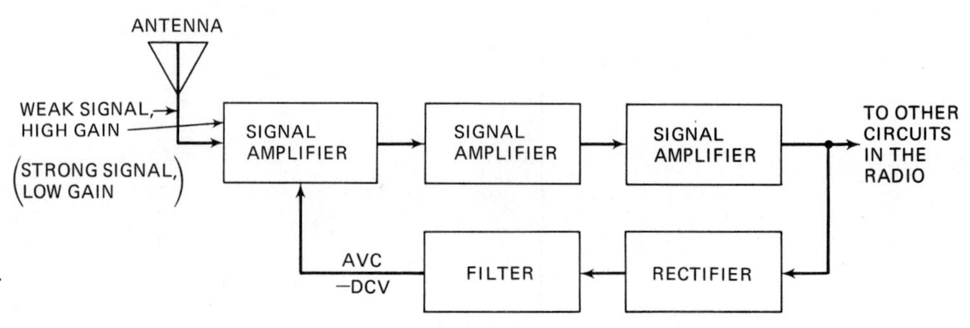

Fig. 7-11 AVC bias in a radio receiver.

voltage. Electrons that strike the grid return to the cathode through this resistor. The voltage drop across the grid resistor is the grid bias.

6. Automatic bias, which is also called *cathode bias,* or *self-bias,* is obtained when the cathode current flows through a cathode resistor.
7. Grid-leak bias, which is also called *signal bias,* is obtained by using the input signal to charge the input coupling capacitor. When the signal goes negative the capacitor discharges through the grid resistor, producing the bias.
8. There are two very important points to remember about grid-leak bias: (*a*) Grid current must flow for a short period of time during each positive half cycle to charge the capacitor. (*b*) Loss of input signal means loss of grid-leak bias.
9. Normally a loss of bias means an excessive plate-current flow. Many types of tubes may be damaged if the bias is lost.
10. An automatic-volume-control circuit is used to bias a radio receiver amplifier. This bias works to keep the output sound constant, even though the received signal strength may vary.

How Is Bias Obtained in Bipolar Transistor Circuits?

Bipolar transistors are different from tubes and field-effect transistors in a number of important ways. Both tubes and FETs are voltage-operated devices. This means that the input signal must be a voltage that controls the plate or the drain current. Furthermore, in the normal operation of tubes and FETs there is usually no current flow in the grid or gate.

Base-to-emitter current *must* flow in a bipolar transistor circuit in order for collector current to flow. In other words, the transistor operates in such a way that the no-signal idling current in the collector circuit is obtained only if there is a current flow in the base-to-emitter circuit.

What Is Simple Bias?

A simple base bias circuit for a bipolar transistor is shown in Fig. 7-12. Although an NPN transistor is shown in the illustration the same basic circuit

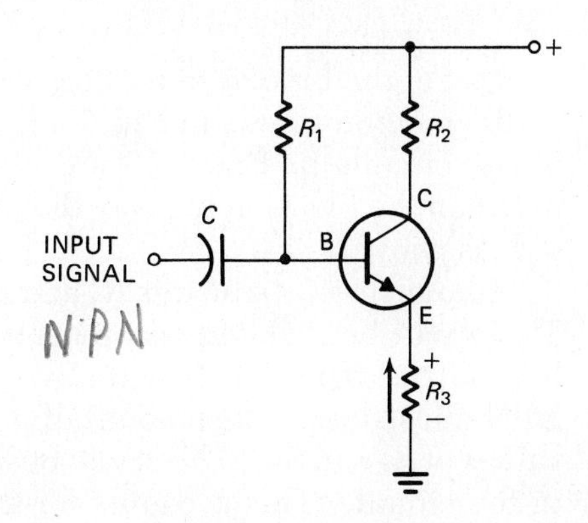

Fig. 7-12 Simple bias.

is used for PNP transistors. The only difference is in the polarity of the applied voltages (negative on both base and collector for PNP). (The same thing is true for all the bipolar transistor circuits discussed in this chapter; that is, NPN transistors are shown, although PNP transistors can also be used if the voltage polarities are reversed.)

The base of the transistor in Fig. 7-12 is connected to the positive supply voltage through R_1. The collector is connected to the positive supply voltage through R_2. A complete path for current exists through R_3, into the emitter of the transistor, and then through both R_1 and R_2. In normal operation the base current which flows through R_1 must be much smaller than the idling collector current. Here, forward bias is provided by the emitter-to base-current through R_3 and R_1.

The input signal through coupling capacitor C increases and decreases the base voltage. This in turn causes the base current to increase and decrease. Varying the base current causes the collector current also to vary, but by an amplified amount.

Resistor R_3 does *not* bias the transistor circuit. It is called an *emitter stabilization resistor.* Its purpose is to protect the transistor from excessive current in the collector circuit. If, for some reason, the temperature of the transistor should increase to a fairly high degree, then the resistance of the transistor would decrease. This could cause the collector current to increase to a very high value. Increasing the collector current would cause more heat in the transistor and cause an additional drop in its resistance. Very quickly the transistor would be ruined by the high collector current. The loss of a transistor due to high temperature, as just described, is called *thermal runaway.*

When R_3 is connected into the emitter circuit any large increase in dc will cause an increase in emitter voltage. In the NPN transistor shown it makes the emitter more positive with respect to ground. In an NPN transistor the emitter must be negative with respect to the base in order for collector current to flow. Making the emitter more positive has the same effect as making the base less positive. This reduces the collector current and prevents thermal runaway. You will study more about the emitter resistor in the chapter on amplifiers (Chapter 13).

What Is Voltage Divider Bias?

Most of the bipolar transistors that you find in amplifier circuits are biased by the method shown in Fig. 7-13. This is called *voltage divider bias.* Both of the circuits in Fig. 7-13 are electrically identical. The only difference between the two is in the way that the circuits are drawn. (An NPN transistor only is shown.)

In both circuits resistors R_1 and R_2 are connected directly across battery E. There will be a voltage drop across each resistor, so that the voltage at the base of the transistor is actually some positive value above ground (for an NPN transistor). The amount of positive voltage at the base depends on the values of R_1 and R_2. The arrow beside R_2 shows the direction of electron flow in the circuit. This accounts for the positive voltage on the base, which in turn causes the needed base-emitter forward-bias current.

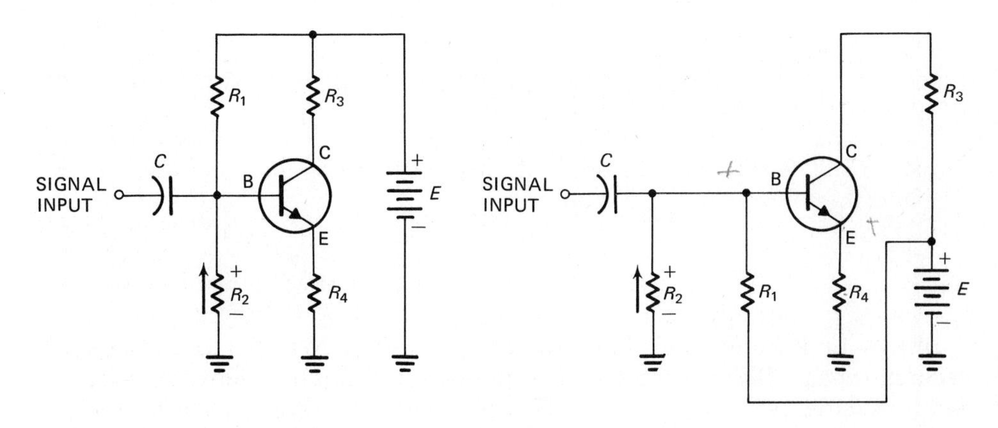

Fig. 7-13 Two ways to draw the same voltage-divider-bias circuit.

Resistor R_3 is a collector load resistor. The signal voltage will be developed across this resistor in an amplifier circuit. Resistor R_4 is the protective resistor, similar to R_3 in Fig. 7-12.

As we said, the input signal delivered through capacitor C increases and decreases the base voltage. This varies the base current and therefore varies the collector current through R_3 in an amplified manner.

How Are Separate Supplies Used to Bias Bipolar Transistors?

In the circuits shown in Figs. 7-12 and 7-13 a single power supply is used to obtain both the base bias and the collector voltage. It is also possible to use two separate supplies for this purpose.

Figure 7-14 shows how an NPN transistor can be powered by using one battery for the base circuit and one for the collector circuit. The input signal is delivered through capacitor C and developed across resistor R_1. Battery E_1 places a positive voltage on the base with respect to the emitter. Therefore base-to-emitter current flows. Resistor R_2 is the emitter stabilization resistor, mentioned earlier.

The collector current is obtained with a positive voltage on the collector of the transistor. This current flows through load resistor R_3. In an amplifier the output signal voltage will also be developed across this resistor.

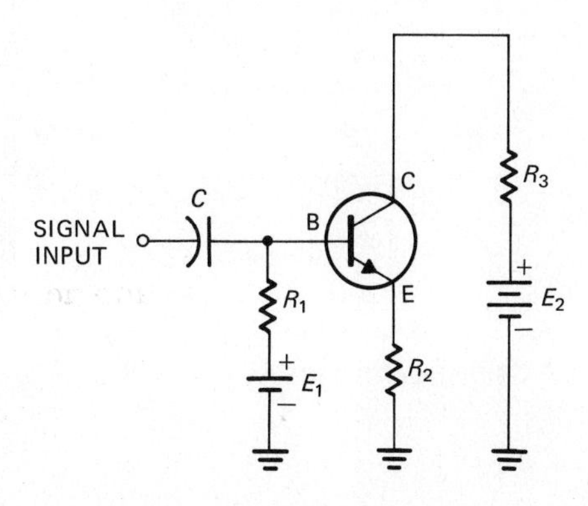

Fig. 7-14 Battery bias for an NPN transistor.

The disadvantage of the circuit in Fig. 7-14 is that two different batteries are needed, so it is not a circuit that is used often. However, you will see this arrangement in some basic hobby circuits.

Two separate power supplies can be used instead of the two batteries shown in Fig. 7-14. As with the batteries, this connection would be expensive and therefore is seldom used.

How Is AVC Bias Used with Bipolar Transistors?

Automatic-volume-control bias can be used with transistors as well as with vacuum tubes. However, there is an important difference between the tube and transistor AVC bias circuits. To understand this difference refer to Fig. 7-15.

Figure 7-15*a* is a graph of the grid voltage of a vacuum tube versus the gain of that tube. When the grid bias voltage is set at −1 volt the gain of the amplifier is at point *a*; and when the bias is set at −2 volts the gain is at point *b*. The important fact about the graph in Fig. 7-15*a* is that as you make the grid bias voltage more negative, the gain decreases.

Figure 7-15*b* shows an example of a graph of the base bias current of a bipolar transistor versus its gain. As the base bias current is increased from 0 to 2 milliamperes, the gain of the transistor increases. When the bias is increased beyond 2 milliamperes the gain of the transistor begins to decrease. This means that the gain of the transistor will increase as you increase the base bias current up to a point. After that point the gain decreases as you continue to increase the base bias current.

If a transistor in a radio is AVC-biased in such a way that a decrease in base bias current causes a decrease in gain, the condition is known as *reverse AVC*. This is shown in the shaded part of the curve in Fig. 7-15*b*. If an increase of bias current causes a decrease of gain, this is called *forward* AVC.

A comparison of tube bias and bias for bipolar transistors can now be made. There are six methods of biasing tubes: *battery bias*, *power-supply bias*, *contact bias*, *cathode bias* (also called *self-bias* or *automatic bias*), *grid-leak bias*, and *AVC* (or *AGC*) *bias*.

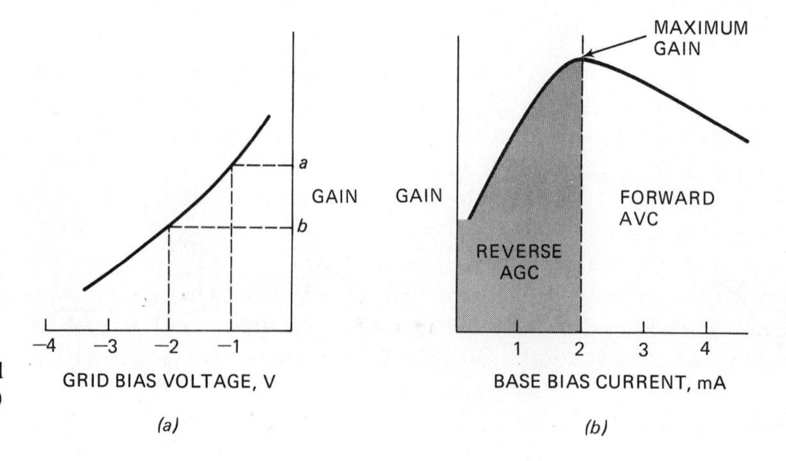

Fig. 7-15 Comparison of bias and gain for (*a*) tube amplifier and (*b*) transistor amplifier.

Transistors can be biased with *battery bias, power-supply bias, simple bias, voltage-divider bias,* and *AVC* (or *AGC*) *bias.* Bipolar transistors are *not* biased by the following methods: contact bias, self-bias, or grid-leak bias.

Summary

1. In most vacuum-tube circuits there is no current flow in the grid circuit.
2. In a bipolar transistor circuit current *must* flow in the base circuit.
3. Separate batteries can be used to obtain base bias and collector voltages.
4. Separate power supplies can be used to obtain base bias and collector voltages.
5. Base bias voltage can be obtained from a single resistor connected between the base and power supply. This is called *simple bias.*
6. A voltage divider can be used to obtain the required base voltage for biasing a transistor.
7. A transistor can be biased with an AVC (or AGC) voltage.
8. If a decrease in AVC base bias current causes a decrease in the gain of the transistor amplifier, this is called *reverse AVC.*
9. If an increase in AVC base bias current causes a decrease in gain of the transistor amplifier, this is called *forward AVC.*
10. Transistors are not biased by the following methods: contact bias, self-bias, or grid-leak bias.
11. There are six methods of biasing tubes and only five methods of biasing a transistor.

How Are Field-Effect Transistors Biased?

Many of the bias circuits that you have studied for tubes and bipolar transistors are also used for field-effect transistors. Important exceptions are grid-leak bias and contact bias. Neither of these bias circuits is used for biasing FETs.

Figure 7-16 shows some examples of N-channel FET bias circuits. The circuit illustrated in Fig. 7-16*a* can be used only with certain field-effect transistors. Here, the FET has no bias, and this system can be used only in circuits where the input signal is very small. Here again, you must not be confused by the layout of the circuit. It looks very much like a contact bias circuit for vacuum tubes. However, remember that contact bias depends upon a small amount of grid-current flow, and in JFETs and MOSFETs no current is presumed to flow in the gate circuit. Remember that all JFETs have a depletion region around the base. This depletion region, illustrated in Fig. 7-17, is present whether or not there is an input signal.

The very small amount of input signal in the circuit of Fig. 7-16*a* reaches the gate of the JFET through coupling capacitor C, and it is developed across the gate resistor R_G. This signal voltage increases and decreases the size of the depletion region around the gate and therefore controls the drain-current flow.

It must be repeated that the input signal must be very small for the kind of

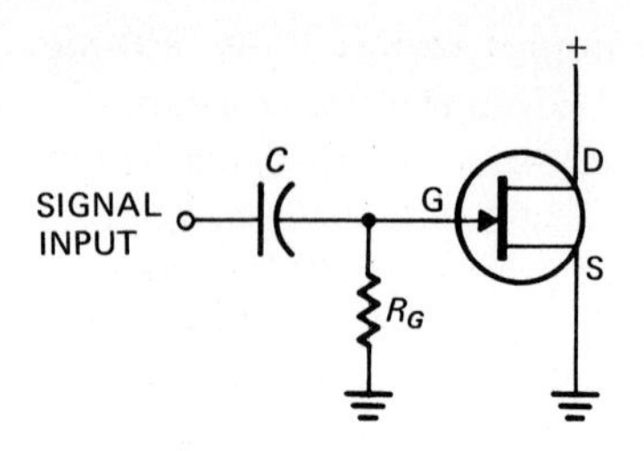

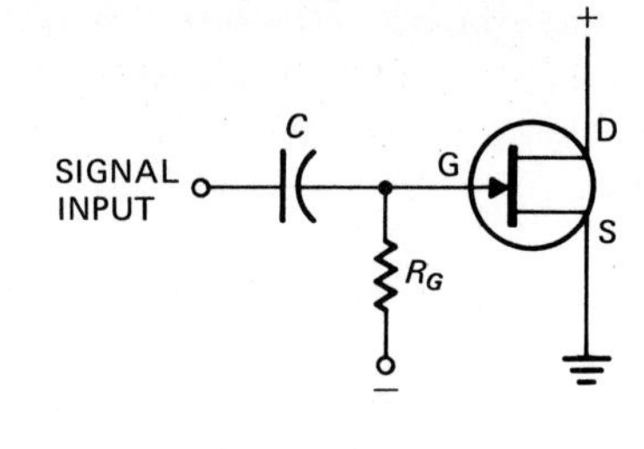

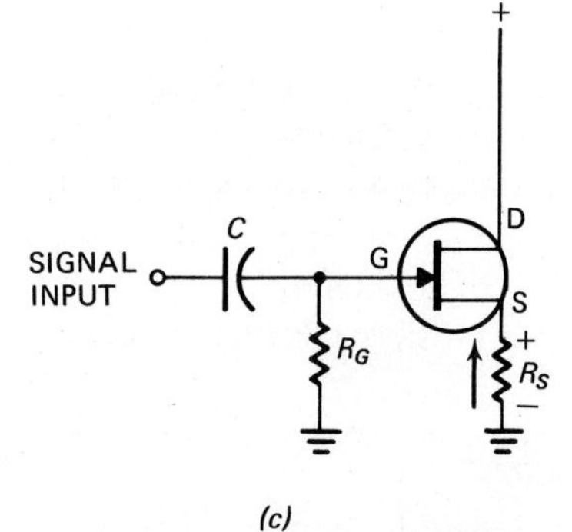

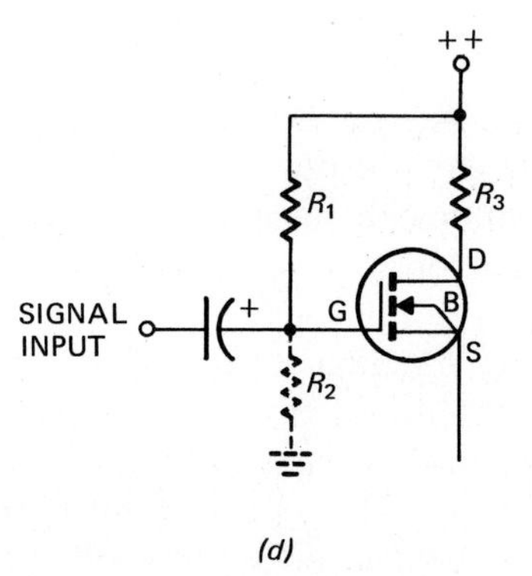

Fig. 7-16 Examples of FET bias circuits. (*a*) No bias; (*b*) battery or power-supply or AGC bias; (*c*) self-bias; (*d*) voltage-divider bias for enhancement-type MOSFET.

bias shown in Fig. 7-16*a*. If the input signal is too large, then signal peaks will reverse-bias the gate junction, and the transistor can be destroyed.

If some vacuum tubes are operated with no bias voltage, they may be damaged because of excessive plate-current flow. If a bipolar transistor is operated with no bias, the transistor is cut off, because it is necessary for base-current to flow in order to get a collector-current flow. Only an FET transistor can be operated with no bias as in Fig. 7-16*a*. This circuit works with both N- and P-channel JFETs and depletion MOSFETs.

Figure 7-16*b* shows an FET connected for an external type of bias. The

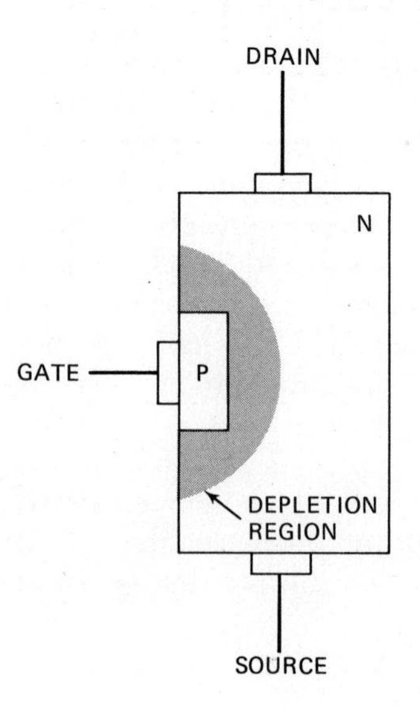

Fig. 7-17 Depletion region in a JFET.

negative bias voltage may come from a separate battery, from a separate power supply, or from the AVC voltage in a receiver.

Figure 7-16*c* shows a self-bias circuit for an FET. This circuit is also called *automatic bias* or *source bias*. Current flowing through the source resistor, indicated by an arrow, makes the source of the FET positive with respect to ground. The gate is at dc ground potential through resistor R_G. Therefore the source is positive with respect to the gate. (This is another way of saying that the gate is negative with respect to the source.) This is the required bias for an N-channel JFET.

For enhancement-type MOSFETs the gate voltage must be the same polarity as the drain voltage. In this way it is similar to bipolar transistors in which the base and collector have the same voltage polarities. Figure 7-16*d* shows how an enhancement MOSFET can be biased. An N-channel device is shown, and both the gate and the drain must have a positive voltage for proper operation. The drain receives its positive voltage through load resistor R_3. The gate obtains its positive bias voltage through gate resistor R_1.

Instead of a single resistor, as shown in Fig. 7-16*d*, it is possible to use a voltage divider (similar to Fig. 7-13). For a voltage divider both R_1 and R_2 are used.

In Table 7-1 the different types of bias used for amplifying devices are listed.

Summary

1. For very low amplitude input signals an FET can be operated with no bias.

Table 7-1 Uses of Various Types of Bias

Type of Bias	Used with Vacuum Tubes?	Used with Bipolar Transistors?	Used with JFETs and Depletion-type MOSFETs?	Used with Enhancement-type MOSFETs?
AVC (or AGC)	Yes	Yes	Yes	Seldom
Battery	Yes	Yes	Yes	Yes
Contact	Yes	No	No	No
Grid leak	Yes	No	No	No
No bias	No (tube may be damaged)	No (transistor cutoff)	Yes Low signal amplitude only	Possible but not used
Power supply	Yes	Yes	Yes	Yes
Self	Yes	No	Yes	No
Simple	No	Yes	No	Yes
Voltage divider	Yes	Yes, often	Seldom	Yes

2. The no-bias circuit works for both JFETs and depletion MOSFETs.
3. An FET may be biased with a separate battery.
4. An FET may be biased with a separate power supply.
5. An AVC (or AGC) voltage can be used to bias an FET.
6. Self-bias can be used with an FET. A source resistor is used, and its operation is similar to the use of a cathode bias resistor for tube operation.

PROGRAMMED REVIEW QUESTIONS

(Instructions for using this programmed section are given in Chapter 1.)

We will review the important concepts of this chapter. If you have understood the material, you will progress easily through this section. Do not skip this material because some additional theory is presented.

1. In the circuit of Fig. 7-18 the JFET is self-biased by source resistor R_2. Current flow through this resistor makes the source
   A. positive with respect to the gate. (Proceed to block 9.)
   B. negative with respect to the gate. (Proceed to block 17.)

2. *Your answer to the question in block 7 is* **B**. *This answer is wrong. A bipolar transistor does not have a contact-current flow, so it cannot be biased by this method.* Proceed to block 14.

3. *Your answer to the question in block 12 is* **A**. *This answer is wrong. When there is no bias on a bipolar transistor, the transistor is cut off. Therefore it cannot be used as an amplifier for very weak signals.* Proceed to block 27.

4. *Your answer to the question in block 21 is* **C**. *This answer is wrong. A short circuit between the emitter and the base of a bipolar transistor will turn the transistor* OFF. Proceed to block 24.

5. *The correct answer to the question in block 16 is* **B**. *The term* ***quiescent current*** *is sometimes used in place of idling current.* Here is your next question.

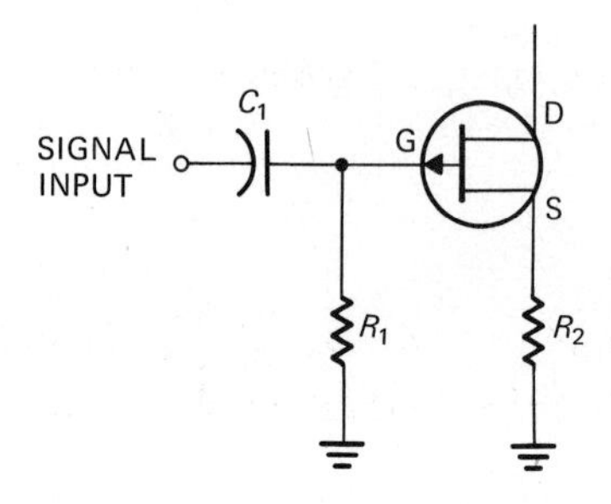

Fig. 7-18 What is the polarity of the source with respect to the gate?

Increasing the base current of a bipolar transistor will cause
A. an increase in collector current. (Proceed to block 13.)
B. a decrease in collector current. (Proceed to block 19.)

6. *Your answer to the question in block 18 is* **B.** *This answer is wrong. Figure 7-1 shows the polarities of bias voltages for all the popular amplifying components. Study this illustration, then* proceed to block 22.

7. *The correct answer to the question in block 13 is* **B.** *The source resistor is used for self-bias, which is also called* ***automatic bias.*** *It does the same job as the cathode bias resistor in a tube circuit.* Here is your next question.
Contact bias is used for
A. MOSFET circuits. (Proceed to block 25.)
B. bipolar transistor circuits. (Proceed to block 2.)
C. tube circuits. (Proceed to block 14.)

8. *Your answer to the question in block 14 is* **B.** *This answer is wrong. Figure 7-13 shows how voltage-divider bias is used with an NPN transistor. Any bias that can be used with an NPN transistor can also be used with a PNP transistor. Of course, the polarities of the voltages must be changed.* Proceed to block 21.

9. *Your answer to the question in block 1 is* **A.** *This answer is wrong. The JFET shown in Fig. 7-18 is a P-channel type. With a P-channel JFET the source must be negative with respect to the gate.* Proceed to block 17.

10. *Your answer to the question in block 17 is* **A.** *This answer is wrong. Figure 7-5 shows the grid bias circuit in which the bias voltage is obtained with a C battery. Study this circuit, then* proceed to block 18.

11. *Your answer to the question in block 22 is* **A.** *This answer is wrong. Study Fig. 7-1, then* proceed to block 16.

12. *The correct answer to the question in block 24 is* **A.** *A short circuit from cathode to grid will remove the grid bias. The plate current will become excessive, and the tube may be damaged if operated without bias.* Here is your next question.
Which of the following may be operated as a class A amplifier with no bias for amplifying very weak signals?
A. Bipolar transistor. (Proceed to block 3.)
B. JFET. (Proceed to block 27.)

13. *The correct answer to the question in block 5 is* **A.** *An increase in base current can be obtained by increasing the voltage on the base.* Here is your next question.

Which of the following is a method of obtaining a dc bias voltage? (Refer to Fig. 7-19.)

A. Emitter current flowing through an emitter resistor, as shown in Fig. 7-19*a*. (Proceed to block 26.)

B. Source current flowing through a source resistor, as shown in Fig. 7-19*b*. (Proceed to block 7.)

14. *The correct answer to the question in block 7 is* **C.** *Contact bias depends upon the fact that some electrons strike the grid and must be returned to the cathode. The flow of these electrons is called* **contact current.** *It does not occur in bipolar transistors or MOSFETs.* Here is your next question.

Is this statement true or false? Voltage-divider bias can be used with PNP transistors.

A. True. (Proceed to block 21.)

B. False. (Proceed to block 8.)

15. *Your answer to the question in block 21 is* **A.** *This answer is wrong. A short circuit between the emitter and the base of a bipolar transistor will turn the transistor* OFF. Proceed to block 24.

16. *The correct answer to the question in block 22 is* **B.** *The correct polarities are shown in Fig. 7-1. You must know the proper voltages for each component. Once you know what voltage* **should** *be at each point you can measure the voltage with a voltmeter. If the polarity of the voltage is found to be wrong, then you have found a likely cause of trouble.*

Here is your next question.

When the control electrode of a tube or transistor has only the bias voltage or current, the output current through the device is called the

A. reverse current. (Proceed to block 23.)

B. idling current. (Proceed to block 5.)

17. *The correct answer to the question in block 1 is* **B.** *For the P-channel device of Fig. 7-18, the source must be negative with respect to the gate.*

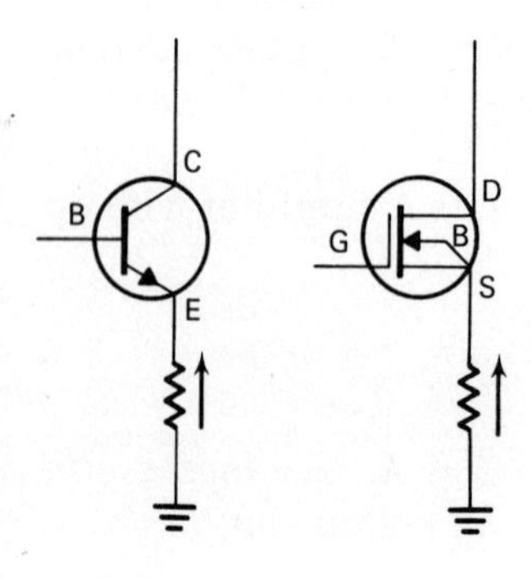

Fig. 7-19 Which is a bias circuit?

Current flows from the source toward ground through R_2. Thus the source is negative with respect to ground. The gate is at ground potential, so the gate is positive with respect to the source. This is the correct voltage polarity. Here is your next question.
Which of the following is a battery used for obtaining grid bias?
A. An A battery. (Proceed to block 10.)
B. A C battery. (Proceed to block 18.)

18. *The correct answer to the question in block 17 is* **B**. *An A battery is used as a filament power supply.* Here is your next question.
Which of the following requires a positive bias voltage on its control electrode?
A. P-channel JFET. (Proceed to block 22.)
B. N-channel JFET. (Proceed to block 6.)

19. *Your answer to the question in block 5 is* **B**. *This answer is wrong. No matter which type of bipolar transistor you have (NPN or PNP), the collector current will increase when you increase the base current. This assumes that the collector is connected to a power supply.*
Proceed to block 13.

20. *Your answer to the question in block 24 is* **B**. *This answer is wrong. Grid bias is needed in a tube to limit the plate current. A short circuit between cathode and grid will eliminate the grid bias voltage.*
Proceed to block 12.

21. *The correct answer to the question in block 14 is* **A**. *Figure 7-20 shows a PNP transistor with voltage-divider bias.* Here is your next question.
In the circuit of Fig. 7-21 ammeter M_1 measures the collector current

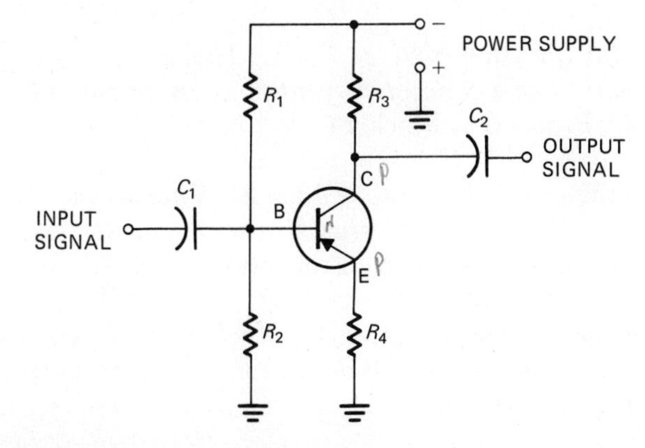

Fig. 7-20 Voltage divider bias (R_1 and R_2).

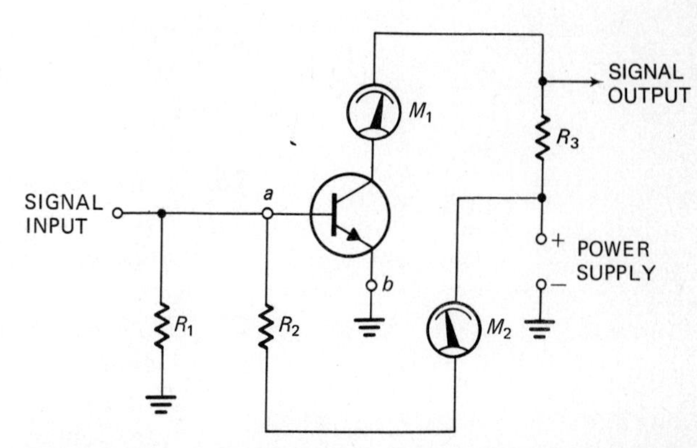

Fig. 7-21 How will a short circuit from point *a* to point *b* affect currents?

and ammeter M_2 measures the base current. If you connect a wire from point *a* to point *b*,

A. the transistor will be destroyed because of excessive collector current. (Proceed to block 15.)
B. the collector current will be zero and the transistor will not be destroyed. (Proceed to block 24.)
C. nothing will happen. (Proceed to block 4.)

22. *The correct answer to the question in block 18 is* **A.** *A positive bias voltage on the gate is required for a P-channel JFET.* Here is your next question.
Which type of MOSFET requires a positive voltage on its gate and a positive voltage on its drain?
A. N-channel depletion MOSFET. (Proceed to block 11.)
B. N-channel enhancement MOSFET. (Proceed to block 16.)

23. *Your answer to the question in block 16 is* **A.** *This answer is wrong. There has been no discussion of any kind of reverse current with regard to output current.* Proceed to block 5.

24. *The correct answer to the question in block 21 is* **B.** *With a short circuit between points a and b there will be no current from emitter to base within the transistor. Instead, all this current will flow through the short circuit. In a bipolar transistor there* **must** *be an emitter-to-base current in order to get a collector current. Therefore this short circuit will shut off the collector current.*

The emitter-to-base short circuit is sometimes used by technicians to determine if the transistor is working in a circuit. Here is your next question.
A short circuit from the cathode to the grid of a vacuum-tube triode
A. may damage the tube due to excessive plate current. (Proceed to block 12.)
B. will have no effect on the operation of the tube. (Proceed to block 20.)

25. *Your answer to the question in block 7 is* **A.** *This answer is wrong. A MOSFET does not have a contact current flow, so it cannot be biased by this method.* Proceed to block 14.

26. *Your answer to the question in block 13 is* **A.** *This answer is wrong. The emitter resistor is used for stabilization against temperature changes. It is not a bias resistor.* Proceed to block 7.

27. *The correct answer to the question in block 12 is* **B.** *Very small signals on the gate will cause the depletion region to vary in size. This in turn will cause the drain current to vary.* Here is your next question.
For a P-channel depletion MOSFET, is the gate normally positive or negative with respect to the source? (Proceed to block 28.)

28. *The gate of a P-channel depletion MOSFET is normally positive with respect to the source.*

You have now completed the programmed questions. The next step is to put some of these ideas to work in laboratory experiments. Proceed to the Experiment section of this chapter.

EXPERIMENT

(The experiment described in this section may be performed on the circuit board described in Appendix C or on a similar laboratory setup.)

Purpose The purpose of this experiment is to show you how to make measurements in a JFET circuit, and also to show you what these measurements tell about the amplifier. In this chapter you learned how the bias voltage for the JFET is obtained.

Theory In Chapter 3 you learned how a JFET works. (JFET is usually pronounced "*jay-fet.*")

The lab experiment you are now going to do uses an N-channel JFET. Figure 7-22 shows the symbol for the JFET and the bias circuit most often used.

The amount of dc current flowing through gate resistor R_1 is so small you can ignore it. Since there is no direct current through R_1, there is no dc voltage across it. This makes the gate 0 volts dc with respect to ground.

Electron current flows from ground through the source resistor R_3, and the source is positive with respect to ground. The source is also positive with respect to the gate. Another way of saying this is that the gate is negative with respect to the source. For an N-channel JFET the gate must be negative with respect to the source, and so the polarity of the bias is correct.

Resistor R_2 is the output load resistor for the circuit. The voltage drop across R_2 causes the voltage at the drain to be less positive than the power-supply voltage.

The input signal comes to the gate through C_1, and the output signal is delivered to the next stage through C_2. These capacitors are in the signal

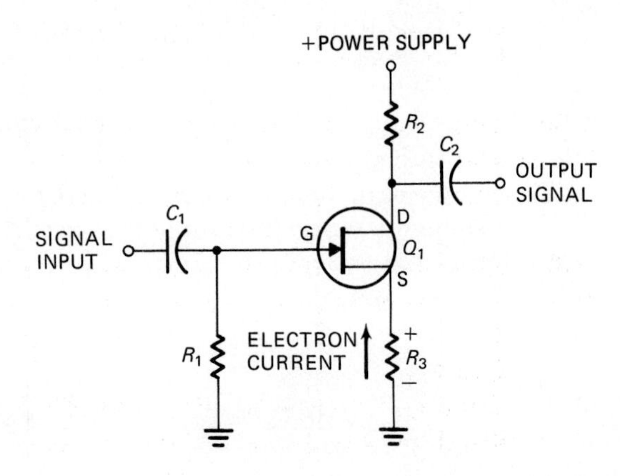

Fig. 7-22 Self-bias circuit for JFET.

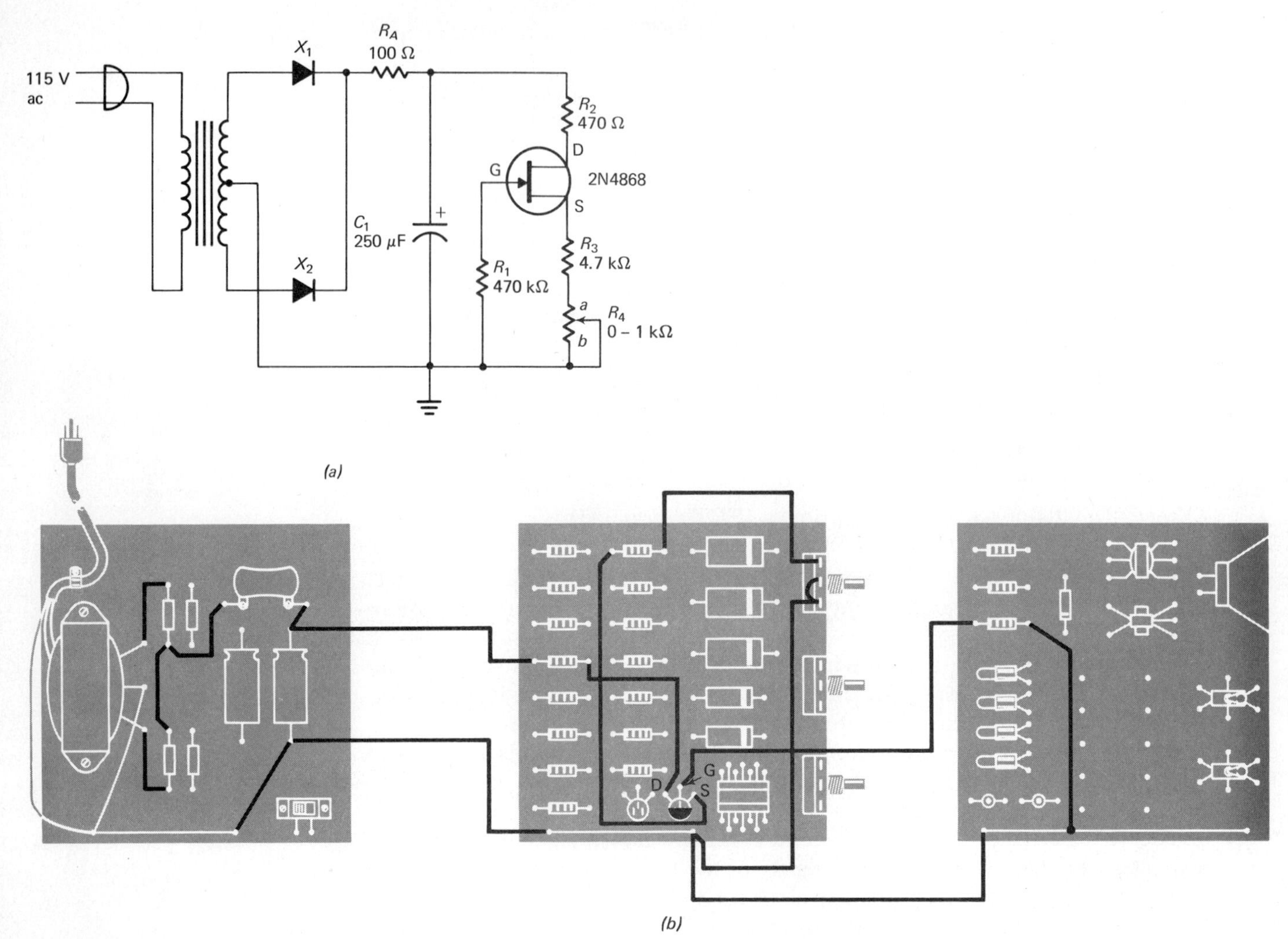

Fig. 7-23 Lab setup. (*a*) **The schematic diagram;** (*b*) **the pictorial diagram.**

path. They have no effect on the dc voltages that you will study in this experiment.

The method of biasing the JFET in the circuit of Fig. 7-22 is called *source bias*. It is also called *automatic bias* or *self-bias*. You will also see this method of bias used for tubes and depletion MOSFETs.

Test Setup Wire the circuit as shown in Fig. 7-23. Figure 7-23*a* shows the schematic and Fig. 7-23*b* the pictorial drawing. This circuit is similar to the one shown in Fig. 7-22.

Two source resistors are used. One (R_4) is variable to permit you to change the bias voltage. When the arm of the variable resistor is at point *a*, resistor R_4 is shorted out of the circuit. In that case, only R_3 sets the bias and the bias voltage will be minimum. Here is the reason:
By Ohm's law:

$$E = IR$$

$$\text{Bias voltage} = \text{source current} \times \text{source resistance}$$

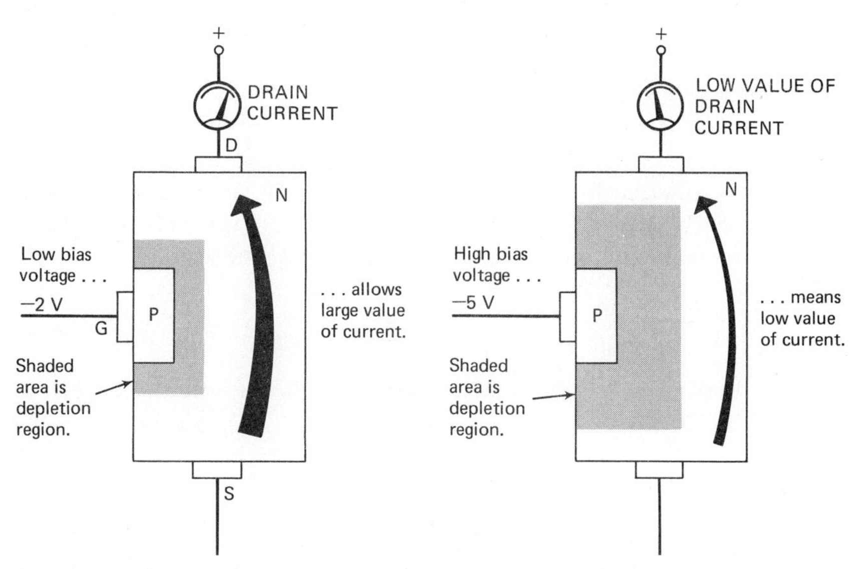

Fig. 7-24 When the bias is minimum the current is maximum.

Fig. 7-25 When the bias is maximum the current is minimum.

When you make the source resistance smaller the bias voltage E also becomes smaller.

When the arm of R_4 is at point b then all the resistance of R_4 is in series with R_3. This is the most source resistance that you can get with this circuit. It is also the most bias voltage that you can get by adjusting R_4.

When the bias voltage is minimum (the arm of R_4 at point a), the current through the JFET should be maximum. Remember that the bias is a reverse voltage across the gate-to-channel PN junction. The lower the bias, the smaller the depletion region and the larger the channel current. This is shown in Fig. 7-24.

When the bias is maximum (the arm of R_4 at point b), the current through the JFET should be minimum. Figure 7-25 shows what happens. The larger bias makes the depletion region larger and decreases the current.

Procedure

step one Construct the circuit shown in Fig. 7-23. Set the variable resistor R_4 to the center of its range.

step two Connect a voltmeter directly between the gate and source as shown in Fig. 7-26. Note that the negative lead of the voltmeter must go to the gate and the positive lead must go to the source.

step three With R_4 set to the center of its range, record the value of source bias shown by the voltmeter. V_G = ________________ volts

step four Adjust R_4 for minimum resistance. (Turn R_4 to its maximum counterclockwise position to obtain the minimum resistance.) Record the voltmeter reading for this setting of R_4. ________________ volts

Did the bias increase or decrease when the resistance was set to the minimum value? ______________
Increase, Decrease

Your bias voltage should decrease when R_4 is set to minimum resistance.

step five Adjust resistor R_4 to its maximum (fully clockwise) resistance. Record the bias voltage as indicated by the voltmeter.

V_G = ______________ volts

Did the bias voltage increase when the resistance was increased?

____
Yes or No

The bias voltage should increase when you increase the source resistance.

step six Set R_4 to the center of its range. Connect the voltmeter between the drain and ground as shown in Fig. 7-27 (V_D).

step seven Record the drain voltage as indicated by the voltmeter.

V_D = ______________ volts

step eight Adjust R_4 for minimum resistance (minimum bias). (Do this by setting R_4 to its maximum counterclockwise position.) Record the drain voltage shown by the voltmeter. V_D = ______________ volts

When the bias is minimum the current through the channel and through drain load resistor R_2 is maximum. This makes the voltage drop across R_2 (V_{R_2}) maximum for this circuit. The voltmeter V_D indicates the supply voltage minus the voltage drop across R_2. Thus the voltmeter reading should be minimum when the bias is minimum. This action is described in Fig. 7-27.

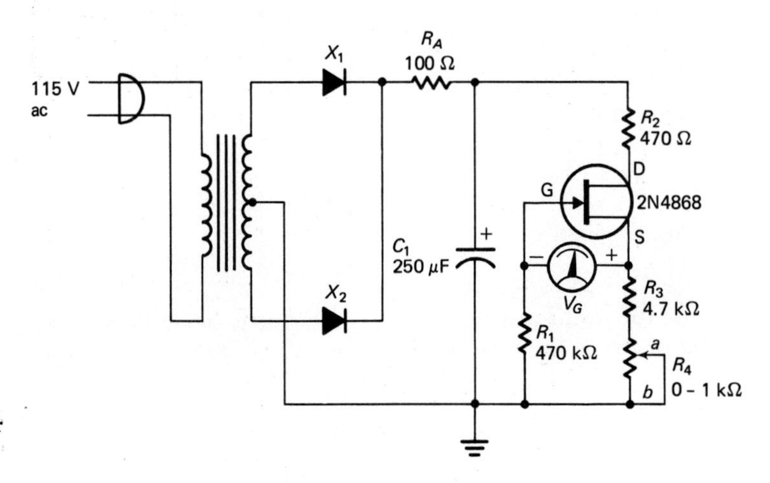

Fig. 7-26 Connection of voltmeter for measuring gate-bias voltage.

Now this is a very important point: From the action just described, you can see that when the gate-to-source bias becomes effectively less negative, the drain voltage becomes less positive.

step nine Adjust resistor R_4 for maximum resistance (maximum bias) by turning the shaft of R_4 fully clockwise. Record the drain voltage as measured by the voltmeter. V_D = ______ volts

step ten Is the collector voltage more positive than the value you recorded in step 8? ______ Yes or No

Your voltage reading should be more positive. This means that the drain voltage becomes more positive when the gate-to-source bias voltage becomes more negative (effectively, since we are actually varying the voltage on the source, while the gate is always at dc ground potential).

step eleven Subtract your voltage reading in Step 8 from your voltage reading in Step 9. The difference between the two voltages is called ΔV_D.

$$\Delta V_D = \text{change in drain voltage}$$
$$= \text{maximum drain voltage} - \text{minimum drain voltage}$$

ΔV_D = ______ volts

step twelve Subtract your voltage reading in Step 4 from your voltage reading in Step 5. The difference between the two voltages is called ΔV_G.

$$\Delta V_G = \text{change in gate voltage}$$
$$= \text{maximum gate voltage} - \text{minimum gate voltage}$$

ΔV_G = ______ volts

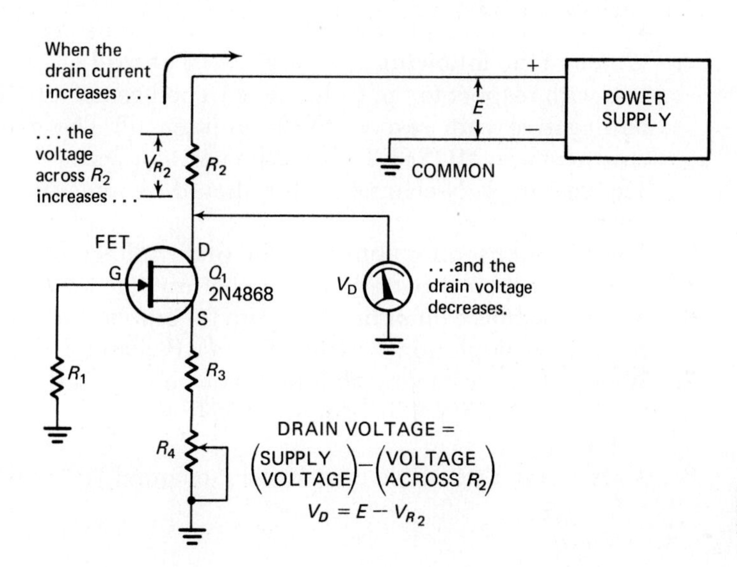

Fig. 7-27 Connection of voltmeter for measuring drain voltage.

step thirteen The voltage gain of the circuit is equal to the change in drain voltage ΔV_D divided by the change in gate voltage that produced it ΔV_G.

$$\text{voltage gain} = \frac{\Delta V_D}{\Delta V_G} = \underline{\hspace{4cm}}$$

The voltage gain of an amplifier tells how much the signal is amplified. The greater the voltage gain, the greater the amount of amplification.

Conclusion A JFET can be biased with resistance in the source circuit. As the source resistance is increased the gate-to-source bias voltage increases. The drain voltage depends upon how much voltage drop there is across the drain resistor. Increasing the drain current increases the voltage drop and makes the drain less positive.

When the circuit of Fig. 7-22 is used as an amplifier, the input signal can be delivered to the gate and the output signal taken from the drain. The amount of amplification is also called the *gain* of the amplifier stage.

It is very important for you to know that a change in gate-to-source voltage will cause a change in drain voltage (and current) in a circuit like the one shown in Fig. 7-22.

SELF-TEST WITH ANSWERS

(Answers with discussions are given at the end of the chapter.)

1. For a P-channel JFET the dc gate bias voltage should be (*a*) positive with respect to the source voltage; (*b*) negative with respect to the source voltage.
2. Which of the following is a current-operated device? (*a*) An enhancement P-channel MOSFET; (*b*) An N-channel JFET; (*c*) A triode tube; (*d*) A bipolar transistor.
3. Contact bias can be used with (*a*) bipolar transistors; (*b*) enhancement MOSFETs; (*c*) depletion MOSFETs; (*d*) some types of tubes.
4. Which of the following is wrong? (*a*) The grid of a tube should be negative with respect to the cathode; (*b*) The base of a PNP transistor should be negative with respect to the emitter; (*c*) The gate of an N-channel enhancement MOSFET should be positive with respect to the drain; (*d*) The gate of an N-channel JFET should be negative with respect to the source.
5. The battery used for obtaining the plate voltage of a tube is called (*a*) an A battery; (*b*) a B battery; (*c*) a C battery; (*d*) an X battery.
6. Automatic bias cannot be used with (*a*) bipolar transistors; (*b*) JFETs; (*c*) N-channel depletion MOSFETs; (*d*) P-channel depletion MOSFETs.
7. Which of the following should never be operated without some type of bias? (*a*) An NPN transistor; (*b*) A PNP transistor; (*c*) A triode tube; (*d*) A JFET.
8. A resistor in the source circuit of a P-channel JFET will make the source

(*a*) positive with respect to common ground; (*b*) negative with respect to common ground.

9. Which of the following devices can be operated as an amplifier of weak signals, even though there is no bias voltage? (*a*) An NPN transistor; (*b*) A PNP transistor; (*c*) A triode tube; (*d*) A JFET.
10. Another name for quiescent (no-signal) current is (*a*) excessive current; (*b*) alternating current; (*c*) idling current; (*d*) no current.

ANSWERS TO SELF-TEST

1. (*a*)—In the experiment you worked with an N-channel JFET. In the lab circuit the gate was negative with respect to the source.
2. (*d*)—Both NPN and PNP transistors are devices in which the base current controls the collector current.
3. (*d*)—Tubes capable of high-voltage amplification are best suited for contact bias. These tubes have very close grid-to-cathode spacing.
4. (*c*)—It is important for you to learn the polarities of the voltages needed to operate the various amplifying devices.
5. (*b*)—Power supplies for plates, collectors, and drains are often called *B supplies.*
6. (*a*)—Emitter resistors are not used for bias. They are used for temperature stabilization.
7. (*c*)—A triode tube may be damaged if it is operated without a grid bias voltage.
8. (*b*)—For a P-channel JFET the source voltage is opposite to the source voltage of an N-channel JFET. In the laboratory experiment you worked with an N-channel JFET.
9. (*d*)—This can be done only with a field-effect transistor.
10. (*c*)

what are voltage and power amplifiers? 8

INTRODUCTION

Amplifiers are normally operated with two different types of voltages, and both are important and necessary.

One of the voltages is the dc operating voltage for getting the amplifying devices into operation. In the previous chapters you have studied these dc voltages. You know the polarities of the voltages required for each type of amplifying device. You also know the methods of obtaining the dc bias voltage for operating each type of amplifier device. These subjects were covered in previous chapters.

In this chapter you will learn about the second voltage that is present in an amplifier. It is the *ac signal.*

Technicians have different ways of referring to amplifiers. If an amplifier is used for making a small signal voltage into a large signal voltage, it is called a *voltage amplifier.* If it changes a signal voltage into a fairly high level signal current for operating a transducer, it is called a *power amplifier.*

The names *class A, class B,* and *class C* refer to the way the amplifying device is biased. As you will learn when you study this chapter, class A amplifiers are not always better than class B amplifiers.

The terms *RC-coupled, impedance-coupled, transformer-coupled,* and *direct-coupled* refer to the way the signal is passed from one amplifier stage to the next. The methods of coupling amplifiers will be studied in a later chapter.

Amplifiers may also be called names according to the type of signal they amplify. Thus you have *audio amplifiers* and *RF amplifiers* for amplifying audio or radio frequency signals.

In this chapter you will learn about the classes of amplifiers and the methods of bias related to these classes. As mentioned before, you will study about the amplifiers in terms of the ac signal. What you learn about classes of amplifiers applies to both voltage amplifiers and power amplifiers.

A typical voltage and power amplifier circuit will be discussed in this chapter. A bipolar transistor circuit is used for the example. As you will see, you already have enough training to understand this two-stage amplifier.

You will be able to answer these questions after studying this chapter.

- ☐ **What are the classes of amplifiers?**
- ☐ **How is the class of operation affected by types of bias?**
- ☐ **How do typical voltage and power amplifiers work?**
- ☐ **How can you tell voltage amplifiers from power amplifiers?**

INSTRUCTION

What Are the Classes of Amplifiers?

In its simplest form an amplifier takes an input signal, increases its amplitude (amplifies it) and delivers an output signal to another stage or transducer. In such an amplifier it is important that the output signal be an exact copy of the input signal as far as the shape of the signal waveform is concerned. (The obvious difference between the input and the output is the amplitude.)

When the output-signal waveform is the same shape as the input-signal waveform it is called a *class A amplifier.* Other classes of amplifiers such as class B, class AB, and class C not only amplify the signal, but they change its shape in some important way. At this time we will define the different kinds of amplifiers according to letter classifications.

What Is a Class A Amplifier?

Figure 8-1 shows the relationship between the input signal and the output signal for a class A amplifier. Below the input signal there is a dotted line marked *cutoff.* If the signal crosses this line the amplifier is cut off for short periods. The signal is not permitted to cross the line for class A operation. As a matter of fact in class A amplification it is usually not permitted to even get close to this cutoff point. Also it is equally important that the height of the positive half cycles not be too great.

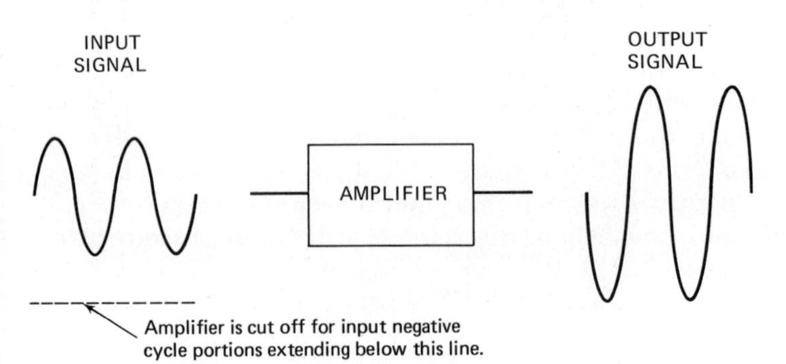

Fig. 8-1 Class A operation.

The output signal is a replica of the input signal except for one important fact: The signal is upside down. Another way of saying this is that the output signal is 180° out of phase with the input signal. When the input signal is going in a positive direction, the output signal is going in a negative direction. When the input signal is going in a negative direction, the output signal is going in a positive direction. The shape of the input and output waveforms are the same except for this fact. The fact that the output signal is upside down is often called *phase inversion.* Later in this chapter you will see why this occurs.

Phase inversion is not an important problem in most amplifiers. For one thing, a second amplifier will invert the phase back so that it is right-side up. This is illustrated in Fig. 8-2. Notice that after the first stage the signal is reversed in phase (upside down), and in the second stage the signal is reversed again. The output of the second stage is *in phase* with the input of the amplifier. When two signals are in phase it means that they both go positive at the same instant and they also go negative at the same instant.

If the circuit is used as an audio amplifier, the phase inversion has no effect on the output sound signal. The ear does not respond to the types of phase inversion described here.

Another important characteristic of a class A amplifier is that there is an output current through the amplifying device at all times. This is true regardless of whether or not there is a signal being amplified.

To summarize, a class A amplifier is one in which the output signal is like the input signal except for two things: it usually has a greater amplitude, and its phase may be inverted. There is always a current flow through the amplifying device in class A operation.

Summary

1. Voltage amplifiers are used for increasing the amplitude of a signal voltage.
2. Power amplifiers are used for changing a signal voltage into a fairly large signal current.
3. Power amplifiers are often used to deliver signals to transducers. Thus an audio power amplifier delivers a signal to a loudspeaker.
4. Amplifiers may be given letter names such as *class* A and *class B*. These

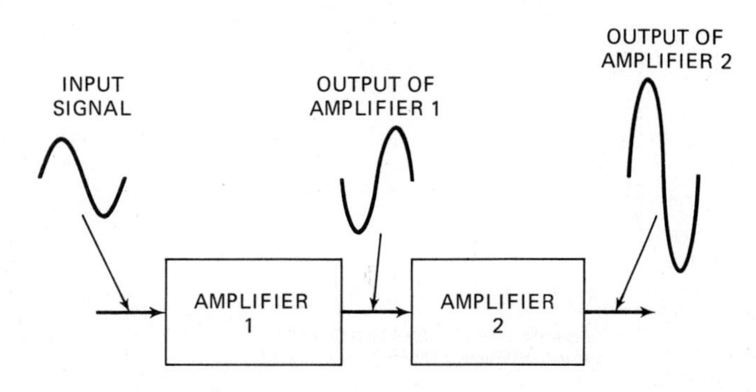

Fig. 8-2 The output signal of amplifier #2 is in phase with the input signal.

names identify how much bias is used and how the input and output signals are related.
5. The method of coupling the signal from one amplifier to another may be used for naming the amplifier. Examples are *direct-coupled amplifiers* and *transformer-coupled amplifiers*.
6. In class A amplifiers the output signal has the same shape as the input signal. However, the signals may be 180° out of phase.
7. In a class A amplifier there is current flowing through the amplifying device at all times. This is true even though there may be no input signal.
8. The input signal in a class A amplifier causes the current through the amplifying device to increase and decrease.

What Is a Class B Amplifier?

In a class B amplifier there is an output signal during only one half cycle of the input signal. This is illustrated in Fig. 8-3. Note that the input signal is above cutoff for one half cycle and below cutoff for the next half cycle. During the half cycle that the signal is below cutoff there is no amplifier output signal.

The output waveform for the class B amplifier shows that it is a series of half cycles. As before, the output waveforms are upside down—that is, they are 180° out of phase with the input signal. Also they are of a greater amplitude than the input signal.

You may wonder if there is any advantage of class B operation over the class A operation of Fig. 8-1. A very important advantage is its greater efficiency. You can think of amplifiers as being like people. If you give them a little time off, they work more efficiently. The fact that the amplifier is cut off for one-half of the input cycle means that it has a resting period. The efficiency of a class B amplifier is much greater than that of a class A amplifier.

Efficiency in an amplifier is a measure of how much signal power it can deliver compared with how much dc power it takes to get the amplifier to work. Mathematically,

$$\text{Percent efficiency} = \frac{\text{amount of output signal power}}{\text{amount of input dc power}} \times 100$$

Some amplifying devices are more efficient than others. However, any given amplifying device (tube, bipolar transistor, or FET) will be more efficient as a class B amplifier compared with the same device operated class A.

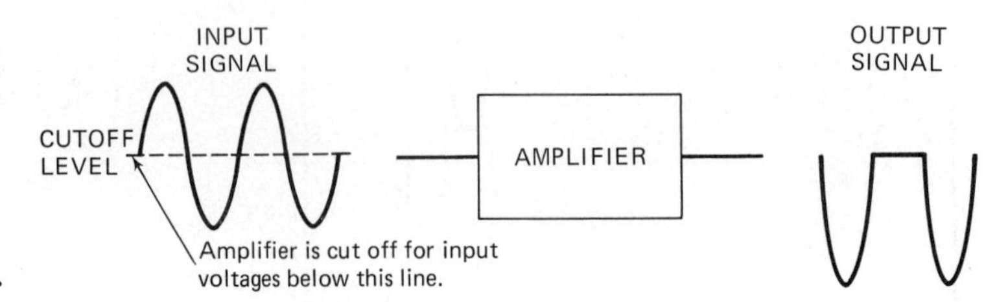

Fig. 8-3 Class B operation.

A class B amplifier may be used where the distortion of the output signal is no problem. Distortion is a measure of how much the output signal differs from the input signal. A highly distorted signal is one in which the shape of the output signal does not resemble the input signal very much. The output signal of a class B amplifier is highly distorted.

Two class B amplifiers can be connected in such a way that one amplifies only the positive half cycle of the signal and the other amplifies only the negative half cycle of the signal. These two efficient amplifiers produce an undistorted output, as shown in Fig. 8-4. The two half cycles have been combined in a circuit component, such as a transformer, so that the output signal contains both half cycles.

There are several types of circuits used to produce the two half waveforms shown in Fig. 8-4. You will study these circuits in a later chapter.

What Is a Class C Amplifier?

A class C amplifier is one in which the output waveform represents considerably less than one-half cycle of the input waveform. This is shown in Fig. 8-5. Note that the cutoff point on the input signal is very high, so there is an output signal only during the shaded portions of the input signal. Again, the output signal is inverted 180°, and it has a much greater amplitude than the portion of the input signal that is above cutoff.

Class C amplifiers are much more efficient than either class A or class B amplifiers. However, they have the disadvantage that the output signal no longer looks anything like the input signal. One example of the use of class C amplifiers is in oscillator circuits, which we will be studying in the next chapter.

What Is a Class AB Amplifier?

You learned that a class A amplifier is one in which both half cycles of the signal are present in the output. A class B amplifier is one in which the output signal contains only one-half cycle of the signal.

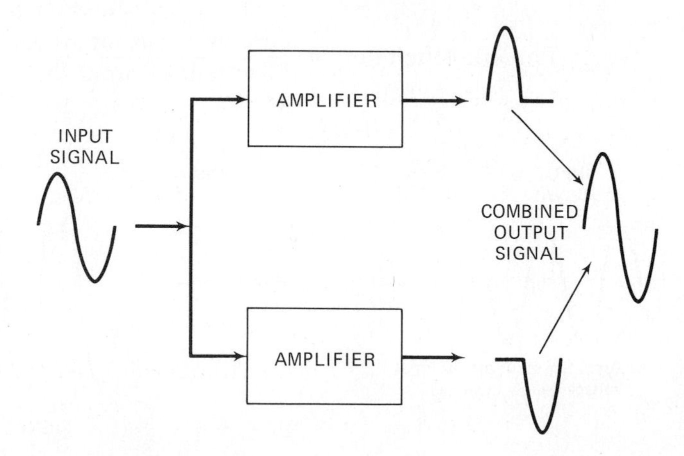

Fig. 8-4 Two class B amplifiers combine to produce a large undistorted output signal.

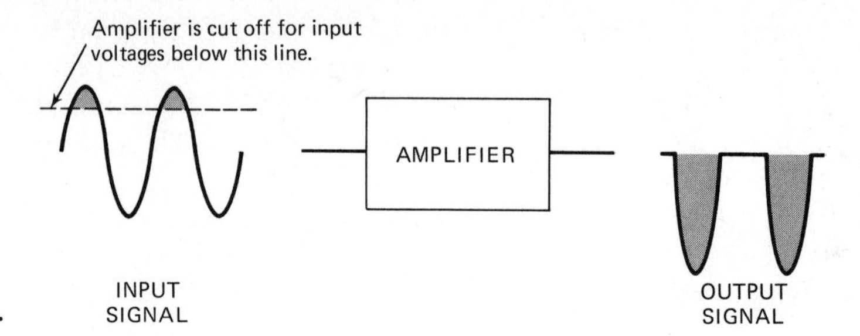

Fig. 8-5 Class C operation.

A class AB amplifier is one in which the output signal occurs for more than half of the input cycle, but less than the full input. In other words, it is somewhere between class A and class B operation.

Figure 8-6 shows an example of class AB operation. Note that the cutoff point permits most of the input signal to produce an output, but the amplifier is cut off for a small part of one-half cycle. Compare the waveform of Fig. 8-6 with those of class A and class B operation.

There are two types of class AB operation which are used only with vacuum tubes. They do not apply to other types of amplifying devices. If the vaccuum tube is operated class AB with no grid current flowing at any time, it is called *class* AB_1. If grid current is permitted to flow during some portion of the input cycle, then it is called *class* AB_2. Remember that these two types of class AB operation do not apply to bipolar transistors or field-effect transistors.

Summary

1. The output of a class B amplifier contains only half cycles of the input signal.
2. The output waveform of a class B amplifier is distorted, but class B operation is more efficient than class A. Two class B amplifiers must be used together (Fig. 8-4) to amplify audio signals.
3. Amplifier efficiency is a measure of how much ac signal power occurs for an amount of dc input power.
4. Mathematically,

$$\text{Percent efficiency} = \frac{\text{amount of output signal power}}{\text{amount of input dc power}} \times 100$$

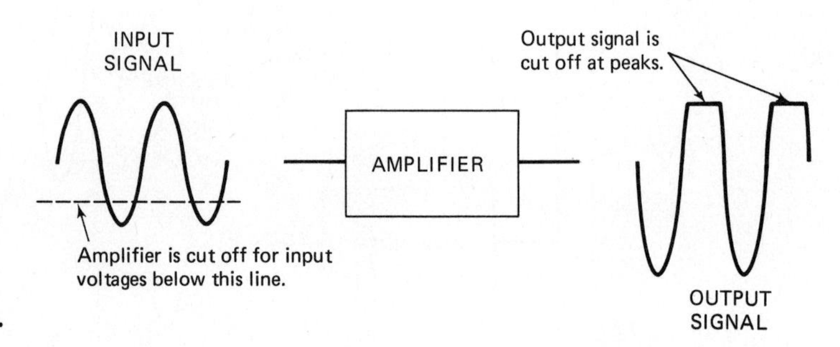

Fig. 8-6 Class AB operation.

5. The output signals from two class B amplifiers can be combined to produce a complete output signal.
6. Class C amplifiers are most efficient, but their output is most distorted. They cannot be used to amplify audio signals.
7. A class AB amplifier has an output signal for more than one-half cycle of input, but less than a full cycle.
8. For vacuum-tube circuits there are two types of class AB operation. With class AB_1 there is no grid current. With class AB_2 there is grid current for a brief time during each cycle of input signal.

How Is the Class of Operation Affected by Types of Bias?

In some classes of operation the amplifying device must be cut off for certain portions of time, and in other classes it must not be cut off at all. The input signal combines with the bias to produce the wanted output signal. The class of amplification is directly related to the amount of bias used. To understand this we will give a few examples.

Figure 8-7 shows a grid-leak-bias circuit. In this circuit capacitor C charges during the positive half cycle of input signal. The charging current flows through the grid circuit as shown by the solid arrow. In reality this charging takes place only during a small portion of the positive half cycle.

As the input signal starts to go negative the capacitor discharges through the grid-leak resistor. The discharge-current path is shown by the dotted arrow in Fig. 8-7. The discharge of the capacitor through the grid-leak resistor produces a negative voltage at the grid. The tube may be shut off during a portion of the input signal when grid-leak bias is used. The obvious conclusion is, then, that you cannot use grid-leak bias for class A operation.

Figure 8-8 shows an FET self-bias circuit. The bias voltage is obtained by current flowing through source resistor R_2. This makes the source positive with respect to ground, and the gate is grounded through R_1. Therefore the gate is negative with respect to the source.

The amount of source bias voltage depends upon the amount of current flow through the amplifying device. This is a very important point. In order to use self-bias with an amplifying device, you must have *current flowing through the device at all times.* Therefore self-bias cannot be used to cut the

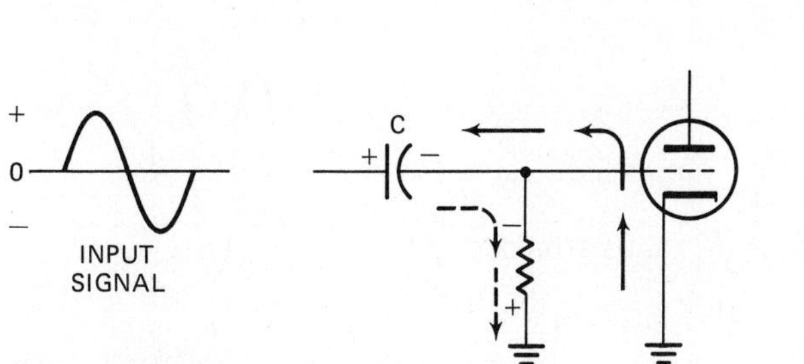

Fig. 8-7 Class A operation cannot be obtained with this bias circuit.

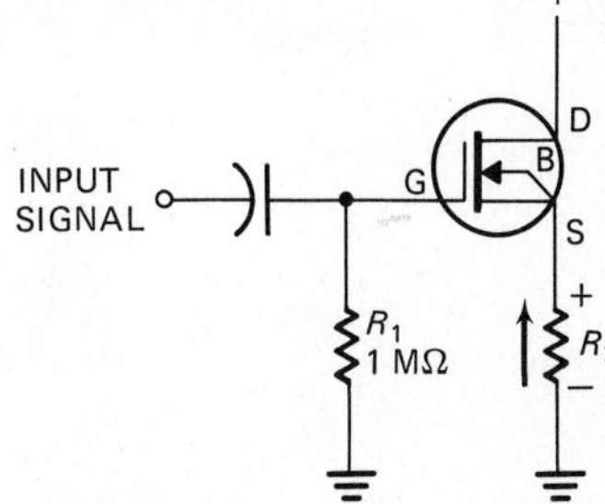

Fig. 8-8 This type of bias is good for class A operation.

amplifier OFF. (If it were possible to use a large enough resistor to shut the amplifier OFF, there would no longer be a voltage drop across the resistor, and the bias would go to zero. This is true for tubes as well as field-effect transistors.

Figure 8-9 shows an amplifier circuit using a bipolar NPN transistor. There is no forward bias for the base circuit. Note in Fig. 8-9*a* that the base is connected to ground through resistor R_1 and the emitter is grounded through R_2. Without any forward bias the transistor cannot conduct. This means that the transistor is normally cut off, without forward base bias.

The input signal, shown in Fig. 8-9*b*, is applied to the circuit through capacitor C_1. The positive half cycle, which is the shaded area, causes the NPN transistor to be forward-biased at the base. Collector current flows during this positive half cycle of input. The output signal is taken across the collector resistor R_3. It is delivered through C_2 to the next stage.

On the negative half cycle of input signal, which is the area not shaded in Fig. 8-9*b*, the base is made negative with respect to the emitter. The emitter-base junction is reverse-biased, and the transistor is biased well beyond cutoff. Therefore the transistor cannot conduct during the negative half cycles of input signal.

The output waveform for the transistor amplifier is shown in Fig. 8-9*c*. Note that the output waveform flows for approximately one-half cycle of the input, so this is a form of class B amplification.

The circuit shown in Fig. 8-9*a* cannot be used for class A operation because the transistor is cut off during the negative half cycles of input.

One important point should be added here. During the negative half cycles of input signal the emitter-base junction is reverse-biased. As you know, the emitter-base junction in a bipolar transistor is like the junction in a diode. Current is not supposed to flow during the times when it is reverse-biased. However, if you apply enough reverse bias, a junction breakdown occurs, and the current rises very rapidly. The voltage at which this breakdown occurs is called the *zener voltage*, and it can quickly destroy a PN junction. Zener diodes are designed to operate in the zener region, and they are not destroyed when the zener voltage is reached. However, most transistors are not designed this way. Therefore, if the input signal is so great that it causes the zener voltage to be reached on the base, then the transistor will be

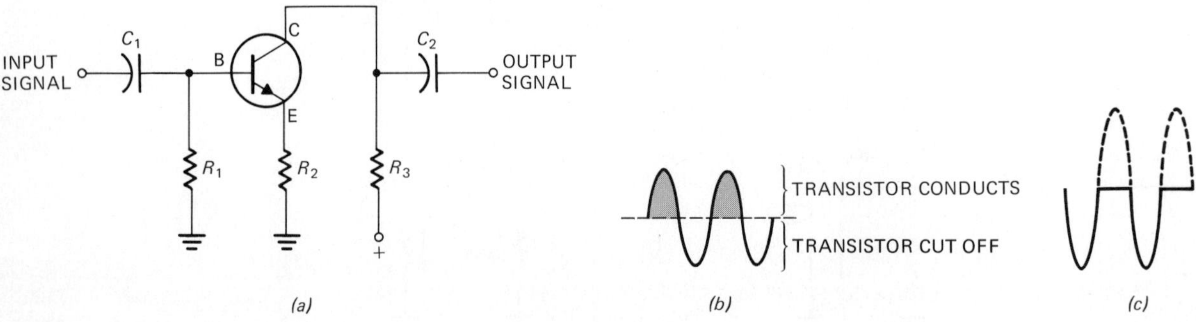

Fig. 8-9 Method of getting class B operation. (*a*) A class B circuit. (*b*) Input signal. (*c*) Output signal.

destroyed. This is an important point to remember: *The amplitude of the input signal in the circuit of Fig. 8-9a must not be large enough to cause a zener voltage to occur.*

We have given just three examples of how the amount of bias affects the class of amplification. Table 8-1 summarizes all the types of bias and relates them to the class of amplification. Study this table carefully. It is important to know this information because it enables you to analyze a circuit. A good place to start is to determine how the amplifier is biased. Once you know what type of bias is used, you will be better able to judge the class of operation.

Summary

1. The type of bias used with an amplifying device may limit it to certain classes of operation.
2. Self-bias cannot be used for class C operation.
3. Grid-leak bias cannot be used for class A operation.
4. When a bipolar transistor is operated without a forward base bias it is being operated class B.
5. Table 8-1 shows which types of bias can be used for each kind of operation.

How Do Typical Voltage and Power Amplifiers Work?

Figure 8-10 shows an audio voltage-amplifier circuit Q_1 and an audio power-amplifier circuit Q_2. The output of Q_1 is delivered to the input of Q_2. When two amplifiers are connected this way they are said to be *cascaded*.

Table 8-1 Classes of Operation for Each Type of Bias

Type of Bias	Used for Class A?	Used for Class B?	Used for Class C?
AVC (or AGC)	Yes	No	No
Battery	Yes	Yes	Yes
Contact	Yes	No	No
Grid leak	No	No	Yes
No bias (FET)	Yes	No	No
Power supply	Yes	Yes	Yes
Self	Yes	No	No
Simple	Yes	No	No
Voltage divider	Yes	Yes	Yes

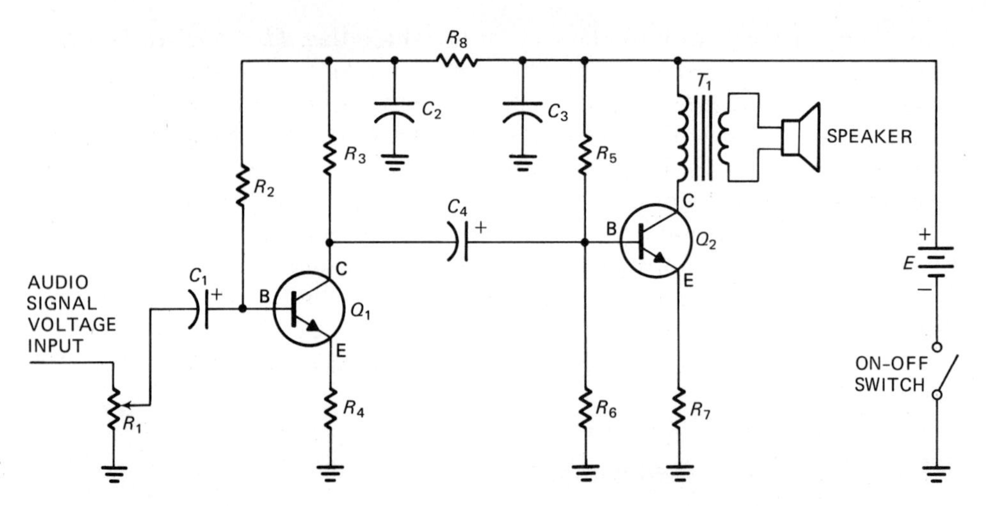

Fig. 8-10 A two-stage audio amplifier.

A voltage amplifier increases the strength of the signal. In the circuit of Fig. 8-10 a weak audio signal voltage is the input of Q_1. The output of Q_1 is an audio signal *voltage* with a higher amplitude.

The input signal to transistor Q_2 is an audio signal with a high voltage amplitude. The output of Q_2 is an audio signal *current* that flows through the primary of audio transformer T_1. Transistor Q_2 is a power amplifier because it changes a signal voltage into a signal current.

A single power supply E is used for both transistor stages. When you are analyzing a transistor circuit a good place to start is to trace the dc current paths and make sure that the polarities of the voltages are correct for each amplifying device. A technician usually makes voltage measurements at various points in the circuit while tracing the dc paths.

You will note that the negative side of E is connected to ground, or common, through the ON/OFF switch. To trace the dc paths you will start at the common (ground) point and go through the devices and back to the positive terminal of the voltage supply.

What Are the DC Paths for Q_2?

Starting at common, go through R_7, through Q_2 (emitter to collector), and through the primary of T_1. Taking this path you can get back to the positive terminal of the voltage supply. Therefore Q_2 (along with R_7 and the primary of T_1) offers a complete direct current path.

The collector of Q_2 is positive with respect to its emitter. This is the proper polarity for an NPN transistor. The transistor will conduct if there is a positive bias voltage on the base of this transistor. Starting at the common point and going through R_6 and R_5, you can again go back to the positive terminal voltage. Resistors R_5 and R_6 form a voltage divider network to obtain the positive bias voltage for the base of the transistor.

The dc paths for Q_2 have now been traced. The proper voltage polarities are on the base and collector.

What Are the DC Paths for Q_1?

Starting at resistor R_4 and going through Q_1 (emitter to collector) and R_3, you come to a positive point in the circuit. However, you must go through R_8 to get to the positive terminal of the supply voltage. The network consisting of R_8, C_2, and C_3 serves two purposes. First, it is a filter circuit to prevent any of the ac variations from transistor Q_2 from getting into the collector circuit of Q_1. This circuit is sometimes called a *decoupling filter.* The second purpose of this circuit is to drop the dc voltage of E to a lower value for operating Q_1. Power amplifiers in many circuits require a larger voltage for their collector than is required for voltage amplifiers. This is the case for the circuit of Fig. 8-10.

In order to use the same power supply for all transistor circuits, a large voltage value is used for the power amplifier and the voltage is dropped to a lower value for operating other transistor circuits. In the circuit of Fig. 8-10 the voltage is dropped to a lower value by R_8.

Transistor Q_1 has a positive voltage applied to its collector circuit, so it will operate if it has the proper base bias voltage. A simple bias circuit, with resistor R_2, is used to provide the positive base voltage for Q_1.

What Is the AC Signal Path for the Circuit?

Having traced the dc paths, we now see that both transistors have the proper polarities of bias and collector voltages. The next step is to trace the ac signal through the circuit.

This is an audio amplifier, and the *input*-signal voltage is developed across variable resistor R_1. If this is part of a phonograph or radio system, R_1 would be called the *volume control.* The closer the arm of R_1 is adjusted toward ground, the lower the signal-voltage amplitude and the lower the volume of sound from the speaker.

The audio signal is coupled to the base circuit of Q_1 through an electrolytic capacitor C_1. An electrolytic capacitor is used here because it has the large capacitance value needed for transistor circuits and it will pass the lower audio frequencies. This is common practice in transistor circuits.

The audio signal voltage from C_1 is developed across resistor R_2. This produces a varying current in the base. The collector current will be an amplified version of the audio base current.

The audio signal voltage across load resistor R_3 will be greater than the input signal to the base of Q_1. Therefore voltage amplification has taken place.

The output of Q_1 is delivered to another coupling capacitor C_4. This is also an electrolytic capacitor. The audio signal is delivered to the base of Q_2, and therefore it controls the collector current.

The collector current flowing through the primary of T_1 causes an expanding and contracting magnetic field around the transformer primary. As a result a voltage is developed across the secondary, and an audio signal and current flow through the loudspeaker. The speaker is a transducer in this circuit. It converts the audio current variations into sound.

Resistors R_4 and R_7 serve as emitter stabilization resistors. They protect the transistor from a thermal runaway.

How Can You Tell Voltage Amplifiers from Power Amplifiers?

The fact that the output of transistor Q_2 delivers its signal to a transducer (through a transformer) means that it must be a power amplifier. Power amplifiers are always preceded by voltage amplifiers. Therefore Q_1 must be a voltage amplifier.

If you were working with the actual circuit, you would have other clues to tell you which is the voltage amplifier and which is the power amplifier. Power amplifier transistors are normally larger than voltage amplifier transistors. Power transistors are usually connected to some form of heat sink to permit them to dissipate more heat during normal operation. Figure 8-11 shows two examples of heat sinks. Their purpose is to conduct heat away from the transistor metal case. This means that the power transistors will operate cooler with the heat sink.

Power amplifiers usually run hotter than the voltage amplifiers. This is also true of vacuum-tube circuits. In fact, power amplifiers are usually one of the hottest components in the circuit. The only other tube or semiconductor component that might rival them for heat would be the power-supply rectifiers.

Voltage amplifiers run cooler because they are not required to produce an output current for operating transducers. Since there is less current flowing in a voltage amplifier, resistors in a voltage-amplifier circuit may have lower wattage ratings.

Both of the amplifiers in Fig. 8-10 are operated class A, but you could not definitely tell this from the circuit layout. The thing that suggests that it is class A operation is the fact that it is an audio amplifier and therefore the signal must not be distorted. Class A operation produces an output signal that looks like the input signal.

Summary

1. A good place to start analyzing an electronic circuit is to trace the dc paths.
2. Each amplifying component must have two dc voltages for its operation.

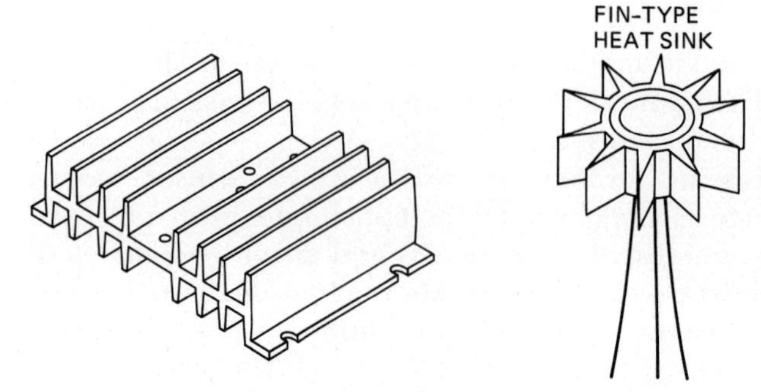

Fig. 8-11 Examples of heat sinks.

There must be a dc voltage for current through the component, and there must be a bias voltage.

3. Electrolytic capacitors are used for coupling the signal from one stage to another in bipolar circuits.
4. Power amplifiers are usually larger than voltage amplifiers because they have to be able to pass a larger current.
5. Power amplifiers usually operate at a higher temperature than voltage amplifiers.
6. A heat sink is often used with a power-amplifier transistor. It helps to conduct the heat away from the transistor.
7. Power-amplifier transistors often have a higher collector voltage than voltage-amplifier transistors.

PROGRAMMED REVIEW QUESTIONS

(Instructions for using this programmed section are given in Chapter 1.)
We will review the important concepts of this chapter. If you have understood the material, you will progress easily through this section. Do not skip this material because some additional theory is presented.

1. Class C operation cannot be obtained with
   A. self-bias. (Proceed to block 9.)
   B. battery bias. (Proceed to block 17.)

2. *The correct answer to the question in block 9 is* **B**. *Making the base more positive causes an increase in base current. If the base current of an NPN transistor increases, then its collector current also increases.* Here is your next question.
   Figure 8-12 shows the input signal and the output signal of an amplifier. The amplifier is being operated
   A. class A. (Proceed to block 15.)
   B. class C. (Proceed to block 24.)

3. *Your answer to the question in block 18 is* **A**. *This answer is wrong. The purpose of a coupling circuit is to pass the desired ac signal.* Proceed to block 12.

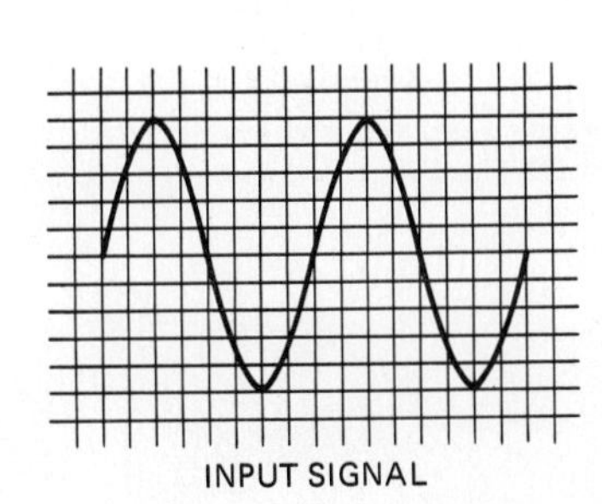

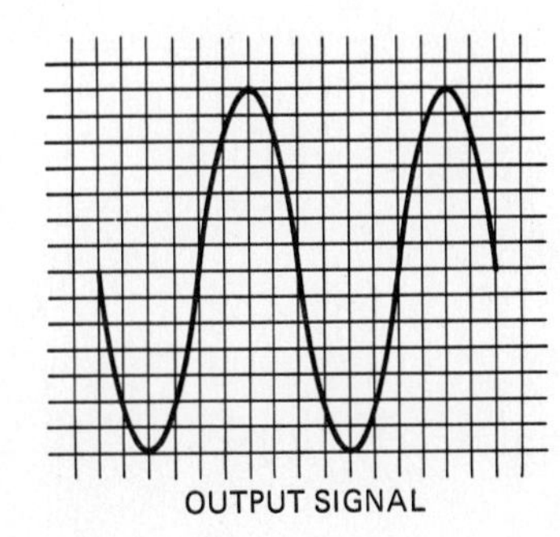

Fig. 8-12 Input and output signals of an amplifier.

4. *The correct answer to the question in block 16 is* **B**. *The output waveform of amplifier 1 in Fig. 8-15 shows that the amplifier is operated class B. Note that the output waveform occurs for one-half cycle. The output of amplifier 2 is less than one-half cycle, and this means that it is a class C amplifier.* Here is your next question. Which of the following would more likely be mounted with a heat sink?
   A. A transistor used in a voltage-amplifying circuit. (Proceed to block 13.)
   B. A transistor used in a power-amplifying circuit. (Proceed to block 22.)

5. *Your answer to the question in block 26 is* **A**. *This answer is wrong. Class B operation is only more efficient than class A operation.* Proceed to block 16.

6. *The correct answer to the question in block 14 is* **A**. *A transducer frequently (but not in every case) requires a larger current than may be supplied by a voltage amplifier. Therefore a power amplifier usually has its output signal delivered to a transducer.* Here is your next question.
   The PNP transistor circuit in Fig. 8-13 is being operated
   A. class A. (Proceed to block 27.)
   B. class B. (Proceed to block 11.)

7. *The correct answer to the question in block 15 is* **B**. *If there is current flowing through the amplifying device at all times, it is a class A amplifier circuit. This is true for both voltage and power amplifiers.* Here is your next question.
   An advantage of using an electrolytic capacitor to couple a signal from one amplifier to another is that the large capacitance allows lower frequencies to be coupled without much loss. A disadvantage is that
   A. electrolytic capacitors cause fires. (Proceed to block 23.)
   B. electrolytic capacitors are polarized. (Proceed to block 14.)

8. *Your answer to the question in block 9 is* **A**. *This answer is wrong. When you increase the positive voltage on the base of an NPN transistor its base and collector currents both increase.* Proceed to block 2.

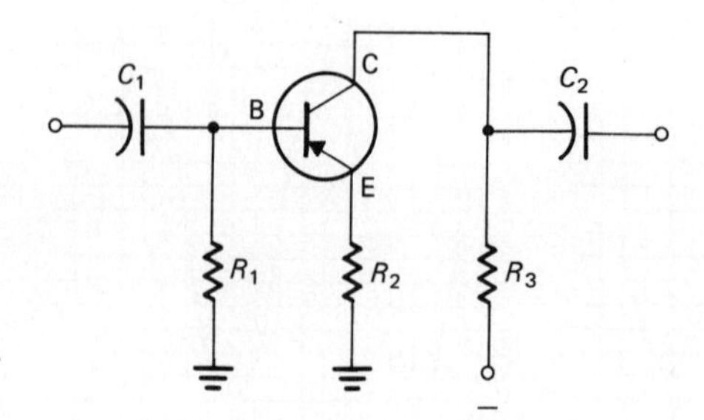

Fig. 8-13 What class of amplification is this?

9. *The correct answer to the question in block 1 is* **A**. *You get self-bias in a vacuum-tube circuit by using a cathode resistor. In a FET circuit a source resistor is used. In both cases the bias occurs when tube or FET current flows through the resistor. However, with class C bias the tube or FET is cut off, so no current flows through the resistor in the cathode or source circuit. There could be no bias voltage if the device were cut off.*

   Only the peaks of the input signal can overcome class C bias and make the amplifying device conduct. Here is your next question.
   If you make the base of an NPN transistor more positive,
   A. the collector current will decrease. (Proceed to block 8.)
   B. the collector current will increase. (Proceed to block 2.)

10. *Your answer to the question in block 15 is* **A**. *This answer is wrong. Either a voltage amplifier or a power amplifier can be operated class A.* Proceed to block 7.

11. *The correct answer to the question in block 6 is* **B**. *There is no dc base bias current flowing in the circuit of Fig. 8-13. The base is connected to ground through resistor R_1, so there is no source for dc bias current. The negative half cycles of input signal will forward-bias the PNP transistor and cause it to conduct. The positive half cycles will hold it in cutoff.* Here is your next question.
    In the circuit of Fig. 8-14 the circuit comprising R_1, C_1, and C_2 serves two purposes. One purpose is to drop the power-supply voltage to a lower value for Q_1. The second purpose is to
    A. couple the signal from Q_1 to Q_2. (Proceed to block 19.)
    B. prevent undesired coupling from one stage to another. (Proceed to block 26.)

12. *The correct answer to the question in block 18 is* **B**. *The dc voltage at the output of one stage may not be the same value as the dc voltage at the next stage. The capacitor prevents direct current from flowing between the two values of dc voltage but passes the ac signal.* Here is your next question.
    Explain how it is possible to use two class B amplifiers to get an undistorted output signal. (Proceed to block 28.)

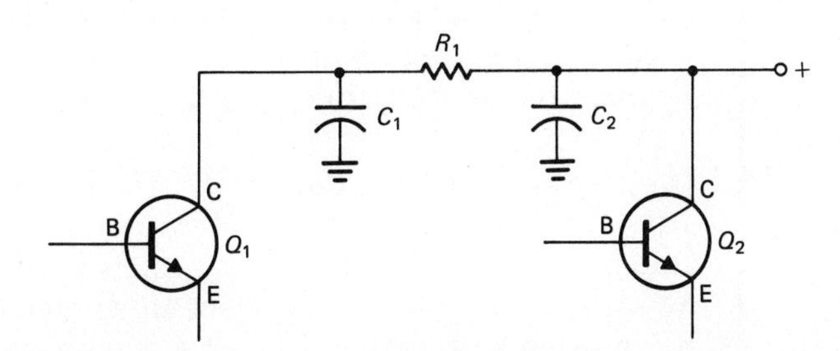

Fig. 8-14 What is the purpose of R_1, C_1, and C_2?

13. *Your answer to the question in block 4 is* **A**. *This answer is wrong. Voltage amplifiers increase the amplitude of the signal voltage, but they do not normally conduct large currents. A small amount of heat is usually generated in voltage amplifiers compared with the amount of heat generated in power amplifiers. Heat sinks are not used with voltage amplifiers.* Proceed to block 22.

14. *The correct answer to the question in block 7 is* **B**. *A capacitor will pass high frequencies more easily than it will pass low frequencies. However, as a general rule you can say that the larger the capacitor, the lower the frequency it can pass. Electrolytic capacitors have a large capacitance value in a small package, and so they are able to pass lower audio frequencies. However, they are polarized and must be connected properly with regard to + and − voltages.* Here is your next question.

One way to locate the power stage in a circuit is to

A. determine which amplifier has its output connected to a transducer. (Proceed to block 6.)

B. determine which amplifier has its output connected to a transformer. (Proceed to block 20.)

15. *The correct answer to the question in block 2 is* **A**. *There is an output signal at all times during the input signal. This is class A operation.* Here is your next question.

Which of the following statements is correct?

A. A class A amplifier must be a voltage amplifier. (Proceed to block 10.)

B. A class A amplifier can be either a voltage amplifier or a power amplifier. (Proceed to block 7.)

16. *The correct answer to the question in block 26 is* **B**. *Class C operation is more efficient than class A or class B. The amplifying device is cut off for more than one-half cycle with class C operation. As a general rule, the greater the off time, the more efficient the operation.* Here is your next question.

Which of the amplifiers in Fig. 8-15 is operating Class C?

A. Amplifier 1. (Proceed to block 25.)

B. Amplifier 2. (Proceed to block 4.)

17. *Your answer to the question in block 1 is* **B**. *This answer is wrong. Batteries are made in a wide choice of voltages. It is always possible to get a battery with a large enough voltage to cut off current flow in a tube, bipolar transistor, or FET.* Proceed to block 9.

18. *The correct answer to the question in block 22 is* **B**. *That is the definition of class A operation.* Here is your next question.

One purpose of the coupling capacitor in *RC*-coupled amplifiers is to

A. pass a dc voltage and reject an ac signal. (Proceed to block 3.)

B. pass an ac signal and reject a dc voltage. (Proceed to block 12.)

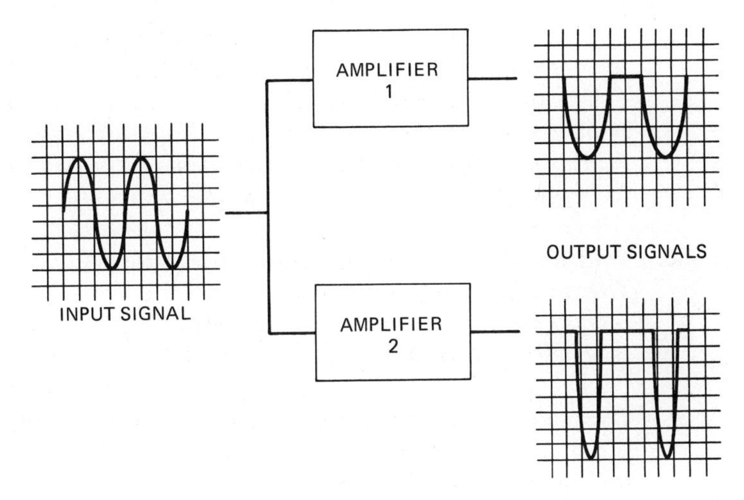

Fig. 8-15 Which amplifier is operating class C?

19. *Your answer to the question in block 11 is* **A**. *This answer is wrong. The circuit is used for filtering. It prevents signals from passing. It is not used for coupling.* Proceed to block 26.

20. *Your answer to the question in block 14 is* **B**. *This answer is wrong. The power amplifier for the circuit discussed in this chapter had its output connected to a speaker through a transformer. However, many voltage amplifiers also have their output signals connected to the next stage through a transformer. Also, many power amplifiers do not deliver their signal to the transducer through a transformer.* Proceed to block 6.

21. *Your answer to the question in block 22 is* **A**. *This answer is wrong. A class B amplifier is more efficient than a class A amplifier.* Proceed to block 18.

22. *The correct answer to the question in block 4 is* **B**. *Power transistors are usually constructed in such a way that internal heat is conducted to a metal surface. However, the metal surface is usually too small to permit all the heat to be radiated into the air. A heat sink simply increases the amount of surface area from which the heat can be dissipated.*

In order for the heat sink to work it must be mounted firmly against the metal surface of the transistor. Many companies use a silicon gel to improve the heat sink. Here is your next question. The term *class A amplifier* always means

A. that it is more efficient than a class B amplifier. (Proceed to block 21.)
B. that there is current flowing through the amplifying device at all times—even when no signal is being amplified. (Proceed to block 18.)

23. *Your answer to the question in block 7 is* **A**. *This answer is wrong.*

There is no evidence that electrolytic capacitors cause fires. Proceed to block 14.

24. *Your answer to the question in block 2 is* **B**. *This answer is wrong. Study the input and output waveforms of Fig. 8-12. There is an output signal at all times when there is an input.* Proceed to block 15.

25. *Your answer to the question in block 16 is* **A**. *This answer is wrong. The output signal from amplifier 1 is a half cycle. Review the waveforms in this chapter for the different classes of amplification, then* proceed to block 4.

26. *The correct answer to the question in block 11 is* **B**. *This decoupling filter has been discussed in the chapter. It is used between* Q_1 *and* Q_2 *in Fig. 8-10.* Here is your next question.
Which of the following is more efficient?
A. Class B operation of amplifiers. (Proceed to block 5.)
B. Class C operation of amplifiers. (Proceed to block 16.)

27. *Your answer to the question in block 6 is* **A**. *This answer is wrong. For class A operation collector current must flow at all times, even when there is no input signal. The transistor in the circuit of Fig. 8-13 is cut off when there is no input signal.* Proceed to block 11.

28. *One of the class B amplifiers amplifies the positive half cycles of signal and the other amplifies the negative half cycles of signal.*
You have now completed the programmed questions. The next step is to put some of these ideas to work in laboratory experiments. Proceed to the Experiment section of this chapter.

EXPERIMENT

(The experiment described in this section may be performed on the circuit board described in Appendix C or on a similar laboratory setup.)

Purpose In this experiment you will show that both ac and dc voltages (and currents) can exist in a circuit at the same time. Also you will show that the common point for the ac signal and the common point for the dc voltage may be at different places in the circuit.

You will also show that an amplifying component can be connected three ways in a circuit.

PART I

Theory There are three important signal points in all amplifiers. They are shown in Fig. 8-16.

The *signal input* point is where the signal enters the amplifier. The *signal output* point is where the signal leaves the amplifier. The *common* point is the place where the signal voltage is considered to be 0 volts. The common point is often called *ground*, but this term may be misleading. The common point may be ground as far as the signal is concerned, but for some circuits it may have a positive or negative dc voltage.

Figure 8-17 shows how it is possible for a point to be at 0 volts as far as the signal is concerned and also have a dc voltage. Four resistors (R_1, R_2, R_3, and R_4) are connected across a dc voltage source. The ac input signal is across resistors R_2 and R_3. The ac output signal is across R_3 only, since point *d* is at ac ground potential because of the low impedance of capacitor C_3. Capacitor C_1 isolates the ac generator from the dc voltage. This capacitor passes the ac signal to point *b*, but prevents the dc voltage at point *b* from getting to the generator.

Capacitor C_2 delivers the signal to the ac output terminal. At the same time it prevents the dc voltage at point *c* from getting to the ac output terminal.

Capacitor C_3 causes point *d* to be grounded or common for ac only. Note that this is not the dc common point. Ac voltage measurements would be taken with respect to point *d*, but dc voltage measurements would be taken with respect to point *e*.

The circuit of Fig. 8-17 shows why you must treat the dc and ac signal voltages separately. The two voltages may not have the same common point. Also the ac signal input is not at the same point as the dc input to the circuit. You must learn to treat dc and ac as two separate and distinct voltages (and currents) in a circuit.

Figure 8-18 shows how a voltmeter can be connected so that it will measure only ac voltages.

The voltmeter in Fig. 8-18*a* will measure both ac and dc voltages. However, most multimeters will indicate a voltage value when their function switch is set for ac voltage and their probes are connected to a dc voltage. This means that a dc voltage will affect the ac voltage reading when a measurement is made in a circuit that has both dc and ac voltages.

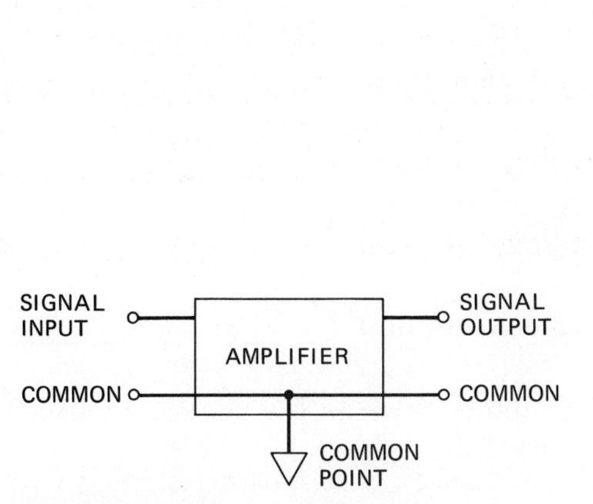

Fig. 8-16 Important signal points.

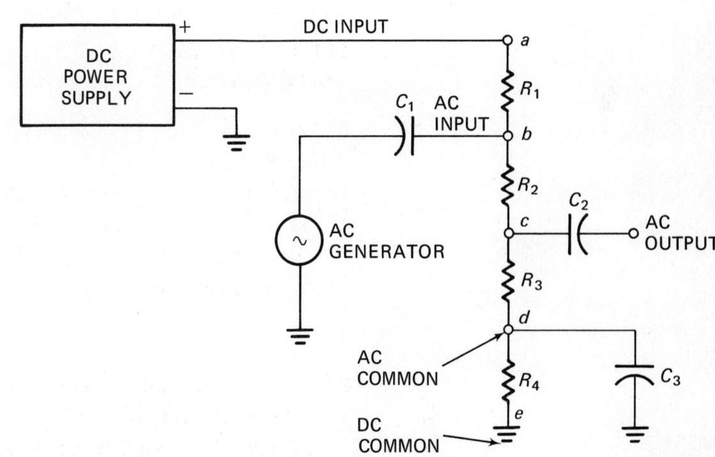

Fig. 8-17 A circuit with both dc and ac voltages.

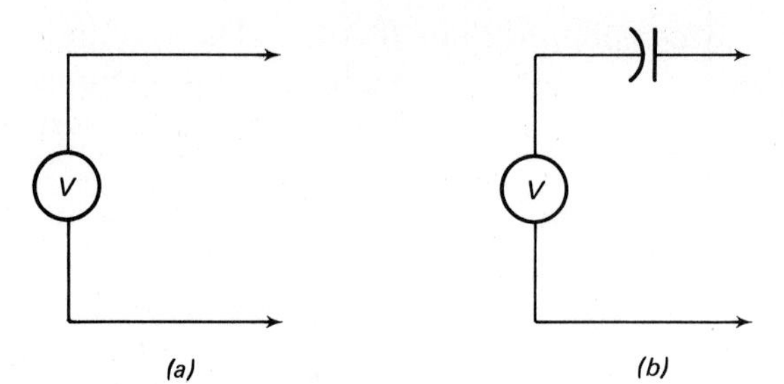

Fig. 8-18 Voltmeters for measuring (*a*) ac or dc and (*b*) ac only.

To prevent this, a capacitor can be connected in series with the voltmeter as shown in Fig. 8-18*b*. When you do this, you must use a capacitor that is large enough in value that it does not drop the ac voltage. Keep in mind that larger values of capacitance have less reactance and therefore produce less voltage drop.

Test Setup The circuit for Part I of this experiment is shown in Fig. 8-19. Figure 8-19*a* shows the schematic diagram, and Fig. 8-19*b* shows the pictorial drawing. Diode X_1 and capacitor C_1 form a dc supply. The dc voltage at point *a* is about 9 volts.

The center tap of the secondary is connected to common. The ac signal is taken from the transformer secondary at point *y*. Capacitors C_2 and C_3 are connected in parallel, so their capacitance values add. The ac signal is connected from point *y* to point *b* through C_2 and C_3.

Point *d* in the voltage divider is the ac common point. There will be a dc voltage at point *d*, but there will be no ac voltage at that point because of the very low impedance of capacitor C_5.

Procedure

step one Wire the circuit of Fig. 8-19.

step two Using the dc voltmeter, measure the dc voltages at the following points.

All dc voltages are to be measured with respect to dc common point *e*.

dc voltage at point *a* ________________ volts dc

dc voltage at point *b* ________________ volts dc

dc voltage at point *c* ________________ volts dc

dc voltage at point *d* ________________ volts dc

step three Using the ac voltmeter shown in Fig. 8-19, measure the voltage between points *d* and *e*. Record the value.

________________ volts ac

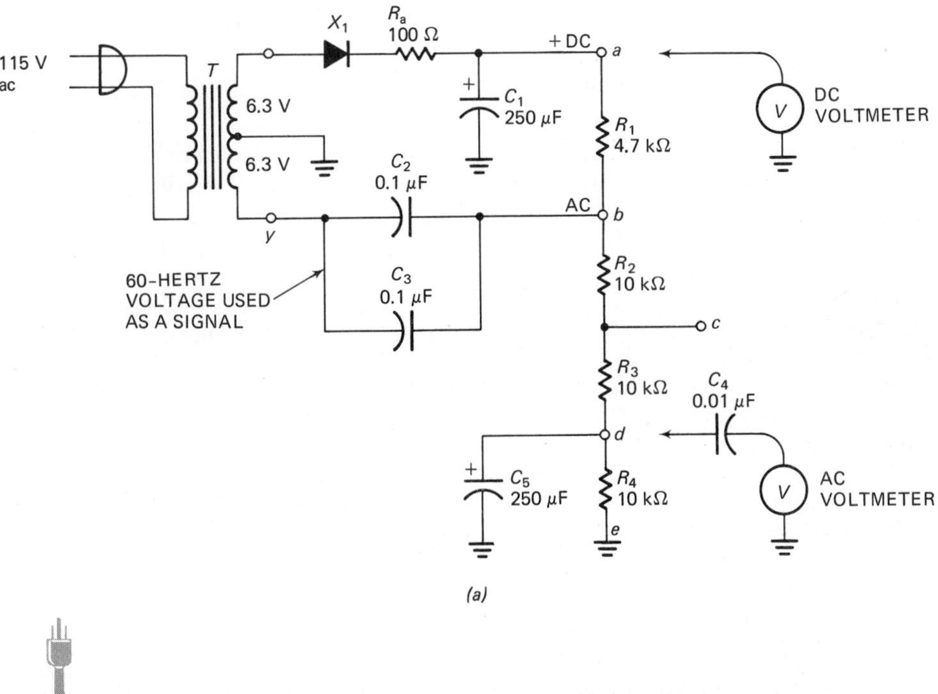

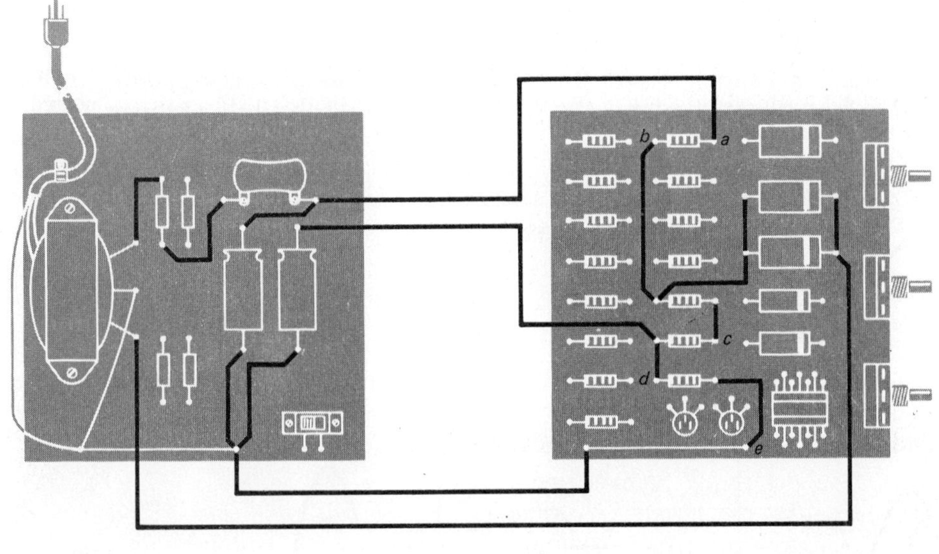

Fig. 8-19 Test setup for Part I of experiment. (*a*) Schematic diagram. (*b*) Pictorial drawing.

Your measurement in Step 3 should be close to 0 volts. Capacitor C_5 acts almost as a short circuit for an ac signal across R_4. Point *d* is the common point for ac signals.

step four Connect the common lead of the ac voltmeter to point *d*. All ac voltage measurements will be taken with respect to point *d*. Measure and record the ac voltages at the following points.

ac voltage at point *a* __________________ volts ac

ac voltage at point *b* __________________ volts ac

ac voltage at point *c* ______________ volts ac

ac voltage at point *y* ______________ volts ac

step five Your ac voltage measurement at point *a* should be almost 0 volts. Which component causes this voltage to be almost 0 volts?

____

Your answer should be C_1. This large capacitance value acts almost as an ac short circuit from point *a* to the common point. The common point at *d* is reached through C_1 and C_5.

Conclusion You measured an ac voltage and a dc voltage at points *b* and *c*. This means that both kinds of voltage can exist at a point at the same time.

At point *d* you measured a dc voltage but no ac voltage. Point *d* is the common point for the ac signal. In practice the dc voltage at the ac common point may be a positive or negative value.

PART II

Theory There are three possible ways to connect amplifier devices. We are considering three amplifying components: tubes, bipolar transistors, and FETs. The three possible ways to connect an amplifier are shown in Fig. 8-20. A bipolar transistor is shown, but you could substitute a tube or FET for

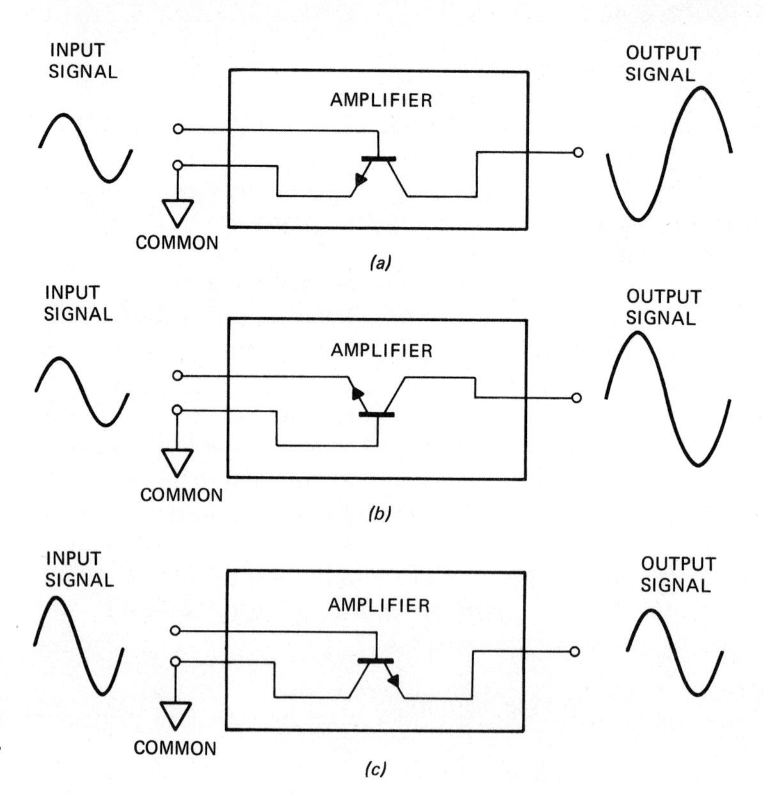

Fig. 8-20 **Amplifiers.** **(*a*) Common-emitter.** **(*b*) Common-base.** **(*c*) Common-collector.**

each type of circuit. Thus a common-base circuit would be a common-grid (or grounded-grid) circuit if a tube were used. It would be a common-gate (or grounded-gate) circuit if an FET were used.

The common-emitter circuit of Fig. 8-20*a* is the one you will most often work with. The input signal is to the base, and the output signal is taken from the collector. This type of amplifier circuit has a high voltage gain. The output signal is 180° out of phase with the input signal.

The common-base circuit of Fig. 8-20*b* has the input signal to the emitter, and the output signal is taken from the collector. There is a high voltage gain and no phase inversion between the input and output signals. You will find common-base amplifiers used in high-frequency electronic systems.

The common-collector circuit of Fig. 8-20*c* is also known as an *emitter follower.* The input signal is to the base and the output signal is taken from the emitter. There is no voltage gain with this type of circuit. It is used for impedance matching. The input impedance is high, and the output impedance is low. (An explanation of impedance matching was given in Chapter 4.) There is no phase inversion between the input and output signals in emitter-follower circuits.

The amplifiers of Fig. 8-20 have components and dc voltages that are not shown. Only the input and output signal points are given along with the ac common point.

In this experiment you will show that a bipolar transistor can be used as a common-emitter, common-base, or common-collector amplifier. You will be using a voltage amplifier, but the same types of circuits can also be used with power amplifiers.

Test Setup Figure 8-21 shows the circuit for this experiment.

A full-wave rectifier is used for a power supply The rectifiers are X_1 and X_2. The filter is C_1.

Simple bias is used for the transistor circuit. Two resistors (R_1 and R_2) are used in the bias circuit. When R_2 is adjusted to its minimum value, the resistance of the base circuit consists only of R_1.

If it were not for R_1, it would be possible to adjust R_2 so that the base of the transistor is at $B+$. This would produce excessive base current and destroy the transistor. Therefore resistor R_1 protects the transistor by limiting the base current.

We will simulate an input signal at the base by varying R_2. This will cause the base voltage to change. This is what an input signal normally does—it varies the base voltage.

The input signal to the emitter, when required, will be produced by varying R_4.

Procedure

step one Wire the circuit shown in Fig. 8-21. Figure 8-21*a* shows the schematic diagram and Fig. 8-21*b* shows the pictorial drawing. Set R_2 and R_4 to the center of their adjustment.

step two Measure and record the dc voltage at point A with respect to common (ground). V_A = ________________ volts dc

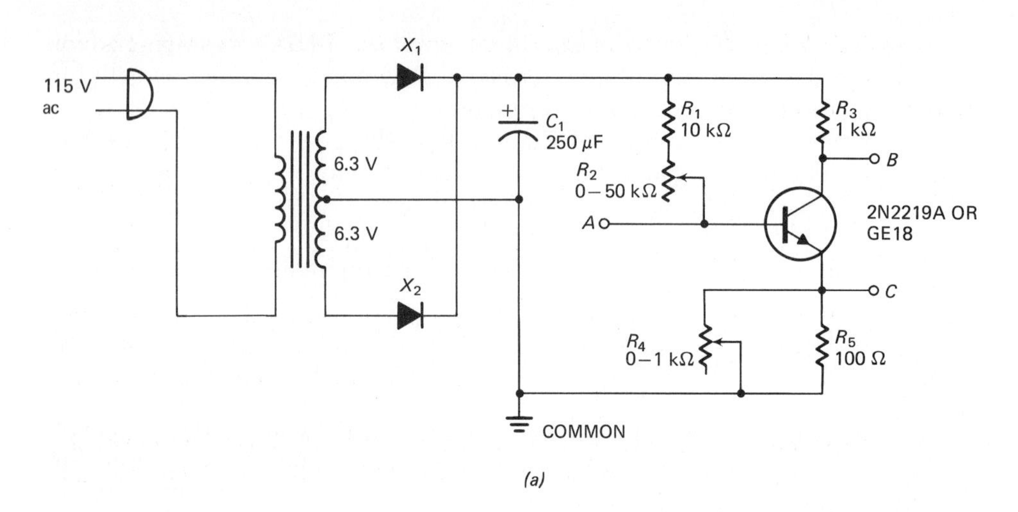

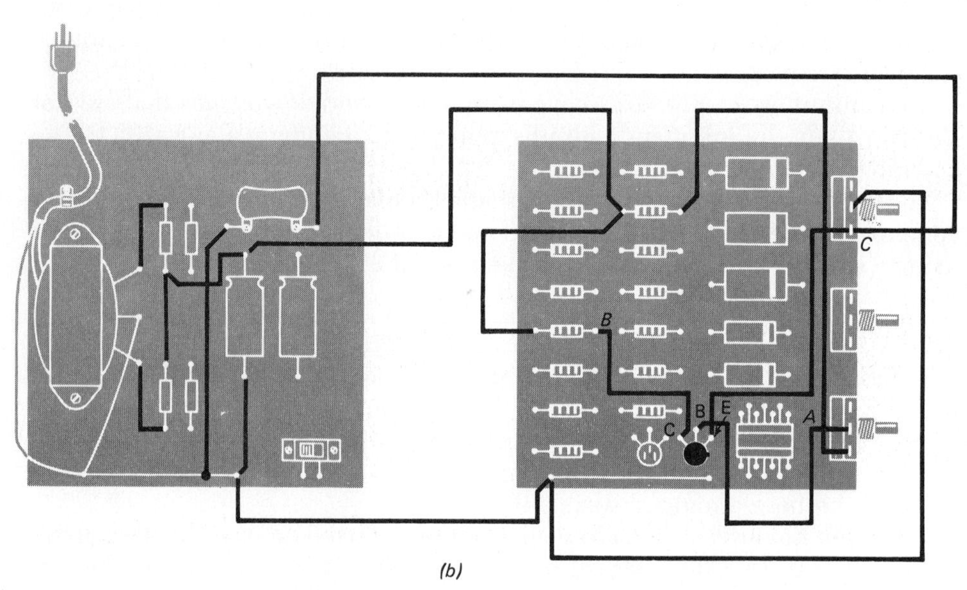

Fig. 8-21 Test setup for Part II of experiment. (*a*) Schematic diagram. (*b*) Pictorial drawing.

step three Adjust R_2 to its full clockwise position. Did the voltage at point A increase or decrease? ________________

If you have wired the circuit correctly, the voltage at point A should increase when you turn the shaft of R_2 clockwise.

step four Return the shaft of R_2 to the center of its range. Measure and record the collector voltage at point B.

V_B = ________________ volts dc

step five Adjust R_2 to its full clockwise position. Did the voltage at point B increase or decrease? ________________

The collector voltage should decrease. This is a very important point. When the base voltage became more positive, the collector voltage became less positive. In other words, the output change is the reverse of the input change.

The reason for the reversal is easy to understand. When the base is made more positive, the base current increases. This causes the collector current to increase. The voltage across R_3 increases when the collector current increases.

When the voltage drop across R_3 increases point B becomes less positive.

step six Return the shaft of R_2 to the center of its rotation. Measure and record the emitter voltage at point C.

V_C = ________________ volts dc

step seven Turn the shaft of R_2 clockwise to make the base voltage of the transistor more positive. Does the emitter voltage as measured at point C increase or decrease? ________________

The emitter voltage becomes more positive when the base becomes more positive. There is no phase inversion in this circuit.

step eight Return the shaft of R_2 to the center of its rotation. Measure and record the emitter voltage at point C.

V_C = ________________ volts dc

step nine Turn the shaft of R_4 clockwise. Does the emitter voltage increase or decrease? ________________

If you wired the circuit correctly, the emitter voltage will become more positive as the shaft of R_4 is turned clockwise.

step ten Return the shaft of R_4 to the center of its rotation. Measure and record the collector voltage at point B.

V_B = ________________ volts dc

step eleven Turn the shaft of R_4 fully clockwise. Does the collector voltage increase or decrease? ________________

The collector voltage should increase.

Conclusion When the base voltage is changed both the collector and emitter voltages also change. The collector voltage becomes less positive when the base is made more positive. This means that the output signal at the collector is 180° out of phase with the base signal. When the input signal is at the base and the output signal is taken from the collector as here, the amplifier is of the common-emitter type.

A change in base voltage changes the emitter voltage, but there is no phase inversion. If the input signal is at the base and the output signal is at the emitter, the amplifier is a common-collector type.

An input signal to the emitter produces an output signal at the collector. There is no phase inversion. When the input signal is at the emitter and the output signal is at the collector the amplifier is a common-base type.

SELF-TEST WITH ANSWERS

(Answers with discussions are given at the end of the chapter.)

1. Figure 8-22 shows the input signal and output signal for a certain amplifier. This amplifier is operated (*a*) class A; (*b*) class B; (*c*) class AB; (*d*) class C.
2. In a tube amplifier circuit you cannot obtain class C operation with (*a*) power-supply bias; (*b*) battery bias; (*c*) grid-leak bias; (*d*) cathode bias.
3. Bipolar transistors in receivers are seldom (if ever) operated (*a*) class A; (*b*) class B; (*c*) class C; (*d*) class AB.
4. Which of the following is used to keep power bipolar transistors operating at a safe temperature? (*a*) A thermocouple; (*b*) A heat sink; (*c*) A hot wire; (*d*) A tracer.
5. Which of the following is used to prevent the signal of one amplifier from getting into another amplifier through a common power-supply connection? (*a*) A NOT circuit; (*b*) a NOR circuit; (*c*) A cleaning circuit; (*d*) A decoupling filter.
6. Which of the following circuits is called an amplifier even though it does not have any voltage gain? (*a*) A common-cathode amplifier; (*b*) A common-collector amplifier; (*c*) A common-base amplifier; (*d*) A common-source amplifier.
7. Is the following statement true or false? It is possible to have an ac voltage and a dc voltage across a resistor at the same time. (*a*) True; (*b*) False.
8. Is the following statement true or false? It is possible to have an alternat-

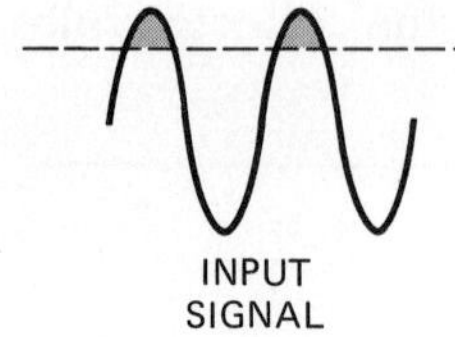

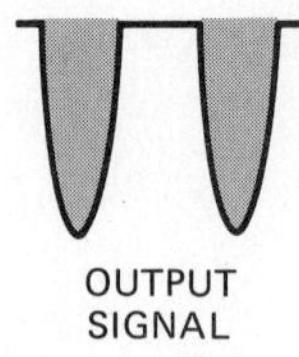

Fig. 8-22 In what class of operation is this amplifier being operated?

ing current and a direct current flowing through a resistor at the same time. (*a*) True; (*b*) False.

9. The input signal to the emitter of a common-base NPN voltage-amplifier circuit is increasing in the positive direction. The collector voltage is (*a*) increasing in the positive direction; (*b*) increasing in the negative direction.

10. Which is more efficient? (*a*) A class A amplifier; (*b*) A class C amplifier.

ANSWERS TO SELF-TEST

1. (*d*)—The waveforms in Fig. 8-22 are the same as the waveforms in Fig. 8-5.

2. (*d*)—To get class C operation the tube must be cut off. You cannot cut a tube off with cathode bias. See Table 8-1.

3. (*c*)—In most receiver circuits the amplifiers are operated class A, class AB, or class B.

4. (*b*)—Figure 8-11 shows examples of heat sinks.

5. (*d*)—Figure 8-10 shows two amplifiers connected to the same power supply. A decoupling filter is used in the power-supply line to the voltage amplifier.

6. (*b*)—A common-collector circuit is also called an *emitter follower.* It does not have a voltage gain. Another way of saying this is that the voltage gain

$$\frac{\text{Output signal voltage}}{\text{Input signal voltage}}$$

is less than 1.0.

7. (*a*)—This was demonstrated in the experiment.

8. (*a*)—The ac voltage and dc voltage across a resistor are obtained with alternating- and direct-current flow.

9. (*a*)—This was demonstrated in the experiment.

10. (*b*)—The advantage of class A amplifiers over class C amplifiers is that the class A amplifier does not distort the signal waveform.

how do oscillators work? 9

INTRODUCTION

An *oscillator* can be defined as a circuit that converts dc to ac. This is a very broad definition, and it covers all types of oscillators used in electronic circuitry. Oscillators are used in all receivers and transmitters and in many types of test equipment.

Figure 9-1 shows the basic parts of an electronic oscillator that produces a sine-wave voltage. A power supply is needed for operating the amplifier, which is a very important part of the oscillator circuit. The amplifying component may be a tube or transistor. The oscillator usually has some form of tuned circuit to determine the frequency.

There is a positive feedback signal employed in all oscillator circuits. Positive feedback is also called *regenerative feedback*. With this type of feedback a small part of the output signal from the amplifier is fed back to the input. You will understand this better when you study the oscillator circuitry in this chapter.

In the basic block diagram of Fig. 9-1 the input to the system is a dc voltage from a power supply and the output from the system is a sine-wave (ac) voltage.

All oscillators used in electronics can be divided into two different types: *sinusoidal* and *nonsinusoidal*. These names refer to the type of waveform of the output. Figure 9-2 shows some of these waveforms. If the output waveform is a sine wave, as shown in Fig. 9-2*a*, the oscillator is said to be a sinusoidal oscillator. If the output is a square wave, a sawtooth, or any other waveform of the type shown in Fig. 9-2*b*, the oscillator is called a nonsinusoidal oscillator. You will often see the term *relaxation oscillator* used to refer to the types that produce waveforms like the ones shown in Fig. 9-2*b*.

In addition to being identified by the type of waveform, oscillators may also be identified by the frequency of oscillation. Thus you have *audio oscillators* that produce an audio-frequency waveform. *RF oscillators* produce a radio-frequency sine-wave output.

In some cases the oscillator circuit is named for its inventor. Thus you have *Armstrong, Hartley*, and *Colpitts* oscillators.

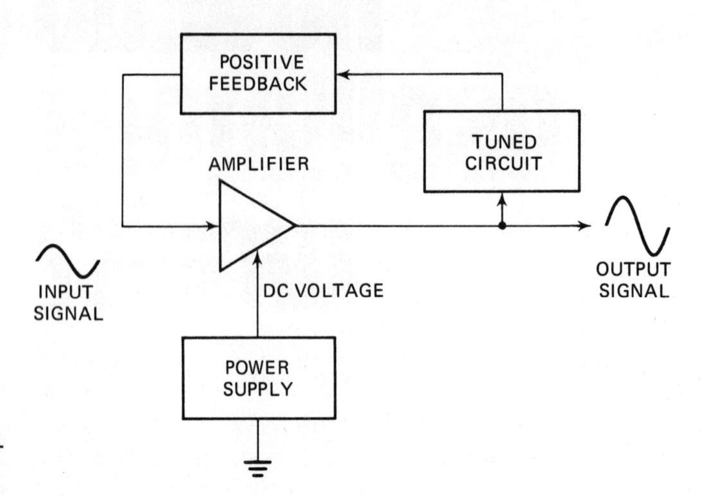

Fig. 9-1 Block diagram of an oscillator.

In this chapter you will study oscillator circuits and learn to identify some of the more important types of oscillators.

You will be able to answer these questions after studying this chapter.

- ☐ **What are the amplifier configurations?**
- ☐ **What is the flywheel effect?**
- ☐ **What are examples of sine-wave oscillator circuits?**
- ☐ **How does a relaxation oscillator work?**
- ☐ **What is an example of a relaxation oscillator circuit?**

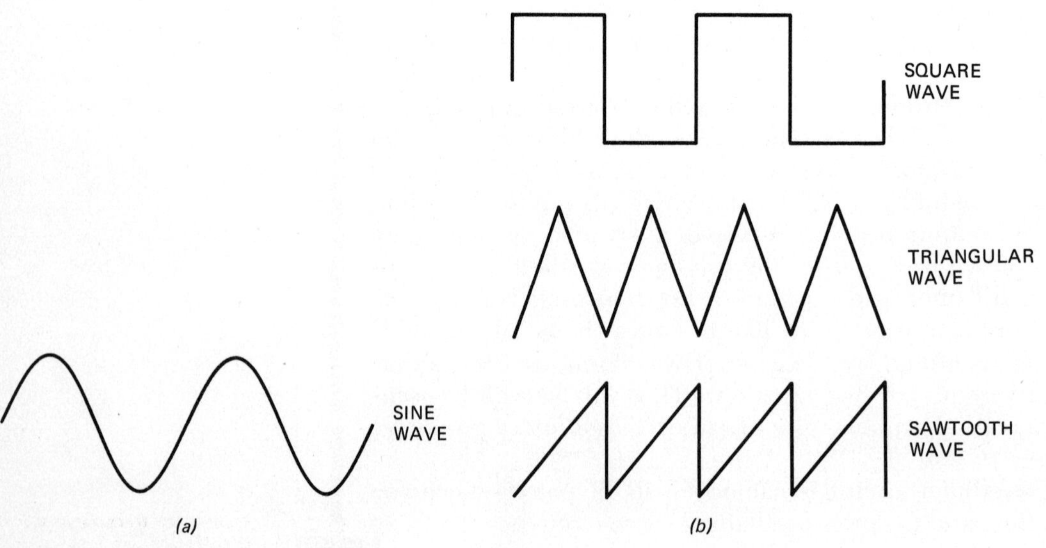

Fig. 9-2 Oscillator waveforms. (*a*) Sinusoidal. (*b*) Nonsinusoidal.

INSTRUCTION

What Are the Amplifier Configurations?

In the Experiment section of Chapter 8 you learned a very important principle related to amplifier circuits: The amplifier is a three-terminal device. The input signal is delivered to one terminal, the output signal is taken from another terminal, and the third terminal is common to both. This concept is important in understanding how an oscillator circuit works.

The three basic forms of amplifiers, called *amplifier configurations*, are shown in Fig. 9-3. In Fig. 9-3*a* the input signal is delivered to the control electrode: the base of a bipolar transistor, gate of an FET, or control grid of a tube. The output signal is taken from the dc output terminal. This is the collector or drain or plate. The dc input terminal is common to both the input and output signals. The dc input terminal is the emitter or source or cathode.

Amplifiers like the ones shown in Fig. 9-3*a* are called *conventional*. They have a high *voltage gain*. This means that the output-signal voltage is much larger in amplitude than the input voltage. Also the output signal is 180° out of phase with the input signal. This means that when the input signal is

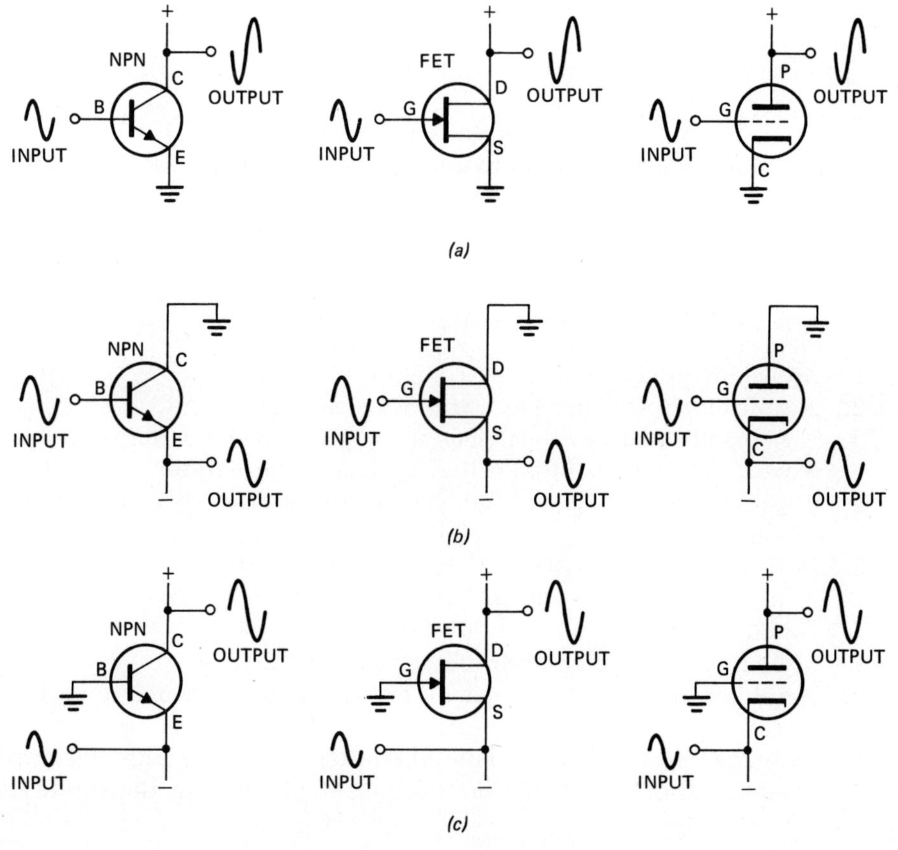

Fig. 9-3 Signal flow in basic amplifiers. (*a*) Conventional. (*b*) Follower. (*c*) Common-control electrode.

going positive the output signal is going negative. Likewise, when the input signal is going negative the output signal is going positive.

Figure 9-3*b* shows a common-collector circuit. Here the input signal is at the base or gate or control grid. The output signal is taken from the emitter or source or cathode. This type of amplifier circuit is called a *follower.* Thus the bipolar transistor circuit is an emitter follower, the FET circuit is a source follower, and the tube circuit is a cathode follower. There is no voltage gain possible with a follower circuit. In other words, the output signal always has less amplitude than the input signal. The output signal is in phase with the input signal. This means that both go positive at the same time and both go negative at the same time.

The follower circuit seems to be useless as an amplifier. However, it has a very important use in electronics. It is used for *impedance matching.* The circuit has a high input impedance and a low output impedance.

Figure 9-3*c* shows amplifiers with a common-control electrode. With this type of circuit the input signal is delivered to the emitter or source or cathode. The output signal is taken from the collector or drain or plate. The input and output signals are in phase, and the circuit has a high voltage gain. This type of circuit is often used as a high-frequency amplifier.

The amplifiers of Fig. 9-3 are sometimes given names that tell which electrode is common or grounded. The bipolar transistor circuit in Fig. 9-3*a* may be called a *common-emitter* or *grounded-emitter* amplifier. As another example, the bipolar transistor circuit of Fig. 9-3*b* is also called a *common-collector* amplifier.

In the introduction of this chapter we noted that an oscillator circuit employs an amplifier with positive feedback. This means that the feedback signal from the output of the amplifier must be returned to the input in phase with the input signal. In order to understand these feedback circuits it is important to know the phase of the output signal in relation to the input signal. That is why Fig. 9-3 is so important.

Summary

1. An oscillator is a circuit that changes dc to ac.
2. A sinusoidal oscillator has a sine-wave voltage output.
3. A nonsinusoidal oscillator has an output waveform that is not a sine wave. This type is often called a *relaxation oscillator.*
4. A sine-wave oscillator circuit has a dc power supply, an amplifier, a tuned circuit, and a positive feedback circuit.
5. Positive feedback is also called *regenerative* feedback.
6. Amplifiers are three-terminal circuits in which one of the electrodes is common.
7. The amplifier configurations are shown in Fig. 9-3.
8. Amplifiers are sometimes named to identify the common or ground terminal.
9. Only the conventional (grounded emitter, source, or cathode) amplifier has an output signal that is 180° out of phase with the input signal.

10. Follower circuits have no voltage gain. The equation for the voltage gain A_v of an amplifier is

$$A_v = \frac{\text{output signal voltage}}{\text{input signal voltage}}$$

What Is the Flywheel Effect?

Many sine-wave oscillators use an *LC* circuit like the one shown in Fig. 9-4. This circuit can produce an oscillating current.

In Fig. 9-4*a* a permanent magnet is moved past inductor *L* so that its magnetic field moves through the coil. According to *Faraday's law,* anytime a magnetic field cuts across a conductor, a voltage is induced. Therefore the motion of the magnet produces a voltage across *L*. This voltage causes capacitor *C* to charge, and the circuit is now ready to begin its oscillation. (It is important to note that the same effect could be obtained by connecting a voltage directly across *C* using a battery.)

In Fig. 9-4*b* the capacitor has begun to discharge. The discharge current flows from minus to plus on the capacitor as shown by the arrow. This current flows through coil *L* and sets up a magnetic field around the coil.

Once the capacitor is completely discharged the current tries to stop. At this time the magnetic field around *L* begins to collapse and induces a voltage which keeps the current going. This is shown in Fig. 9-4*c*. Note that the voltage polarity across the coil is such that it keeps the current flowing in the same direction as it was going. The result is that the capacitor will now be charged in the opposite direction as it was in Fig. 9-4*a*.

When the capacitor is fully charged, the current in Fig. 9-4*c* will stop flowing. The magnetic field around the coil has collapsed completely. Now the capacitor begins to discharge. The discharge current is shown in Fig. 9-4*d*. This current sets up a magnetic field around *L* again.

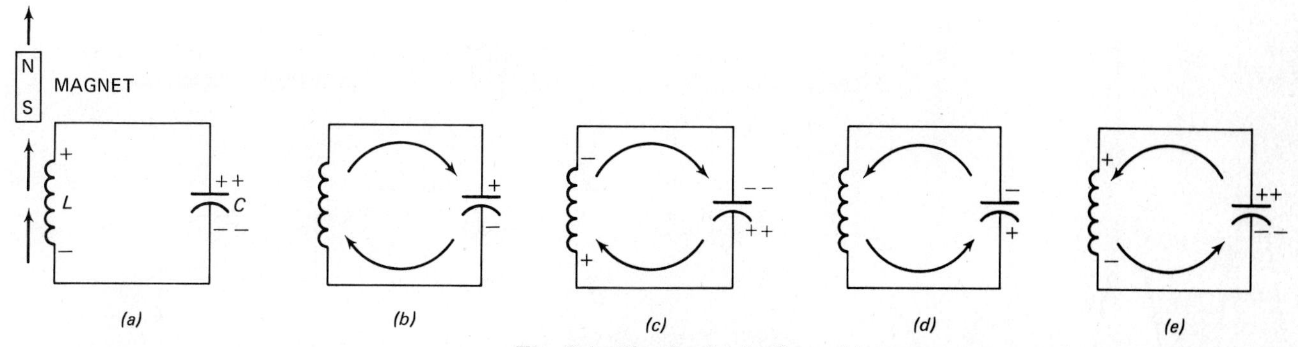

Fig. 9-4 The flywheel effect. (*a*) Moving magnet induces a voltage which charges the capacitor. *C* is shown in charged state. (*b*) Capacitor discharges. *C* is shown partially discharged. (*c*) Countervoltage charges capacitor. *C* is shown in recharged state. (*d*) Capacitor discharges. *C* is shown in partially discharged state. (*e*) Countervoltage charges capacitor. *C* is shown in charged state.

Once the capacitor has become fully discharged, the current tries to stop. However, the magnetic field around the coil collapses and keeps the current going to charge the capacitor in the opposite direction. This is shown in Fig. 9-4*e*.

The above chain of events is called the *flywheel effect.* The capacitor is charged first in one direction and then in the other. The voltage across the capacitor is an ac voltage. If you were able to look at this voltage waveform, it would be a sine wave.

If there were no losses in the capacitor and coil, this sine-wave voltage would go on and on forever. In reality there are losses in the circuit, due mostly to the resistance of the coil wire. There is also some loss in the dielectric of the capacitor. For each cycle of operation the losses take their toll.

The final result is a waveform that looks like the one in Fig. 9-5. It is a sine wave except that each peak decreases as time increases. The waveform is known as a *damped wave.* The lower the amount of loss, the longer the oscillations will continue.

In order to keep the oscillations going it is only necessary to supply a small amount of energy to the system to replace the energy that is lost due to the losses. Figure 9-6 shows how energy can be returned to the system using a transformer winding. The pulses at the primary of this transformer arrive at just the right moment so that they add to the energy supplied by the coil when the field collapses.

Figure 9-7 shows the circuit for a *ringing oscillator.* The *LC* oscillator circuit, which is often called a *tank circuit,* is in the cathode lead for the triode. The triode is normally cut OFF by battery *E*. Input pulses at the grid turn the tube ON at just the right moment so that the cathode current adds to the coil current. The additional energy supplied to the circuit prevents the oscillations from dying out.

A very important point to remember is that *the LC circuit provides the oscillation.* The tube serves as a switch to supply energy to the tank circuit at the right moment.

You will hear terms like *oscillator tube* and *oscillator transistor.* Do not

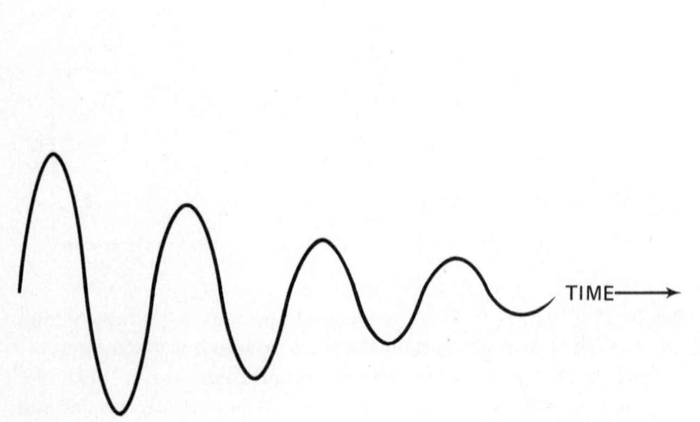

Fig. 9-5 Damped wave.

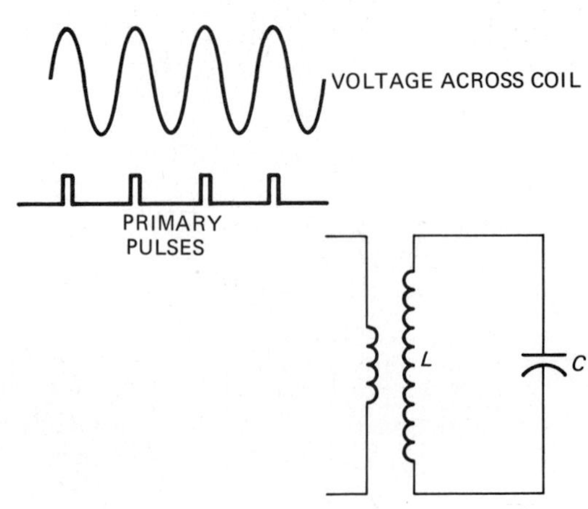

Fig. 9-6 One way to replace energy in a tank circuit.

let these terms confuse you. The oscillation occurs in the tank circuit, *not* in the amplifying component.

What Determines the Frequency of Oscillation?

For every combination of L and C in the circuit of Fig. 9-4, there is one frequency of oscillation, called the *resonant frequency*. The following equation gives the value of frequency:

$$f_r = \frac{1}{2\pi\sqrt{LC}}$$

where f_r = the frequency of oscillation
L = the inductance of the coil, in henrys
C = the capacitance of the capacitor, in farads

From this equation you can see that the resonant frequency can be changed by changing L or C. If the circuit is used in an oscillator, then the oscillator frequency can be tuned with a variable inductance or a variable capacitor.

Summary

1. The LC circuit that determines frequency in an oscillator is often called a *tank circuit.*
2. Once an oscillating current is started in the tank it will continue to flow back and forth for some time. This is called the *flywheel effect.*
3. Losses in a tank circuit cause the amplitude of the flywheel current to decrease with each cycle. The result is a damped wave.
4. In an oscillator circuit the amplifier releases energy to the tank circuit periodically. This replaces energy losses.
5. In an oscillator circuit it is the LC combination that provides the sine-wave voltage.
6. The frequency of oscillation for an oscillator tank circuit is given by the equation

$$f_r = \frac{1}{2\pi\sqrt{LC}}$$

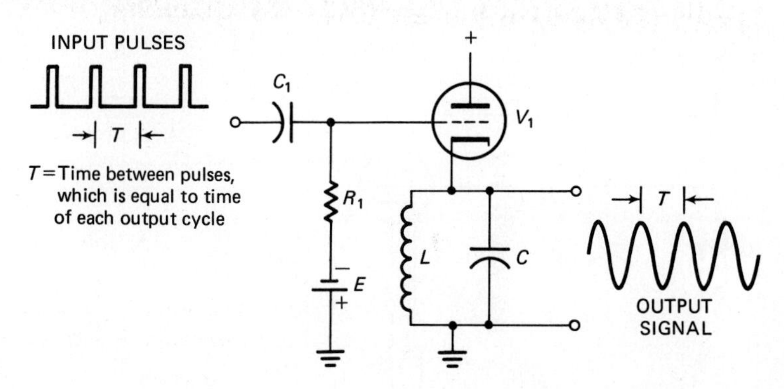

Fig. 9-7 A ringing oscillator.

7. The equation for frequency shows that it does not depend upon the power-supply voltage or the amount of amplification.

What Are Examples of Sine-Wave Oscillator Circuits?

Sine-wave oscillators used in electronic circuits have the basic form shown in Fig. 9-1. When you are analyzing an oscillator circuit you should look for each of the sections shown in this block diagram.

As a general rule, if the tuned circuit has a variable capacitor in it, then the oscillator is capable of generating radio-frequency (very high frequency) signals. If the tuned circuit consists of fixed inductors and capacitors, or fixed capacitors and resistors, then the oscillator is usually an audio-frequency (or lower-frequency) type. There are some exceptions to these rules. For example an audio frequency generator that is used for testing audio circuits will have a variable capacitor frequency adjustment.

What Is an Armstrong Oscillator?

Figure 9-8 shows an Armstrong oscillator that uses a vacuum-tube amplifier. The plate voltage for this amplifier, V_1, is delivered through coil L_1. Grid-leak bias is used, and the bias circuit is made up of R_1 and C_2. You will remember that with this type of bias there is no negative voltage on the grid unless there is an input grid signal.

Coils L_1, L_2, and L_3 in the circuit are actually parts of a single transformer, with L_1 being the primary and L_2 and L_3 being secondaries. The feedback for the oscillator circuit is obtained between windings L_1 and L_2. The output signal is taken from winding L_3.

The tuned circuit for the oscillator is composed of C_1 and L_2. This circuit sets the frequency of the oscillator.

Now that we have discussed each of the parts of the oscillator circuit, it is

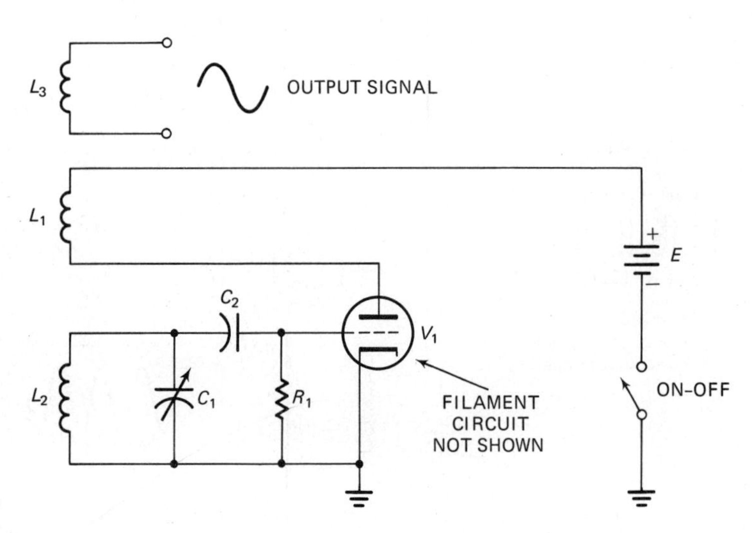

Fig. 9-8 Armstrong oscillator.

possible to explain its operation. When the ON/OFF switch is turned to the ON position a positive voltage is applied to the plate of V_1, and it begins to conduct heavily. Remember that grid-leak bias is used and there is no input signal, so there is now no bias on the grid. The rapidly increasing plate current through L_1 causes an expanding field that cuts across L_2 and induces a voltage in it.

The voltage induced across L_2 starts the flywheel current flowing in the tank circuit. Capacitor C_2 couples the sine-wave voltage across C_1 to the grid-leak circuit.

The tube acts as a valve in the circuit. At exactly the right moment a positive voltage is delivered through C_2 to the grid to permit the tube to conduct. When it conducts, the expanding magnetic field around L_1 induces a voltage across L_2. This voltage adds energy to the tank circuit to replace lost energy.

The circuit of Fig. 9-8 is an RF oscillator. Note that C_1 is a variable capacitor across L_2. In some RF oscillators and amplifiers the tank-circuit capacitor is a fixed value and the inductance value is changed. In either case it is normal for many RF oscillators to be tuned.

What Is the Difference between Series-fed and Shunt-fed Circuits?

Figure 9-9 shows two versions of the Armstrong oscillator using transistors. The difference between the circuits is in the fact that in Fig. 9-9*a* the dc collector current flows through part of the tuned circuit L_1, but in Fig. 9-9*b* the dc collector current does not flow through any part of the tuned circuit. The distinction between these two oscillators is important. The one in Fig. 9-9*a* is said to be *series-fed*, and the one in Fig. 9-9*b* is said to be *shunt-fed*. Remember this important point: Series-fed oscillators have their amplifier current flowing through the tuned circuit, while shunt-fed oscillators do not.

The various parts of the oscillators are easily found in Fig. 9-9. In both circuits a positive voltage is fed to the collector circuit and a positive base volt-

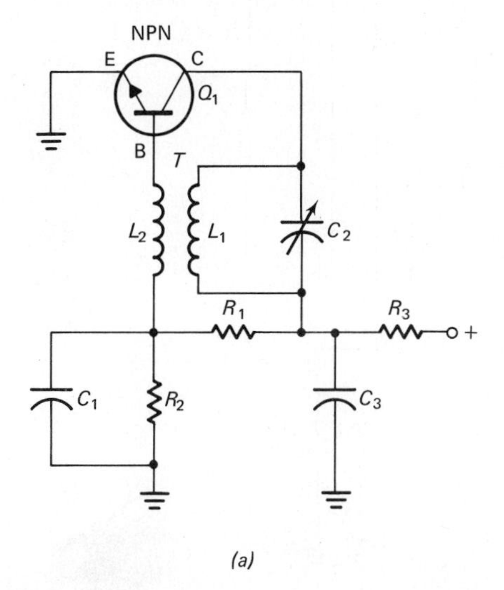

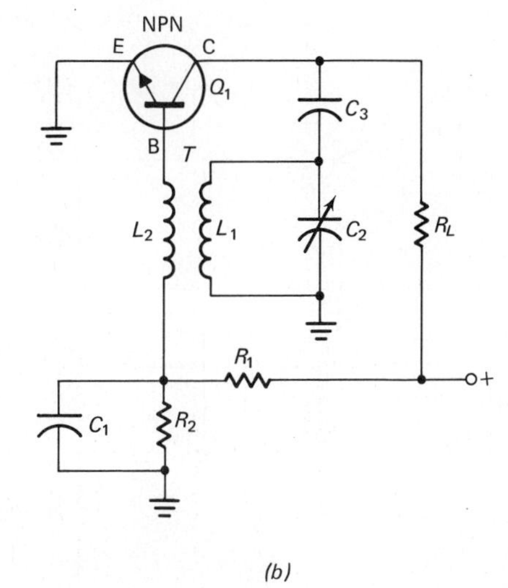

Fig. 9-9 Transistor oscillators. (*a*) Series-fed. (*b*) Shunt-fed.

age is obtained with a voltage divider consisting of R_1 and R_2. The dc bias voltage is fed through the secondary of the transformer L_2. Both amplifiers will work because they have positive collector voltages and positive base voltages.

The positive collector voltage for the circuit of Fig. 9-9*a* is delivered through tuned circuit coil L_1. Since the dc collector current must flow through L_1, it is a series-fed oscillator.

In Fig. 9-9*b* the dc collector current flows through the load resistor R_L. No part of the dc collector current can flow through L_1 because it is blocked by capacitor C_3. This is a shunt-fed oscillator.

The feedback signal for both oscillators is through transformer T. The tuned circuit consists of L_1 and C_2, and L_1 is the primary of both feedback transformers. Capacitor C_1 holds the junction of R_1 and R_2 at ac ground potential. The output signal in both oscillators is a pure sine wave at radio frequency.

Most sine-wave oscillators can be designed as either series-fed or shunt-fed, but we will not show both examples for every oscillator discussed.

What Is a Hartley Oscillator?

Figure 9-10 shows another very popular sine-wave oscillator. In this circuit, amplifier Q_1 is biased with a voltage divider (R_1 and R_2), and the collector voltage is obtained through one-half of coil L marked L_P. The oscillator is ready to operate because it has the proper dc voltages. The feedback voltage passes through capacitor C_1, which delivers the signal developed across L_S to the base of the transistor. The tuned circuit consists of capacitor C_2 across both halves of the winding of L.

Coil L is an autotransformer. The primary of the transformer is L_P and the secondary is L_S. When the varying collector current flows through L_P, its magnetic field cuts across L_S. A voltage is induced across L_S which is fed back to the transistor base through capacitor C_1. Capacitor C_1 also serves as a dc blocking capacitor. It prevents the positive voltage of the battery from being applied to the base of the transistor through L_S. Capacitor C_3 across the

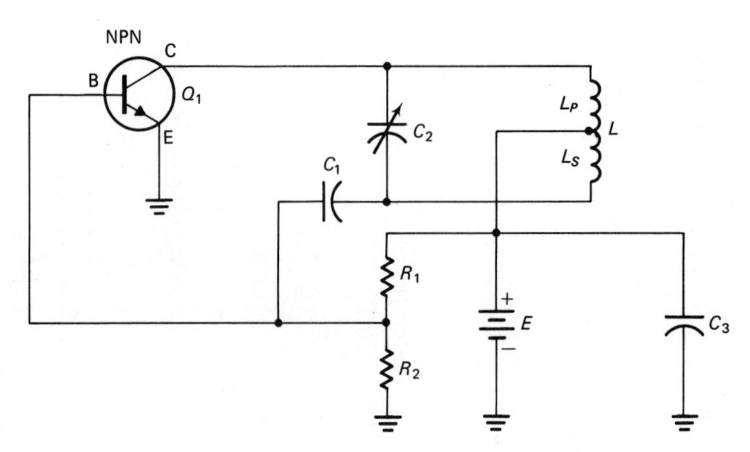

Fig. 9-10 A Hartley oscillator.

battery is used to hold the center tap of autotransformer T at ac ground potential.

The Hartley oscillator is similar in operation to the Armstrong oscillator. The primary difference is that the primary and secondary windings of the Hartley are parts of an autotransformer instead of being two separate windings.

What Is a Colpitts Oscillator?

Figure 9-11 shows a Colpitts oscillator circuit. Bias for the oscillator is obtained with voltage divider R_1 and R_2. R_3 is used for emitter stabilization. Capacitor C_1 across R_3 maintains the emitter at a steady dc potential. (You will study more about capacitors across emitter resistors in a later chapter.)

Capacitors C_3 and C_4 are grounded where they connect. This actually forms an ac split (or tap) of the tuned circuit, somewhat similar to the tapped coil of the Hartley oscillator. The feedback signal is the portion across only C_4, which is fed to the base. The tuned circuit for the oscillator consists of C_3 and C_4 across L.

Again, we have identified the amplifier section (transistor circuit), the power-supply voltages to the amplifier section, the feedback path, and the tuned circuit for the oscillator.

What Is an RC Phase-Shift Oscillator?

All the oscillator circuits you have studied so far use an LC tuned circuit to set the oscillator frequency. The oscillator circuit shown in Fig. 9-12 uses capacitor and resistor combinations to set the frequency. This type operates only at audio frequencies.

First we will look at the common-emitter amplifier portion of the oscillator. Resistors R_3 and R_4 form a voltage divider for base bias, while the collector voltage is obtained through load resistor R_5. The transistor is properly biased, and it has the proper collector voltage, so it will amplify.

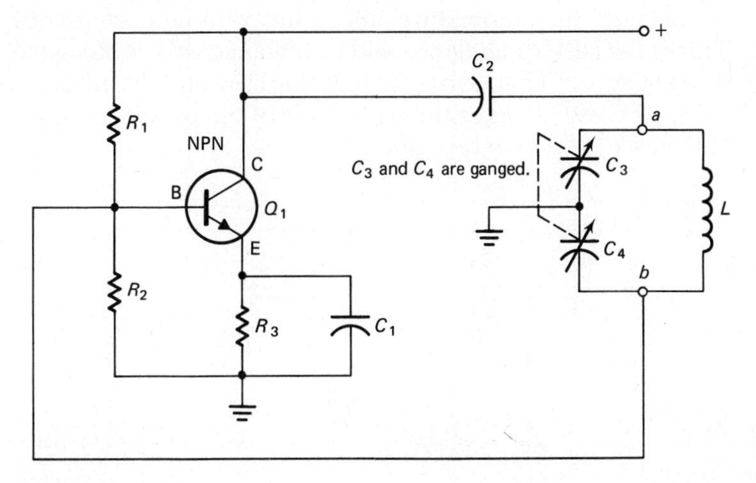

Fig. 9-11 A Colpitts oscillator.

The feedback circuit is obtained by taking the output signal across R_5 and connecting it to the base through a phase shift network.

You will remember that the collector voltage in a common-emitter amplifier is 180° out of phase with the base voltage. If the collector were fed back to the base directly, it would cancel the input signal. In an oscillator *the feedback signal must be in phase with the signal at the base.* To accomplish this the feedback network is in three sections: C_1R_1, C_2R_2, and C_3R_3. Each section shifts the phase 60°, so the total phase shift for the three sections is 180°.

To summarize, the collector voltage is 180° out of phase with the base voltage. The phase-shift network shifts the collector voltage another 180°, making a total of 360° phase shift. This is the same as having 0° phase shift or no phase shift at all. Thus the signal to the base is in phase with the signal at the collector, and the circuit will oscillate.

What Is a Crystal Oscillator?

Earlier in this chapter you learned that the resonant frequency of an *LC* tank is given by the equation

$$f_r = \frac{1}{2\pi\sqrt{LC}}$$

When the tank is used in an oscillator circuit, as in Fig. 9-9, the equation is still valid. However, it becomes more difficult to determine the true values of L and C.

In a practical circuit there is capacitance between the conductors. Also there is an input capacitance to the amplifying component. The conductors have a certain amount of inductance, especially at very high frequencies. Wiring and input capacitance, and the inductance of conductors, is distributed throughout the circuit. You will hear such terms as *distributed capacitance* or *distributed inductance* and *wiring capacitance* or *wiring inductance.* They refer to capacitance and inductance that is not in the form of individual components. The term *lumped components* refers to resistors, capacitors, and inductors that you buy at the parts supply house.

Changes in temperature affect the resonant frequency of an oscillator. This is because capacitance and inductance values change when the temperature changes. This is true for both lumped and distributed components. Because of these changes the frequency of the oscillator signal *drifts*—that is, it changes slowly over a period of time.

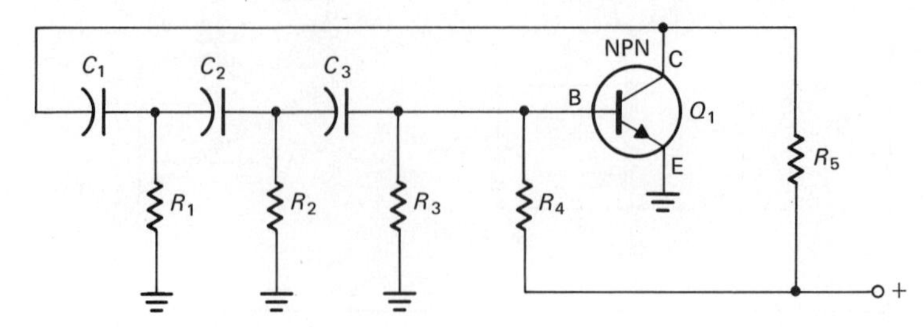

Fig. 9-12 Phase-shift oscillator.

In some cases, such as in transmitters, the frequency of oscillation must be held to very close tolerance. It must not be permitted to drift over a long period of time, and it must not have quick momentary changes.

It is possible to design an *LC* oscillator that has a steady frequency, but it is easier to use a piezoelectric crystal to hold the frequency to an exact value.

In some applications the crystal is placed in a heated container called a *crystal oven*. This container holds the temperature of the crystal constant to make the frequency more constant. Even without a crystal oven the crystal-oscillator frequency is steadier than can normally be obtained with *LC* circuitry.

What Does the Crystal Do?

Piezoelectric crystals are made of quartz, tourmaline, or barium titanate materials. The secret to the operation of crystals is in their piezoelectric effect. This means that these materials change their shape when a voltage is placed across them. Furthermore, if their shape is changed, they produce a voltage between their surfaces.

The crystal of piezoelectric material is cut to a very exact size so that its mechanical rate of vibration is an exact frequency. The relationship between mechanical vibration and frequency is illustrated in Fig. 9-13. A stick is clamped to a table so that one end of the stick is free to vibrate. If you pluck the stick, it will vibrate at a very low frequency when it is clamped as shown in Fig. 9-13*a*. In other words, when there is a large amount of stick free to vibrate, the frequency is low. When you clamp it as shown in Fig. 9-13*b*, the frequency of vibration is much higher.

You can make a basic rule from the illustration in Fig. 9-13: *The smaller the mass, the higher the frequency of vibration.* The crystal used in crystal oscillators has a very much smaller dimension than the stick shown in Fig. 9-13, but it does have a natural period of vibration which is determined by its physical size.

The symbol for a crystal is shown in Fig. 9-14. Figure 9-15 shows a circuit that imitates a crystal. (A circuit that imitates another circuit or device is called an *equivalent circuit*.) If you had perfect resistors and capacitors and coils, you could build a circuit like the one shown in Fig. 9-15, and it would do exactly the same thing as the crystal of Fig. 9-14. However, components

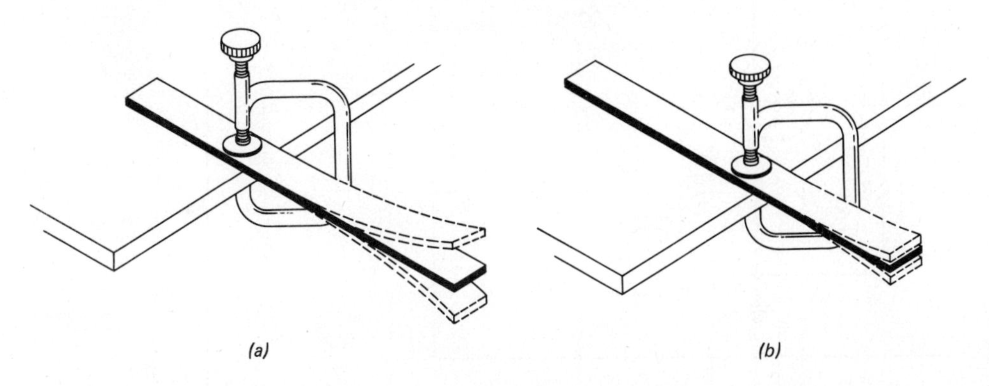

Fig. 9-13 Vibrating meter stick. (*a*) Low frequency. (*b*) High frequency.

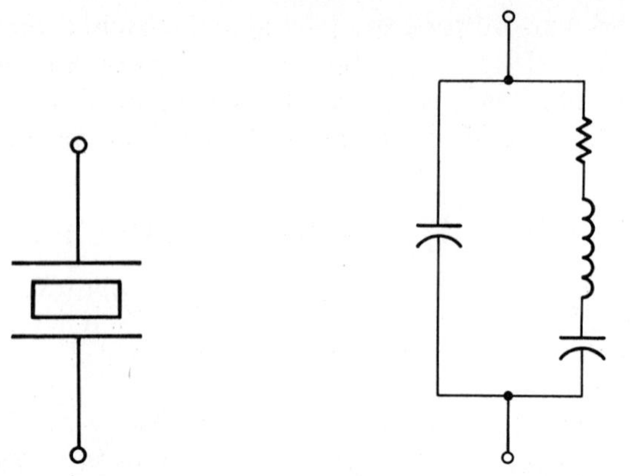

Fig. 9-14 Symbol for a crystal.

Fig. 9-15 Circuit that imitates a crystal.

are not perfect. Capacitors have leakage and inductors have loss of flux lines or incomplete linkages. Resistors are not simply resistors because they also have capacitance and, at high frequencies, inductance. The crystal has the characteristics of the circuit in Fig. 9-15, but with almost perfect component parts.

The important thing for you to remember is that the crystal vibrates and produces an electrical signal at a very steady fixed frequency.

What Is an Example of a Crystal Oscillator Circuit?

Figure 9-16 shows how the crystal can be used in a Colpitts oscillator circuit. (Compare with Fig. 9-11.) Again, we will identify the parts of the oscillator.

The amplifier is Q_1. Bias for the base is obtained with voltage divider R_1 and R_2. R_3 is the emitter stabilization resistor. Capacitor C_2 places the emitter at ac ground potential.

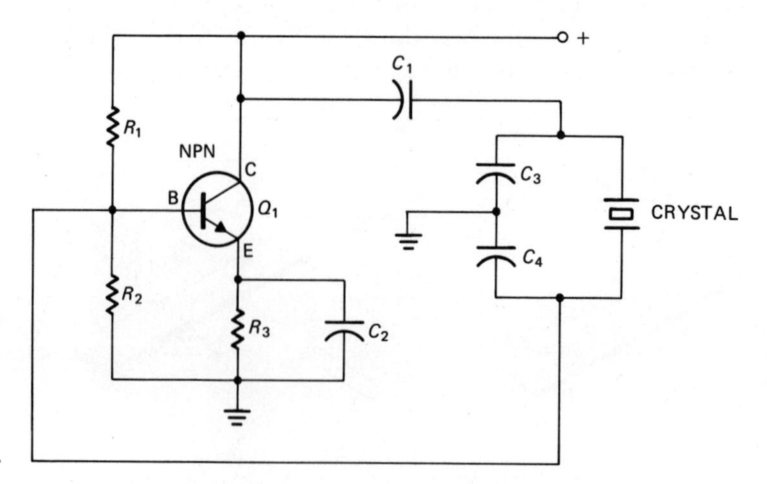

Fig. 9-16 Crystal oscillator circuit.

The feedback path is through coupling capacitor C_1, through the crystal circuit, and back to the base of the amplifier. The feedback voltage is the voltage across C_4. The crystal in the circuit vibrates at a steady frequency and produces a sine-wave oscillation voltage similar to that of an LC tuned circuit.

How Does a Relaxation Oscillator Work?

You have studied sine-wave oscillators and learned that most of them have LC tank circuits or crystals. All such oscillators have an amplifier that replaces energy in the tank circuit or crystal.

A *relaxation oscillator* depends upon the charge and discharge of voltage across a capacitor or the buildup and decay of current in an inductor. The principle of operation for a relaxation oscillator that uses a capacitor is shown in Fig. 9-17.

Figure 9-17*a* shows the basic sections of the relaxation oscillator. A dc power source E is needed to charge the capacitor C. The discharge circuit causes the capacitor to discharge periodically.

Figure 9-17*b* shows a relaxation oscillator that uses mechanical switches for the charging and discharging circuits. When switch A is closed and switch B is open, the capacitor charges through resistor R. This resistor limits the charging current, so it takes time for the voltage across the capacitor to rise to the desired peak voltage. Figure 9-17*c* shows the rising voltage across C at this time.

After the voltage across the capacitor has reached the desired peak value, switch A is opened and switch B is closed. This causes the capacitor to dis-

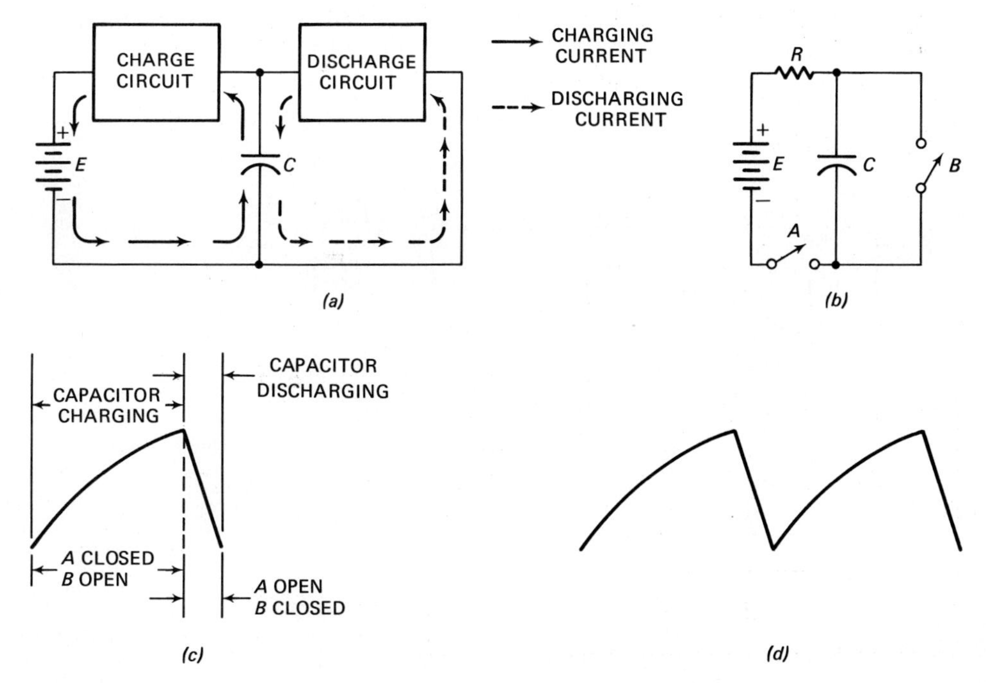

Fig. 9-17 Relaxation oscillator. (*a*) Basic parts. (*b*) Simple circuit. (*c*) One cycle. (*d*) Two cycles.

charge rapidly. This is also shown by the waveform in Fig. 9-17*c*. You will note that the time for discharge is very short compared to the amount of time required for charging.

If you could operate the switches very rapidly, you could produce a sawtooth waveform across *C* like the waveform shown in Fig. 9-17*d*. In practice electronic devices are used to control the charging and discharging of the capacitor. In practical applications an amplifying component such as a tube, transistor, or FET is used to perform the required switching.

As mentioned before, the increasing and decreasing of current through a coil could also be used to make a relaxation oscillator.

What Is an Example of a Relaxation-Oscillator Circuit?

Figure 9-18 shows one example of a relaxation oscillator. This circuit is known as a *blocking oscillator*. Such oscillators are used in TV sets and radar. (Additional relaxation oscillators will be taken up in a later chapter.) This type of oscillator generates a series of pulses, not a continuous sine wave.

The circuit of Fig. 9-18 shows the same basic kind of transformer feedback coupling as used in an Armstrong oscillator, but there is no *LC* tuned circuit. Increasing collector current causes regenerative feedback through capacitor C_1 to the base of the transistor and drives the base positive. This causes the transistor collector current to become saturated (maximum current). It also quickly charges capacitor C_1 negative at the base end. After saturation is reached the current through T_1 primary does not increase further and there is no induced feedback voltage into T_1 secondary. This causes the positive base voltage to drop and the collector current to decrease. Now the decreasing current through T_1 primary induces a negative voltage feedback which is applied to the base and quickly cuts off the transistor. Because of the prior charging of C_1, which was negative at the base end, the transistor is held cut off until C_1 can discharge enough to permit the transistor to again conduct. At

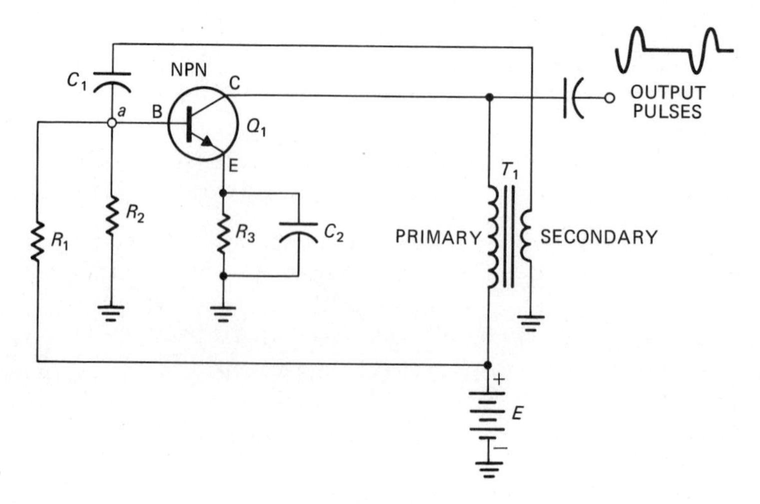

Fig. 9-18 A blocking oscillator.

this time the cycle again repeats, and the entire process continues until power is removed. The length of time C_1 takes to discharge its negative voltage is a function of the size of C_1 and of R_2, through which it discharges. As shown in Fig. 9-18, the output of the blocking oscillator is a continuous series of positive and negative pulses. Either polarity may be utilized to trigger other circuits, at the pulse repetition rate (PRR).

Summary

1. As a general rule, if there is an *LC* circuit in an oscillator that is variable, it is an RF oscillator.
2. There are exceptions to the above rule. Fixed-frequency oscillators are used in many transmitter circuits.
3. In an Armstrong oscillator the feedback signal goes from the primary to the secondary of a transformer.
4. In a series-fed oscillator the direct current through the amplifying device flows through some part of the tuned circuit.
5. In a shunt-fed oscillator circuit the direct current through the amplifying device does not flow through any part of the tuned circuit.
6. The Hartley oscillator has an inductive feedback, but it differs from the Armstrong oscillator in that an autotransformer is used.
7. In a Colpitts oscillator the feedback is determined by the two series capacitors across the tank coil.
8. The physical size of a piezoelectric crystal determines the frequency of its oscillation.
9. The equivalent circuit of a crystal is similar to the tuned circuit used in *LC* oscillators.
10. In special applications the piezoelectric crystal may be located in an oven to keep the temperature control very precise. This keeps the crystal vibrating at an exact frequency.
11. A relaxation oscillator produces a nonsinusoidal waveform. Relaxation oscillators operate on the principle of charge and discharge of capacitors, or increasing and decreasing of current through inductors.
12. A blocking oscillator is an example of a relaxation oscillator. It uses a transformer feedback circuit, and its output is a series of pulses.

PROGRAMMED REVIEW QUESTIONS

(Instructions for using this programmed section are given in Chapter 1.)

We will review the important concepts of this chapter. If you have understood the material, you will progress easily through this section. Do not skip this material because some additional theory is presented.

1. Figure 9-19 shows a
   A. special form of sawtooth wave. (Proceed to block 17.)
   B. damped wave. (Proceed to block 9.)

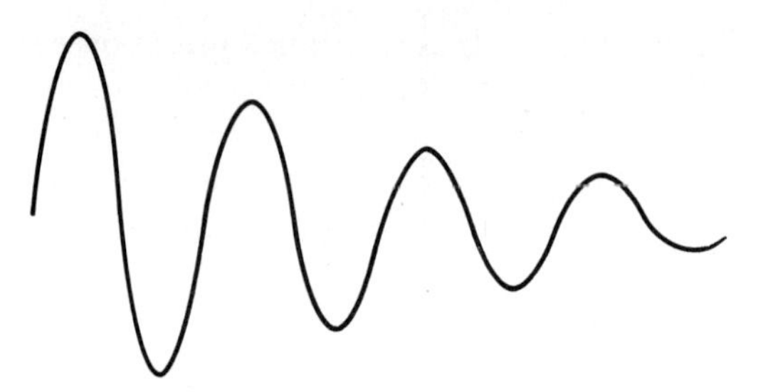

Fig. 9-19 What kind of waveform is this?

2. *The correct answer to the question in block 8 is* **B**. *The equation for resonant frequency*

$$f_r = \frac{1}{2\pi\sqrt{LC}}$$

does not involve the gain of the amplifier or the amount of voltage delivered to the amplifier. This equation shows that the resonant frequency is dependent only upon inductance and capacitance. Here is your next question.

The term *flywheel effect* refers to

A. oscillation in a relaxation oscillator. (Proceed to block 11.)

B. oscillation in an *LC* circuit. (Proceed to block 16.)

3. *The correct answer to the question in block 24 is* **A**. *A conventional amplifier is also known as a* **common-cathode, common-emitter,** *or* **common-source** *circuit. The input signal is at the grid, base, or gate, and the output signal is taken from the plate, collector, or drain. The output signal is 180° out of phase with the input signal in conventional amplifiers.* Here is your next question.

When current through the amplifying component also flows through the oscillator tank circuit, the oscillator is said to be

A. series-fed. (Proceed to block 15.)

B. shunt-fed. (Proceed to block 26.)

4. *Your answer to the question in block 8 is* **A**. *This answer is wrong. The equation for resonant frequency does not involve the gain of an amplifier. This is another way of saying that the resonant frequency is not directly affected by gain.* Proceed to block 2.

5. *Your answer to the question in block 15 is* **A**. *This answer is wrong. A common-emitter amplifier has a high voltage gain.* Proceed to block 12.

6. *The correct answer to the question in block 9 is* **B**. *A transistor must be forward-biased in order to start it into conduction.*

A tube will start to conduct when there is no bias on the grid, so a tube-oscillator circuit usually has grid-leak bias. When the circuit is first energized, the tube starts to go to saturation and the oscillating-circuit cycle is begun.

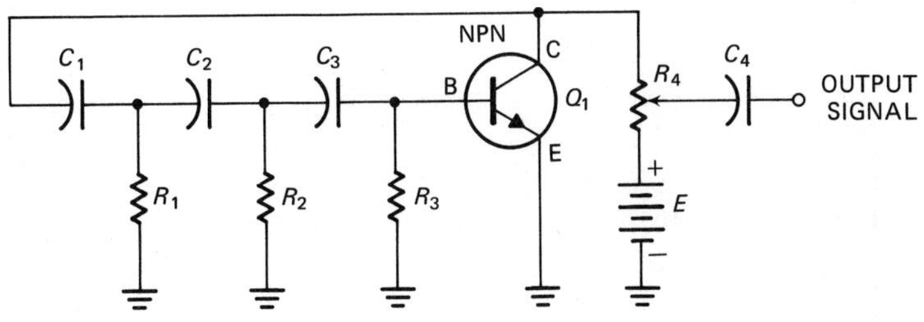

Fig. 9-20 What is wrong with this circuit?

When a bipolar transistor is used, it is necessary to forward-bias the transistor in order to start it into conduction. Unless it starts into conduction, oscillations do not occur. Here is your next question.

In a bipolar-transistor oscillator circuit the feedback signal at the base must be

A. 180° out of phase with the signal at the collector. (Proceed to block 13.)
B. in phase with the signal at the collector. (Proceed to block 19.)

7. *Your answer to the question in block 27 is* **A**. *This answer is wrong. Study the parts of a sine-wave oscillator circuit as shown in Fig. 9-1, then* proceed to block 14.

8. *The correct answer to the question in block 13 is* **A**. *The output waveform of a relaxation oscillator may be a square wave, a sawtooth wave, a series of pulses, or almost any other nonsinusoidal waveform.* Here is your next question.

The resonant frequency of a sine-wave oscillator is best changed by

A. changing the gain of the amplifier. (Proceed to block 4.)
B. changing the amount of inductance or capacitance in the tank circuit. (Proceed to block 2.)

9. *The correct answer to the question in block 1 is* **B**. *For a damped wave each new cycle has a lower amplitude than the preceding cycle.* Here is your next question.

The circuit of Fig. 9-20 will not work. Why?

A. The battery voltage is reversed. (Proceed to block 25.)
B. There is no forward bias for Q_1. (Proceed to block 6.)

10. *Your answer to the question in block 24 is* **B**. *This answer is wrong. An FET source follower is actually a common-drain circuit. The output signal taken at the source is in phase with the input signal at the gate.* Proceed to block 3.

11. *Your answer to the question in block 2 is* **A**. *This answer is wrong. Figure 9-4 shows the principle of the flywheel effect.* Proceed to block 16.

12. *The correct answer to the question in block 15 is* **B**. *A common-collector circuit has no voltage gain. In other words, the output-signal voltage is less than the input-signal voltage.* Here is your next question.
Another name for a common-collector circuit is
A. emitter-follower. (Proceed to block 18.)
B. conventional amplifier. (Proceed to block 22.)

13. *The correct answer to the question in block 6 is* **A**. *The feedback is always regenerative in an oscillator circuit.* Here is your next question.
A nonsinusoidal waveform is produced by
A. a relaxation oscillator. (Proceed to block 8.)
B. an Armstrong oscillator. (Proceed to block 23.)

14. *The correct answer to the question in block 27 is* **B**. *A tank circuit (tuned circuit) determines the resonant frequency of most sine-wave oscillator circuits.* Here is your next question.
A crystal oven is used to
A. deliver heat energy for operating the crystal. (Proceed to block 20.)
B. hold the temperature of the crystal constant so the oscillator frequency will not vary. (Proceed to block 24.)

15. *The correct answer to the question in block 3 is* **A**. *This is a definition of a series-fed oscillator.* Here is your next question.
Which of the following amplifiers has no voltage gain?
A. Common emitter. (Proceed to block 5.)
B. Common collector. (Proceed to block 12.)

16. *The correct answer to the question in block 2 is* **B**. *The flywheel current in a tank circuit is an oscillating current as shown in Fig. 9-4.* Here is your next question.
The feedback circuit of a blocking oscillator is like the feedback circuit of
A. a phase-shift oscillator. (Proceed to block 21.)
B. an Armstrong oscillator. (Proceed to block 27.)

17. *Your answer to the question in block 1 is* **A**. *This answer is wrong. A sawtooth waveform is shown in Fig. 9-17. This does not look like the waveform shown in Fig. 9-19.* Proceed to block 9.

18. *The correct answer to the question in block 12 is* **A**. *Bipolar follower circuits have their collector grounded. Their output signal at the emitter is in phase with their input signal at the base. There is no voltage gain, and they are used mostly for matching a high impedance to a low impedance.* Here is your next question.

You will remember that the equation for resonant frequency of a tank circuit is

$$f_r = \frac{1}{2\pi\sqrt{LC}}$$

Will the resonant frequency increase or decrease when you hold the inductance constant and increase the capacitance? After you have determined the answer, proceed to block 28.

19. *Your answer to the question in block 6 is* **B**. *This answer is wrong. The feedback in an oscillator must be regenerative. This means that the feedback signal must be fed back in phase with the base signal. Unless the base and feedback signals are in phase, oscillation cannot occur.* Proceed to block 13.

20. *Your answer to the question in block 14 is* **A**. *This answer is wrong. Electrical energy is used for operating a crystal. It comes from the amplifying device.* Proceed to block 24.

21. *Your answer to the question in block 16 is* **A**. *This answer is wrong. Phase-shift oscillators use an* RC *feedback network.* Proceed to block 27.

22. *Your answer to the question in block 12 is* **B**. *This answer is wrong. A conventional bipolar amplifier has a common or grounded emitter.* Proceed to block 18.

23. *Your answer to the question in block 13 is* **B**. *This answer is wrong. An Armstrong oscillator produces a sine-wave output.* Proceed to block 8.

24. *The correct answer to the question in block 14 is* **B**. *A crystal requires electrical energy, not heat energy. The purpose of the oven is to hold the crystal temperature constant. This is necessary so it can hold the frequency to an exact value.* Here is your next question. Which of the following amplifiers has an output signal that is 180° out of phase with its input signal?

A. Conventional amplifier. (Proceed to block 3.)
B. Source follower. (Proceed to block 10.)

25. *Your answer to the question in block 9 is* **A**. *This answer is wrong. The transistor is an NPN type, and its collector is connected to the positive side of the voltage source. This is the proper polarity for an NPN transistor.* Proceed to block 6.

26. *Your answer to the question in block 3 is* **B**. *This answer is wrong. Learn the definitions of series-fed and shunt-fed oscillators, then proceed to block 15.*

27. *The correct answer to the question in block 16 is* **B.** *Both the Armstrong oscillator and the blocking oscillator use a transformer for the feedback voltage. The difference between the two circuits is that the Armstrong oscillator has a tuned circuit to determine the frequency. In the blocking oscillator the repetition frequency is determined by the rate at which the input capacitor discharges.* Here is your next question.

The *LC* tuned circuit that determines the frequency in an oscillator circuit is called

A. a feedback network. (Proceed to block 7.)

B. a tank circuit. (Proceed to block 14.)

28. *Increasing the size of a capacitor will increase the denominator of the equation for resonant frequency. When you make the denominator of a fraction larger, you make the value of the fraction smaller. Therefore increasing the capacitance will decrease the resonant frequency.*

By similar reasoning, increasing the inductance would also decrease the resonant frequency. Obviously, the reverse is also true. Making either the capacitance or the inductance smaller will increase the resonant frequency.

You have now completed the programmed questions. The next step is to put some of these ideas to work in laboratory experiments. Proceed to the Experiment section of this chapter.

EXPERIMENT

(The experiment described in this section may be performed on the circuit board described in Appendix C or on a similar laboratory setup.)

Purpose In this experiment you will build a full-wave rectifier, a UJT relaxation oscillator, and a power amplifier. You will show that the frequency of oscillation is set by resistor-capacitor combination. This is typical for nonsinusoidal oscillators.

As part of this experiment you will learn some valuable methods of finding trouble in electronic circuits.

You will build the power supply, oscillator, and amplifier separately. Then you will combine them into a working unit. This can be called a *modular* approach to circuitry. In Chapter 11 the idea of a modular construction will be further extended.

Theory There are a number of components that are useful because of their *breakover* behavior. They will not conduct current until the voltage across their terminals reaches a certain minimum value.

A neon lamp is a good example of a two-terminal breakover component. It has a breakover point that is usually somewhere between 50 and 100 volts, depending upon the type of lamp. When the voltage across the lamp terminals is lower than the breakover point (called the *firing potential* of the lamp), there is no current flow. When the voltage across the lamp exceeds the firing potential, the lamp glows and conducts current. In a later chapter you will learn how this feature makes it possible to use a neon lamp in a relaxation oscillator.

Three-layer diodes (called *diacs*) and four-layer diodes are examples of solid-state two-terminal breakover components. The difference between them is that the diac can conduct in either direction, but a four-layer diode can conduct in only one direction. In both diodes there is no current flow until the voltage across their terminals reaches the breakover point.

The unijunction transistor (UJT) shown in Fig. 9-21 is an example of a three-terminal breakover component. The symbol in Fig. 9-21*a* shows that it has an emitter and two base junctions, but it does not have a collector. When the voltage between the emitter and base 1 reaches a certain minimum value, then current flows through the UJT from base 1 to base 2.

The graph of Fig. 9-21*b* shows the voltage between the emitter and base 1 on one axis and the current from base 1 to base 2 (through the UJT) on the other axis. Note that the voltage must be increased to point *a* before a large amount of current can start to flow. After current starts to flow, the voltage to maintain a large current drops to a lower value (point *b* on the curve).

Figure 9-22 shows how a UJT can be used as a relaxation oscillator. This circuit has two separate branches. One branch (R_1 and C_1) produces a ramp voltage (part of sawtooth wave) at point *a*. A ramp voltage is one that increases steadily as the capacitor charges. The other branch (R_2, Q_1, and R_3) discharges the capacitor when the voltage reaches the breakover point.

Assume the circuit has just been turned ON. Capacitor C_1 starts to charge through R_1. The solid arrows show the path of charging current. The voltage at point *a* increases as the capacitor charges. Figure 9-23 shows the waveform produced at point *a*.

When the voltage across the capacitor in Fig. 9-22 reaches the breakover

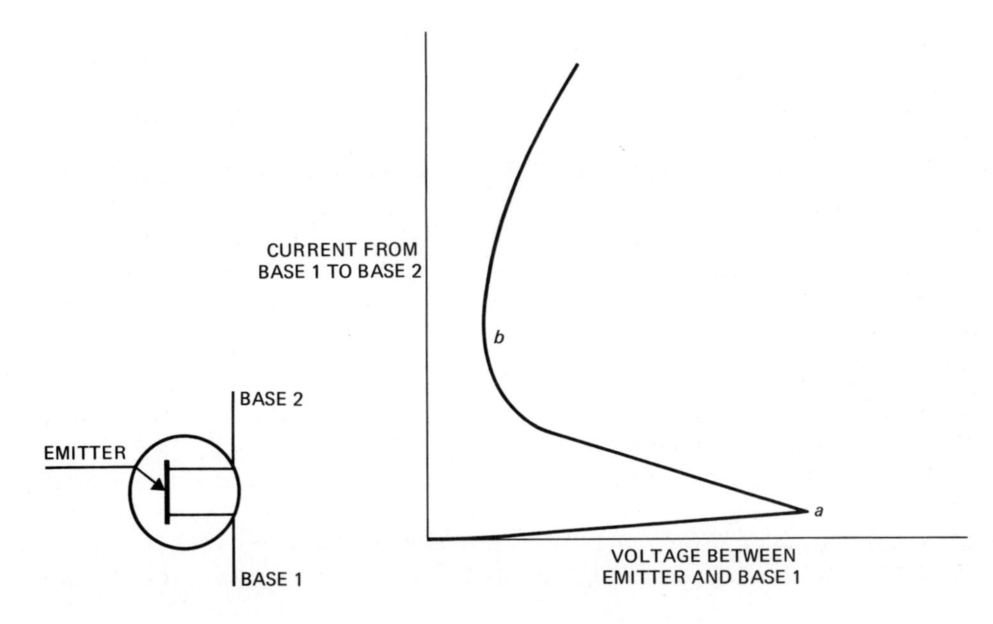

Fig. 9-21 The unijunction transistor. (*a*) Symbol. (*b*) Characteristic curve.

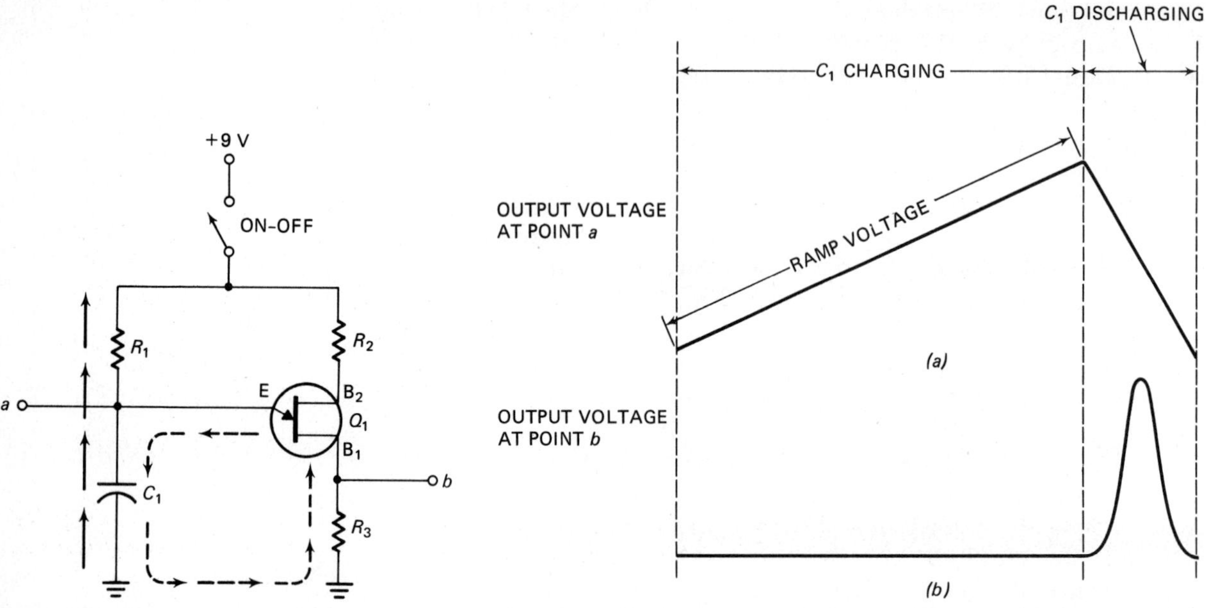

Fig. 9-22 UJT relaxation oscillator.

Fig. 9-23 One cycle of output waveform for the circuit of Fig. 9-22. (*a*) Sawtooth at point *a*. (*b*) Pulse at point *b*.

point, the UJT conducts through R_2 and R_3. At the same time, capacitor C_1 discharges through R_3 and the emitter to base 1 junction. The discharge-current path is shown with dotted arrows in Fig. 9-22.

Since current flows through R_3 for only a short period of time while C_1 is discharging, the voltage across R_3 consists of short pulses. This is the waveform at point *b* in the circuit as shown in Fig. 9-23.

Figure 9-24 shows several cycles of waveform for the oscillator circuit. The pulse output is not strong enough to operate a transducer directly, so you will use a power amplifier to drive the speaker in your experiment. The UJT oscillator will supply the input signal to the power amplifier.

PART I

Test Setup Figure 9-25 shows the power-supply circuit. Figure 9-25*a* shows the test setup, and Fig. 9-25*b* shows the pictorial drawing.

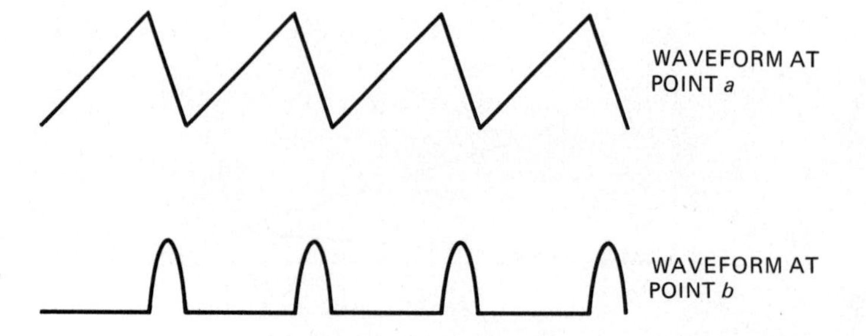

Fig. 9-24 Several cycles of waveform for the circuit of Fig. 9-22.

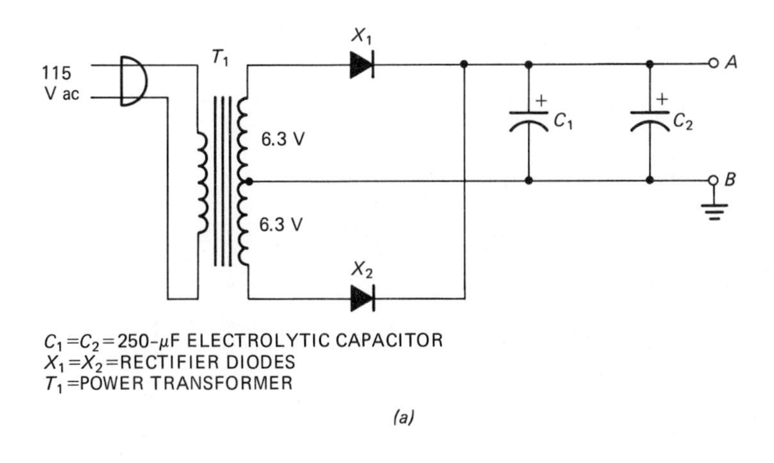

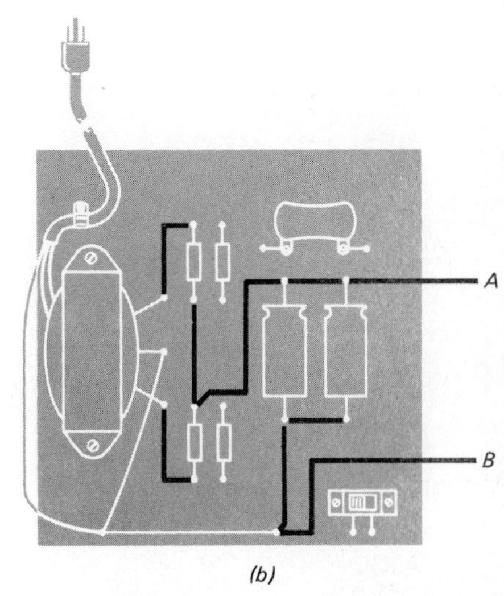

Fig. 9-25 Power-supply setup. (*a*) **Test setup.** (*b*) **Pictorial drawing.**

Procedure

step one Wire the circuit shown in Fig. 9-25. What type of power supply is this? ________________

Your answer should be a full-wave rectifier.

step two Measure the dc voltage across the output. This measurement can be taken directly across one of the filter capacitors. Record the voltage here. ________________

Your value should be about 9 volts. The exact value depends upon the amount of voltage on the power line and the resistance of the diodes.

step three The circuit has two 250-microfarad electrolytic filter capacitors connected in parallel. What is the capacitance of the parallel combination? ________________ microfarads

Your answer should be 500 microfarads. When capacitors are connected in parallel, their capacitance values add.

step four If diode X_2 were removed from the circuit of Fig. 9-25, what type of power supply would you have? ________________

Your answer should be half-wave rectifier. The circuit would work, but the voltage would drop more when a load is connected to terminals A and B. In other words, the regulation is not as good. Also there would be more ripple.

PART II

Test Setup Figure 9-26 shows the power-amplifier circuit. Figure 9-26*a* shows the test setup, and Fig. 9-26*b* shows the pictorial drawing.

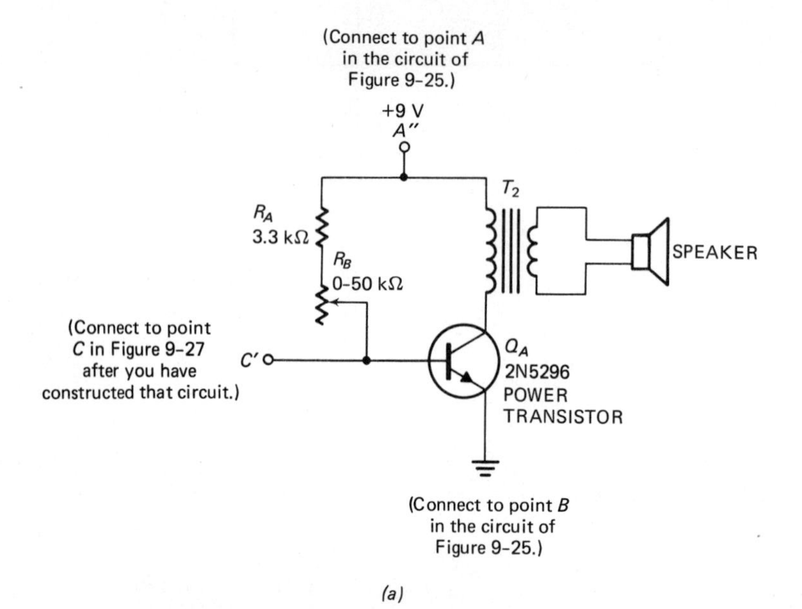

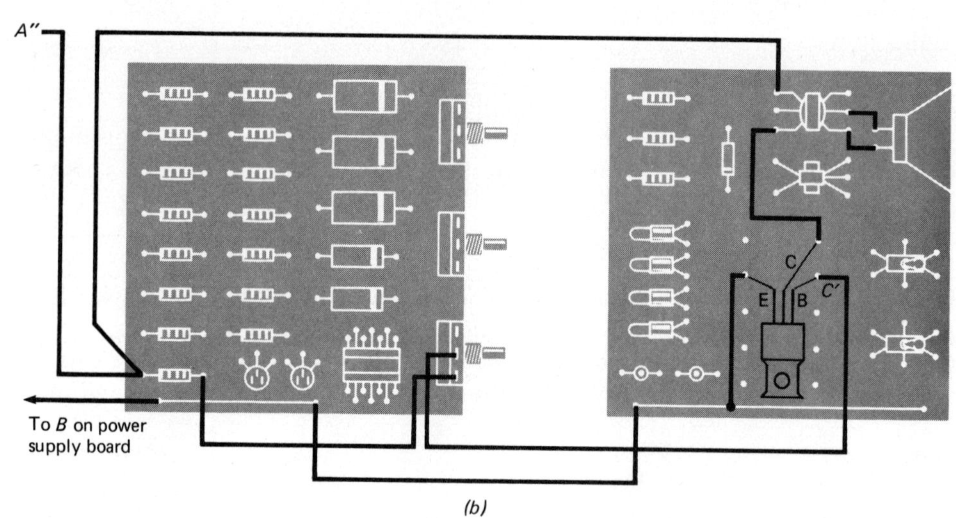

Fig. 9-26 **Power-amplifier setup. (*a*) Test setup. (*b*) Pictorial drawing.**

Procedure

step one Wire the circuit shown in Fig. 9-26. This is a power amplifier with simple bias. Be sure the power-supply circuit is not energized when you connect the power amplifier to it. This is a basic rule of good lab practice:

Do not connect or disconnect components in a circuit with the power ON.

The surges of current (called *transients*) can destroy a component.

In this section you are going to perform an experiment that requires you to disconnect a component with the energy applied. This is an exception to the rule. Even so, you should not connect and disconnect the circuit more than a few times to demonstrate the point being made.

step two Connect the power supply as indicated.

step three Adjust R_B to the center of its range.

step four A good way to check the power amplifier is to momentarily open the emitter-circuit lead. If current is flowing from emitter to collector and through the speaker, you will hear a clicking noise when you make or break the emitter lead. Try it. Do you hear a sound in the speaker?

____
Yes or No

Your answer should be yes. If you do not hear the sound, check the collector voltage. It should be almost as high as the power-supply voltage. Also check the base voltage. It should be a positive value, and the value should change when you vary the setting of R_B. If these voltages are all right, change the transistor first, then the speaker, to isolate the trouble.

step five Another quick test for the transistor is to momentarily short the emitter to the base.

Be sure you short *only* the emitter and base leads!

When the emitter is shorted to the base, the current through the transistor stops suddenly. This makes a clicking noise in the speaker. Try it. Do you hear a sound in the speaker?

____
Yes or No

Your answer should be yes.

PART III

Test Setup Figure 9-27 shows the UJT oscillator. Figure 9-27*a* shows the test setup, and Fig. 9-27*b* shows the pictorial drawing.

Procedure

step one Wire the circuit shown in Fig. 9-27. This is a UJT oscillator like the one you studied in the theory section. Connect the oscillator to the power supply and power amplifier as shown. Figure 9-28 shows the complete assembly for this experiment. Figure 9-28*a* shows the test setup, and Fig. 9-28*b* shows the pictorial. Adjust the resistors for a low-volume output tone in the speaker.

step two Do you hear a tone in the speaker?

____
Yes or No

Your answer should be yes. If you have a good ear for musical sounds, you

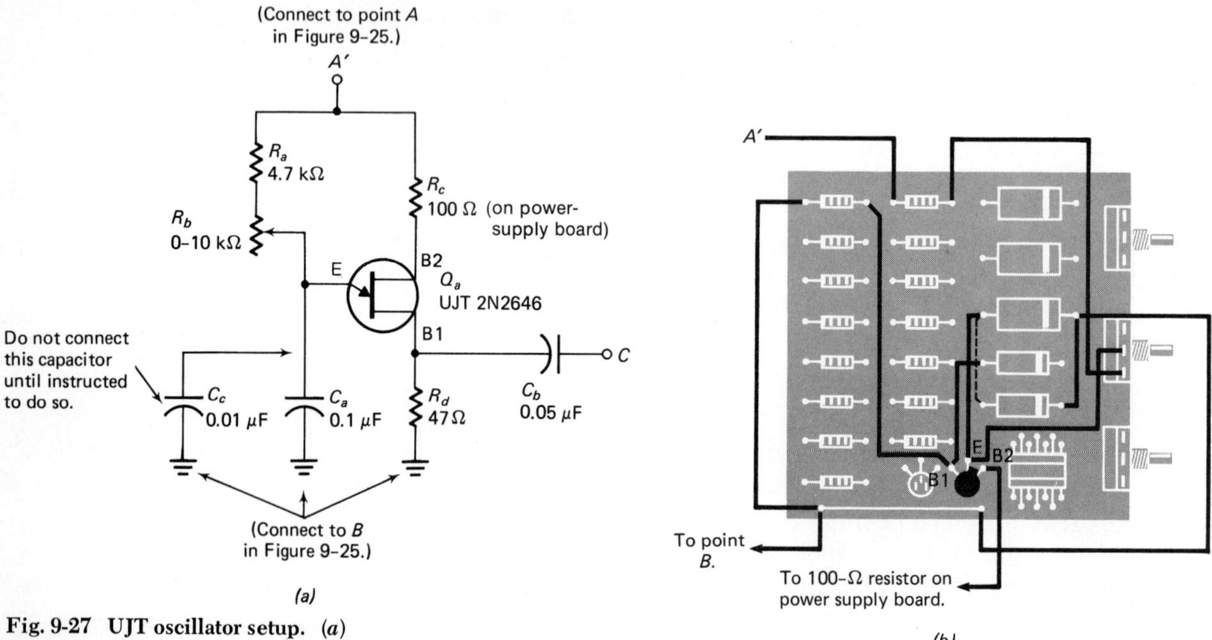

Fig. 9-27 UJT oscillator setup. (*a*) **Test setup.** (*b*) **Pictorial drawing.**

will no doubt recognize the fact that the tone is not pure. This is because it is produced by a nonsinusoidal oscillator.

step three Connect capacitor C_c in parallel with C_a. What effect does this have on the tone? _______________ Higher or Lower

The frequency of the tone should be lower. It takes longer to charge the two capacitors. Remember, their capacitance values are added when they are connected in parallel.

Since it takes longer for the capacitors to charge, each cycle will take a longer period of time. The time for one cycle is called the *period*, and it is represented by the letter T. The frequency is represented by the letter f. The period and frequency are related by the equation

$$f = \frac{1}{T}$$

Making T larger will cause f to be smaller.

step four Disconnect C_c.

step five When you make R_b smaller, will the frequency increase or decrease? Try it. _______________ Increase or Decrease

Your answer should be increase. With a smaller value for R_b the charging cur-

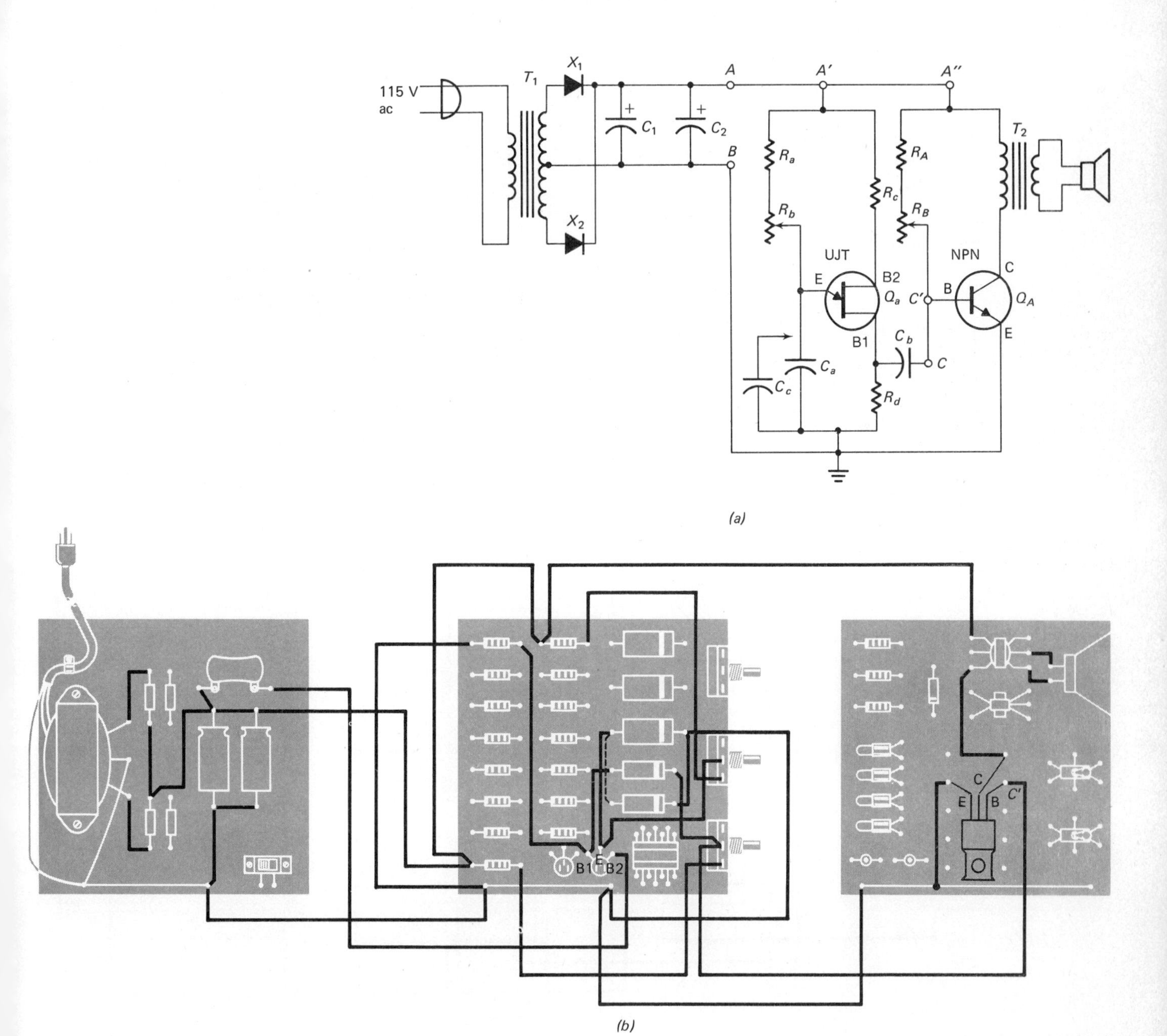

Fig. 9-28 The completed circuit. (*a*) Test setup. (*b*) Pictorial drawing.

rent is greater and the time required to charge the capacitor is less. The shorter the time for one cycle, the higher the frequency.

step six According to the theory section there should be an output signal from the *emitter* of the UJT. Connect the emitter output to the power amplifier as shown in Fig. 9-29. Figure 9-29*a* shows the test setup, and Fig. 9-29*b*

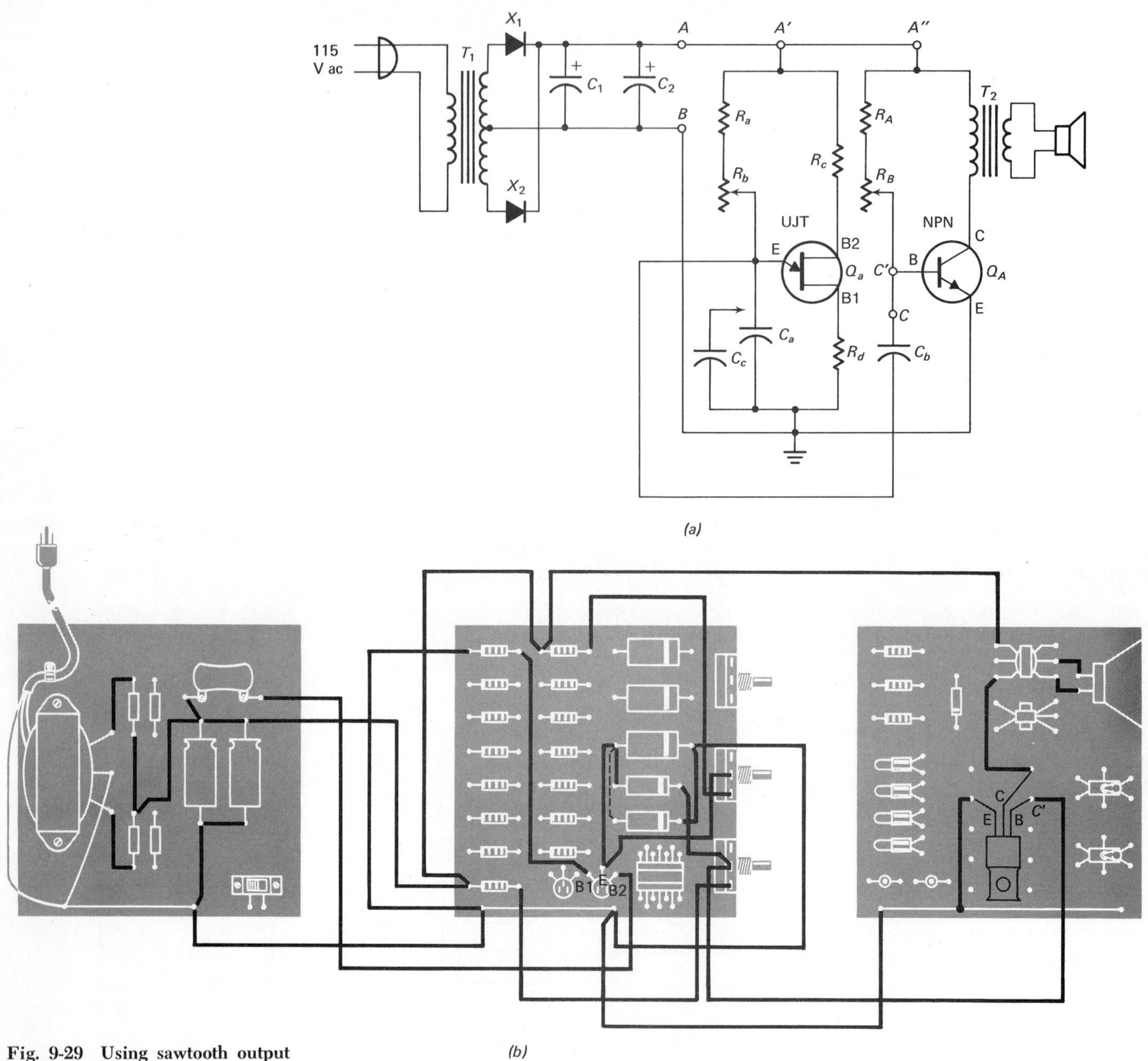

Fig. 9-29 Using sawtooth output for an audio signal. (*a*) Test setup. (*b*) Pictorial drawing.

(b)

shows the pictorial drawing. Energize the circuit. Do you hear a sound in the speaker? ____________ Yes or No

The answer should be yes. The sound will be different from the one you get with the output from base 1. The reason is that the waveforms from the two points are different.

Conclusions A unijunction transistor can be used in a relaxation oscillator. The frequency of oscillation depends upon the amount of time required for the capacitor to charge to the breakover point. This time can be changed by changing the value of either the resistor or the capacitor.

There is an output signal at the emitter of the UJT and at base 1. Although not demonstrated in the experiment, there is also an output at base 2, but this output is seldom used. It has the same waveshape as the one shown in Fig. 9-24 for point *b*, but it is inverted (upside down).

SELF-TEST WITH ANSWERS

(Answers with discussions are given at the end of the chapter.)

1. Feedback in an oscillator circuit *must* be (*a*) from plate to cathode; (*b*) from gate to drain; (*c*) regenerative; (*d*) degenerative.
2. A sawtooth waveform is produced by (*a*) a Colpitts oscillator; (*b*) a Hartley oscillator; (*c*) a crystal oscillator; (*d*) a UJT oscillator.
3. Which of the following amplifiers has an output that is out of phase with the input? (*a*) Common-drain; (*b*) Common-emitter; (*c*) Common-base; (*d*) Common-gate.
4. To increase the resonant frequency of an *LC* tank circuit you would (*a*) increase the inductance; (*b*) decrease the inductance.
5. Which of the following is not true regarding a ringing oscillator? (*a*) It has an *LC* tank circuit; (*b*) It has a feedback network; (*c*) It has an amplifier; (*d*) It requires a dc voltage for its amplifier.
6. Which of the following oscillators uses a transformer in its feedback network? (*a*) Armstrong; (*b*) Colpitts; (*c*) Crystal; (*d*) UJT.
7. When the dc current through the amplifying component also flows through the tuned circuit, the oscillator is said to be (*a*) series-fed; (*b*) shunt-fed.
8. Which of the following sine-wave oscillators does not use an *LC* tank circuit? (*a*) Armstrong; (*b*) Hartley; (*c*) Colpitts; (*d*) Phase-shift.
9. A blocking oscillator is similar to an Armstrong oscillator in the sense that they (*a*) both use only transistors; (*b*) both produce sine waves; (*c*) both have transformer feedback; (*d*) both have tank circuits.
10. An oscillator that produces a nonsinusoidal waveform by charging and discharging a capacitor is called (*a*) a capcharger; (*b*) a full-wave oscillator; (*c*) a relaxation oscillator; (*d*) a half-wave oscillator.

ANSWERS TO SELF-TEST

1. (*c*)—The question asks which type of feedback *must* be used. The only type which will sustain oscillations is regenerative feedback.
2. (*d*)—All other oscillators mentioned produce a sine wave.
3. (*b*)—A common-emitter amplifier is an example of a conventional amplifier. Its output at the collector is 180° out of phase with its input at the base.

4. (*b*)—Remember that decreasing the denominator of a fraction increases the fraction's value. The equation

$$f_r = \frac{1}{2\pi\sqrt{LC}}$$

shows that the frequency f_r will increase if either L or C is decreased.
5. (*b*)—Figure 9-7 shows an example of a ringing oscillator. It has no feedback circuit.
6. (*a*)—See Figure 9-8.
7. (*a*)—This is the definition of a series-fed oscillator.
8. (*d*)—Figure 9-12 shows the circuit for a phase-shift oscillator. It has no tank circuit.
9. (*c*)
10. (*c*)—None of the other terms have any meaning in electronics.

what are integrated circuits and operational amplifiers? 10

INTRODUCTION

In electronics we make a distinction between two kinds of systems: *analog* and *digital*. An analog system is one that produces a steady output of some type which is proportional to the system input. The input (or output) can be either mechanical or electrical.

A good example of an analog device is the speedometer in a car. It uses a needle or pointer as the output device. The deflection of the needle depends on the speed of the car. You interpret the speed by reading the position of the needle on a scale behind the needle. The important thing is that the needle position depends on the speed of the car. Therefore this is an analog system.

Another important feature of an analog system is that whenever there is an input there is always an output. The output is always directly related to the input. If you change the speed of the car, the speedometer pointer shows the change in speed by deflecting to a new number.

A digital system is one in which the output is a series of digits or numbers. They give you a specific value. A good example is the odometer in an automobile. The odometer tells the total number of miles that the car has been driven. If the odometer reads 13156 you know that the car has been driven 13,156 miles. You do not need the added step of first seeing the position of the pointer and then finding the number behind the pointer, as with the speedometer.

Figure 10-1 shows how multimeters can be made with either an analog or a digital readout. The analog meter of Fig. 10-1*a* has a needle that deflects to a voltage or resistance value. To read the meter you line up the number or mark behind the needle.

The digital meter of Fig. 10-1*b* does not require any pointer and scale. You can read the number directly.

As a general rule, meters with an analog readout are less expensive, but the digital meters are more accurate. Thus price and accuracy are *tradeoffs*. You cannot have a low price and high accuracy at the same time.

Most of the early designs for electronic computers were analog types, and many analog computers are still being made. They produce an output which depends on the value of input.

(a)

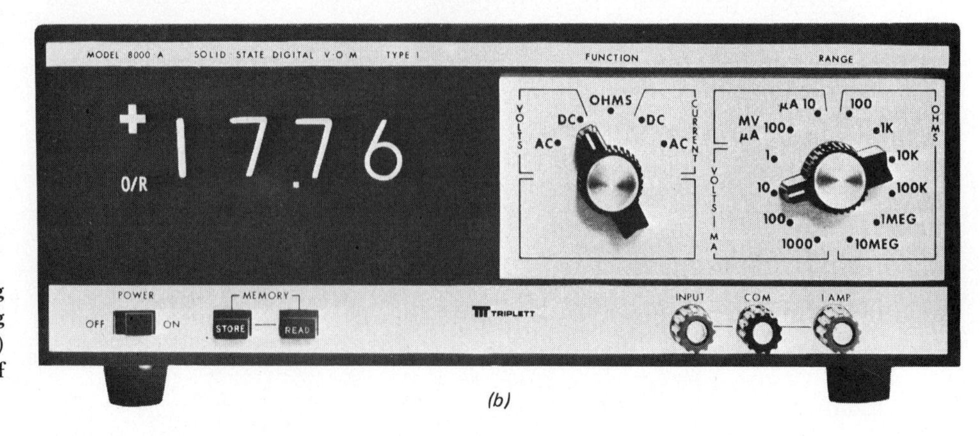

(b)

Fig. 10-1 Multimeters with analog and digital readouts. (*a*) An analog readout. (Courtesy of Triplett.) (*b*) A digital readout. (Courtesy of Triplett.)

When designing these computers, engineers found it necessary to use amplifiers for performing mathematical *operations* such as multiplication, division, addition, and subtraction. They are used with amplifiers that have an extremely high gain. To perform math operations this amplifier must use the proper amounts of feedback. In other words, by connecting some part of the output signal back to the input the amplifier is able to perform the mathematical operations.

The basic amplifier is the same for all the math operations, and it has become known as an *operational amplifier*, or *op amp*. Of course, the earlier versions were vacuum-tube types. This means that they were quite large. They wasted a lot of energy because the tube filaments had to be kept heated at all times—even though the amplifier was only used once in a while. Today, most operational amplifiers are *integrated circuit* (*IC*) types. An integrated circuit has many transistors, resistors, and capacitors, all mounted or engraved on a single chip of silicon. In addition to the saving in space, they operate with low power and have a high reliability. Figure 10-2 shows a typical operational amplifier circuit schematic. The pictorial drawing shows you the actual size of this op amp when it is made as an integrated circuit.

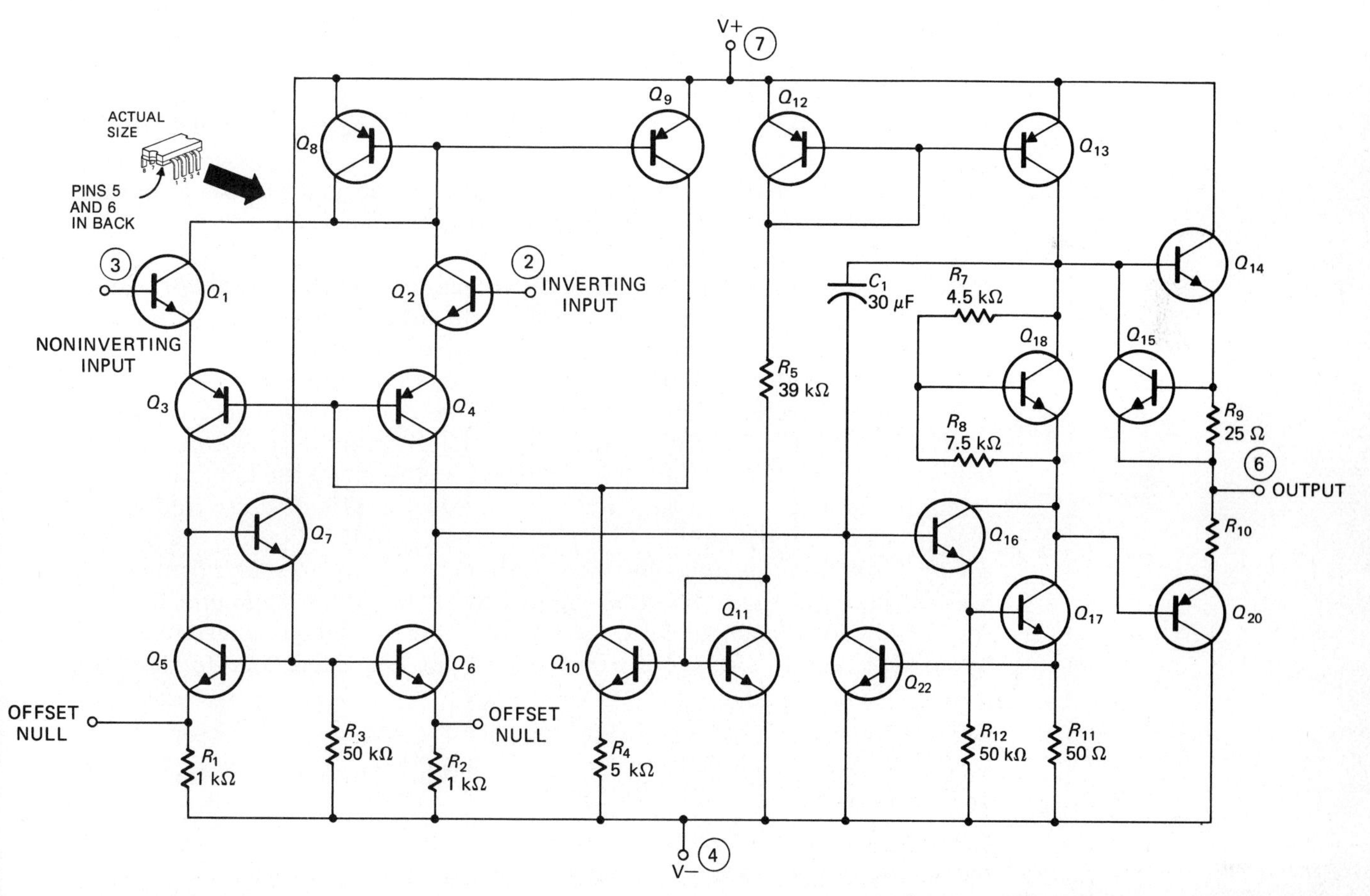

Fig. 10-2 A typical operational amplifier schematic diagram.

The operational amplifier is now being used in many circuits other than analog computers. They are still used for mathematical operations, but they are also used as amplifiers, oscillators, comparators, and in many other types of specialized circuits.

In this chapter you will study some of the basic principles of integrated circuits. Also you will study some of the uses of operational amplifiers and learn about their circuitry.

You will be able to answer these questions after studying this chapter.

☐ **What are integrated circuits?**

☐ **What do MSI and LSI mean?**

☐ **How are integrated circuits made?**

☐ **What is the difference between a linear and a digital IC?**

☐ **What is an operational amplifier?**

☐ **What is frequency compensation in op amps?**

INSTRUCTION

What Are Integrated Circuits?

By the early 1960s solid-state electronics had a firm grasp on the electronics industry. Transistors were used in a wide variety of circuits. They were rapidly replacing vacuum tubes.

It was about 1964 that the idea of the integrated circuit came about. An integrated circuit (or IC) is simply a solid-state chip of silicon on which a large number of individual circuit components have been mounted or etched. Thus, one integrated circuit may contain several complete amplifier stages.

All the connections between the circuit components are made automatically when the IC is manufactured. Therefore it is not necessary for internal soldering connections to be made. This is one reason for the better reliability of an integrated circuit compared with one made with *discrete* (individual) components wired together to make a circuit. Reliability experts say that many of the troubles in hand-wired equipment are due to solder connections which do not properly bond or connect the leads of the components. Also there is always a chance of incorrectly soldering one lead of a terminal to another lead, and this requires bothersome troubleshooting time to straighten the circuit out.

What Do MSI and LSI Mean?

Integrated circuits are often identified according to the number of *active components* they have mounted on a single chip. Active components are used for generating or amplifying signals. A transistor is an active component, but resistors and capacitors are *passive components.* Passive components do not produce or amplify voltages or signals.

If there are less than 50 active components, the IC is called a *M*edium-*S*cale *I*ntegrated circuit, or simply *MSI*. If there are more than 50 active components the IC is called a *L*arge-*S*cale *I*ntegrated circuit, or *LSI*.

Integrated circuits are called *monolithic* if the complete circuit is made out of a single chip of silicon. If the circuit consists of an IC with some additional tiny discrete components attached to it, it is called a *hybrid IC*. The hybrid and monolithic circuits are about the same size, but the monolithic type is less expensive to make. The hybrid type is used mostly for very high frequency work and high-power work.

By using special methods it is possible to change the discrete components on a hybrid IC. You could never do this with a soldering iron, but there are special machines that can do it. Since some of the components can be changed, the hybrid IC could be repaired if it were found to be defective during manufacture.

Another advantage of the hybrid circuit is that the circuit can be changed more easily. For example, a transistor can be added or removed to change the circuit. With a monolithic IC there is no way to change the circuit once it is made.

If a monolithic IC is found to be defective during manufacture, it must be thrown away. There is no way to repair it.

How Are Integrated Circuits Made?

Figure 10-3 shows a simplified procedure for making a monolithic integrated circuit. The various steps for making a single transistor are shown. This transistor may be part of a complete integrated circuit. For example, it might be Q_3 in the circuit of Fig. 10-2. Although you are being shown how one transistor on the IC is made, all the components are made in a similar way, and all the components are made at the same time.

As shown in Fig. 10-3*a*, the integrated circuit is built on a piece of insulating material called the *substrate*. The surface of the IC material is covered with a photosensitive layer. When this layer is exposed to light and developed (like a camera film), it becomes hard.

The film prevents the light from hitting the surface in some areas but permits light to pass through in other areas. During exposure the film is against the surface.

The next step, as shown in Fig. 10-3*b*, is to remove the unexposed photosensitive material. This is done by washing the surface with a spray. Then, in Fig. 10-3*c*, an acid spray removes all the N-type material that was under the unexposed photosensitive area. Note that this leaves an island of N-type material.

The block is covered with a photosensitive material again, and then a new film, or *mask*, is used to cover the complete chip. This is shown in Fig. 10-3*d*. Only the area to be exposed will get light through this mask.

In Fig. 10-3*e* almost the complete surface has been hardened by exposure to light. Only a small area on the N-material island is unexposed. A P-type gas is then directed to the surface. This gas seeps, or diffuses, into the surface that was unexposed. All other areas are protected by the hardened photosensitive material.

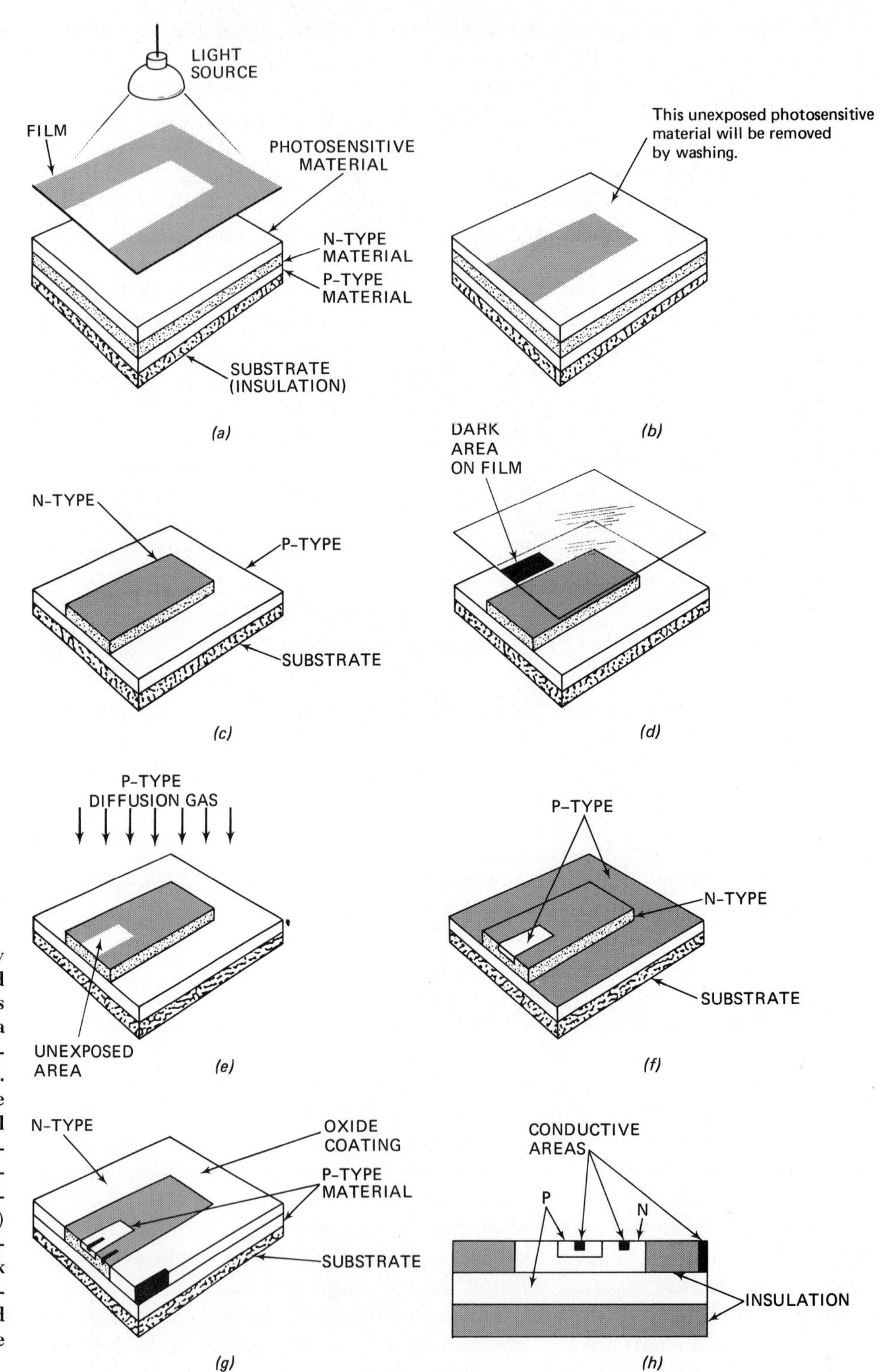

Fig. 10-3 PNP transistor made by IC method. (*a*) A film is placed on the surface. Then the surface is exposed to light. (*b*) Exposed area is hardened and the rest of the photosensitive material is removed. (*c*) Acid spray removes all N-type material not exposed. (*d*) Material covered with photosensitive material and exposed again. (*e*) Unexposed area is diffused with P material. Dark area is a resist. (*f*) Resist is removed. (*g*) Oxide insulation area is added. Black areas are conductor material for adding leads. (*h*) Side-view shaded area is insulation. Black areas are conductors for attaching leads.

In Fig. 10-3*f* the photosensitive material has been removed. Note the P-type block of material that is now in the N-material island.

An oxide coating is used to cover the complete area, as shown in Fig. 10-3*g*. Highly conductive areas are added to the P and N materials. As shown in the side view (Fig. 10-3*h*), a PNP transistor has been formed and is ready for the attachment of the leads.

What Is the Difference between a Linear and a Digital IC?

Integrated circuits are also classified as being *linear* or *digital*. A linear integrated circuit is used in places where it is desired to amplify signals without distorting them. An audio amplifier or an RF amplifier in a radio system are examples of linear circuits. With a linear IC you always have an output signal that is a replica of, or is directly related to, the input signal. Sometimes linear integrated circuits are referred to as analog circuits, but the term *linear* is now preferred.

Digital integrated circuits are used with pulse circuitry and logic circuitry. The amplitude of the output signal for a digital IC may not be directly related to the input signal. Also there may sometimes be no output from a digital IC even though there is an input signal. Digital circuitry is used in computers and in electronic systems that control machinery.

One of the best examples of linear integrated circuits is the *operational amplifier*. This is a very high gain amplifier which may be used for many different purposes in electronic systems. The operational amplifier will be discussed in the next section.

Summary

1. There are two kinds of electronic systems: analog and digital.
2. An analog system produces a voltage or current that depends on an input quantity. There is always an output for an analog circuit whenever there is an input.
3. Digital circuits produce a readout in digits or numbers. You do not have to interpret a digital output.
4. An integrated circuit is made on a single chip of silicon. The base of an integrated circuit is called a *substrate*.
5. Some integrated circuits have individual parts mounted externally on the IC block. These are called *hybrid* types.
6. Some integrated circuits are made from a single chip with no additional components mounted. These are called *monolithic*.
7. If an IC has less than 50 active components, it is called a *M*edium-*S*cale *I*ntegrated circuit (MSI). If there are more than 50 active components, it is called a *L*arge-*S*cale *I*ntegrated circuit (LSI).

What Is an Operational Amplifier?

The symbol for an operational amplifier is shown in Fig. 10-4. The symbol shows two input terminals—one is called the *inverting input terminal*, and

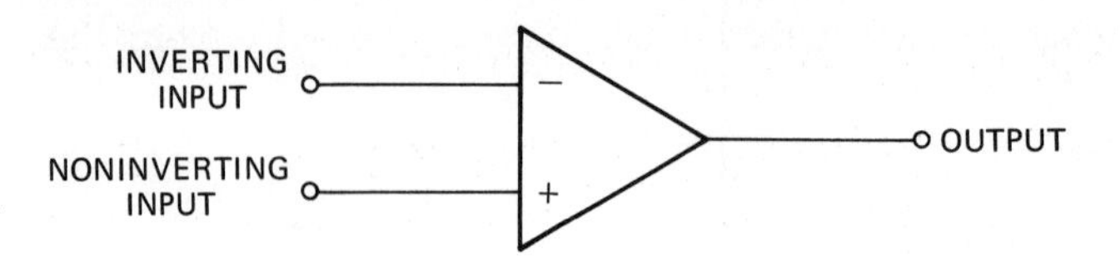

Fig. 10-4 Symbol for an operational amplifier.

one is called the *noninverting input terminal.* There is one output terminal. The inverting input terminal is always marked with a negative sign at the terminal entrance to the operational amplifier. The positive terminal indicates the noninverting input. These two inputs are used for different purposes depending upon the type of circuit.

Obviously not all the connections to the op amp are shown on its symbol. You must also have a dc voltage to get the circuit into operation. The dc voltage input terminals are not normally shown in the op amp symbol.

There are many different ways to design the op amp circuit. However, if it is a typical op amp with an extremely high gain, it does not really matter to you what components are used internally.

Figure 10-5 shows a block diagram of a typical operational amplifier. This does not represent all op amps. As mentioned before, it is possible to use many different circuit designs to get the high gain needed. The input to the operational amplifier is normally to a two-stage differential amplifier. This is how the inverting and noninverting input terminals are obtained. If a signal is applied to the inverting input terminal, the output signal is out of phase with the input. There is no phase difference between the input and output terminals when the input signal is applied to the noninverting terminal.

The output of the differential amplifier goes to a Darlington amplifier, which produces a very high gain. [Remember that a Darlington amplifier is sometimes called a beta squared (β^2) amplifier because its gain is so enormously high.] The Darlington amplifier delivers the amplified signal to an output amplifier. The output amplifier may also be called a power amplifier, but actually, it can deliver only a modest amount of power. It is not to be compared with the high-power amplifiers that you studied in Chapter 8.

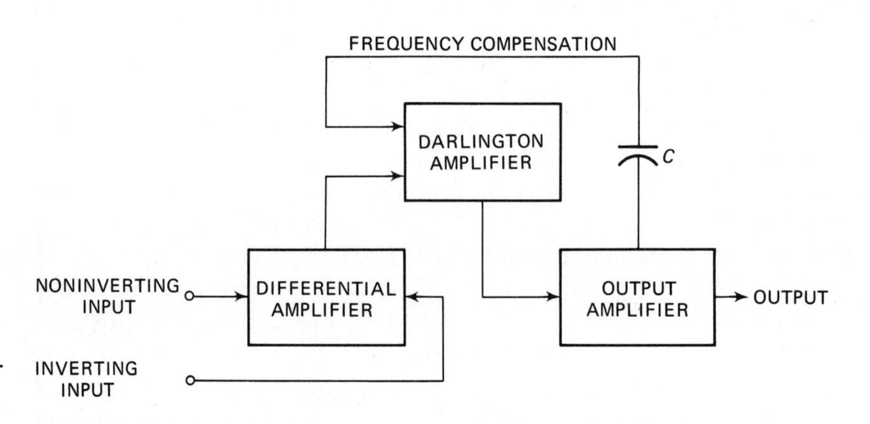

Fig. 10-5 Block diagram of a typical op amp.

What Is Frequency Compensation in Op Amps?

There is a feedback line in the block diagram of Fig. 10-5. This feedback line is marked *frequency compensation.* It determines the frequency response of the operational amplifier.

Capacitor C in the feedback circuit returns higher-frequency signals to the Darlington amplifier, but it opposes the feedback of low frequencies. The overall result is that the gain of high-frequency signals is decreased. In fact the reduction in gain *must be produced at a fixed rate for the op amp.* The fall-off of high-frequency gain is called the *roll-off.*

Figure 10-6 shows the open-loop frequency response of a typical op amp. Remember, this is the open-loop response, which means that there is no external feedback circuit connected.

Not all op amps have an internal frequency compensation circuit. When external compensation is required, the manufacturer will show how the circuit is connected and will also give data showing the component values needed for different frequency-response values.

Figure 10-7 shows a 709 op amp. This type must be compensated with external circuitry. As shown in Fig. 10-7*a*, R_1, C_1, and C_2 provide the feedback to get the desired frequency response. Figure 10-7*b* shows the different curves for each combination of component values. Later in this chapter it will be shown that the open-loop frequency response has a direct effect on the closed-loop gain and frequency response of the amplifier.

What Kind of Power-Supply Circuit Is Used with an Op Amp?

Most operational amplifier symbols do not show how power-supply voltages are applied. Figure 10-8 shows how an operational amplifier is normally powered. Two power supplies are used—one for the positive dc input and the other for a negative dc input. This is the recommended method for getting most operational amplifiers into operation.

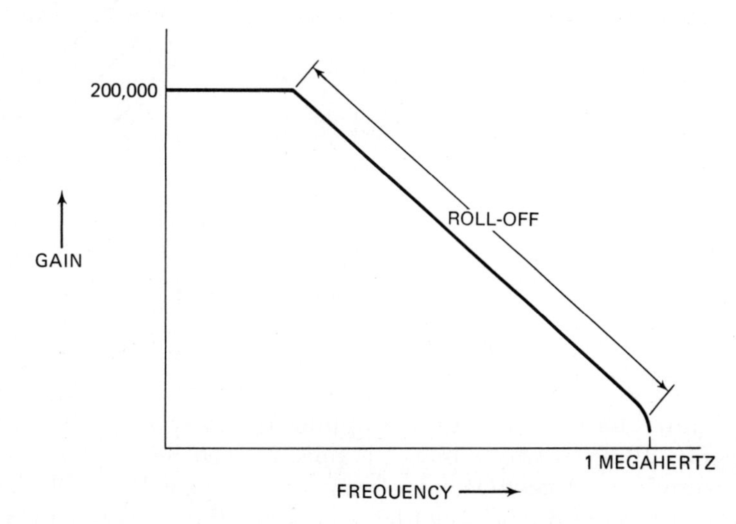

Fig. 10-6 Open-loop response of a typical op amp.

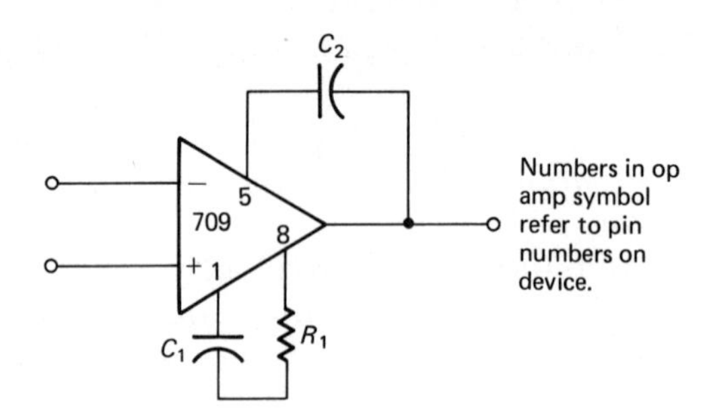

CLOSED-LOOP GAIN	VALUES OF COMPONENTS			CURVE IN FIG. 10-7*b*
	C_1	C_2	R_1	
1000	10 pF	3 pF	0 Ω	*A*
100	100 pF	3 pF	1.5 kΩ	*B*
10	500 pF	20 pF	1.5 kΩ	*C*
1	5000 pF	200 pF	1.5 kΩ	*D*

(a)

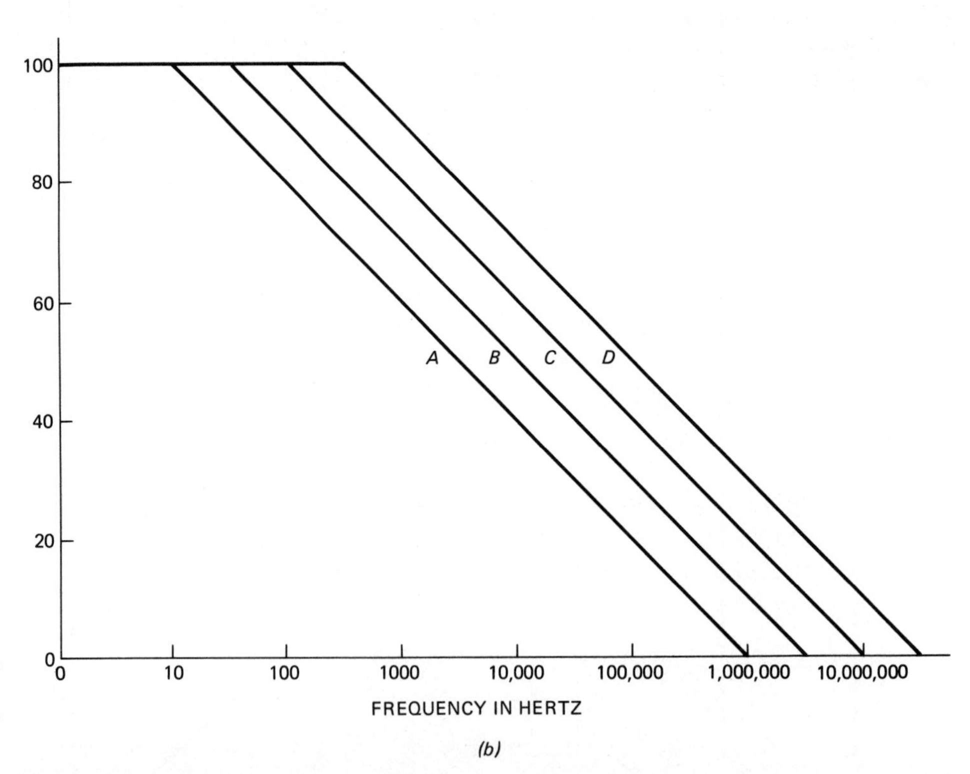

(b)

Fig. 10-7 A 709 op amp. (*a*) The op amp with external circuitry. (*b*) Different curves for each combination of component values.

In some cases you can obtain the positive and negative dc voltages required for the operational amplifier by using a circuit like the one shown in Fig. 10-9. A single power supply is used here. The output of the power supply is connected across two resistors in a voltage-divider network. These resistors are R_1 and R_2 with a junction at the center (point *a*) connected to the

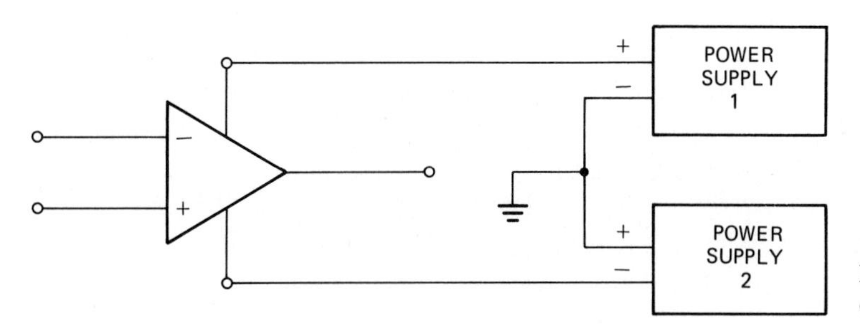

Fig. 10-8 Plus and minus voltages obtained with two power supplies.

common. You cannot go from a positive voltage to a negative voltage without going through zero. If R_1 and R_2 are of equal values, then the voltage at point *a* should be 0 volts or halfway between the positive and negative values.

The 30-volt power supply connected in this manner gives +15 and −15 volts output. Notice that the dc voltages that are shown are applied to the operational amplifier at two terminals, which are not normally shown on the symbol.

How Is an Op Amp Used in an Amplifier Circuit?

Figure 10-10 shows a typical op amp connected as an inverting amplifier. Note that the input signal and output signal are 180° out of phase. This means that when the input signal goes positive, the output signal goes negative, and when the input signal goes negative, the output signal goes positive. Resistor R_L is the output load resistance for the operational amplifier circuit. Resistor R_2 is a feedback resistor which connects part of the inverted output signal to the op amp input. Since part of the output signal is being delivered to the input, and the output signal is 180° out of phase with the input, the overall result will be to reduce the input signal.

Summary

1. An op amp is a high-gain amplifier with two input-signal terminals and one output-signal terminal.
2. When a signal is delivered to the terminal marked with a negative sign, the amplifier output signal will be inverted. In other words, the output signal will be 180° out of phase with the input signal.
3. If a signal is delivered to the terminal with a positive sign, the amplified output signal will be in phase with the input signal.

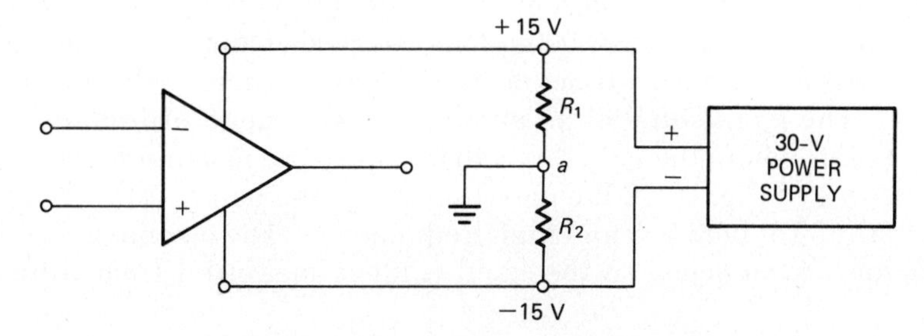

Fig. 10-9 Plus and minus voltages obtained with one power supply.

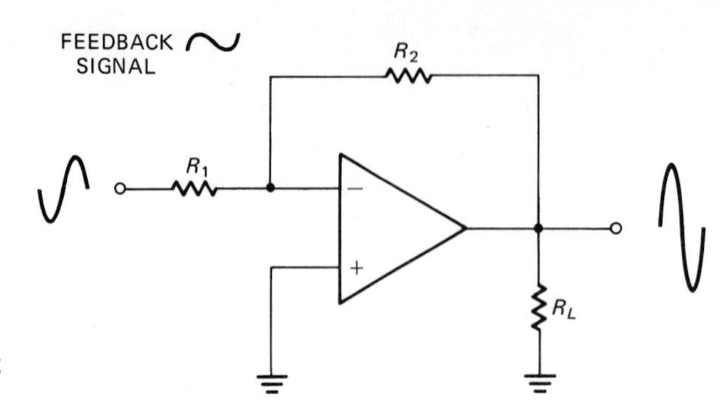

Fig. 10-10 Op amp inverting amplifier.

4. Some of the characteristics of operational amplifiers are very high gain, differential input, and linear roll-off.
5. The roll-off for some op amps is accomplished with an internal capacitor. For other op amps an external compensation circuit must be used.
6. An op amp is normally operated with a positive and negative supply. In some cases, however, a single supply can be used to get the positive and negative dc voltages required.

What Determines the Closed-Loop Gain of the Op Amp Circuit?

The voltage gain of the operational amplifier shown in Fig. 10-10 is approximately equal to the ratio of R_2 to R_1. In other words, you can write a simple equation.

$$\text{Voltage gain} = -\frac{R_2}{R_1}$$

The negative sign indicates that the output is inverted. The load resistance R_L does not normally affect the gain of the op amp circuit, provided its resistance is high enough that it does not draw a large amount of current from the amplifier.

The voltage-gain equation shows that it is quite a simple matter to design an op amp circuit to get any gain you want. However, there is a very important tradeoff between the gain of the operational amplifier and its bandwidth. You can understand this compromise or tradeoff by referring to the following response curves. Figure 10-11 shows the open-loop response curve of an operational amplifier. The circuit of Fig. 10-10 could be used to obtain this response curve if you remove resistors R_1 and R_2. The curve shows that the open-loop gain is 200,000. This means that an input-signal voltage would be amplified 200,000 times at the output if there were no feedback resistor.

The *bandwidth* of an amplifier is the range of frequencies between the points where the gain drops to 0.7 times the maximum gain. In the response curve of Fig. 10-12 the dark line shows the bandwidth. In this case the gain falls off at both low and high frequencies. The op amp gain does not fall off at low frequencies, so the bandwidth is measured from 0 hertz to the point

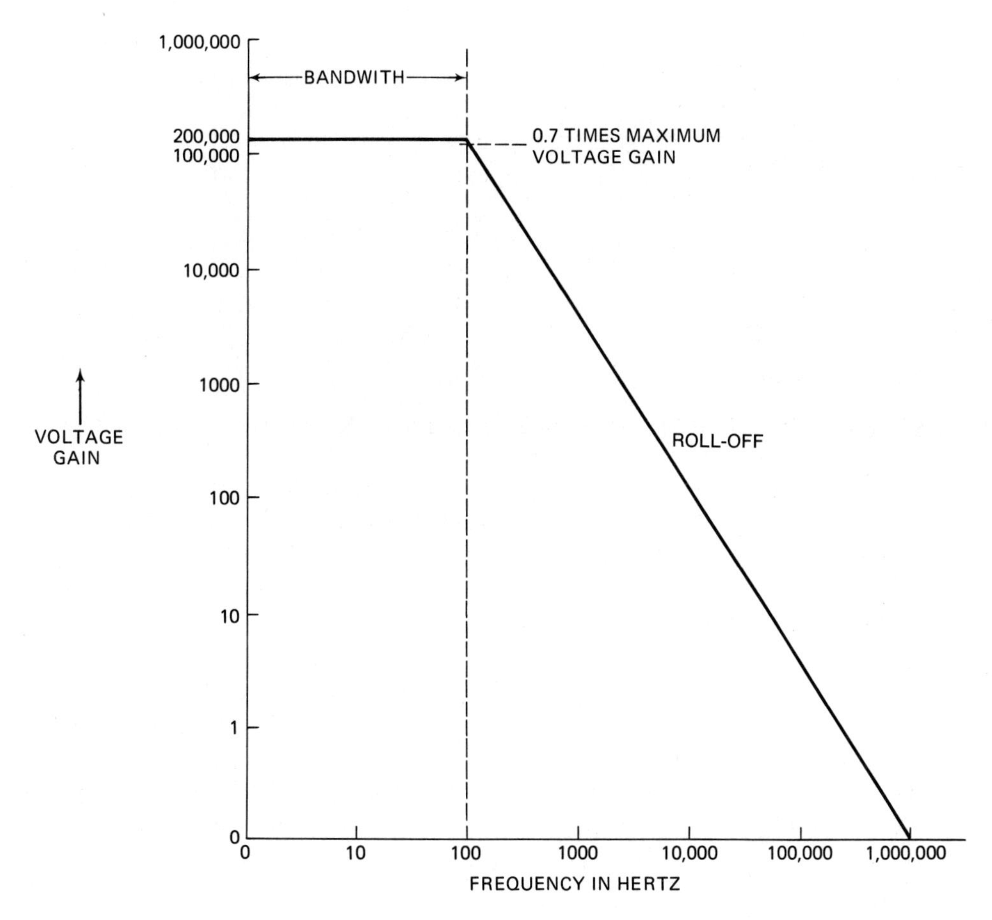

Fig. 10-11 Frequency response of an op amp used without feedback.

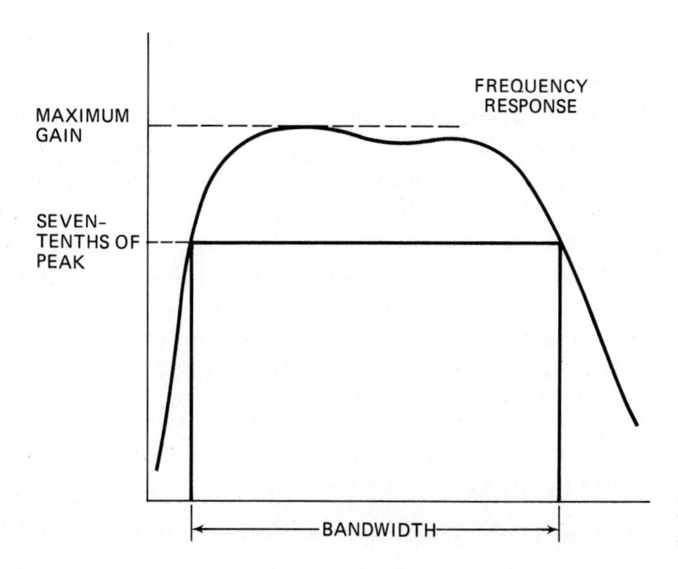

Fig. 10-12 Frequency response and bandwidth of an amplifier.

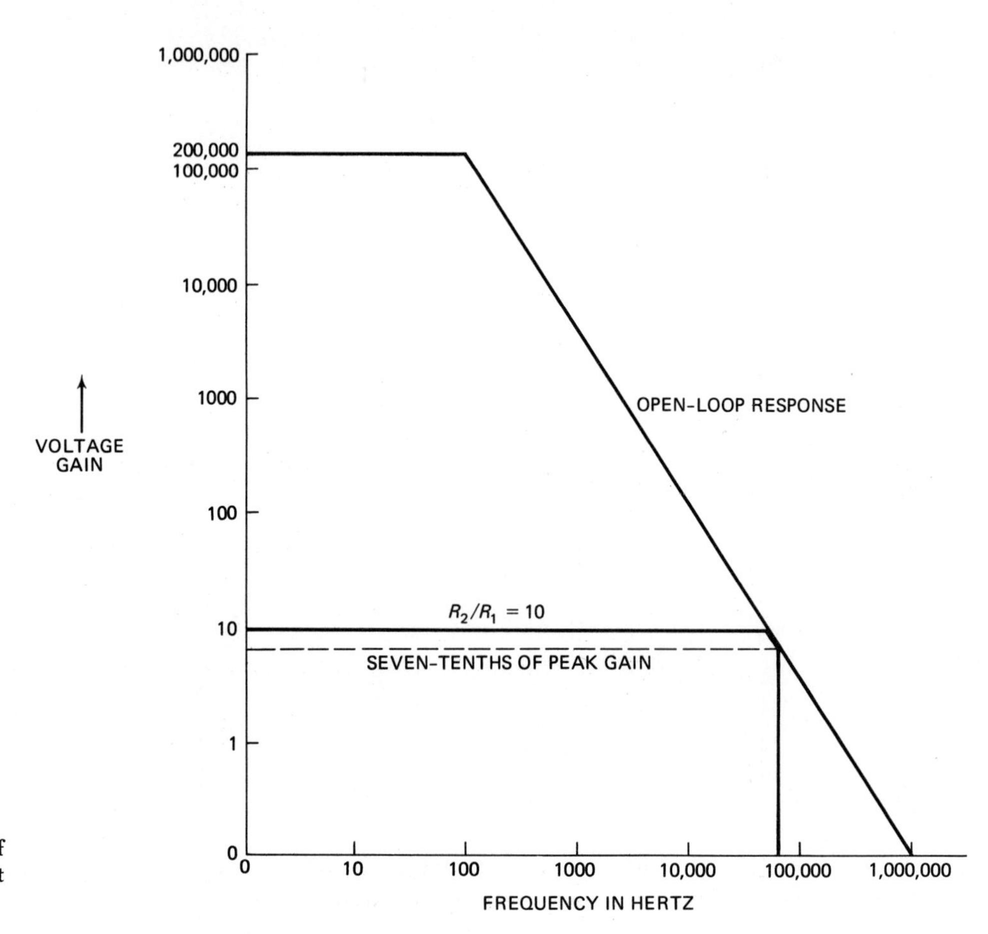

Fig. 10-13 Frequency response of an op amp. The feedback circuit gives it a gain of 10.

where the gain drops to 0.7 times the maximum value for each amplifier. The op amp bandwidth is narrow when it is operated as an open-loop circuit. The dotted line on the curve of Fig. 10-11 shows the bandwidth. Even though there is a lot of gain beyond the point where the bandwidth is marked, by definition the bandwidth is still taken only to the point where the gain falls off to 0.7 times its maximum value.

Figure 10-13 shows the response of the operational amplifier when it is used with R_1 and R_2. Since this is a negative feedback situation, the result is to decrease the gain to a value of 10. This is the maximum gain. The bandwidth is shown with a dotted line.

Figure 10-14 shows the response of the op amp with a gain of 5. The gain has been reduced, but the bandwidth has been increased.

From the illustration of Fig. 10-14 you can see that if you choose a wide bandwidth for your amplifier, then you will have to be satisfied with a lower gain. This is true of all amplifiers, not just operational amplifiers.

Problem Using an operational amplifier with a curve of Fig. 10-11, design an amplifier that has a gain of 1000. What values of R_1 and R_2 are needed? What is the bandwidth of the circuit?

Solution The gain is set by R_2/R_1. For a gain of 1000, $R_2/R_1 = 1000$. The

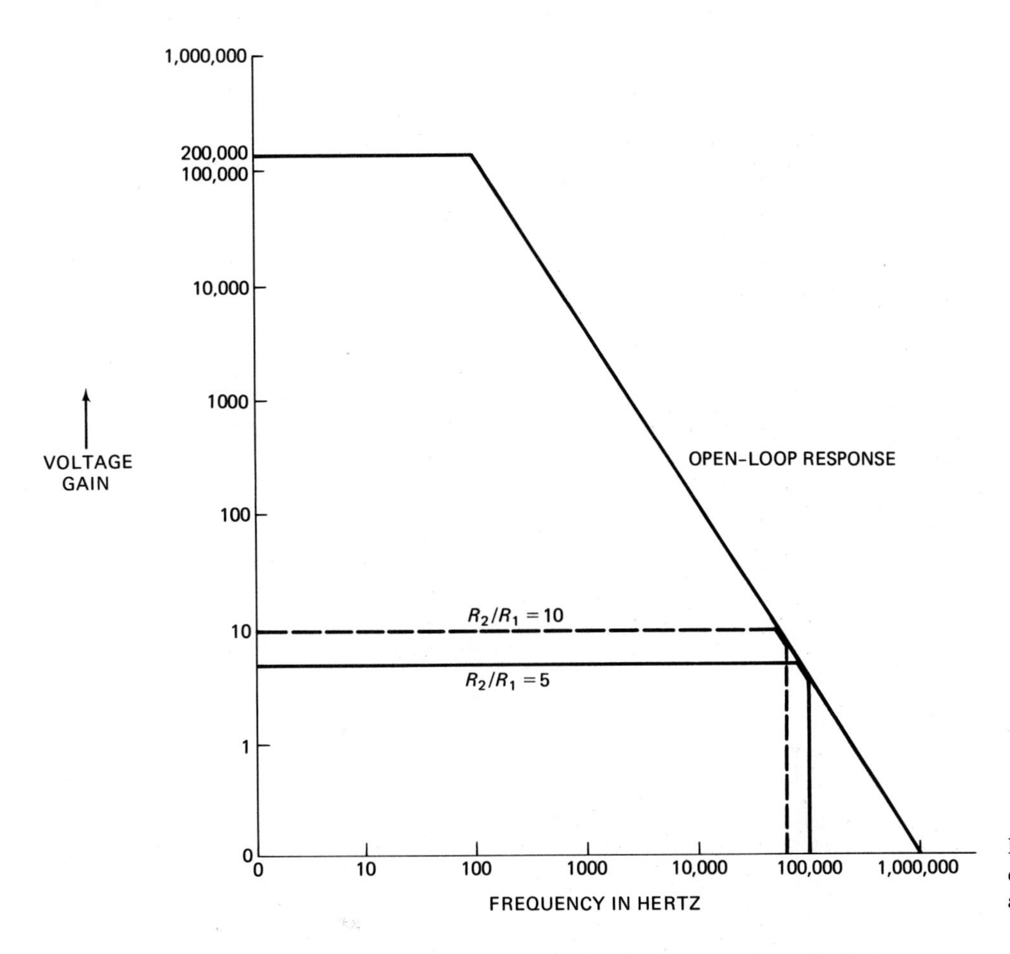

Fig. 10-14 Comparison of frequency response for a gain of 10 and a gain of 5.

usual procedure is to assume an input resistance R_1 and then find the corresponding feedback resistance R_2. The input resistance is usually some value between 1000 ohms and 5000 ohms, although other values are possible. Let $R_1 = 1500$ ohms (or 1.5 kilohms). Then

$$\frac{R_2}{R_1} = 1000$$

By substitution of 1500 for R_1,

$$\frac{R_2}{1500} = 1000$$

There is a basic rule of mathematics that says you can multiply both sides of an equation by the same amount. So we will multiply both sides of the above equation by 1500 in order to get rid of the denominator.

$$\cancel{1500} \times \frac{R_2}{\cancel{1500}} = 1000 \times 1500$$

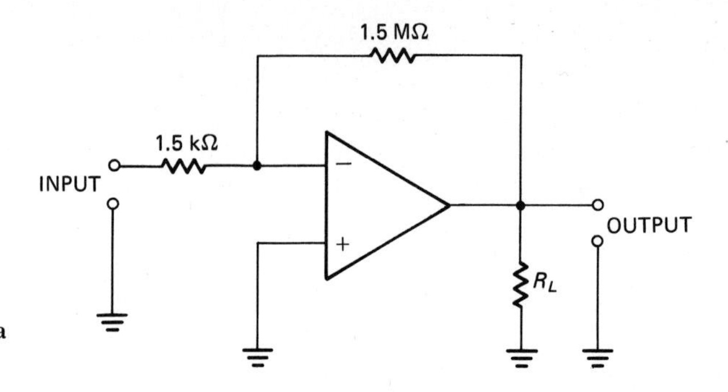

Fig. 10-15 Op amp circuit with a gain of 1000.

The 1500 values cancel, and you get

$$R_2 = 1000 \times 1500$$
$$= 1{,}500{,}000$$
$$\text{or } R_2 = 1.5 \text{ megohms}$$

Figure 10-15 shows the op amp circuit for a gain of 1000.

Figure 10-16 shows that the bandwidth, which is measured to the point

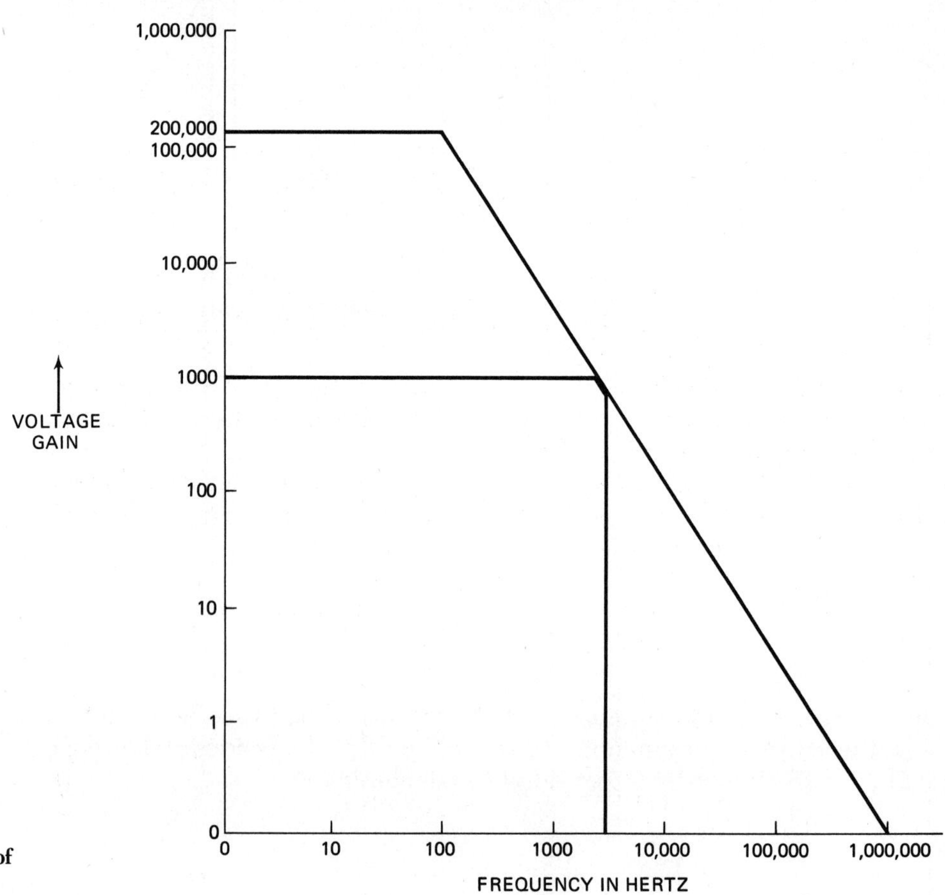

Fig. 10-16 Frequency response of an op amp with a gain of 1000.

where the gain falls to 0.7 times its maximum value, is about halfway between 1000 and 10,000 hertz. This would be approximately 3000 hertz. The amplifier would not be suitable for high-fidelity audio work, but it would work all right as an *intercom.* (An intercom is an amplifier system that permits you to talk to people in another location and permits them to talk to you.)

The value of R_L in the circuit of Fig. 10-11 is not critical. The manufacturer of the op amp may set a minimum value for the circuit so that you will not try to draw too much current.

How Is the Op Amp Used as a Noninverting Amplifier?

Figure 10-17 shows how an operational amplifier can be connected for a noninverting output. The input-signal terminal is delivered to the terminal marked with a +. The output signal is exactly in step, or in phase, with the input signal, but it has a higher amplitude. The gain of this noninverting amplifier is still fixed by the feedback network comprising R_2 and R_1. For all practical purposes the gain of the inverting and noninverting amplifier is the same under most conditions. Mathematically, the gain of the noninverting amplifier is given by the equation

$$\text{Voltage gain} = \frac{R_1 + R_2}{R_1}$$

If R_2 is large compared to R_1, then this equation is approximately R_2/R_1.

The gain equations given for the inverting and noninverting amplifiers are both simplified versions of longer equations. It can be shown that these equations are accurate enough for most work.

How Is an Op Amp Used as a Voltage Follower?

In your study of voltage amplifiers you learned that an amplifier can be connected in a follower configuration. Examples are cathode follower, emitter follower, and source follower.

Figure 10-18 shows an NPN transistor connected as an emitter follower. The input signal is delivered to the base, and the output signal is delivered

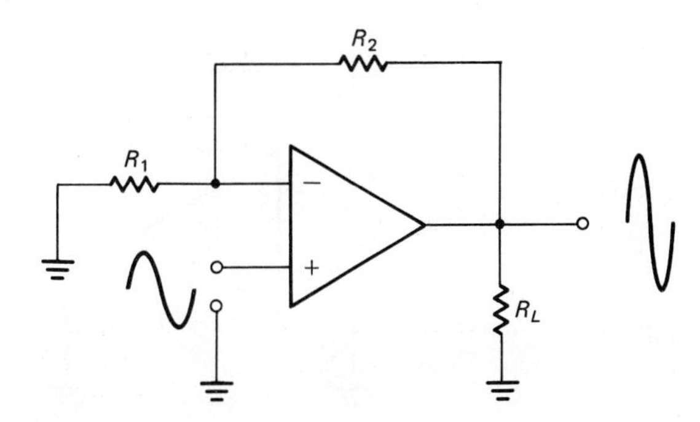

Fig. 10-17 Op amp noninverting amplifier.

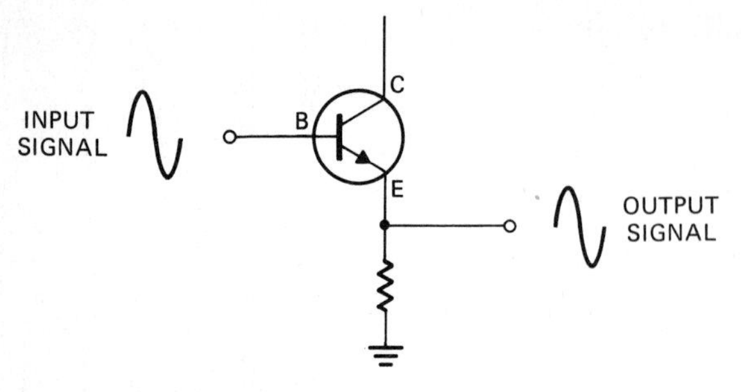

Fig. 10-18 An emitter follower.

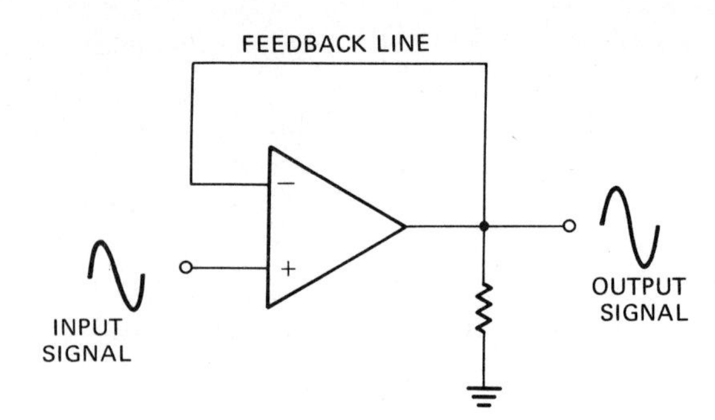

Fig. 10-19 An op amp voltage follower.

from the emitter. This type of amplifier has a high input impedance and a low output impedance. The output impedance is about equal to the emitter resistance. There is no phase inversion in any follower circuit.

Figure 10-19 shows how an operational amplifier can be connected to operate like an emitter follower. The gain of an emitter follower, if it is an ideal circuit, is approximately equal to 1. Under no conditions can the gain ever be greater than 1 due to the degenerative action of the feedback line.

What Are Some Other Types of Op Amp Circuits?

One reason the op amp is very popular is because it is easy to design an amplifier circuit. All you have to do is decide the gain and bandwidth you want, remembering that these are tradeoffs.

Another reason for the popularity of the op amp is the fact that it can be used in a wide variety of circuits. A few of the many examples will be discussed.

The circuit of Fig. 10-20 is called a *summing amplifier*. The output signal is equal to the sum of the input signals. This circuit can be used to perform basic addition and subtraction. To get the circuit to subtract, the number to be subtracted—called the *subtrahend*—is represented by a signal that is 180° out of phase with the number being subtracted from—called the *minuend*. Figure 10-21 shows a basic addition and subtraction problem, along with the voltage used at the input of the comparator and the output voltage that represents the answer.

The output signal might be larger than 0.13 volt due to the gain of the op

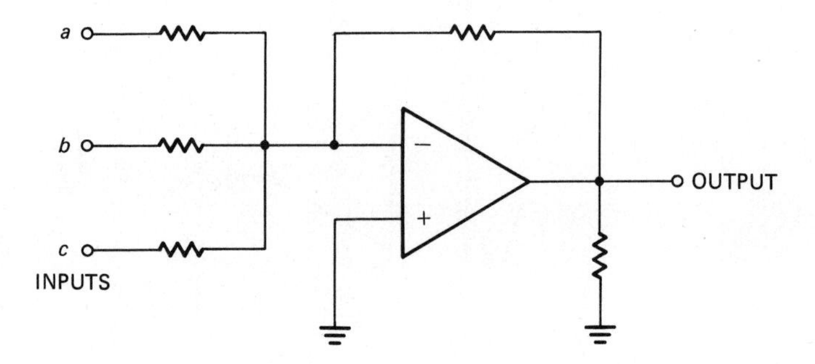

Fig. 10-20 Basic summing amplifier.

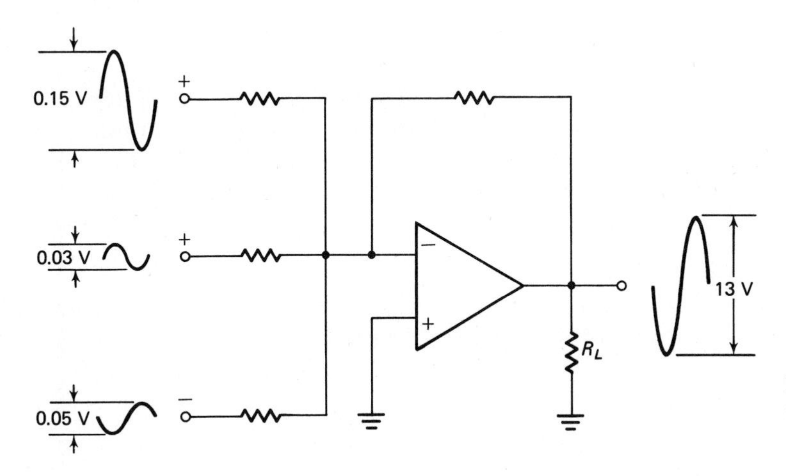

Fig. 10-21 Summing amplifier used for solving the problem 15 + 3 − 5 = 13.

amp. Thus if the gain is 100, the output will be 13 volts, so the answer can be read directly from the voltmeter. Only the magnitude of the ac voltage at the output is used for the answer, so the fact that it is 180° out of phase with the positive inputs can be disregarded.

Figure 10-22 shows the circuit for a low-pass filter. All frequencies above the cutoff point are reduced, or *attenuated.* The cutoff point is considered to be the place where the amplitude drops to 0.7 times the maximum value.

The op amp can be used as a *high-pass filter* that will pass all frequencies above the cutoff point. It can also be used as a band-pass filter to pass only a narrow band of frequencies or as a band-rejection filter to reject a narrow band of frequencies.

Figure 10-23 shows how the op amp can be used as a *comparator.* When the input voltage is less than the reference voltage, the output of the op amp is high. When the input voltage is greater than the reference voltage, the output of the op amp is low.

Manufacturers also make special integrated-circuit comparators that will

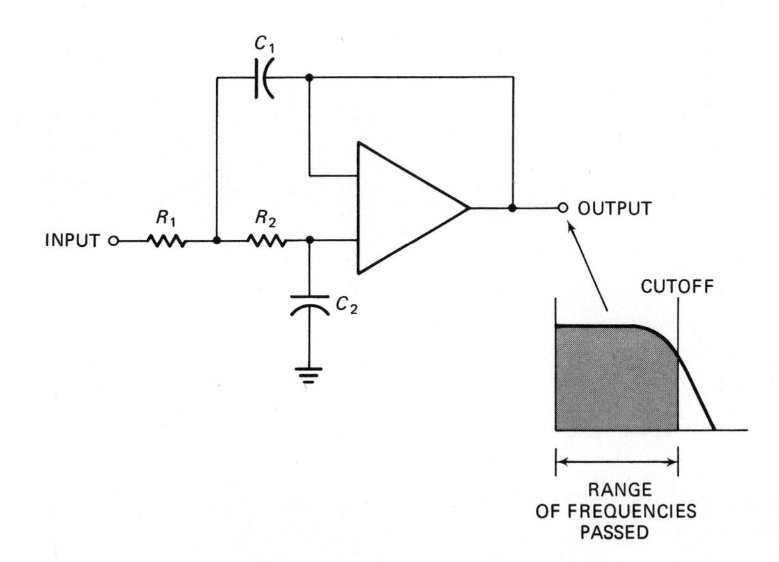

Fig. 10-22 The op amp as a low-pass filter.

switch more quickly, but the op amp can be used where the switching time is not important.

Figure 10-24 shows one of many possible oscillator circuits that you can make with an op amp. This one produces a square-wave output signal. Others will produce a sine wave, pulse, or any other oscillator waveform.

Summary

1. The voltage gain of an op amp used as an inverting amplifier is equal to the ratio of the feedback resistor and input resistor. Thus

$$\text{Voltage gain} = -\frac{R_2}{R_1}$$

2. The negative sign in the voltage-gain equation means that the output is 180° out of phase with the input.
3. Bandwidth is measured to the point where the gain drops to 0.7 of the maximum gain value.
4. Bandwidth and gain are tradeoffs in all types of amplifiers.
5. When an op amp is used as a noninverting amplifier, its voltage gain is given by the equation

$$\text{Voltage gain} = \frac{R_1 + R_2}{R_1}$$

6. An op amp can be used as a voltage follower. This circuit has a high input impedance, low output impedance, wide bandwidth, and voltage gain of approximately 1.0.
7. Op amps can be used in many applications, such as summing amplifiers, followers, and comparators.

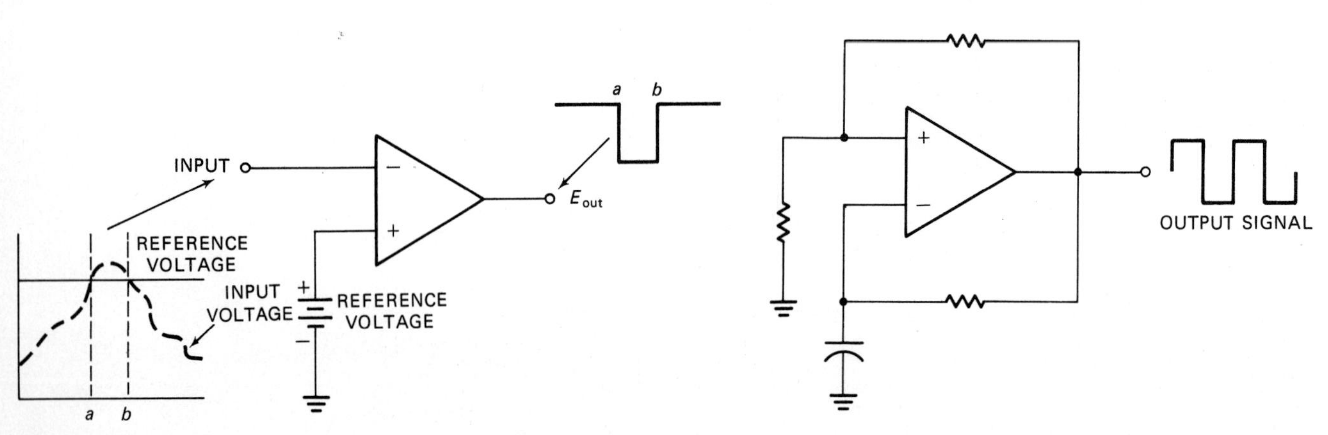

Fig. 10-23 Whenever the input voltage rises above the reference voltage, the output voltage switches to a lower value.

Fig. 10-24 Op amp oscillator circuit.

PROGRAMMED REVIEW QUESTIONS

(Instructions for using this programmed section are given in Chapter 1.) We will review the important concepts of this chapter. If you have understood the material, you will progress easily through this section. Do not skip this material because some additional theory is presented.

1. In Fig. 10-25 the approximate voltage gain of the amplifier circuit is about
A. 2. (Proceed to block 9.)
B. 1. (Proceed to block 17.)

2. *The correct answer to the question in block 17 is* **A.** *To get this answer, find the gain value of 100 on the vertical axis. Take seven-tenths of this, or 70. Move to the right at 70 until you touch the graph, then move down to the frequency value. You will get a value of about 20,000 hertz.*
The bandwidth is greater than 10,000 hertz because it is always measured to the point where the voltage gain drops to about seven-tenths of maximum.
Here is your next question.
The maximum rate of change of amplifier output voltage is called the
A. slew rate. (Proceed to block 20.)
B. reaction. (Proceed to block 8.)

3. *Your answer to the question in block 20 is* **B.** *This answer is wrong. There is no such thing as a Smith plot used for op amp frequency and gain plots.* Proceed to block 13.

4. *The correct answer to the question in block 23 is* **B.** *The substrate is usually made of silicon. It is the foundation for the integrated circuit.* Here is your next question.
An important characteristic of op amps is the *roll-off.* This refers to
A. the rate at which the gain decreases with an increase in frequency. (Proceed to block 21.)
B. the rate at which the voltage across the load decreases when the input voltage decreases. (Proceed to block 26.)

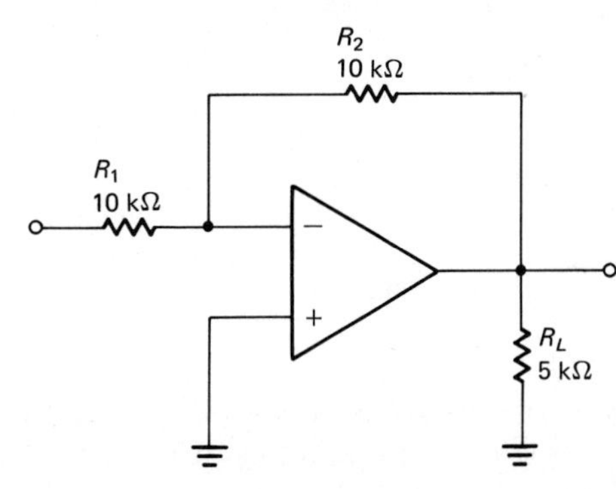

Fig. 10-25 What is the gain of this op amp?

5. *Your answer to the question in block 21 is **B**. This answer is wrong. An open-loop gain of 175 is too low for most op amp circuits.* Proceed to block 10.

6. *The correct answer to the question in block 18 is **B**. Not all op amps have built-in protection against latchup, so external circuitry must be used to prevent it. It is an advantage to have the protection built in. Once latchup occurs, the op amp cannot amplify the input signal.* Here is your next question.
   Which of the following is a popular package for integrated circuits?
   A. DIP. (Proceed to block 23.)
   B. DUP. (Proceed to block 15.)

7. *Your answer to the question in block 24 is **A**. This answer is wrong. Frequency response and bandwidth both refer to the range of frequencies that can be amplified by a given amplifier. The greater the frequency response, the greater the bandwidth. There is no tradeoff between the two.* Proceed to block 14.

8. *Your answer to the question in block 2 is **B**. This answer is wrong. The maximum rate of change of the op amp output voltage is not called reaction.* Proceed to block 20.

9. *Your answer to the question in block 1 is **A**. This answer is wrong. The gain is directly related to the ratio of R_2 to R_1.* Proceed to block 17.

10. *The correct answer to the question in block 21 is **A**. Many op amps have an open-loop gain of 1 million. Others have gain values as low as 20,000.* Here is your next question.
    In which of the following circuits could you use a linear integrated circuit?
    A. An audio amplifier. (Proceed to block 24.)
    B. A switching network. (Proceed to block 19.)

11. *Your answer to the question in block 23 is **A**. This answer is wrong. A subtrahend is the number that you subtract from another number as shown here:*

153	minuend
−26	subtrahend
127	remainder or difference

    Proceed to block 4.

12. *Your answer to the question in block 18 is **A**. This answer is wrong. When latchup occurs in an op amp it can no longer act as an amplifier.* Proceed to block 6.

13. *The correct answer to the question in block 20 is* **A.** *The op amp Bode plot is useful in designing an amplifier with a given gain or bandwidth.*

You should not get the idea that Bode plots are used only with op amps. Actually they are used for all types of amplifier circuits. Here is your next question.

A sine-wave signal is applied to both the inverting and noninverting input terminals at the same time. The output of the op amp should be

A. twice as large as it would be if the signal were applied to only one of the input terminals. (Proceed to block 25.)

B. no signal. (Proceed to block 18.)

14. *The correct answer to the question in block 24 is* **B.** *Anything you do to increase the bandwidth will automatically decrease the gain. The product of the bandwidth and gain is a constant value which is called the gain-bandwidth product (GBP). Mathematically,*

$$\text{GBP} = \text{gain} \times \text{bandwidth}$$

or

$$\text{Gain} = \frac{\text{GBP}}{\text{bandwidth}}$$

From this simple equation you can see that an increase in bandwidth will result in a decrease in gain. Likewise, a decrease in bandwidth causes an increase in gain. Here is your next question.

On the input leads of an op amp the + and − signs show

A. where the positive and negative power-supply voltages are to be applied. (Proceed to block 22.)

B. which terminal is for an inverting input and which is for a noninverting input. (Proceed to block 16.)

15. *Your answer to the question in block 6 is* **B.** *This answer is wrong. There is no such thing as a DUP package for integrated circuits.* Proceed to block 23.

16. *The correct answer to the question in block 14 is* **B.** *When a signal is delivered to the input terminal marked with a minus sign, the amplified output terminal will be 180° out of phase with the input. This simply means that when the input signal is on the positive half cycle, the output signal is on the negative half cycle. Also, when the input goes into the negative half cycle, the output goes into the positive half cycle.* Here is your next question.

Can an op amp be operated from a single power supply?

____
Yes or No

Proceed to block 28.

17. *The correct answer to the question in block 1 is* **B.** *The approximate voltage gain A_v of the circuit in Fig. 10-25 is found as follows:*

$$A_v = \frac{R_2}{R_1} = \frac{10 \text{ kilohms}}{10 \text{ kilohms}} = 1$$

Here is your next question.
If an op amp has the characteristic curve shown in Fig. 10-26, what bandwidth will be obtained with an amplifier gain of 100?

A. About 20,000 hertz. (Proceed to block 27.)
B. About 100 hertz. (Proceed to block 27.)

18. *The correct answer to the question in block 13 is* **B.** *The common mode rejection (CMR) of an op amp should be high. This means that the op amp will reject any signal that is common to both terminals.*
Here is your next question.
A company which makes a certain op amp says that it will not latch up. Which of the following is true?

A. This is a disadvantage. It means that the op amp cannot hold a current value, even for a short period of time. (Proceed to block 12.)

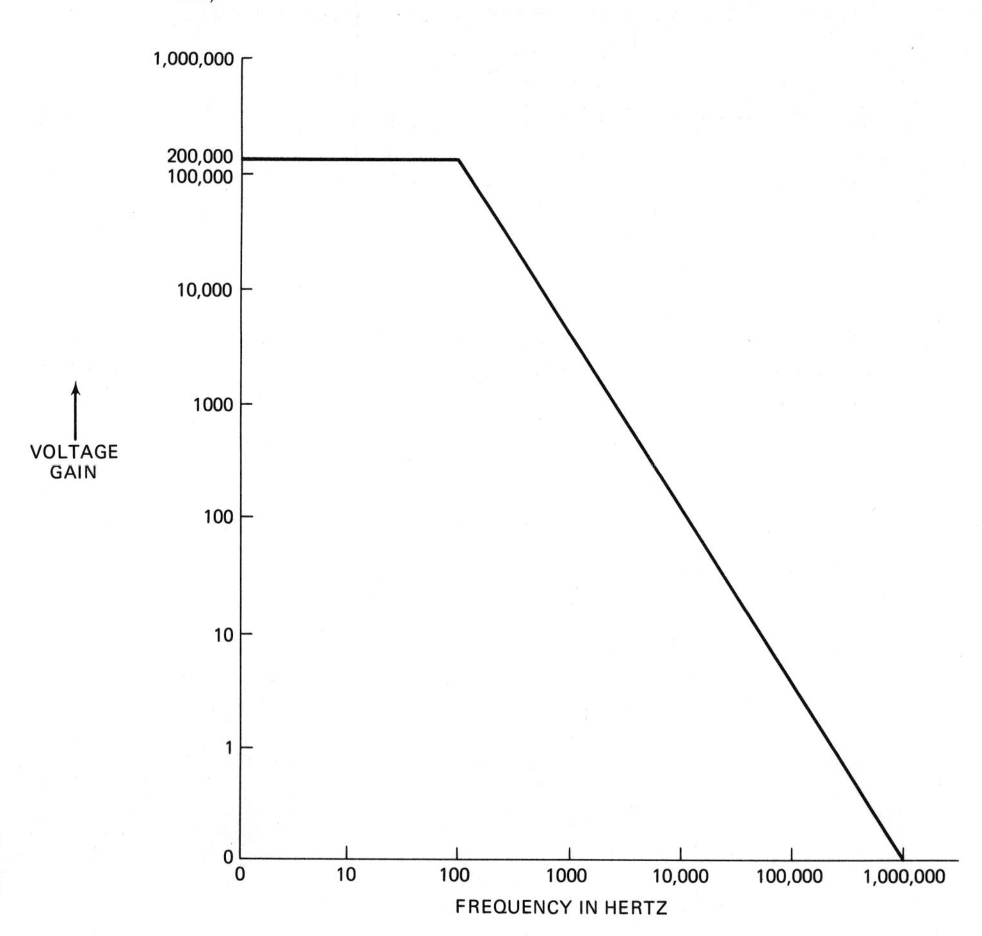

Fig. 10-26 What bandwidth will be obtained with an amplifier gain of 100?

B. This is an advantage. It means that the op amp will not become locked in saturation. (Proceed to block 6.)

19. *Your answer to the question in block 10 is* **B.** *This answer is wrong. In a switching network the output signal is not amplified. Also, the output is* ON *or* OFF, *but there is no condition between these states.* Proceed to block 24.

20. *The correct answer to the question in block 2 is* **A.** *The slew rate of an op amp is usually measured in volts per microsecond.* Here is your next question.
A graph of amplifier gain versus frequency, such as the one shown in Fig. 10-27, is called
A. a Bode plot. (Proceed to block 13.)
B. a Smith plot. (Proceed to block 3.)

21. *The correct answer to the question in block 4 is* **A.** *A graph of the gain versus frequency is called a Bode plot. A Bode plot of an operational amplifier response clearly shows the roll-off.* Here is your next question.
Which of the following is a typical value of open-loop gain for an op amp?
A. 200,000. (Proceed to block 10.)
B. 175. (Proceed to block 5.)

22. *Your answer to the question in block 14 is* **A.** *This answer is wrong. The op amp symbol does not usually show power-supply connections. The + and − are not for determining the polarities of the power-supply connection.* Proceed to block 16.

23. *The correct answer to the question in block 6 is* **A.** *The letters DIP stand for* Dual Inline Package. (See Fig. 10-2.) Here is your next question.

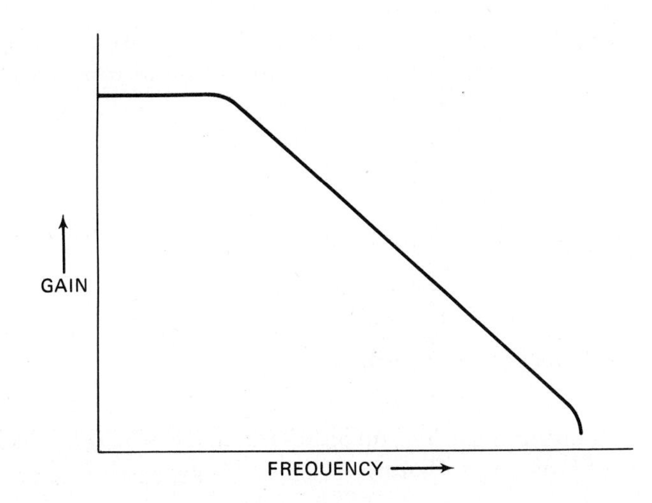

Fig. 10-27 What type of plot is this?

The word *monolithic* means "one stone." An integrated circuit is monolithic because it is made on one chip of silicon. The base of the integrated circuit is called
A. the subtrahend. (Proceed to block 11.)
B. the substrate. (Proceed to block 4.)

24. *The correct answer to the question in block 10 is* **A.** *The output of an audio amplifier is an amplified version of the input signal. This is typical of linear circuits.* Here is your next question.
Which of the following are tradeoffs when you are designing an op amp circuit?
A. Frequency response and bandwidth. (Proceed to block 7.)
B. Frequency response and gain. (Proceed to block 14.)

25. *Your answer to the question in block 13 is* **A.** *This answer is wrong. There should be no output when a signal is applied to both the inverting and noninverting terminals at the same time.* Proceed to block 18.

26. *Your answer to the question in block 4 is* **B.** *This answer is wrong. The rate at which the output voltage decreases when the input voltage decreases is related indirectly to the frequency response of the amplifier, but this is not called roll-off.* Proceed to block 21.

27. *Your answer to the question in block 17 is* **B.** *This answer is wrong. Find the gain value of 100 on the vertical axis. What is the frequency of the amplifier when the gain is 100?* Proceed to block 2.

28. *Yes. An op amp can be operated with a single power supply in most cases, but this is not a recommended way to do it. The manufacturer says to use separate positive and negative power supplies. Values of +15 volts and −15 volts are typical for the positive and negative power-supply leads.*
You have now completed the programmed questions. The next step is to put some of these ideas to work in laboratory experiments. Proceed to the Experiment section of this chapter.

EXPERIMENT

(The experiment described in this section may be performed on the circuit board described in Appendix C or on a similar laboratory setup.)

Purpose This experiment gives you experience in connecting an operational amplifier into a circuit.

Theory An oscillator is a high-gain amplifier with regenerative feedback. To determine if the 741 op amp will amplify a signal, it is only necessary to connect it as an oscillator.

Figure 10-28 shows the basic oscillator circuit used in this experiment. Note that the feedback circuit (R_1 and R_2) returns the output signal to the noninverting-amplifier input. As you know from your study of oscillator circuits in Chapter 9, oscillation occurs when the output signal of an amplifier is returned to the input in phase with the input signal. At the noninverting input terminal the input and output signals are in phase.

When the circuit is first energized, the output voltage (at pin 6) rises. This positive-going voltage goes to pin 3 where it is amplified. The output at pin 6 goes further positive as it is amplified, and the op amp goes into saturation very quickly. If an amplifier is saturated, its output can no longer change.

Capacitor C_1 starts to charge to the positive output voltage. The voltage at pin 2 rises with the capacitor charged.

When the voltage across the capacitor at pin 2 becomes more positive than the voltage at pin 3, the op amp output goes negative very quickly. Then C_1 starts to discharge. When the capacitor voltage is less than the voltage at pin 3, the output goes positive again.

The overall result is that the output voltage is a square wave. This is shown on the illustration. The square wave produces sound in the speaker. The oscillator is called a *multivibrator* because its output is a square wave and its frequency depends upon the rate of charging and discharging a capacitor.

Test Setup Figure 10-29 shows the test setup. Figure 10-29*a* shows the schematic and Fig. 10-29*b* shows the laboratory connection. This circuit consists of a single power supply like the one shown in Fig. 10-9, except that the voltage value is different, and an op amp oscillator similar to the one shown in Fig. 10-28.

Procedure

step one Connect the circuit as shown in Fig. 10-29. The op amp should oscillate, and the output at the speaker should be an audio tone.

step two Set R_2 and R_3 to the middle of their range.

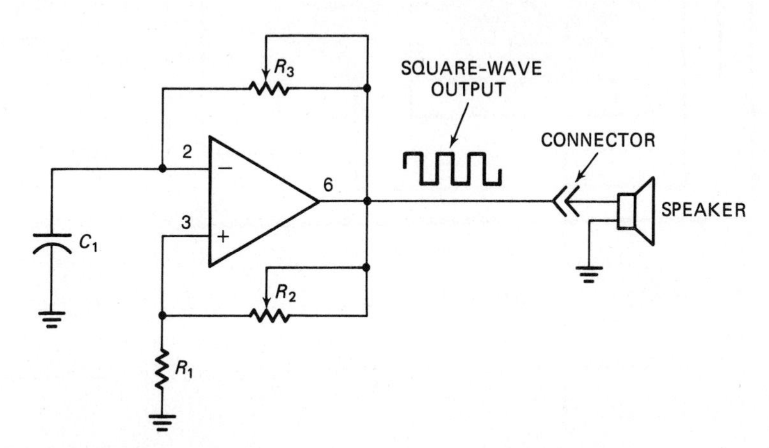

Fig. 10-28 Basic op amp oscillator circuit.

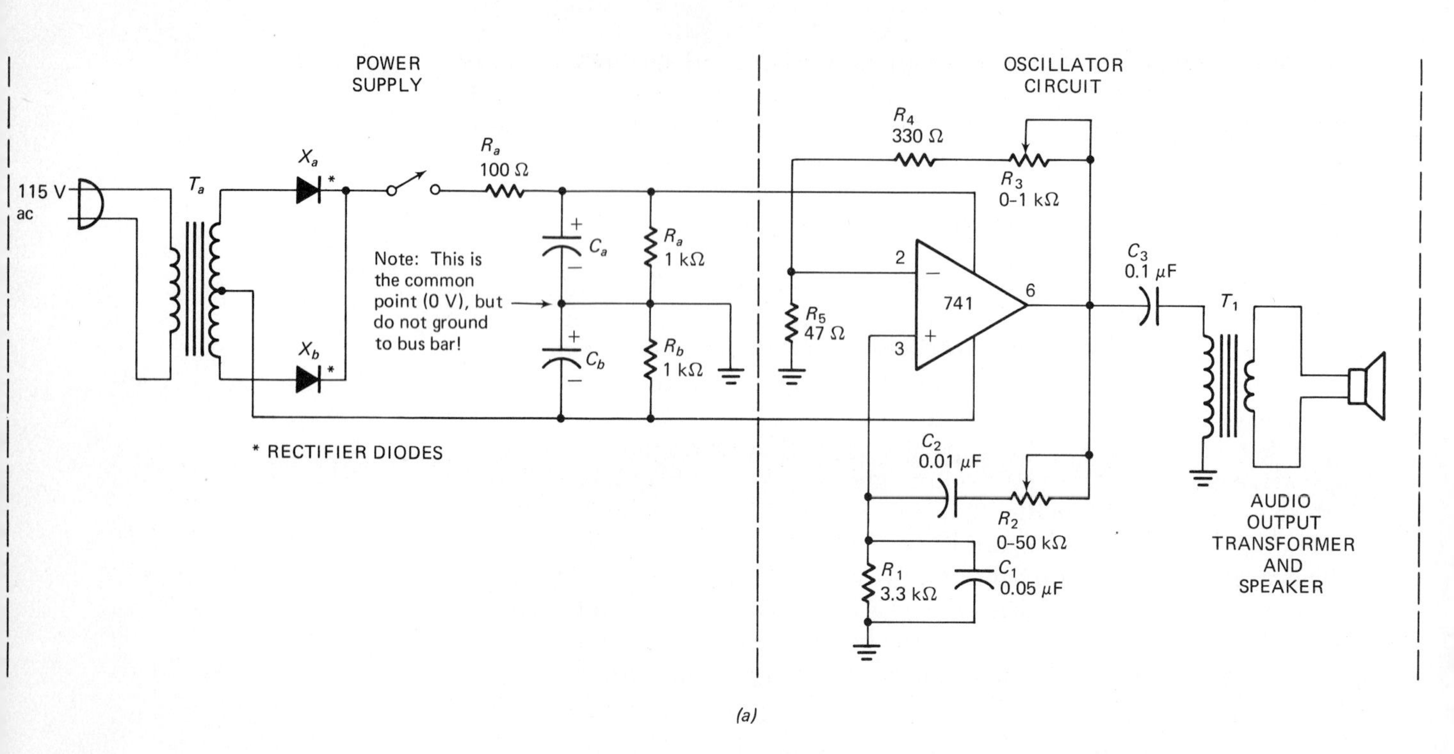

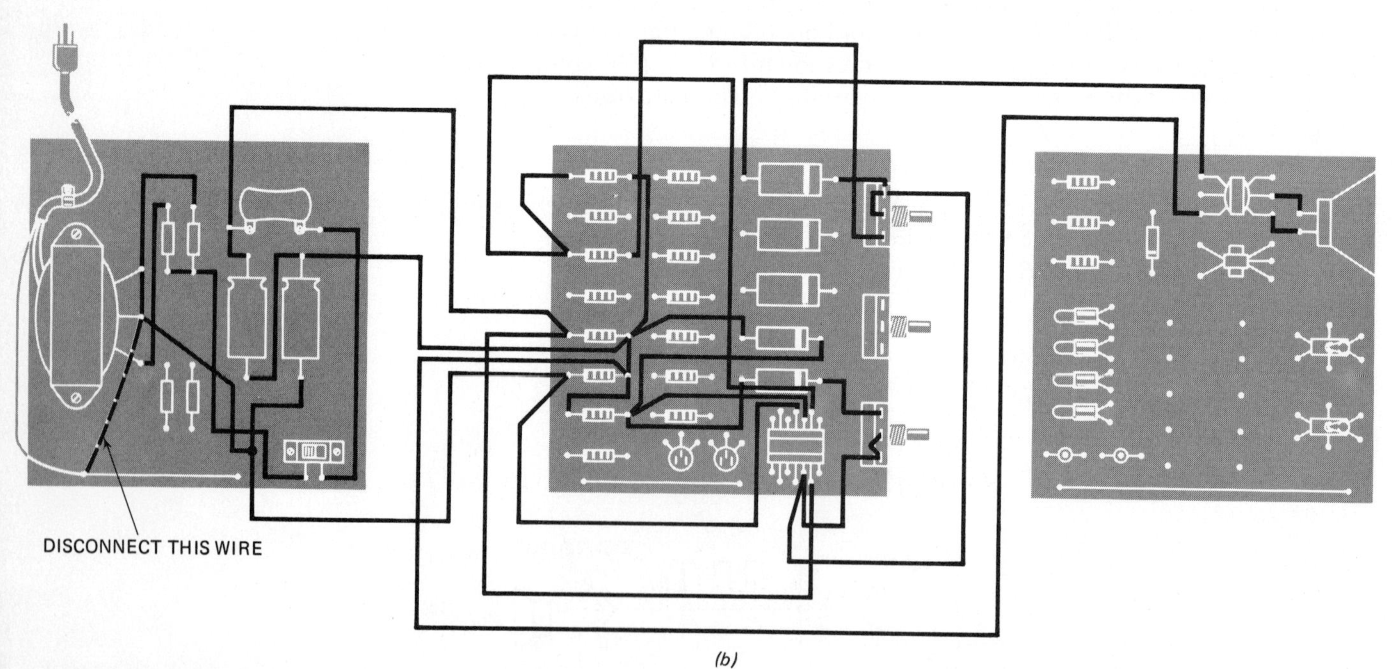

Fig. 10-29 Circuit for the experiment. (a) The schematic. (b) The laboratory connection.

step three Measure the positive and negative power-supply voltage with respect to ground. Record the values here. ______________ volts

______________ volts

The supply voltage should be about 4.5 volts positive and about 4.5 volts negative for a total of about 9 volts between the positive and negative terminals. If the voltage is much less than this, the operational amplifier may not work properly.

step four Is there an audio tone in the speaker? ________________ Yes or No

Your answer should be yes.

step five Adjust R_2. What effect does this have on the output?

____

The adjustment of R_2 changes the amount of time that it takes for C_2 to charge. Therefore it changes the time per cycle or period T of each cycle. The frequency f of the output signal is related to the period by the equation

$$f = \frac{1}{T}$$

Therefore, changing R_2 changes the period, and this in turn changes the frequency of the sound.

step six Does the frequency increase or decrease when the resistance of R_2 is increased? ________________ Increase or Decrease

Increasing the resistance of R_2 will increase the period T and decrease the frequency f.

step seven What effect does R_3 have on the output of the circuit?

____

Resistor R_3 controls the amount of feedback to the noninverting input. The gain of the amplifier depends upon the ratio of $R_3 + R_4$ to R_5. Changing the value of R_3 will change the volume of the sound.

If R_3 is made too small, then the feedback will be reduced to the point where the gain is too small. At this point the amplifier cannot keep the oscillation going.

Conclusion The op amp will oscillate in the circuit of Fig. 10-29. This means that it will amplify in a circuit with a single power supply.

SELF-TEST WITH ANSWERS

1. LSI refers to an integrated circuit with (*a*) more than 5 passive components; (*b*) more than 50 active components; (*c*) more than 500 active components; (*d*) more than 25 passive components.

2. Which of the following is *not* a requirement for an operational amplifier? (*a*) It must be an integrated circuit; (*b*) It must have a very high gain; (*c*) It must have a noninverting input terminal; (*d*) It must have an inverting input terminal.
3. Which of the following is true? (*a*) Increasing the bandwidth of an amplifier is accomplished by increasing the gain; (*b*) Increasing the gain of an amplifier reduces its bandwidth.
4. Is this statement correct? The + and − signs on the op amp symbol tell where to connect the positive and negative leads from the dc supply. (*a*) Yes; (*b*) No.
5. A certain op amp circuit has two input terminals: A and B. Whenever the voltage of A is higher than the voltage of B, the op amp has a low output. Whenever the voltage of A is lower than the voltage of B, the op amp has a high output. This circuit is called (*a*) an analyzer; (*b*) a summing amplifier; (*c*) a closed-loop network; (*d*) a comparator.
6. With reference to subtraction in a summing amplifier, (*a*) it cannot be done; (*b*) the number to be subtracted is fed to the output terminal and the difference is taken from the − terminal; (*c*) the number to be subtracted is 180° out of phase with numbers to be added; (*d*) the number to be subtracted is fed to the + terminal and the difference is taken from the − terminal.
7. A certain op amp is connected so that it has a voltage gain of 1, a high input impedance, and a low output impedance. This type of circuit is called (*a*) an oscillator; (*b*) a follower; (*c*) an open-loop circuit; (*d*) a comparator.
8. A mercury thermometer uses a column of mercury beside a column of numbers. You read the temperature by sighting along the top of the mercury column. This is (*a*) an analog device; (*b*) a digital device.
9. The nonconducting base of an integrated circuit is called the (*a*) foundation; (*b*) slab; (*c*) substrate; (*d*) bottom.
10. A capacitor on the op amp integrated circuit determines the rate at which gain decreases with an increase in frequency. This is (*a*) not possible because capacitors cannot be made on integrated circuits; (*b*) internal frequency compensation; (*c*) a dropout capacitor; (*d*) roll-on.

ANSWERS TO SELF-TEST

1. (*b*)—This is an accepted definition of LSI.
2. (*a*)—Op amps were used in the 1940s, before there were transistors or integrated circuits. They were vacuum-tube versions of the circuits used today.
3. (*b*)—Bandwidth and gain are tradeoffs.
4. (*b*)—The + and − terminals show the noninverting and inverting terminals for the input signal.
5. (*d*)—It does not matter which input is marked A and which is marked B.
6. (*c*)—See Fig. 10-21.

7. (*b*)—This circuit has the same characteristics as an emitter follower.
8. (*a*)
9. (*c*)
10. (*b*)—Not all operational amplifiers have built-in frequency compensation.

how is feedback used in amplifiers? 11

INTRODUCTION

Several people can look at the same car, and each might describe it in a different way. One person may say it's a sedan, another may say it's a blue car, another may say it's a family car.

If the same people are looking at the same house, they may describe it in different ways. It can be a frame house, a one-story house, and a white house all at the same time.

Amplifiers may also have different names, and one amplifier can be described in several different ways. For example, it can be called a voltage amplifier, a class A amplifier, and a common-emitter amplifier. All these names can be used for the same amplifier. From your study of previous chapters you know what these names mean.

Amplifiers may also be called by names that refer to the method of getting the signal from one stage to another. Thus, you may have *RC*-coupled, impedance-coupled, transformer-coupled, or direct-coupled amplifiers. You will study about these circuits in this chapter.

Another way of naming amplifiers is according to the type of signal they are designed to amplify. Examples are audio amplifiers, video amplifiers, and RF amplifiers.

Regardless of what the amplifier is called, or what it is used for, it has at least two important features: *gain* and *bandwidth*. *Voltage gain* is the output-signal voltage divided by the input-signal voltage. *Power gain* is the output-signal power divided by the input-signal power.

Bandwidth is the range of frequencies the amplifier can handle.

Gain and bandwidth are tradeoffs. Anything you do to increase the gain of an amplifier will decrease its bandwidth. On the other hand, decreasing the gain will increase its bandwidth.

A common way to change or control amplifier gain is to use feedback. In this chapter you will learn how gain is increased or decreased by the use of feedback.

You will be able to answer these questions after studying this chapter.

- ☐ **How are signals coupled from one amplifier to another?**
- ☐ **What is high-frequency (peaking) compensation?**
- ☐ **What is transformer coupling?**
- ☐ **What is direct coupling?**
- ☐ **How is feedback used in amplifiers?**
- ☐ **What is current feedback and how is it obtained?**
- ☐ **What is voltage feedback and how is it obtained?**
- ☐ **What is low-frequency compensation?**

INSTRUCTION

How Are Signals Coupled from One Amplifier to Another?

So far you have learned that all the amplifying components require a dc voltage. This includes a dc bias voltage and a voltage across the component. You have learned how these dc voltages affect the operation of the amplifiers. Our concern in this section is with the path of the *signal flow* from one amplifier to another.

What Is RC Coupling?

Figure 11-1 shows two *RC- (resistance-capacitance) coupled* amplifiers. In Fig. 11-1*a* the output signal from amplifier 1 is developed across the load resistor R_L. This load resistor is shown connected between the amplifier and a positive supply voltage for operating the amplifying device. The amplifying device could be a tube, an NPN transistor, or an FET in this example.

The signal voltage across R_L is coupled to resistor R_a by coupling capacitor C_b. The capacitor permits the signal to pass from the first amplifier to the next. At the same time it prevents the positive dc voltage at the output of the first amplifier from passing to the input of the next stage. The input signal to amplifier 2 is the signal voltage developed across R_a.

Capacitor C_a is a power-supply filter capacitor. It also prevents the signal of one amplifier from getting into the other amplifier, which is connected to the same power supply.

Figure 11-1*b* shows an example of two amplifiers with *RC* coupling. In this circuit, the bias for Q_1 and Q_2 is obtained with voltage divider networks (R_1 and R_2, and R_7 and R_a). Both amplifiers have emitter stabilization resistors (R_4 and R_8).

The load resistor for amplifier Q_1 is resistor R_L. This resistor has the same job as R_L in Fig. 11-1*a*. The coupling capacitor between the amplifier circuits is capacitor C_b. The input signal to Q_2 is developed across R_a.

In the experiment for Chapter 8 you learned that a circuit may have both an

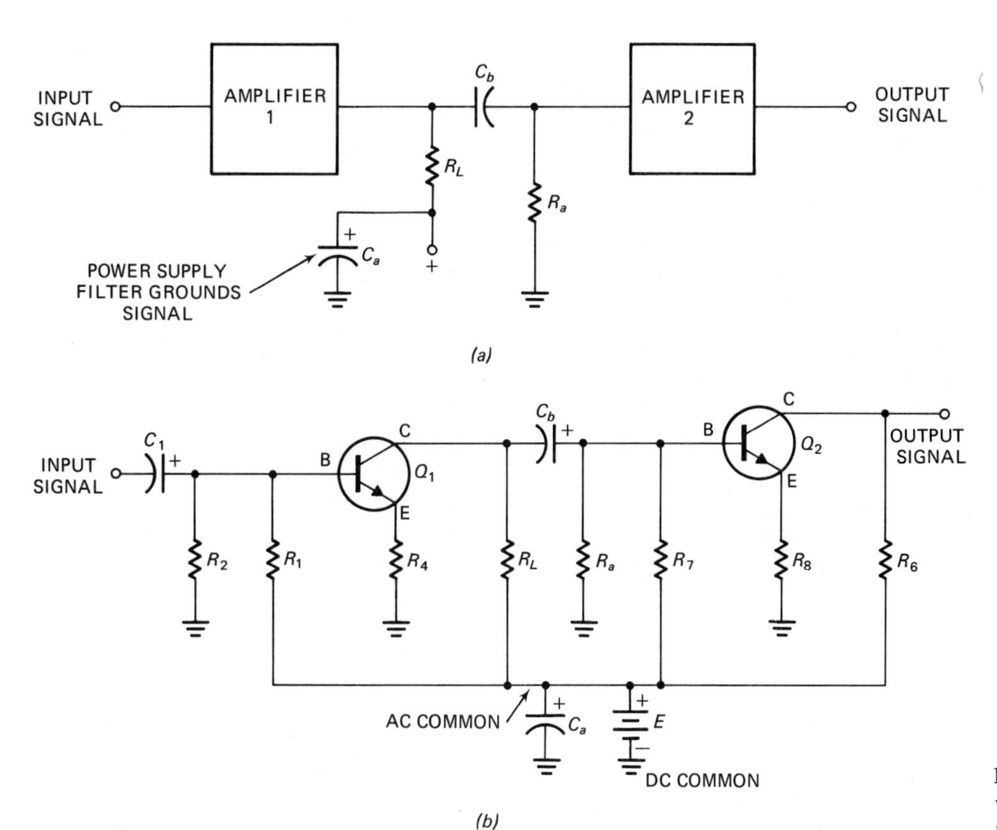

Fig. 11-1 ***RC*** **coupling. (*a*) Network. (*b*) Example with bipolar transistors.**

ac and a dc voltage at the same time. Also you learned that the common point for the ac voltage may be at a different point from the dc common point.

In the circuit of Fig. 11-1*a* capacitor C_a also acts as a decoupling filter. This capacitor causes the positive dc voltage point to be grounded only as far as ac is concerned. Since this is the ac common point, there can be no chance of a signal from one amplifier getting into another amplifier connected to this point. Remember, the ac voltage is 0 volts at the ac common point.

RC-coupled circuits are very popular because of their low cost, but they also have a serious disadvantage. You will remember that a capacitor is a component that opposes any change in voltage across its terminals. The opposition that a capacitor offers to alternating current is called *capacitive reactance,* and it is given by the equation

$$X_C = \frac{1}{2\pi fC}$$

There is an important point shown by this equation. It shows that for a given capacitor C, the higher the frequency, the lower the reactance (or opposition) the capacitor offers. On the other hand, the lower the frequency, the higher the opposition to alternating-current flow.

Returning now to the coupled circuit between the two amplifiers, you can see that the high frequencies will be coupled from amplifier 1 to amplifier 2

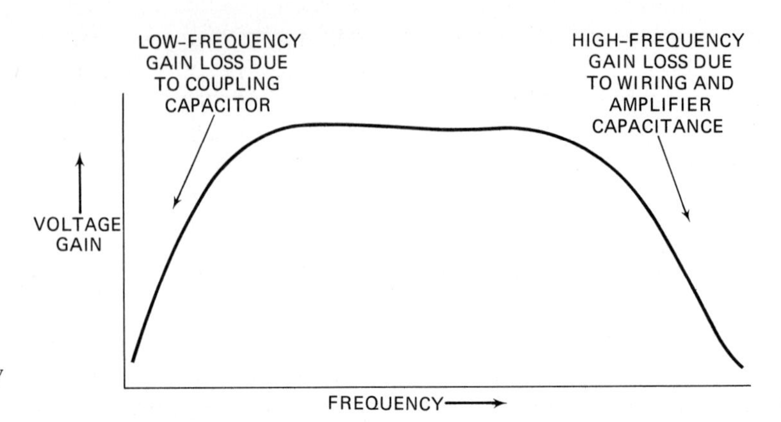

Fig. 11-2 Amplifier frequency response.

with almost no opposition, but the very low frequencies will see a great opposition. The lowest possible frequency—a dc or 0-hertz voltage—cannot be coupled at all.

To summarize, *RC* coupling is inexpensive but has a poor frequency response for the very low frequencies.

Figure 11-2 shows a typical response curve for *RC*-coupled amplifiers. Note that the gain falls off at low frequencies because of the coupling capacitor. The gain also falls off at high frequencies due to the input capacitance and distributed capacitance in the circuit. Figure 11-3 shows these capacitances in dotted lines.

Capacitor C_x in Fig. 11-3 represents the wiring and parts capacitance to ground (chassis) in the circuit. You will recall that a capacitor is made of any two conductors separated by an insulator. The wiring and parts for the circuit are separated from the chassis or common point by insulation (or air), so they act like any capacitor. At low frequencies this capacitance is so small it can be ignored. At high frequencies much of the signal can be lost due to wiring and parts capacitance to the chassis.

Capacitor C_y in Fig. 11-3 is the capacitance between the control electrode (grid, base, or gate) and the common electrode (cathode, emitter, or source). This capacitance is always present. It has the effect of shorting part of the

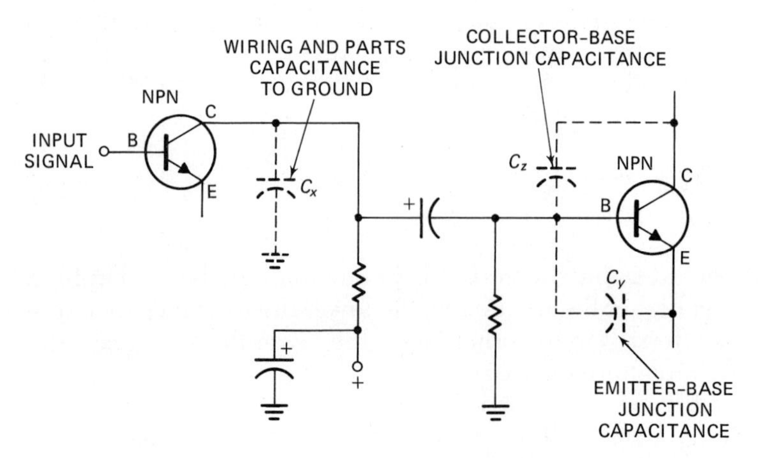

Fig. 11-3 Two reasons for loss of high-frequency gain.

high-frequency signal to ground. The same effect is caused by C_z, the capacitance of the control electrode to the output electrode (plate, collector, or drain).

The loss of signal due to C_x and C_y in Fig. 11-3 cannot be avoided. This is what causes the response curve of the RC-coupled amplifier, as shown in Fig. 11-2, to drop off at the higher frequencies.

What Is High-Frequency (Peaking) Compensation?

One way to improve the gain of amplifiers at high frequencies (above about 100 kilohertz) is to add an inductor in the output circuit. This is called a *peaking coil*, and it is shown in Fig. 11-4. Now the load for amplifier 1 consists of L and R_1 in series. At low frequencies L has almost no reactance, so there is almost no voltage developed across it.

At high frequencies the reactance of L is high. Therefore there will be a voltage drop across L and also across R_1. This means that the voltage delivered by coupling capacitor C_b to input resistor R_2 is greater at high frequencies than it would be without L.

The circuit in Fig. 11-4 is called a *high-frequency compensation circuit.* Such circuits are commonly found in the video amplifiers of TV sets (see Chapter 14) and oscilloscopes.

High-frequency compensation circuits may be in forms other than the one shown in Fig. 11-4. The inductor may be connected in series with the coupling capacitor. In some cases two inductors may be used, one in series with R_1 and one in series with the coupling capacitor.

What Is Transformer Coupling?

It has been pointed out that the coupling capacitor in RC-coupled amplifiers prevents the dc voltage at the output of amplifier 1 from reaching the input circuit of amplifier 2.

A transformer is another circuit component that will pass an ac voltage but not pass a dc voltage. *Transformer coupling* between amplifiers is illustrated in Fig. 11-5.

In Fig. 11-5*a* the transformer symbol indicates that it is an air-core type.

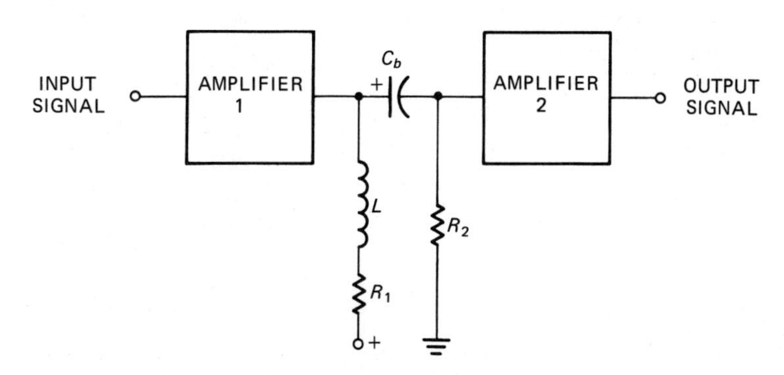

Fig. 11-4 High-frequency compensation.

This means that the amplifier is used only for radio frequencies (RF). Since the primary of the transformer and the secondary are both inductors, either the primary or the secondary, or both windings, of the transformer can be tuned. This is a strong advantage of transformer coupling in radio-frequency circuits. As a general rule you will find that most RF amplifiers have at least

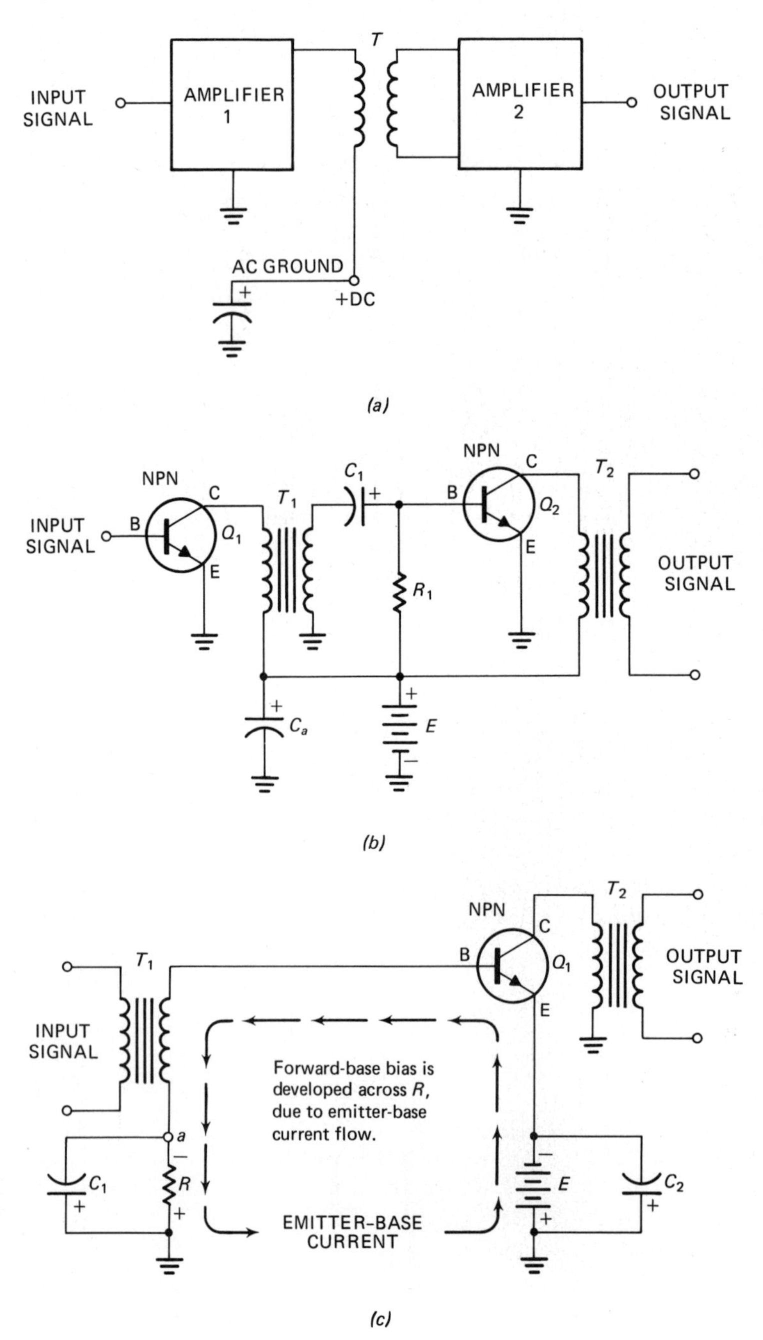

Fig. 11-5 Transformer coupling. (*a*) Coupling between radio-frequency amplifiers. (*b*) Audio-frequency circuit with isolation capacitor C_1. (*c*) Audio-frequency circuit without isolation capacitor.

one tuned circuit in their coupling networks. This tuned circuit will usually be part of an impedance-coupled or transformer-coupled circuit.

Figure 11-5*b* shows a typical audio-frequency transformer-coupled amplifier circuit. Both transistors are NPN types, so they require a positive voltage on their collectors and a positive voltage on their bases with respect to their emitters.

The output signal from transistor Q_1 is delivered to Q_2 through coupling transformer T_1 and coupling capacitor C_1. Note that the primary of the transformer T_1 goes to the positive voltage power supply to provide the positive voltage on the collector of Q_1. In transistor Q_2 the base obtains its positive voltage through resistor R_1, and the collector obtains its positive voltage through the primary of transformer T_2.

Suppose the coupling capacitor C_1 were omitted in the circuit of Fig. 11-5*b*. That would mean that the base would be grounded through the low resistance of the transformer T_1 secondary winding. As you know, transistor Q_2 could not operate class A with its base grounded. It must have a forward base bias voltage. The purpose of capacitor C_1 is to prevent the base voltage of Q_2 from being grounded.

Not all transformer-coupled circuits have a need for such a capacitor. Figure 11-5*c* shows another NPN transistor circuit with transformer coupling, but the coupling capacitor is not needed in this stage.

Note that in the circuit of Fig. 11-5*c* both the collector and the base (for ac only) are connected to ground through the transformer windings and the emitter is connected to a negative voltage. For forward bias it does not make any difference if you make the emitter negative and the base zero or make the base positive and the emitter zero. As far as the NPN transistor is concerned, either type of connection will provide proper forward bias. Also, it does not make any difference (for collector current flow) if you make the collector zero and the emitter negative or the collector positive and the emitter zero. In either case the collector and the base are both positive with respect to the emitter as required for the NPN circuit. The advantage of the circuit, however, is that a coupling capacitor, like C_1 of Fig. 11-5*b*, is not needed. This also improves the low-frequency response of the amplifier.

Resistor R in the circuit of Fig. 11-5*c* is needed to limit the emitter-base current to the desired value. Capacitor C, in parallel with R, maintains the bottom of the secondary winding (point *a*) at ac (audio) ground potential. Therefore the full audio signal is delivered to the base.

What Is Direct Coupling?

None of the methods of coupling audio (or video) amplifiers described so far can be used to amplify a very wide range of frequencies. *RC* coupling has poor very-low-frequency audio response and a limited (above audio) high-frequency response. High-frequency compensated coupling has the same low-frequency response but improved high-frequency response (to 4 megahertz or more). Transformer coupling can be used for a range of frequencies set by the transformer design but generally is not used over a very wide range of frequencies.

The fifth method of coupling, called *direct coupling* (shown in Fig. 11-6), has the advantage that it has the best possible low-frequency response, down to 0 hertz. However, this type of coupling still has basically the same limited high-frequency response as an uncompensated *RC*-coupling circuit. There is just a straight piece of wire connecting amplifier 1 and amplifier 2 in Fig. 11-6*a*. In order to do this, great care must be taken in the design of the amplifier circuit. It is no longer possible to isolate the normally higher dc collector voltage—9 volts—of the first stage from the normally lower voltage of the second base input.

Figure 11-6*b* shows how direct coupling may be obtained with NPN bipolar transistors. The load resistor for Q_1 is R_3. The dc collector voltage of Q_1 is also the base voltage for Q_2. The arrows show the dc collector current path for Q_1 and Q_2 and the base current flow for Q_2. Note that two currents flow through R_3.

The emitter resistors have different values. The emitter of Q_2 is more positive than the emitter of Q_1. (The dc voltages shown in squares are with respect to ground. The voltage *across* each transistor is the same.)

Transistor Q_1 has an emitter-base voltage of 0.8 volts, and transistor Q_2 also has an emitter-base voltage of 0.8 volts. The emitter-to-collector voltage for both transistors is 7 volts, as shown in Fig. 11-6.

The operating voltages of both transistors are the same, even though they have different voltages with respect to ground. However, the important point is that the transistors both work correctly. Their operation is based on

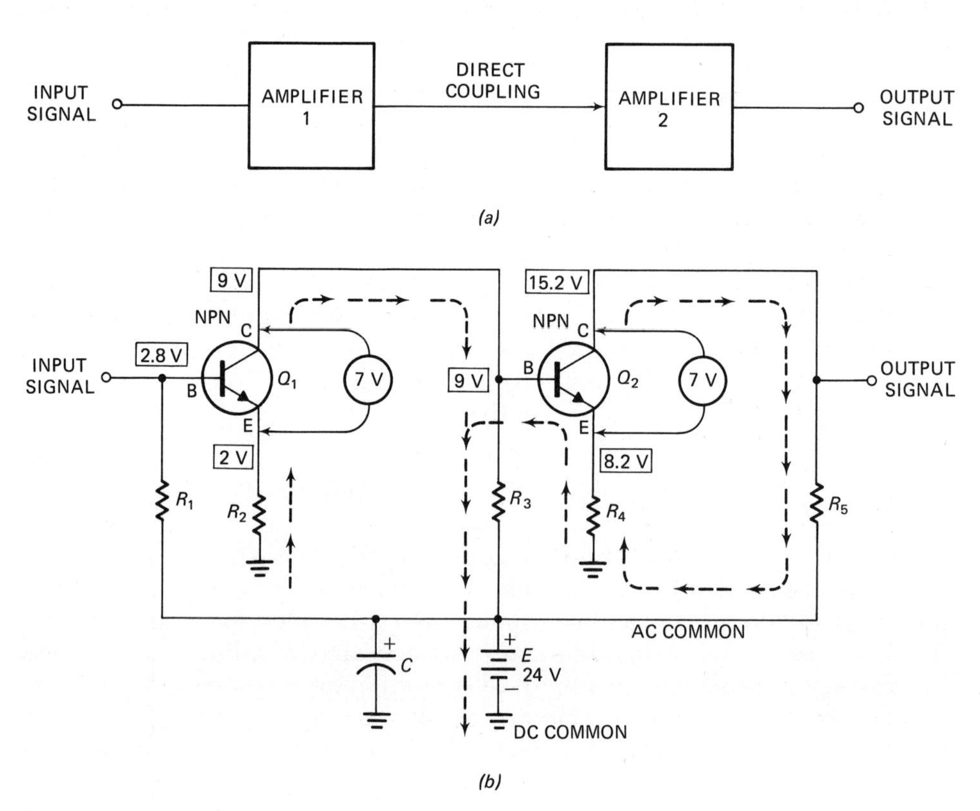

Fig. 11-6 Direct coupling. (*a*) Between amplifiers. (*b*) Between bipolar transistors.

emitter-base and emitter-collector voltages and currents, not upon the voltages on the electrodes with respect to ground.

There are wiring capacitances in direct-coupled circuits. Also the amplifying components have input and output capacitances. This means that there is a practical limit to the high-frequency response of direct-coupled amplifiers, as previously stated.

Figure 11-7 shows how amplifier components can be direct-coupled, even though their output electrodes would normally require a voltage of opposite polarity to that applied to their control electrodes.

In this application vacuum tubes are used, but a similar idea would work for JFETs or depletion MOSFETs.

There is a resistance voltage divider (R_1, R_2, R_3, and R_4) between the +500-volt mark and ground. This voltage divider has taps at the 0-volt, 10-volt, 210-volt, 300-volt, and 500-volt points. The cathode voltage of tube V_1 is +10 volts, and the grid is at 0 volts. This is the same as saying that the control grid is 10 volts negative with respect to the cathode. Thus V_1 is properly biased for operation.

The plate of V_1 is directly connected to the grid of V_2. The voltage at this point (plate and grid) is +200 volts. However, the cathode of the tube is connected to a +210-volt point in the voltage divider. Again, with the cathode at +210 volts and the grid at +200 volts, the grid is actually 10 volts negative with respect to the cathode, and the tube is properly biased. The overall result is that the plate of the first tube can be directly connected to the grid of the next tube and that proper operating conditions do exist for both tubes. Note that the plate-to-cathode voltage is 190 volts for both tubes.

Summary

1. An amplifier may be known by several different names. It can be a class A audio voltage amplifier, and at the same time it can be a direct-coupled or common-emitter amplifier.
2. The coupling capacitor in RC-coupled amplifiers passes the ac signal. At the same time it prevents the dc voltage of one amplifier from reaching the next stage.

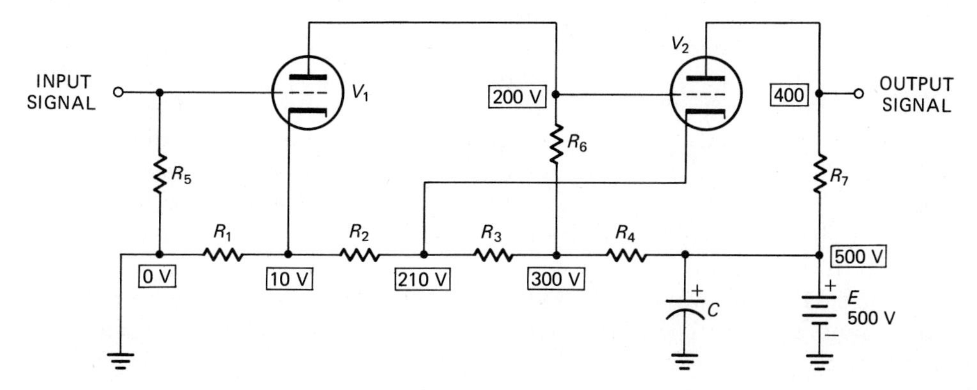

Fig. 11-7 Direct-coupled tube amplifiers.

3. The coupling capacitor in *RC*-coupled circuits tends to cause poor low-frequency response. This can be offset in some circuits by using a large value of coupling capacitance.
4. High-frequency compensation circuits can produce a better high-frequency response than simple *RC* coupling. They use peaking coils.
5. Transformer coupling is often used in RF circuits, as well as in audio circuits. The primary or secondary winding (or both) of an RF transformer can be tuned with a capacitor so that only one frequency or a narrow range of frequencies will pass.
6. Direct coupling gives the best low-frequency response, but special care must be taken in their design.
7. One disadvantage of direct coupling is that the power supply must provide a higher voltage. (There is an exception to this, called a *complementary amplifier circuit*. It will be discussed later in this chapter.)

How Is Feedback Used in Amplifiers?

In Chapter 9 you learned that a small amount of output signal from an amplifier can be fed back to the input and cause the amplifier circuit to oscillate. One of the most important conditions for oscillation is that the feedback signal must be in phase with the input signal. This type of feedback is called *positive* or *regenerative*.

There is another way to feed a signal back to the input of an amplifier, and it is called *negative* or *degenerative* feedback. With negative feedback the output signal is returned to the input in such a way that it subtracts from the input signal. In other words the feedback signal is 180° out of phase with the input signal.

Figure 11-8 shows the two types of feedback in block diagram form. In Fig. 11-8*a* part of the output signal from the amplifier is fed through a feedback network, and it combines with the input signal. These two signals are said to be in phase because they both go positive at the same time and they both go negative at the same time.

Negative feedback is shown in Fig. 11-8*b*. Part of the output signal is passed through a feedback network and combined with the input. However, the feedback signal goes negative when the input signal is going positive, and it goes positive when the input signal is going negative. Thus the feedback signal subtracts from the input signal. One important result is a loss of gain.

Figure 11-9 shows how the signals are combined for the two types of feedback. In Fig. 11-9*a* a positive-feedback signal is shown in dotted lines. It is combined with the input signal, which is shown in solid lines. The combination of the two signals is a simple addition. In other words, the amplitude of the peak input signal a simply adds to the amplitude of the peak feedback signal b. The result of adding the two signals is a waveform that has exactly the same shape but a higher amplitude $(a + b)$. (The negative half cycle is increased in exactly the same way.)

Figure 11-9*b* shows how a negative-feedback signal b combines with the input signal a. Again, the feedback signal is shown with dotted lines and the input signal is shown with a solid line. On the positive half cycle of input

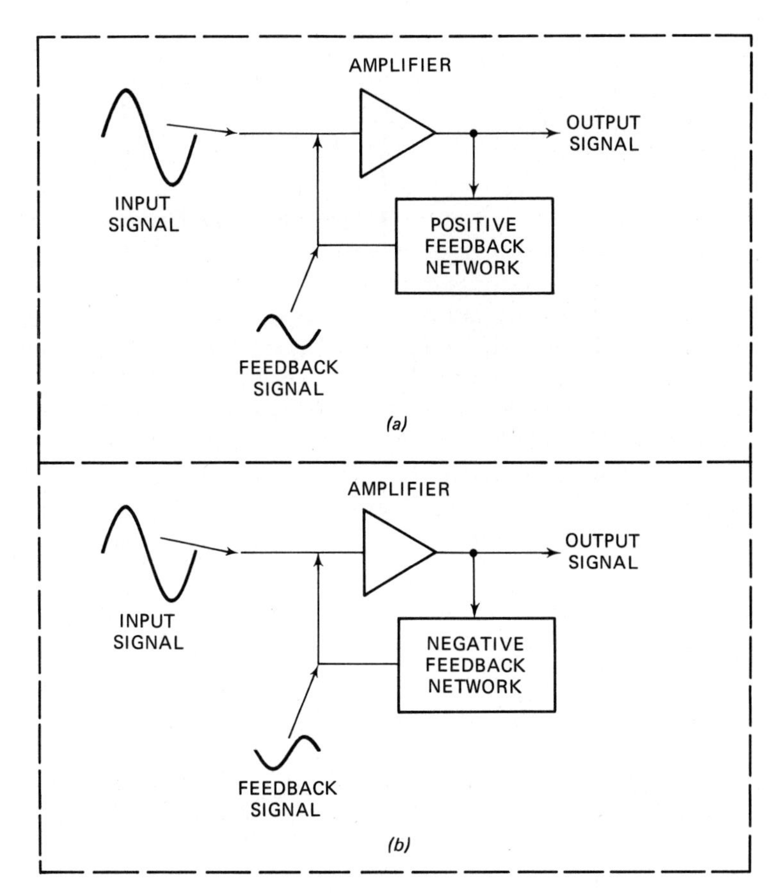

Fig. 11-8 Types of feedback. (*a*) Positive. (*b*) Negative.

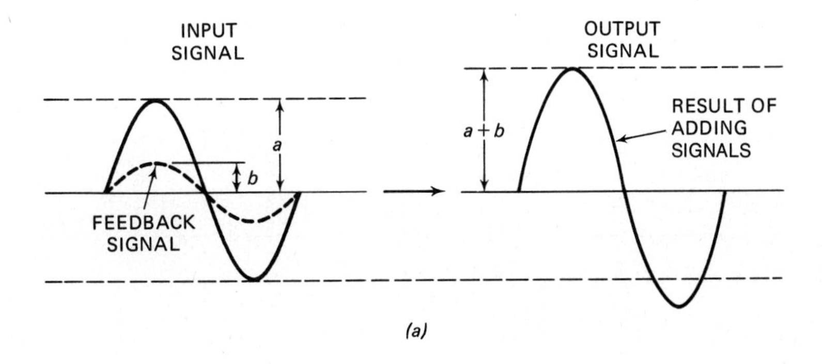

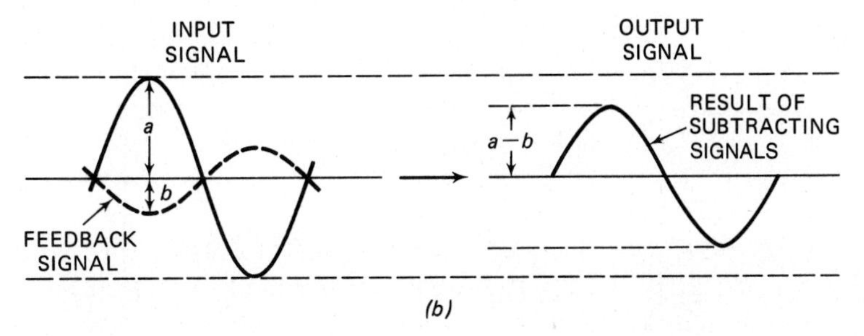

Fig. 11-9 Result of combining input and feedback signals. (*a*) With positive feedback. (*b*) With negative feedback.

signal the amplitude a is in the opposite polarity to the feedback signal b. The two signals subtract from one another. The result of subtracting these signals is that the output signal is of smaller amplitude $(a - b)$. (The negative half cycle of the output signal is reduced the same as the positive portion.)

How Is Positive Feedback Used in Amplifiers?

In Chapter 9 you learned that oscillation is one result of positive feedback. If there is enough signal from the feedback amplifier to replace the energy lost in an *LC* oscillating circuit during each cycle, then the circuit will oscillate. The overall circuitry is called an *oscillator.*

Positive feedback is used for purposes other than oscillation. In some high-frequency amplifiers it is hard to get a large amount of signal gain due to capacitive losses in the circuit. To increase the gain of the circuit, a very small amount of the output signal is fed back to the input of the amplifier to reinforce or add to the input signal. This results in a much larger gain than could be obtained without the feedback.

An example of this type of circuit is shown in Fig. 11-10. In this circuit transistor Q_1 serves as an amplifier for the RF input signal. A separate battery E_2 is used for base bias. The collector voltage is supplied by E_1. The input signal is passed through C_1 to the base, and the output signal is taken through coupling capacitor C_2. Resistor R_1 limits the amount of base bias current in the circuit.

A portion of the output signal is fed back through transformer T to the base of the amplifier. This transformer has a *dot notation,* which you may see on many commercial schematic drawings. These dots show the points of the transformer at the primary and secondary where the signals are in phase. This means that when the primary signal is going positive at the dot, the secondary is also going positive at the dot. This dot notation is important to technicians. To replace some transformers the secondary must be wired in a certain way.

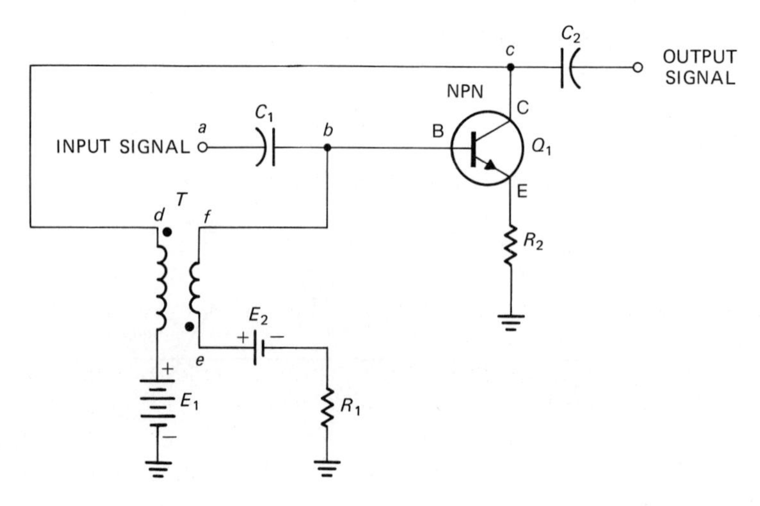

Fig. 11-10 A positive-feedback circuit.

With the dots you can tell from the transformer symbol whether the feedback is positive or negative.

The polarity of the feedback signal in the circuit in Fig. 11-10 is easy to follow. The input signal at point *a* is passed to the transistor base through capacitor C_1. To increase circuit gain the feedback signal must be in phase with this input signal.

Suppose the input signal is going positive at some instant. The positive-going input signal at point *a* causes the base voltage at point *b* to also go positive. Since Q_1 is a conventional amplifier, the output voltage at point *c* is reversed from that of *b*. In other words, when the base of *b* is going positive, the collector voltage at *c* is going negative. This negative-going voltage is delivered to the primary of *T*. Remember that the points with the dots on the transformer are the same polarity. Therefore when the voltage at *d* is going negative, the voltage at *e* is also going negative. If *e* is going negative, then the voltage at the other side of the secondary (point *f*) is going positive.

The positive-going voltage at *f* adds to the positive-going voltage at *a* to increase the signal voltage at *b*.

The same chain of events takes place when the voltage at point *a* is going negative. The overall result is that the voltage at point *f* on the secondary of the transformer is always in phase with the voltage at point *a*. These two voltages combine to produce a larger input voltage than would be obtained with the voltage at *a* alone. Thus the gain of the stage is increased.

You can see immediately that there will be a problem with the circuit of Fig. 11-10. It looks very much like the Armstrong oscillator you studied in Chapter 9. What prevents this circuit from going into oscillation? The answer is that the feedback must be maintained at a very small value. In other words, the feedback signal must be less than the amount of signal required for oscillation. This is accomplished partly by using a step-down transformer for the feedback circuit. As you will see later in this chapter, resistor R_2 is used for emitter stabilization. This resistor also reduces the gain of the stage and therefore serves to reduce the chance of oscillation in the circuit.

You will not see the circuit of Fig. 11-10 used in many systems because it does have a tendency to oscillate. Though factory adjustments are made to prevent oscillation, if someone replaces transistor Q_1 in the circuit, the difference in gain might be enough to drive the circuit into oscillation.

How Is Negative Feedback Used in Amplifiers?

In the circuit of Fig. 11-11 the input signal is passed to the base through C_1, and the output signal is passed through C_2. Resistors R_1 and R_2 form a voltage divider for base bias. The load resistor is R_3, and R_4 is the emitter stabilization resistor.

You will remember from your discussion of amplifier configurations that the input signal can be delivered to the control electrode (base) of the amplifier, and the output signal can be taken from the emitter. This type of circuit is called a *follower*. It is also called a *common-collector*, *common-drain*, or *common-plate* circuit, depending upon the type of amplifying component used.

An important characteristic of a follower is that the output signal is in phase with the input signal. In Fig. 11-11 the signal voltage at the emitter would be the output of the circuit, if it were a follower. Note that the input signal and the signal at the emitter are both in phase. This means that when the base of the transistor becomes more positive, the emitter also becomes more positive, and when the base becomes less positive, the emitter becomes less positive.

The signal at the emitter reduces the gain of the amplifier. In order to increase the collector current in an NPN transistor, you must increase the base current. This is done by making the base more positive with respect to the emitter. If the base in the circuit of Fig. 11-11 starts to go more positive, the emitter also goes more positive. Because of this, the ac base-to-emitter voltage is now less than it would have been if the emitter voltage were held constant. The result is reduced amplifier gain.

What Is Current Feedback and How Is It Obtained?

The overall result of resistor R_4 in Fig. 11-11, then, is to reduce the gain of the stage by permitting the emitter voltage to follow the input voltage. In reality this is a form of feedback in the amplifier. It is called *current feedback* because the feedback voltage is derived from current flowing through the emitter resistor.

You can also get current feedback in vacuum-tube and FET circuits. Current feedback is a form of negative feedback because it causes a loss of gain.

You might wonder what negative feedback accomplishes. After you use an amplifier to obtain gain, you purposely feed back a signal to reduce the gain. You must remember this very important thing about all amplifiers: *There is a tradeoff between gain and bandwidth!* In other words, anything you do to increase the gain of a given amplifier decreases the range of frequencies that it can amplify.

If the amplifier in Fig. 11-11 is to be used for audio signals, then negative feedback has the effect of reducing the gain but increasing the range of audio

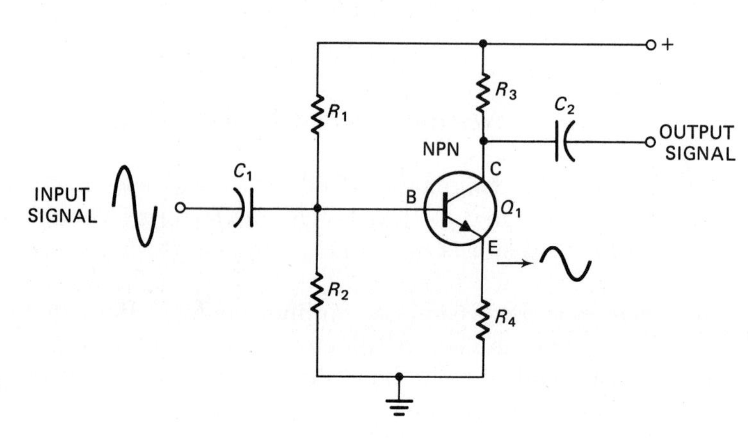

Fig. 11-11 Amplifier with current feedback.

frequencies amplified. In other words, a greater range of audio frequencies can be amplified than would be possible without negative feedback.

It is never possible for the feedback voltage at the emitter of the amplifier to be as large as the input voltage at the base. When you studied the emitter-follower circuits, you learned that the gain of the emitter follower is always less than one. This means that the amplitude of the output signal at the emitter is always less than the amplitude of the input signal at the base. Therefore even though the emitter voltage changes with the base voltage, it cannot change by the same amount. Thus there is always an effective base signal. The amount of current feedback depends upon the resistance of resistor R_4. The greater the resistance of R_4, the greater the amount of feedback.

Current feedback is used in high-fidelity audio circuits and also in radio and television receivers.

From what has been said about current feedback, it is now easier to understand the purpose of R_2 in Fig. 11-10. Resistor R_2 is introducing current feedback and therefore reducing the gain of amplifier Q_1. If the circuit is properly designed, the gain will be reduced sufficiently to prevent oscillation from occurring in the circuit. Of course, the net gain is higher than you would get if you did not use the positive feedback.

What Is Voltage Feedback and How Is It Obtained?

Negative feedback can be obtained with *voltage feedback* as well as with current feedback. An example of how this is done is shown in Fig. 11-12. This is a two-stage *RC*-coupled amplifier circuit. To understand the circuit let us first look at the amplifier circuit. Resistors R_1 and R_2 form a voltage divider for the base bias of Q_1, and C_1 is the coupling capacitor. Resistors R_5 and R_6 form the voltage divider for the base bias voltage of Q_2. Resistors R_3 and R_7 are the load resistors for the two amplifiers.

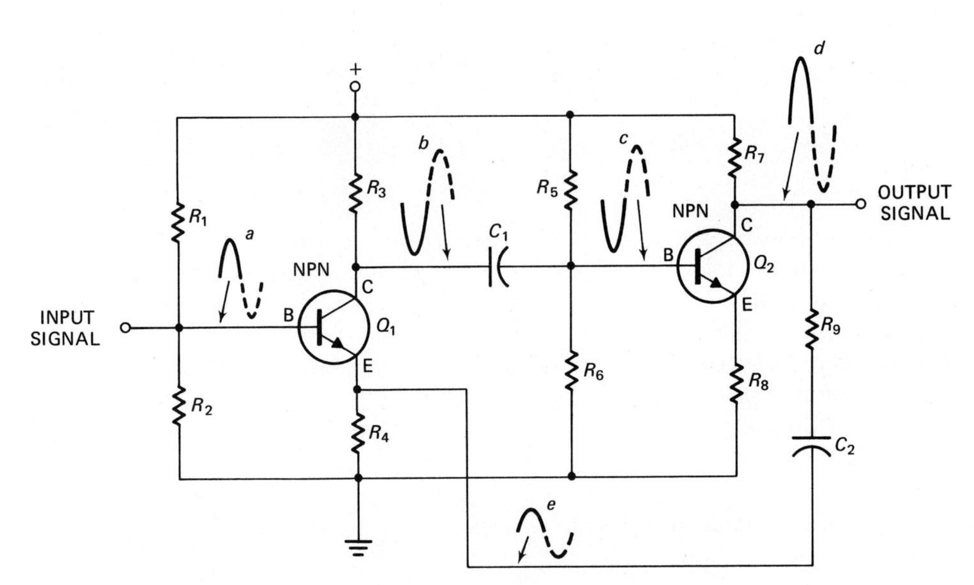

Fig. 11-12 ***RC*-coupled amplifiers with voltage feedback.**

Two emitter stabilization resistors (R_4 and R_8) are used. These resistors will produce current feedback as well as stabilize the circuit against temperature changes.

To understand the voltage-feedback circuit we will look at the waveforms at various points in the circuit. The input signal at the base is a sine-wave voltage. The positive half cycle of this sine wave is shown with a solid line. Since this is a common-emitter amplifier, the output signal, taken at the collector, is 180° out of phase with the input. This is indicated by the solid line for the negative half cycle in waveform *b*.

The waveform *c* is the same as the waveform *b*. It is still also 180° out of phase with *a*. Amplifier Q_2 is in a common-emitter circuit, so the output from Q_2 is also 180° out of phase with its input. This is shown as waveform *d*. Note that waveform *d* is now in phase with waveform *a*. In other words, they both go positive at the same time, and they both go negative at the same time.

Part of the in-phase signal at *d* is returned to the emitter of Q_1 through resistor R_9 and capacitor C_2. Capacitor C_2 is needed to isolate the positive collector voltage at Q_2 from the emitter voltage of Q_1. Resistor R_9 reduces the amplitude of the feedback signal. Note that the feedback signal (marked *e*) is in phase with the input signal of Q_1 (marked *a*). They both go positive and they both go negative at the same time. This feedback signal is injected into the emitter of Q_1. It serves to reduce the amount of change of voltage between the emitter and the base (effective input signal) when the input signal is delivered.

To summarize, the circuit of Fig. 11-12 has two kinds of negative feedback. Current feedback is obtained from the emitter resistors, and voltage feedback is obtained through R_9 and C_2.

Another example of voltage feedback is shown in Fig. 10-10. In this simple operational amplifier the feedback signal is passed through resistor R_2. Note that the feedback signal is out of phase with the input signal, so they subtract. The gain is reduced by the feedback signal. Voltage feedback is always used in a normal op amp circuit—that is, when the op amp is used as an amplifier.

Current feedback is generally used to obtain negative feedback. Voltage feedback is used for both positive and negative feedback.

Summary

1. Positive feedback is also called *regenerative* feedback.
2. Positive feedback is used to increase amplifier gain. It is also used in oscillator circuits.
3. Negative feedback is also called *degenerative* feedback.
4. Negative feedback is used to decrease the gain of an amplifier and thereby increase its bandwidth.
5. A transformer can be used for positive feedback. The dot notation on a transformer symbol shows the points that are in phase.
6. An emitter resistor can be used for negative feedback. The same is true of a cathode resistor or a source resistor.

7. Feedback with an emitter resistor, cathode resistor, or source resistor is an example of current feedback.
8. Current feedback is generally negative feedback.
9. Voltage feedback can be either positive or negative.

How Can Negative Feedback Be Avoided?

Emitter stabilization resistors are included in most bipolar transistor circuits to protect the transistor from thermal runaway. It has just been shown that the emitter resistor can also be used to obtain current feedback to reduce the gain of a stage. In some cases this can be undesirable. If you are trying to get a high gain and wide bandwidth is not needed, you do not want negative feedback.

Figure 11-13 shows two ways in which degenerative feedback can be avoided. In Fig. 11-13*a* the emitter resistor is eliminated from the circuit and the emitter goes directly to ground. This will result in a larger gain than you would get if there were an emitter resistor. However, the disadvantage is that there is no stabilization in the circuit, so the gain will be affected by variations in temperature. In Fig. 11-13*b* the emitter resistor is bypassed by a capacitor. You will remember that a capacitor is a component that opposes any change in voltage on its terminals. The capacitor holds the emitter voltage constant. It does this by charging and discharging as shown in Fig. 11-14. In Fig. 11-14*a* the emitter current has increased. This would normally produce a larger voltage drop across emitter resistor *R*. However, the extra current goes into one plate of the capacitor and out of the other plate as shown by the dotted arrows. Another way of saying this is that as the current starts to increase, the capacitor becomes charged.

In Fig. 11-14*b* the emitter current has decreased. Now the voltage across emitter resistor *R* starts to decrease. As the voltage goes lower than the voltage across capacitor *C*, *C* discharges. (The voltage across capacitor *C* was obtained when it was charged by an increase in emitter current. Now the

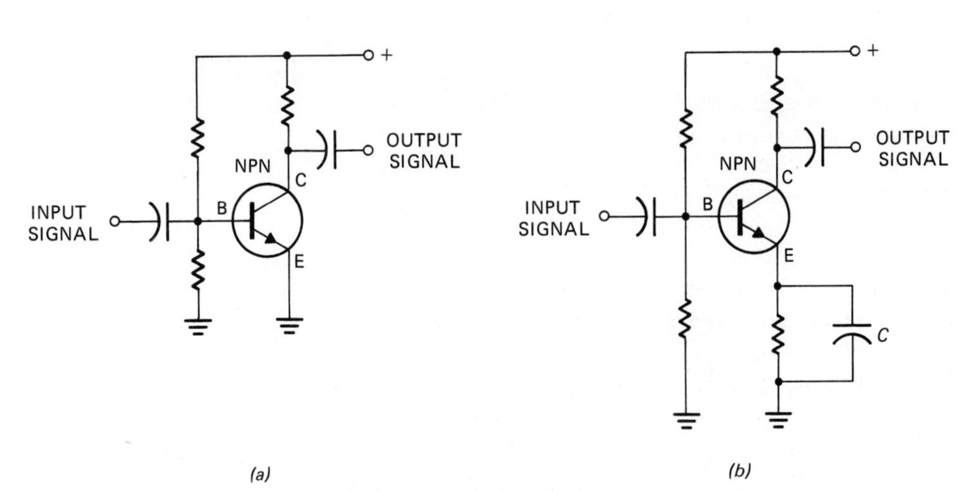

Fig. 11-13 How degeneration is avoided. (*a*) No emitter resistor. (*b*) Bypass capacitor.

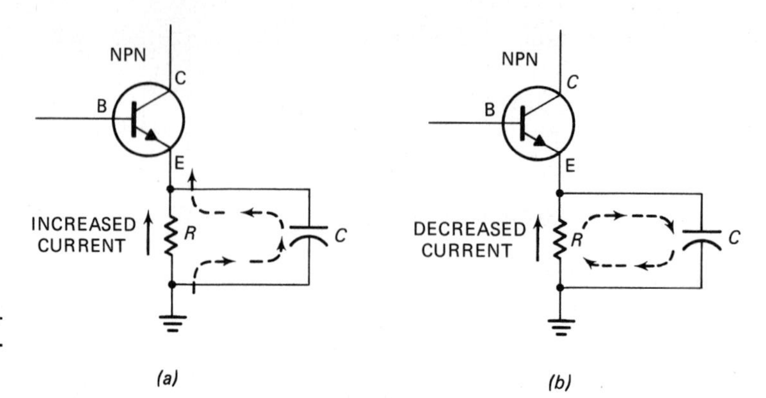

Fig. 11-14 Action of bypass capacitor. (*a*) Charging. (*b*) Discharging.

emitter current has decreased.) The capacitor cannot now hold its charge, so it discharges through resistor *R*, as shown by the dotted arrow.

The discharge current of the capacitor adds to the emitter current so that the voltage across the resistor now remains at the same value as the voltage across the resistor in Fig. 11-14*a*.

Figure 11-15 shows a tube amplifier and an FET circuit with bias resistors. These resistors will cause degeneration unless they are bypassed by a capacitor. Thus for all amplifying components in this type of operation, the resistor at the emitter, cathode, or source can be used for current feedback. The resistor may be bypassed to stop negative feedback and increase gain.

What Is Low-Frequency Compensation?

Figure 11-16 shows a typical *response curve* for the *RC*-coupled amplifier shown in the inset. As with all amplifier-response curves, it shows gain versus frequency. Of interest here is the way the gain falls off at low frequencies. This is shown by the shaded area. The dark dotted line shows a perfect low-frequency response.

The coupling capacitor between the amplifiers is the reason for the poor low-frequency gain. The reactance (opposition) of this capacitor is high at low frequencies.

A low-frequency compensating network increases the low-frequency gain to offset the effect of the capacitor. This is done by making the gain of the first amplifier higher at low frequencies.

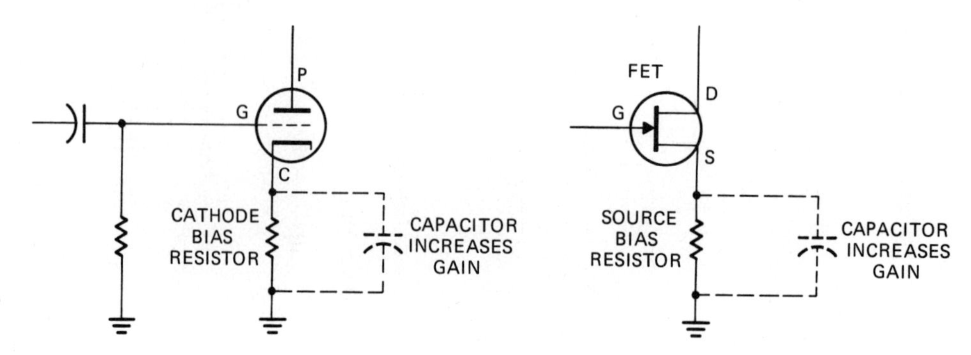

Fig. 11-15 Bias resistors that cause negative feedback.

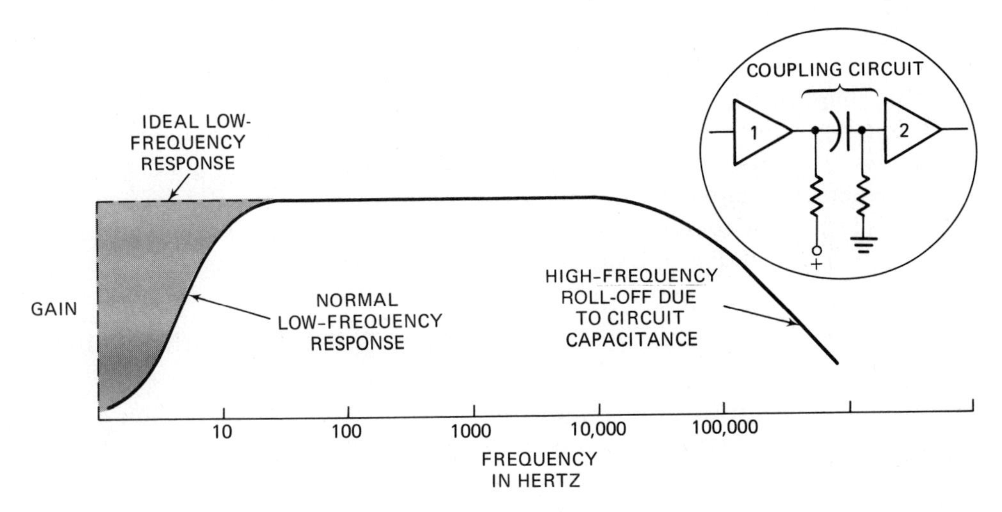

Fig. 11-16 Frequency response of *RC*-coupled amplifiers.

The gain of an amplifier depends upon the value of load resistance. The compensating network causes the amplifier to have a higher load resistance (and improved gain) at low frequencies and a lower load resistance at high frequencies.

Figure 11-17 shows how it works. In this bipolar NPN transistor circuit, the load resistance for Q_1 is series circuit R_1 and R_2. At low frequencies capacitor C_1 has such a high reactance that it is like an open circuit. Therefore the load resistance equals the sum of the resistances of R_1 and R_2.

At high frequencies C_1 has such a low reactance that it is like a short circuit. Since the high frequencies are shorted to ground through C_1, only R_1 is the load resistance for Q_1 at high frequencies.

The result is that the amplifier has a higher gain at low frequencies. This offsets the greater opposition of C_2 at the lower frequencies. The net result is that the low-frequency response is greatly improved.

Summary

1. Negative feedback is not always desired. If the highest gain is needed, then negative feedback must be eliminated.

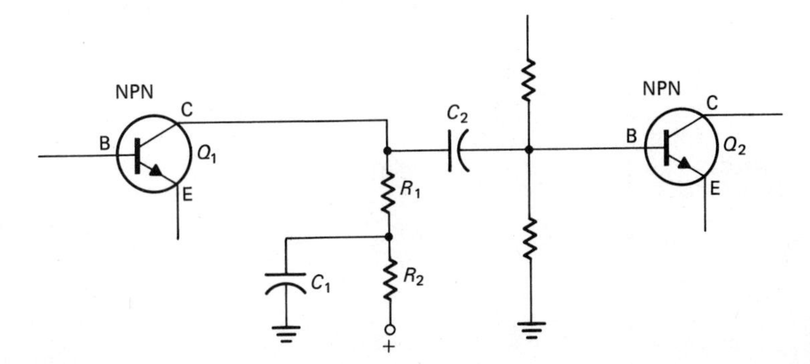

Fig. 11-17 Low-frequency compensating network.

2. A capacitor across an emitter, cathode, or source resistor eliminates negative current feedback.
3. The coupling capacitor in an RC-coupled amplifier causes loss of low-frequency gain.
4. A low-frequency compensating network increases low-frequency gain. The increased gain makes up for the loss of lows in the coupling capacitor.
5. Low-frequency compensating networks are used in transistor and tube amplifiers, in audio high fidelity equipment and TV equipment.

PROGRAMMED REVIEW QUESTIONS

(Instructions for using this programmed section are given in chapter 1.)

We will now review the important concepts of this chapter. If you have understood the material, you will progress easily through this section. Do not skip this material because some additional theory is presented.

1. Which of the following statements is true?
   A. Positive feedback may be used in circuits that are not oscillators. (Proceed to block 17.)
   B. Positive feedback is used only in oscillator circuits. (Proceed to block 9.)

2. *Your answer to the question in block 14 is* **B**. *This answer is wrong. Positive feedback is used in op amp oscillator circuits, but not in op amp amplifier circuits.* Proceed to block 27.

3. *The correct answer to the question in block 26 is* **B**. *Negative feedback reduces the gain of the amplifier. It also increases its bandwidth.* Here is your next question.
   Figure 11-18 shows a simple common-emitter transistor circuit. Which of the following is not a use of emitter resistor R?
   A. Bias the transistor. (Proceed to block 19.)
   B. Stabilize the transistor against temperature changes. (Proceed to block 11.)
   C. Produce feedback. (Proceed to block 24.)

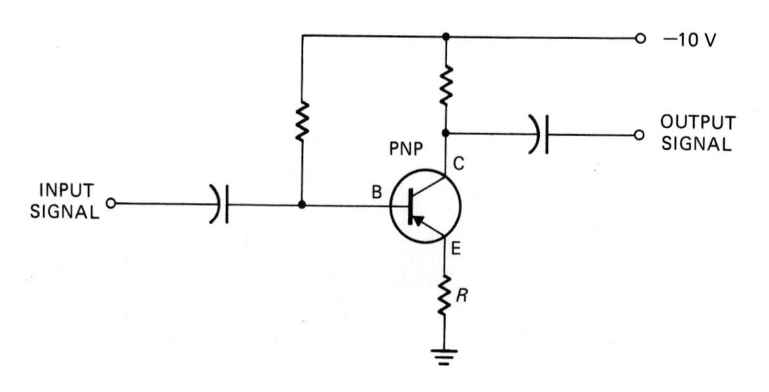

Fig. 11-18 What is the purpose of R?

4. *Your answer to the question in block 27 is* **B**. *This answer is wrong. Review the subject of peaking compensation, then* proceed to block 5.

5. *The correct answer to the question in block 27 is* **A**. *Peaking coils resonate with circuit capacitance to improve high-frequency response.* Here is your next question.
The impedance of a parallel *LC* circuit at the resonant frequency is
A. very high. (Proceed to block 25.)
B. very low. (Proceed to block 18.)

6. *Your answer to the question in block 23 is* **B**. *This answer is wrong. Review the discussion related to Fig. 11-11, then* proceed to block 20.

7. *The correct answer to the question in block 15 is* **B**. *With direct coupling the signal is passed from one amplifier to another through a straight piece of wire. This piece of wire has the best possible low-frequency response. Other methods of coupling use transformers or capacitors. These components, which are also known as* **reactive** *components, tend to oppose either the high frequencies or the low frequencies. Direct coupling can be used with any of the amplifying components you have studied.*

It is much easier to direct-couple bipolar transistors than it is to direct-couple either vacuum tubes or FETs. With bipolar transistors the collector and base both have the same voltage polarity. Neither the tube nor the FET has the same voltage polarity on the control electrode as it has on the dc output electrode. Here is your next question.
Refer to the FET audio amplifier circuit of Fig. 11-19. If electrolytic capacitor *C* is removed from across the source resistor, the gain of this stage will
A. increase. (Proceed to block 22.)
B. decrease. (Proceed to block 12.)

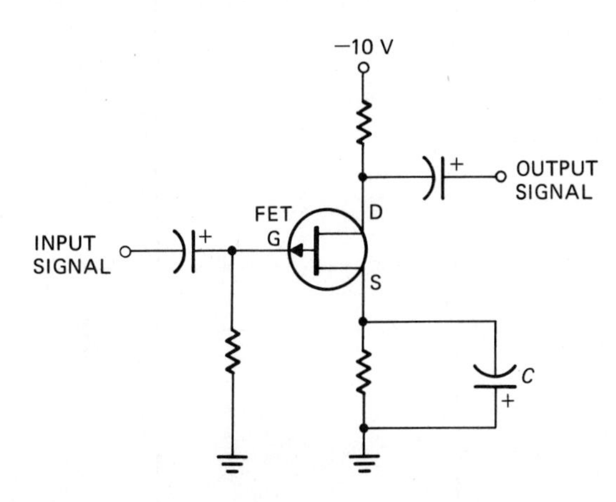

Fig. 11-19 **What happens to the gain when *C* is removed?**

8. *Your answer to the question in block 20 is* **B**. *This answer is wrong. A coupling capacitor can pass high frequencies readily, but it has a great opposition to very low frequencies.* Proceed to block 15.

9. *Your answer to the question in block 1 is* **B**. *This answer is wrong. Although positive feedback is used in oscillator circuits, that is not its only use.* Proceed to block 17.

10. *Your answer to the question in block 19 is* **A**. *This answer is wrong. Audio amplifiers do not use tuned circuits in their input or output stages.* Proceed to block 23.

11. *Your answer to the question in block 3 is* **B**. *This answer is wrong. The question asks which of the three choices is* **not** *a proper use of the resistor. The resistor* **is** *used for bias stabilization against temperature changes.* Proceed to block 19.

12. *The correct answer to the question in block 7 is* **B**. *Removing the bypass capacitor will mean that the source voltage is no longer a constant dc value. Instead, the source voltage will fluctuate with the signal current. This will cause degeneration, or loss of gain.* Here is your next question.
In Fig. 11-20 capacitor *C* serves the following purpose:
A. It improves the frequency response of the amplifier. (Proceed to block 28.)
B. It prevents the base voltage of the transistor from being grounded through the transformer winding. (Proceed to block 14.)

13. *Your answer to the question in block 17 is* **A**. *This answer is wrong. Positive feedback is regenerative feedback, not degenerative feedback.* Proceed to block 26.

14. *The correct answer to the question in block 12 is* **B**. *This question was included to help you test your ability to analyze circuits. The capacitor prevents the dc bias voltage on the base from being grounded*

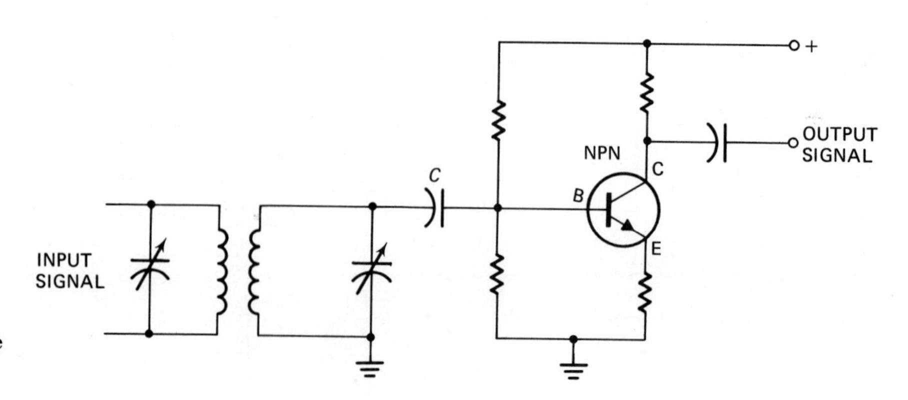

Fig. 11-20 What is the purpose of C?

through the transformer secondary winding. As some capacitors become older, they tend to become leaky—that is, they tend to pass a small amount of direct current. If this happens in the circuit of Fig. 11-20, the direct current flowing through the capacitor will change the base bias voltage. This means that the transistor will no longer be operating properly with the right bias voltage. You must know this in order to understand how to troubleshoot such a circuit. Here is your next question.
Op amps used as amplifiers use
A. negative feedback. (Proceed to block 27.)
B. positive feedback. (Proceed to block 2.)

15. *The correct answer to the question in block 20 is* **A**. *The coupling capacitor can pass higher frequencies, but it cannot pass the very low frequencies. This is because the reactance of a capacitor, which is the capacitor's opposition to alternating-current flow, increases as the frequency decreases.* Here is your next question.
Which of the following methods of coupling will give the best low-frequency response?
A. *RC* coupling. (Proceed to block 21.)
B. Direct coupling. (Proceed to block 7.)

16. *Your answer to the question in block 26 is* **A**. *This answer is wrong. Positive feedback is used to increase the gain of an amplifier. When you increase the gain of the amplifier, you decrease its bandwidth.* Proceed to block 3.

17. *The correct answer to the question in block 1 is* **A**. *Positive feedback may be used to increase the gain of an amplifier. This is done in some high-frequency amplifier circuits.* Here is your next question.
Another name for degeneration is
A. positive feedback. (Proceed to block 13.)
B. negative feedback. (Proceed to block 26.)

18. *Your answer to the question in block 5 is* **B**. *This answer is wrong.* Proceed to block 25.

19. *The correct answer to the question in block 3 is* **A**. *Of the three choices given, bias is the one that is not a proper use of resistor R in Fig. 11-18. This emitter resistor is sometimes confused with the cathode resistor of a tube circuit or the source resistor in an FET circuit. In those cases the resistor is used to produce bias. However, for bipolar transistors this just simply will not work. Resistor R is not a bias resistor.* Here is your next question.
A certain FET amplifier has a tuned circuit in its drain circuitry. This usually indicates that the amplifier is an
A. audio amplifier. (Proceed to block 10.)
B. RF amplifier. (Proceed to block 23.)

20. *The correct answer to the question in block 23 is* **A**. *An unbypassed cathode resistor produces current feedback. An unbypassed emitter resistor in a bipolar circuit and an unbypassed source resistor in an FET circuit also produce current feedback. This is a negative feedback. It reduces the gain of the stage and increases its bandwidth.* Here is your next question.
In an *RC*-coupled amplifier circuit the coupling capacitor causes
A. poor low-frequency response. (Proceed to block 15.)
B. poor high-frequency response. (Proceed to block 8.)

21. *Your answer to the question in block 15 is* **A**. *This answer is wrong. RC coupling produces poor very low frequency response because the coupling capacitor cannot pass dc and very low frequencies.* Proceed to block 7.

22. *Your answer to the question in block 7 is* **A**. *This answer is wrong. The purpose of the capacitor is to prevent degeneration—that is, it prevents loss of gain due to current feedback. If the capacitor is removed, negative current feedback will occur and the gain of this stage will decrease.*

It does not matter if this resistor is in the cathode circuit of a tube, in the emitter circuit of a bipolar transistor, or in the source circuit of an FET. In all cases the bypass capacitor will increase the gain. Removing the bypass capacitor will always decrease the gain. You will demonstrate this point in the Experiment section. Proceed to block 12.

23. *The correct answer to the question in block 19 is* **B**. *Tuned circuits are used to permit one radio frequency, or a narrow band of radio frequencies, to pass through the amplifier stage. At the same time they prevent all other radio frequencies from passing through.* Here is your next question.
An unbypassed resistor in the cathode of a tube amplifier will produce
A. current feedback. (Proceed to block 20.)
B. voltage feedback. (Proceed to block 6.)

24. *Your answer to the question in block 3 is* **C**. *This answer is wrong. The question asks which of the three choices is* **not** *a use of resistor R in the emitter circuit of Fig. 11-18. This resistor* **is** *used to produce feedback.* Proceed to block 19.

25. *The correct answer to the question in block 5 is* **A**. *It is very important to remember that a series LC circuit has a very low impedance at the resonant frequency. In contrast the impedance of a parallel LC circuit is very high at resonance.* Here is your next question.
Figure 11-21 shows a peaking coil wound on a resistor. The ohmmeter measures the resistance. If the peaking coil is good, will the resistance be high or low? (Proceed to block 29.)

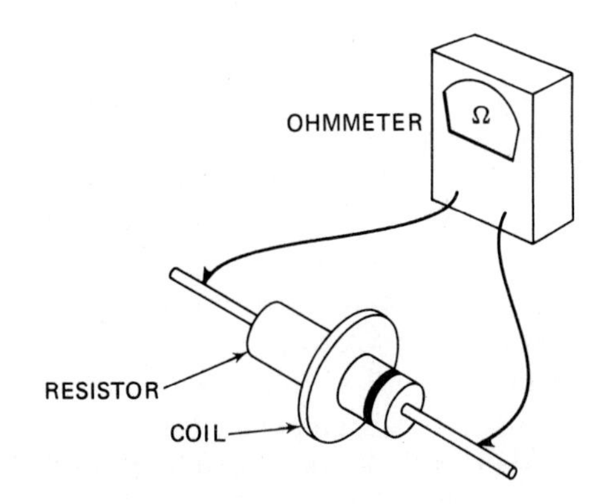

Fig. 11-21 Should the resistance be high or low?

26. *The correct answer to the question in block 17 is **B**. Negative feedback is the same as degenerative feedback. Degeneration is the result of negative feedback.* Here is your next question.
To increase the bandwidth of an amplifier you would use
A. positive feedback. (Proceed to block 16.)
B. negative feedback. (Proceed to block 3.)

27. *The correct answer to the question in block 14 is **A**. The negative voltage feedback reduces the gain but increases the range of frequencies that can be amplified.* Here is your next question.
Peaking compensation is used to improve
A. high-frequency response. (Proceed to block 5.)
B. low-frequency response. (Proceed to block 4.)

28. *Your answer to the question in block 12 is **A**. This answer is wrong. Capacitor C in Fig. 11-20 does not improve the frequency response.* Proceed to block 14.

29. *The answer to the question in block 25 is low. The coil is made of wire, and its dc resistance is very small. You will remember that when resistors are connected in parallel, the resistance of the combination must be lower than the lowest resistor value.*

The peaking coil is actually a coil in parallel with a resistor. The coil resistance is very small, and the parallel resistance is smaller yet. You have now completed the programmed questions. The next step is to put some of these ideas to work in laboratory experiments. Proceed to the Experiment section of this chapter.

EXPERIMENT

(The experiment described in this section may be performed on the circuit board described in Appendix C or on a similar laboratory setup.)

Purpose In this experiment you will show that an electronic module or unit is made by combining circuits.

You will also show the effect of negative current feedback on the gain of an amplifier.

Theory As a technician looking for a trouble in an electronic system, you have a choice of instruments and test procedures. You may be surprised to learn that not all technicians use exactly the same instruments and procedures.

There is a wide difference of opinion about what is the best method of *troubleshooting*—that is, locating a fault in a system. Some technicians prefer to start with a component test. There are many component testers available. Three examples are shown in Fig. 11-22.

A *tube tester* is shown in Fig. 11-22*a*. There are many types of tube testers on the market, but all them can be divided into two classes. One is called an *emission checker.* It simply checks to see if there are enough electrons being emitted from the cathode or filament of the tube. The idea behind the emission checker is that most tube faults occur as a result of low electron emission.

The second kind of tube tester available is the *dynamic checker.* This one tests the tube by putting it in a circuit to see if it will amplify or rectify (depending upon its purpose). Dynamic checkers are more thorough than emission checkers, but they are also more expensive.

Figure 11-22*b* shows an example of a *transistor checker.* There are two types of transistor checkers, but both of them may be in the same instrument. One is called the *out-of-circuit checker.* It checks the transistor for gain and noise. The second type is the *in-circuit checker.* This type of transistor checker usually just tests the transistor to see if it can amplify. It is a very useful instrument because transistors in many circuits are soldered onto a printed circuit board. It is a tedious job to unsolder them from the board in order to test them.

Figure 11-22*c* shows a third kind of component tester, called a *capacitor checker.* It checks for the capacitance value and for *dielectric leakage*—that is, the amount of direct current that flows through the dielectric of the capacitor.

Many technicians prefer to run a series of bench checks with test equipment as a first step in troubleshooting. They perform component checks only after they have narrowed the trouble to one certain circuit area.

In this experiment you are going to build a power supply, an IC oscillator, and a voltage amplifier. In Chapter 9 you constructed a similar circuit. The complete assembly can be called a *module,* which is defined as a combination of basic electronic circuits. The term *unit* is also used for the complete circuit. A combination of modules is called a *system.*

You will find that it is almost impossible not to make a mistake if you try to wire the complete module before testing. One wrong connection can disable the complete module. A more logical approach is to build one circuit at a time and test it before connecting it to other circuits. That is the procedure that will be used here and the one that was used in Chapter 9.

In Part I you will build the power supply. This is the heart of any electronic system, and it must produce the correct voltages or the system cannot work. The power-supply voltages must be rechecked each time you connect

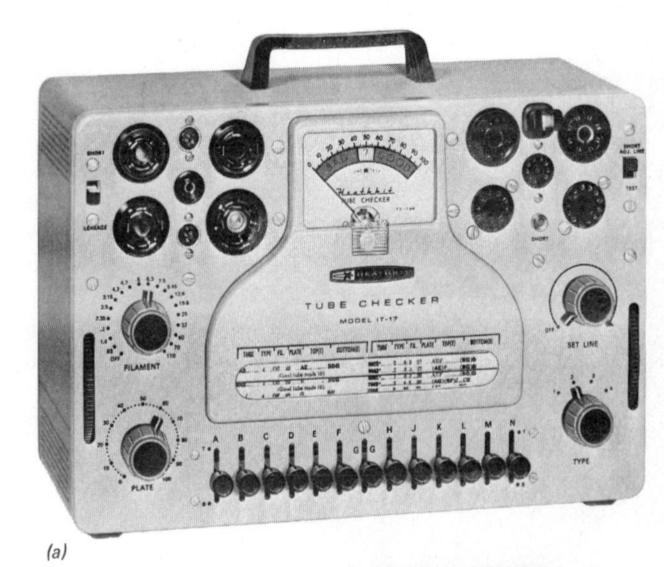

(a)

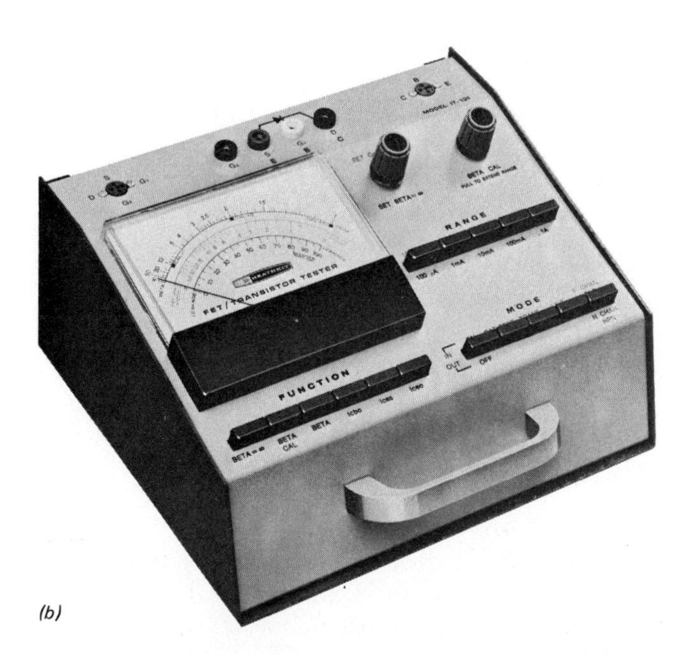

(b)

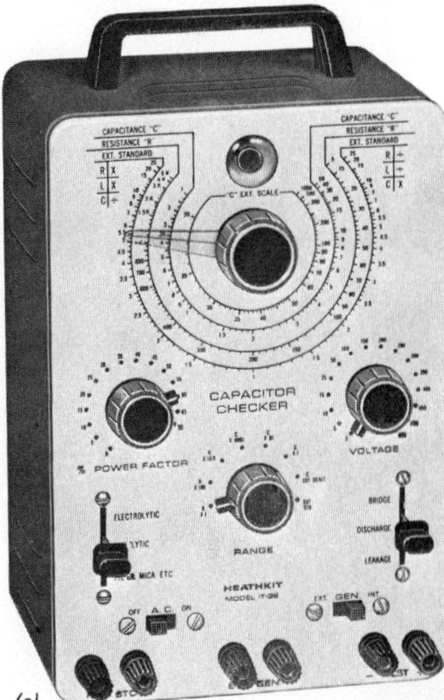

(c)

Fig. 11-22 Component testers. (*a*) Tube tester. (*b*) Transistor checker. (*c*) Capacitor checker. (*Courtesy of The Heath Co.*)

an additional circuit to the power supply. If the power-supply regulation is poor, the voltage will decrease as additional circuitry is connected. The voltage may actually decrease to the point where it can no longer energize the individual circuits. That is why you must check it each time you add a circuit to your module.

In Part II of the experiment you are going to build an integrated-circuit oscillator. Then you will connect this oscillator to the power supply.

In Part III of the experiment you will build a simple transistor amplifier. This amplifier uses simple bias. The transistor will not be operated at anywhere near its maximum capability. As a matter of fact, the gain of this amplifier is purposely made very low by introducing a large amount of negative feedback. You will connect the output of the oscillator circuit to the amplifier and compare the output signals with and without negative feedback.

PART I

Test Setup Figure 11-23 shows the power-supply circuit. Figure 11-23*a* shows the test setup, and Fig. 11-23*b* shows the pictorial drawing. This power supply uses a bridge rectifier circuit. It is connected across the two ends of the 12.6-volt power transformer secondary. The center tap is connected to a common busbar. With this connection one end of the output is positive and the other end is negative. The common point is halfway between the plus and minus voltage values. In other words, the common point is 0 volts.

Resistors R_a and R_b form a voltage divider between the positive and negative voltages. Note that the junction of these two resistors is connected to the

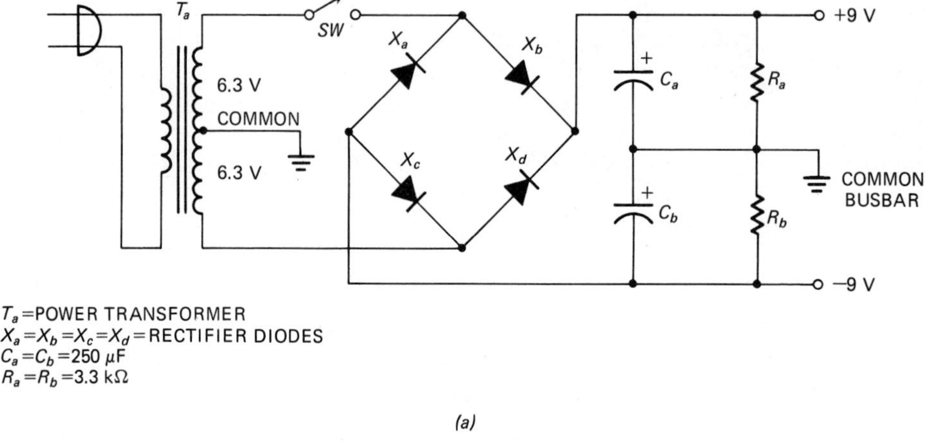

(a)

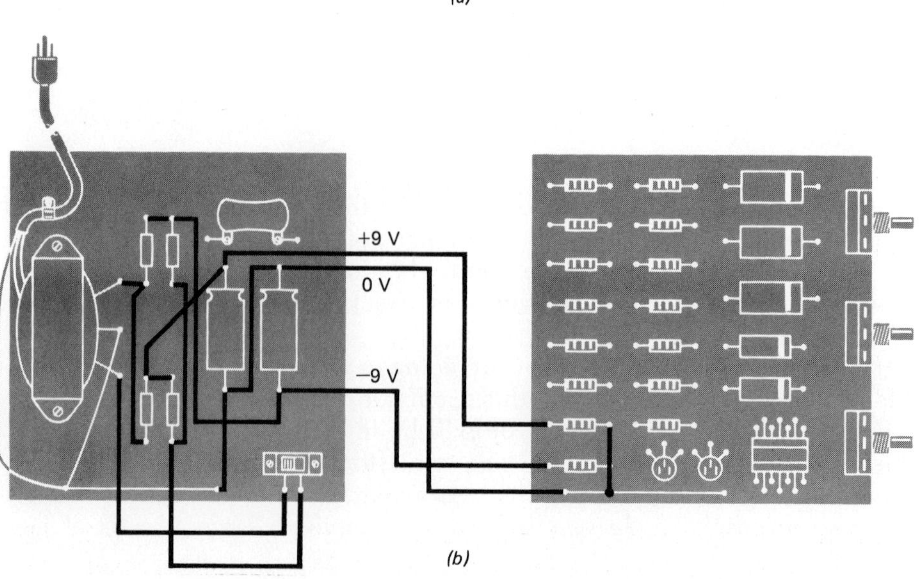

(b)

Fig. 11-23 **Power supply with positive and negative voltage. (*a*) Test setup. (*b*) Pictorial drawing.**

common point for the circuitry. Electrolytic capacitors C_a and C_b are used to filter the output power-supply voltage.

Procedure

step one Construct the power supply shown in Fig. 11-23.

step two Energize the circuit by closing switch *SW*.

step three Measure the +9-volt output with respect to common. To do this the common lead from your voltmeter will be connected to the common point in the circuit, and the + lead will be connected to the +9-volt point. The output voltage should be approximately 9 volts. Record your voltage value here. ______________ volts

If you do not measure about +9 volts, turn switch *SW* OFF right away! Then carefully recheck your wiring.

The exact value of voltage depends upon a number of factors such as the amount of voltage drop across the diodes and the line voltage. That is why we say the voltage should be about +9 volts rather than give an exact value. However, the value should be in the 8- to 10-volt range.

step four Measure the −9-volt output from the power supply with respect to common. If you have a polarity switch on your voltmeter, the common lead from your voltmeter will be connected to the common point in the circuit. The polarity switch should be in the negative position. The voltage will be taken with respect to the −9-volt terminal. If your voltmeter does not have a polarity switch, then the −9 volts must be measured by connecting the common lead to the −9-volt terminal and the positive lead to the common point. Record the value of voltage here. ______________ volts

If you do not measure about −9 volts, turn switch *SW* OFF right away! Then carefully recheck your wiring.

step five What voltage value would you expect to measure between the −9-volt and the +9-volt terminals? ______________

This should be about an 18-volt difference. To measure this voltage, connect the common or negative terminal of the voltmeter to the −9-volt terminal and the positive lead of the voltmeter to the +9-volt terminal. Measure and record the voltage here. ______________

The voltage value should be about 18 volts.

PART II

Test Setup Having wired the power supply, the next step is to wire the oscillator circuit. It is shown in Fig. 11-24. Figure 11-24*a* shows the schematic diagram, and Fig. 11-24*b* shows the pictorial diagram. You constructed a similar circuit in Chapter 10. Positive feedback goes from pin 6 to pin 3. Resistors R_1 and R_2 provide negative ac feedback. This controls the gain.

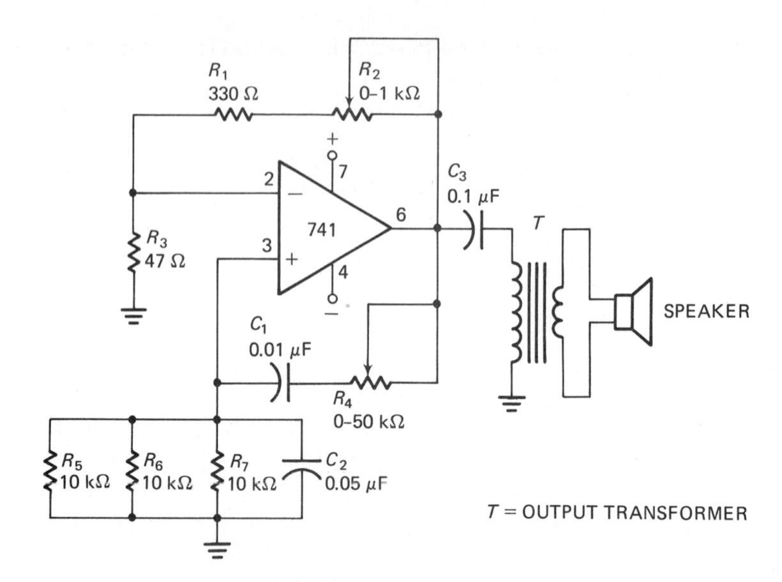

(a)

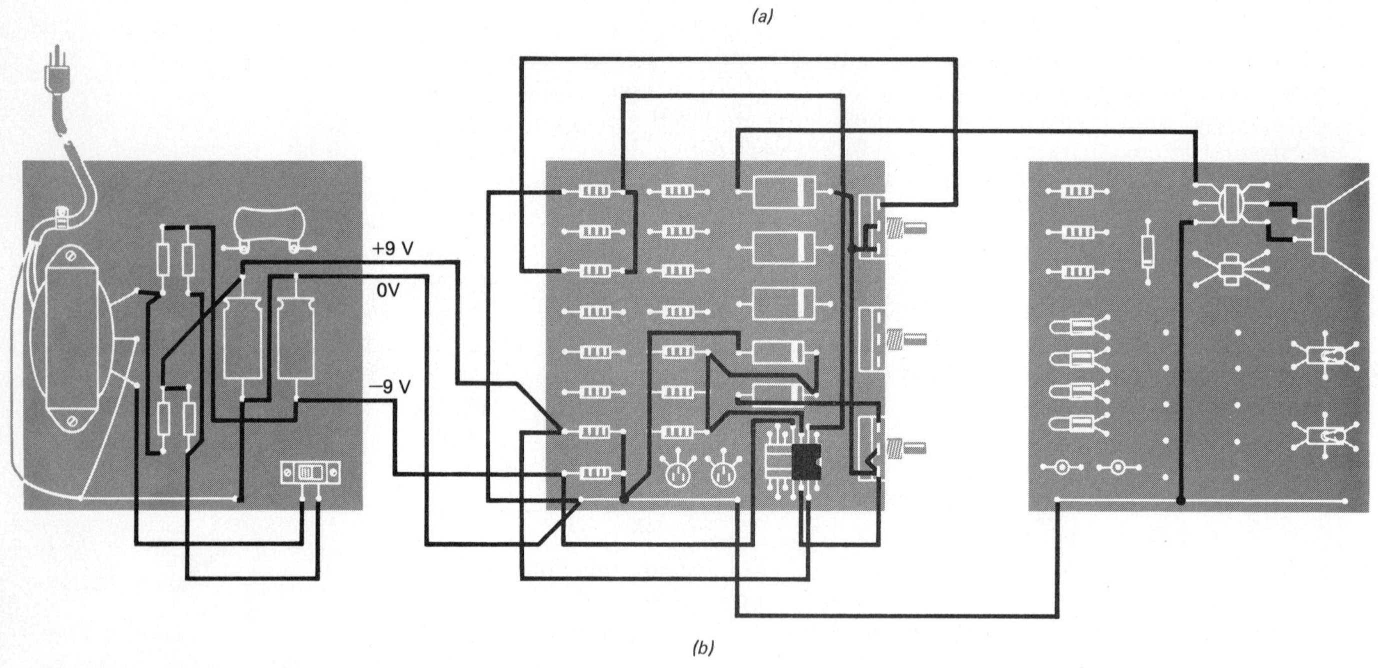

(b)

Fig. 11-24 Op amp oscillator. (*a*) Test setup. (*b*) Pictorial drawing.

Procedure

step one Wire the oscillator circuit shown in Fig. 11-24.

Always switch OFF the power supply before connecting circuits!

step two Connect the positive-voltage terminal (pin 7) of the integrated circuit to the +9-volt terminal of the power supply.

step three Connect the negative-voltage terminal (pin 4) of the integrated circuit to the -9-volt terminal of the power supply. Make sure the common points shown in the oscillator circuit are connected to the common points (busbar) of the power-supply circuit.

step four Measure the positive and negative power-supply voltages with the oscillator circuit connected and energized. The voltages should be about the same values as you measured in Steps 3 and 4 of Part I. (There should be no more than a 10 percent change in the voltage values with the oscillator connected.)

step five With the circuit connected properly you should hear an output sound in the loudspeaker when you close switch *SW*. This will not be a high volume sound because the 741 op amp is not a power amplifier.

Now this is an important point: If the oscillator is not working, then the trouble *must be in the circuit of Fig. 11-24.* It cannot be in the power supply because you have already correctly connected the power-supply circuit and measured the output voltage with the oscillator connected. The fault must be either a faulty component or a wiring error.

step six Adjust R_4 to obtain a tone in the middle audio range.

step seven Revise the circuit as shown in Fig. 11-25. Figure 11-25*a* shows the test setup, and Fig. 11-25*b* shows the pictorial drawing. Variable resistor R_8 and capacitor C_4 replace the speaker in Fig. 11-24. The oscillator produces an output audio signal voltage. The amount of signal voltage is adjustable by resistor R_8. You are going to use this oscillator to inject a signal into an amplifier in the next part of this experiment.

PART III

Test Setup Figure 11-26 shows the amplifier circuit. Figure 11-26*a* shows the test setup, and Fig. 11-26*b* shows the pictorial drawing.

Procedure

step one Wire the circuit as shown in Fig. 11-26. Do not connect the signal from the oscillator at this time, but connect the power supply as shown.

step two Measure the dc voltage from common to the collector of the transistor. This voltage should be less than $+9$ volts. Record the voltage value. ________________

It must be less than 9 volts because there is a voltage drop across resistor R_{101}— the collector resistor—during normal operation of the amplifier. If this voltage drop is not present, it means that the transistor is not conducting and the amplifier will not work.

step three Connect the output from capacitor C_4 of Fig. 11-25 to the input at the base of the amplifier circuit. You have now injected an ac signal into the amplifier. The ac voltmeter connected between the collector and common measures the amplifier output voltage.

step four Adjust variable resistor R_8 in the circuit of Fig. 11-25 to obtain about 1.5 volts of ac signal voltage at the collector of the amplifier. (Capacitor C_{100} is connected at this time.)

step five Momentarily disconnect capacitor C_{100} across emitter resistor

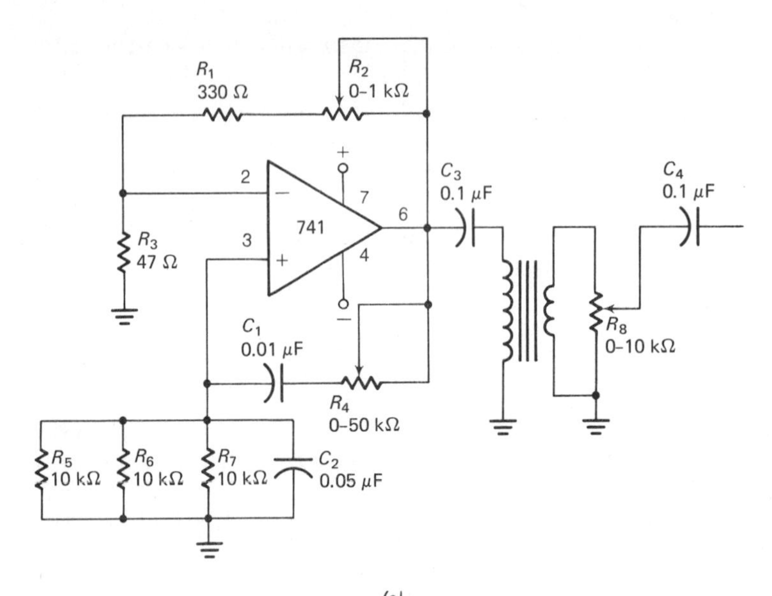

(a)

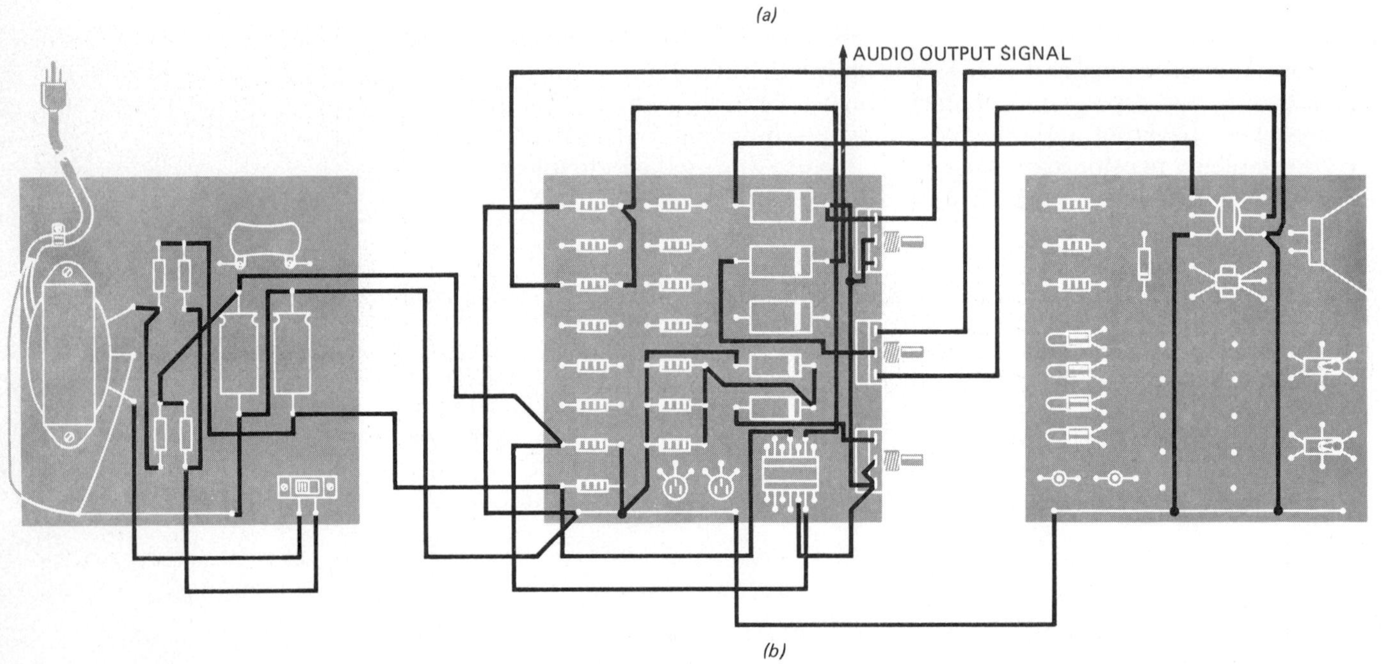

(b)

Fig. 11-25 Oscillator with adjustable output voltage. (*a*) **Test setup.** (*b*) **Pictorial drawing.**

R_{102}. Does this change the ac voltage at the collector? ________________
Yes or No

If you have wired the circuit correctly, and if it is operating properly, disconnecting C_{100} should decrease the collector ac voltage. Without the capacitor there is degeneration (negative feedback). The loss of gain is apparent because the ac collector voltage decreases when the capacitor is removed.

Conclusions You have shown that an electronic module comprises a number of individual electronic circuits. Figure 11-27 shows the complete electronic

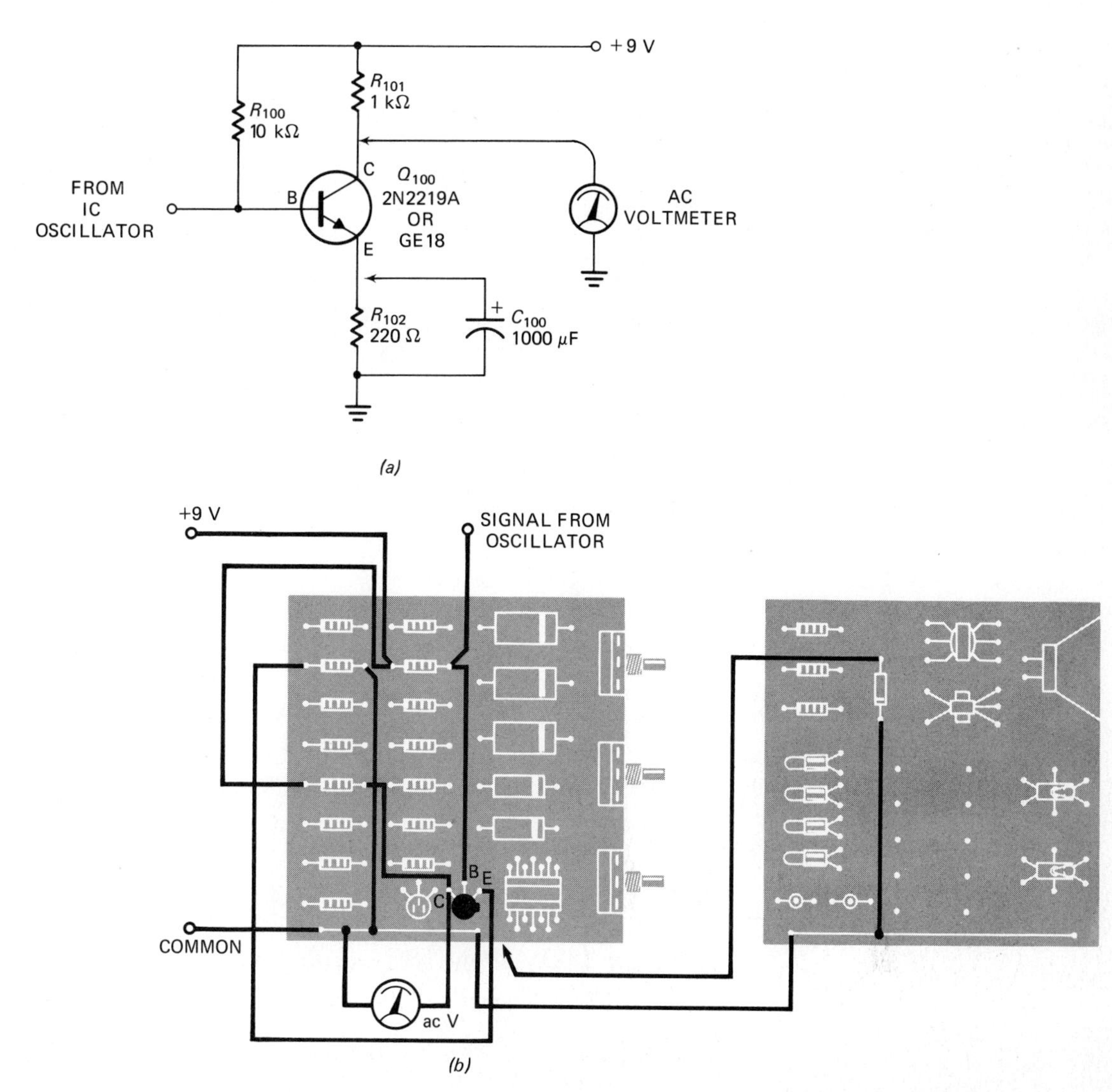

Fig. 11-26 Amplifier circuit. (*a*) Test setup. (*b*) Pictorial drawing.

module with dotted lines separating the circuits. These circuits were built and tested individually, then assembled to make the complete module.

The test procedure that a technician uses for checking out such a circuit is very much the same as the one you used. That procedure is to check each individual circuit, one at a time. The starting point is to check the power supply. This is always a good practice because if the power supply is not operating properly, none of the other circuits can work.

You also learned that degeneration (negative feedback) in an amplifier can be prevented by connecting a capacitor across the emitter (or cathode or source) resistor. This capacitor maintains the emitter at a constant dc value and therefore increases the amount of output signal from the amplifier.

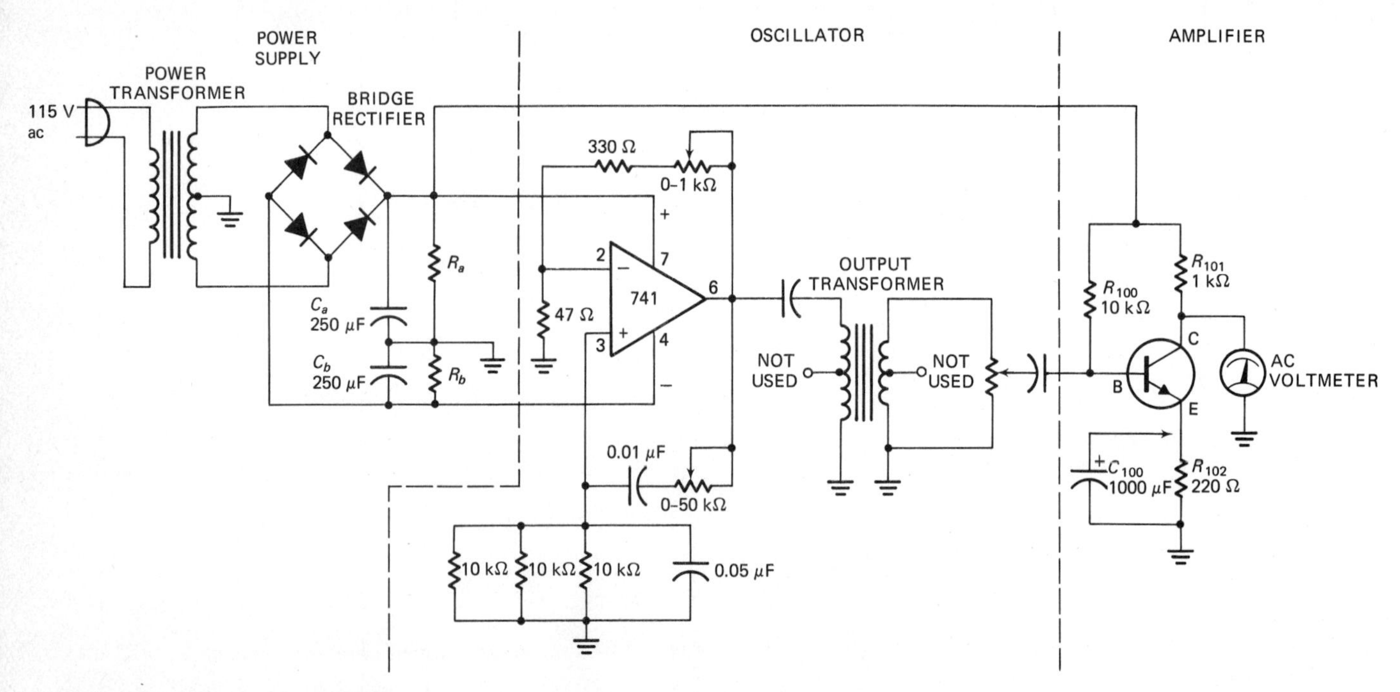

Fig. 11-27 The complete module schematic.

SELF-TEST WITH ANSWERS

(Answers with discussions are given at the end of the chapter.)

1. Figure 11-28 shows *RC*-coupled amplifiers. Poor low-frequency response in this circuit may be caused by (*a*) C_1; (*b*) R_3 and R_4; (*c*) R_1; (*d*) R_5.
2. In the circuit of Fig. 11-28 the gain would be increased by (*a*) increasing the capacitance of C_1; (*b*) increasing the resistance of R_2; (*c*) increasing the resistance of R_5; (*d*) connecting a capacitor across R_5.

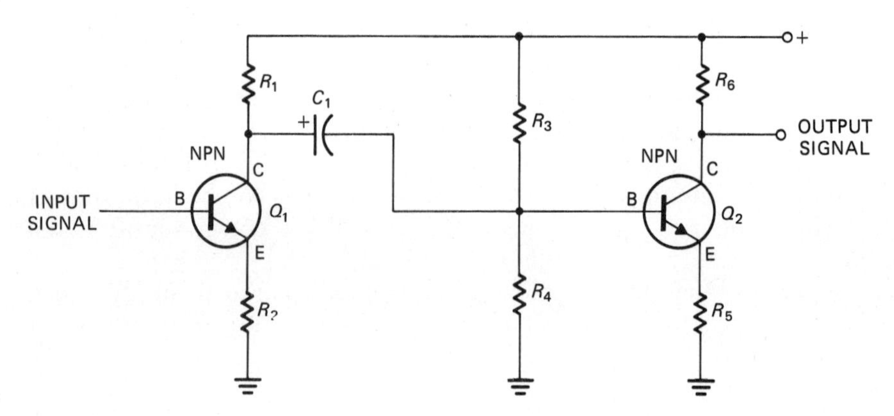

Fig. 11-28 *RC*-coupled amplifiers.

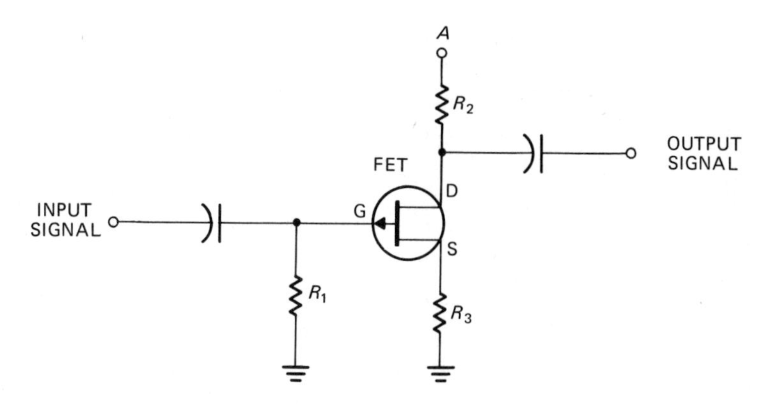

Fig. 11-29 FET amplifier.

3. Peaking coils are used in coupling circuits to (*a*) increase low-frequency response; (*b*) increase middle-frequency response; (*c*) increase high-frequency response; (*d*) remove ripple frequencies from the amplifier output.
4. Positive feedback in an amplifier is (*a*) also called degenerative feedback; (*b*) used only to produce oscillations; (*c*) used only to increase amplifier gain; (*d*) used to either increase gain or produce oscillation, as desired.
5. Two sine-wave voltages have the same frequency, but they are 180° out of phase. Each has an amplitude of 10 volts. If these voltages are combined, the result will be (*a*) 20 volts; (*b*) 0 volts.
6. In the circuit of Fig. 11-29 (*a*) there is no bias; (*b*) there is no feedback; (*c*) there is a positive voltage feedback; (*d*) there is a negative (current) feedback.
7. In the circuit of Fig. 11-29, the dc voltage at point *A* must be (*a*) negative with respect to ground; (*b*) positive with respect to ground.
8. Which of the following types of coupling circuits would most likely have a low-frequency compensating network? (*a*) Direct coupling; (*b*) *RC* coupling; (*c*) Transformer coupling; (*d*) Feedback coupling.
9. A series-resonant *LC* circuit has a (*a*) low impedance; (*b*) high impedance.

ANSWERS TO SELF-TEST

1. (*a*)—Resistors R_3 and R_4 form a voltage divider for bias of Q_2. The collector load resistor for Q_1 is R_1, and R_5 is an emitter stabilization resistor.
2. (*d*)—Increasing the capacitance of C_1 would improve only the very low frequency gain, but not the middle- or high-frequency gain. Increasing the resistance of either R_2 or R_5 would decrease the gain.
3. (*c*)
4. (*d*)—Positive feedback is also called regenerative feedback.
5. (*b*)—When waveforms are 180° out of phase, they subtract. This is shown in Fig. 11-9*b*. If their amplitudes are equal, they cancel and the result is 0 volts.

6. (*d*)—Source resistor R_3 biases the FET. The fact that it is not bypassed with a capacitor means that there is negative (current) feedback.
7. (*a*)—The FET is a P-channel type, so a negative drain voltage is needed.
8. (*b*)—There is no such thing as feedback coupling. Of the other types mentioned, *RC* coupling is the only right answer.
9. (*a*)

how do transmitters and receivers work? 12

INTRODUCTION

Most of the earliest developments in electronics were devoted to sending communications signals and receiving these signals. Much time was spent increasing the *sensitivity* of receivers. (Sensitivity is a measure of how well a receiver can pick up signals.)

The earliest triode tube, which inventor Lee DeForest called the *audion,* was a significant step in improving receiver sensitivity. It made possible the amplification of weak signals and therefore improved long-distance reception.

Once the triode tube was invented, and the idea of amplification was firmly established, other types of amplifying vacuum tubes came onto the scene. These include tetrodes and pentodes, which are more efficient amplifiers in many applications. Until the late 1940s tubes were the only efficient high-frequency amplifiers. Bipolar transistors came in 1948, and they were followed by FETs some time later. Each time a new amplifying device was discovered, there were immediate applications in communications systems.

Before the vacuum-tube amplifier was invented early transmitters used a spark-gap transmission system. This consisted simply of drawing a spark between two electrodes by placing a high voltage across them. Figure 12-1 shows the type of damped waveform produced by spark-gap transmission. It starts with a high-amplitude sine wave, and each successive cycle has a decreased amplitude. There are two main disadvantages to using damped waves for communications. The first is that there is a wide frequency band of transmission. You may have had experience with a car passing your house and producing interference in your television receiver. Each sparkplug in an automobile is a transmitter of damped waves, with the ignition wiring acting as antennas. Since this has such a wide frequency spectrum, it is able to interfere with your television receiver as well as your radio receiver.

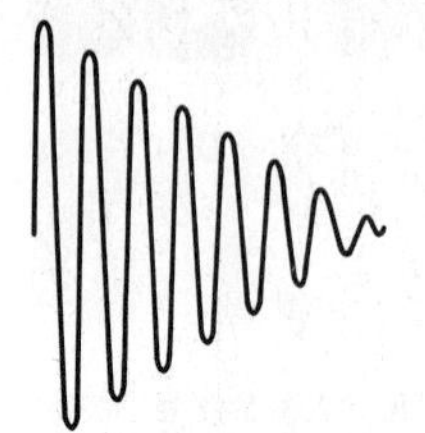

Fig. 12-1 Damped wave used in early radio systems.

A second disadvantage of the damped waveform is the fact that it can be used only for continuous-wave (CW) Morse code transmission. It is completely useless for transmitting voice or television signals.

Today the damped waveform as a method of communication is outlawed by international agreement. These early transmitters have long since been replaced. Modern transmitters and receivers are very accurately tuned to avoid interference between stations on different frequencies.

You will be able to answer these questions after studying this chapter.

- ☐ **What are the sections in a CW transmitter?**
- ☐ **What are frequency-domain and time-domain graphs?**
- ☐ **What types of transmission are being used in radio systems?**
- ☐ **What are the sections of an AM radio transmitter?**
- ☐ **What are the sections of an FM radio transmitter?**
- ☐ **What are the basic sections of a receiver?**
- ☐ **What is an AM superheterodyne receiver?**
- ☐ **What is an FM superheterodyne receiver?**

INSTRUCTION

What Are the Sections in a CW Transmitter?

A CW transmitter is one that produces interrupted continuous waves. Note that these are not damped waves. These waveforms are used for Morse code. This is the simplest kind of transmitter, and its basic parts are shown in Fig. 12-2. The oscillator produces a sine-wave frequency in the radio-frequency range. This frequency is called the *carrier*. In many transmitters the oscillator may be followed by *frequency multipliers,* which increase the oscillator frequency. This permits the oscillator to operate at a lower frequency, where it is more stable.

A telegraph key interrupts the oscillator signal at the point where the signal is delivered to the voltage amplifier. By using this telegraph key it is possible to transmit a series of dots and dashes related to the Morse code.

The voltage amplifier delivers the coded signal to the power amplifier. Most power amplifiers must be preceded by some kind of a voltage amplifier

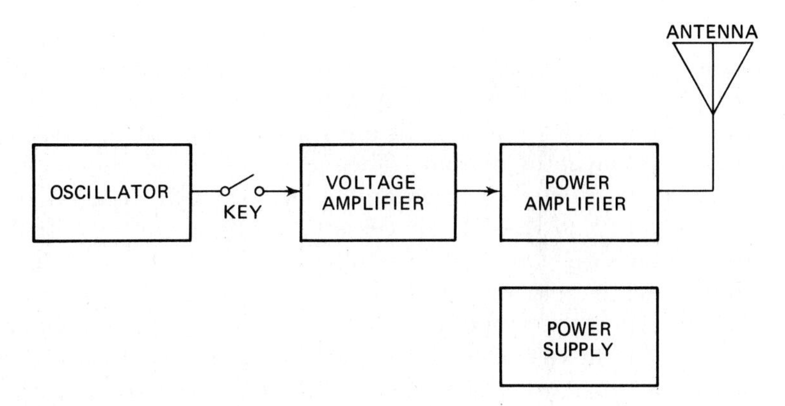

Fig. 12-2 CW transmitter.

because a large input-voltage swing is needed for the power amplifier operation. The power amplifier delivers an RF signal to the antenna, where it is radiated.

A power supply is needed for the CW transmitter circuitry. This can be a battery or an electronic power supply. The greater the amount of radiated power from the antenna, the more power required from the power supply.

What Is the CW Signal?

Figure 12-3 shows how the continuous wave is interrupted to produce dots and dashes for the letter L in Morse code. This illustration is *not* drawn to scale. If the RF-signal frequency is 5 megahertz and a dash takes 0.5 second, then 2.5 million cycles of sine wave would have to be drawn to represent the dash. Obviously this cannot be done in a book illustration, so the sine waves in Fig. 12-3 represent only the RF signal.

Each letter of the alphabet, each number, and each punctuation mark has its own individual code. Although Morse code was one of the first methods of transmitting *intelligence,* it is still being used to some extent today. (The word *intelligence* refers to the message being transmitted in communications.)

What Does a Modulator Do?

The word *modulation* means the combination of two waveforms in such a way that one controls a feature of the other. For example, if the intelligence (audio) wave controls the amplitude of the RF signal (or carrier), then you have amplitude modulation (AM). (See Fig. 12-4.)

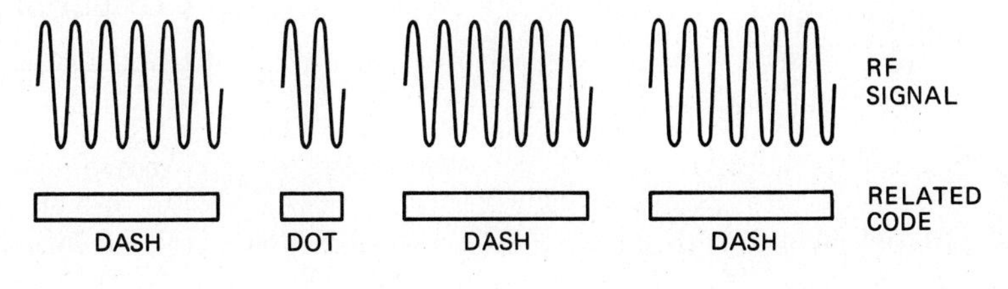

Fig. 12-3 CW signal for the letter L.

Note that the amplitude of the modulated signal increases and decreases at an audio rate. The dotted lines along the outside of the amplitude-modulated signal actually represent the audio signal being transmitted. It is important to note that two audio signals are present: one on the top of the modulated wave and one at the bottom. The outline of the modulated wave is called the *modulation envelope.* In order to combine an RF signal with an audio signal as shown in Fig. 12-4, it is necessary that the modulator circuit be *nonlinear.* For example, if you tried to combine an RF signal and an audio signal in a linear resistor, you would not get modulation.

If you have the proper modulator circuit, then the combination of an audio and an RF signal produces a number of output frequencies. Figure 12-5 shows an example. A 1400-kilohertz RF signal is being modulated by a 1000-hertz audio signal. (A 1-kilohertz signal is the same as a 1000-hertz signal.) When the two signals are combined to produce an amplitude-modulated signal there are actually four output frequencies from the modulator. One of the outputs is the audio 1000-hertz signal, which is immediately discarded by filter circuits. The other three output frequencies are:

- ☐ The radio frequency (1400 kilohertz), or carrier.
- ☐ The radio frequency plus the audio frequency (1401 kilohertz). This is called the *upper sideband.*
- ☐ The radio frequency minus the audio frequency (1399 kilohertz). This is called the *lower sideband.*

In Fig. 12-5 the audio signal is a pure 1000-hertz tone. In practice a large number of audio frequencies may be present during modulation. For example, if a symphony orchestra is playing, all the frequencies of the different instruments will be combined with the radio frequency (carrier).

The presence of sidebands is an important feature of the modulated waveform. We will discuss these sidebands in greater detail, but first it will be necessary to define two kinds of graphs that are used to represent communications signals. They are called *frequency-domain graphs* and *time-domain graphs.* They will be discussed immediately after this summary.

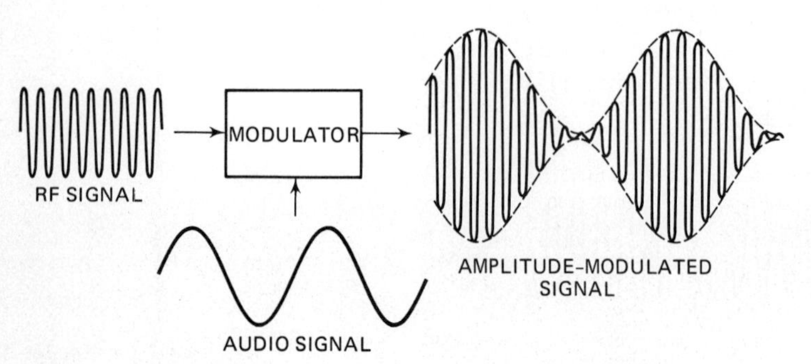

Fig. 12-4 The modulator combines the RF and audio signals.

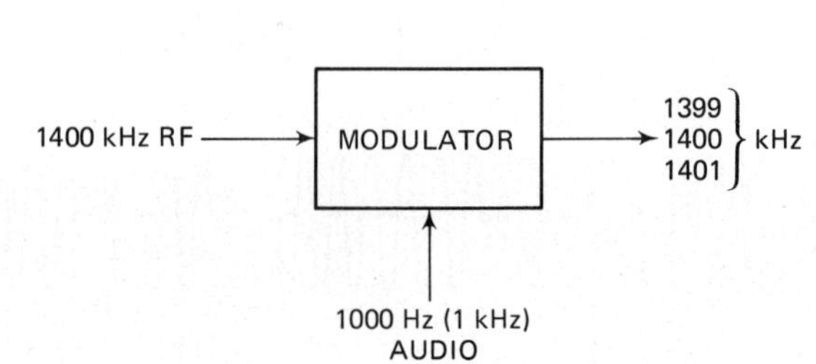

Fig. 12-5 How signals combine in a modulator.

Summary

1. Most of the earliest developments in electronics were in the field of communications—that is, sending and receiving intelligence.
2. The audion was invented by Lee DeForest. It was the first triode tube.
3. Since the audion could amplify weak signals, it greatly increased the sensitivity of receivers. This made reliable long-distance reception possible.
4. Damped waveforms were used in early communications. This type of waveform is now outlawed for use in communications.
5. There are two disadvantages to using damped waveforms in communications: (1) They cover a wide band of frequencies, and (2) they can be used only for CW communications.
6. CW signals consist of dots and dashes used to form letters of the alphabet. This type of communication (called *Morse code*) is still in use.
7. A modulator combines the RF signal of a transmitter with the audio signal. The RF signal is called the *carrier* and the audio signal is called the *intelligence* (or modulation signal).
8. If a carrier with a frequency of 1400 kilohertz is modulated with a 1.0-kilohertz signal in an AM transmitter, then three important waveforms are present in the output: the 1400-kilohertz carrier, a 1401-kilohertz signal, called the *upper sideband,* and a 1399-kilohertz signal, called the *lower sideband.*

What Are Frequency-Domain and Time-Domain Graphs?

Figure 12-6 shows three different ways of representing waveforms that are used in communications. In Fig. 12-6*a* the two waveforms are easily identified. One has a larger amplitude and a lower frequency. The other waveform actually has twice the frequency and half the amplitude of the first. If you look along the axis marked *frequency,* you will see that the further you move away from the zero line, the greater the frequency of the signal. This is how you know the small amplitude signal has a higher frequency.

If you were to look at this three-dimensional graph of Fig. 12-6*a* from the viewpoint of point *A*, you would see a *time-domain graph.* This is shown in Fig. 12-6*b*. The time-domain illustration shows the amplitude versus the time. It also shows the frequency relationship between the two signals. The higher frequency crosses the axis a greater number of times than the lower frequency.

It would be very difficult if not impossible to show a modulated waveform with a time-domain graph. This is especially true if there are more than one or two modulating frequencies. The *frequency-domain graph* of Fig. 12-6*c* shows the two waveforms of Fig. 12-6*a* as viewed from point *B*. Since you are looking at these waveforms on end, you see only a straight line that represents the amplitude. In other words, this is a graph of amplitude versus frequency. (The portion of the display below the frequency line has been eliminated in the illustration. This is the normal procedure for making a frequency display graph.)

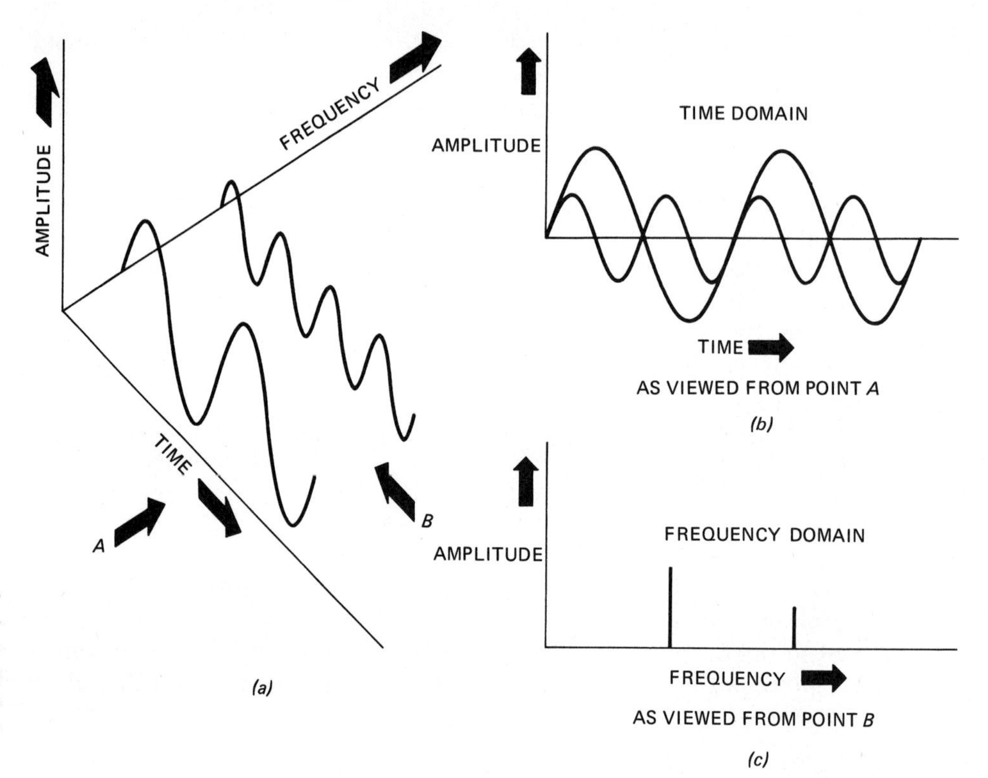

Fig. 12-6 **Frequency- and time-domain graphs. (*a*) Three-dimensional display of two waveforms. (*b*) The same two waveforms on a time-domain graph. (*c*) The same two waveforms on a frequency-domain graph.**

The further you move to the right on the graph of Fig. 12-6*c*, the higher the frequency. This illustration shows you that the first waveform is twice the amplitude of the second and the second waveform is twice the frequency of the first. The frequency-domain illustration is much more convenient to use when discussing sidebands and modulation in transmitters.

What Types of Transmission Are Being Used in Radio Systems?

Now that we have discussed two different ways of showing a graph of radio waves, it is easier to illustrate the types used in radio communications. Figure 12-7 shows you some of the more important examples.

Figure 12-7*a* shows the frequency-domain display for the signals in Fig. 12-5. The carrier is at 1400 kilohertz. This carrier, you will recall, is being modulated by a 1000-hertz (1-kilohertz) audio signal. The combination of carrier and audio signal consists of the carrier, the carrier plus the 1000-hertz audio, and the carrier minus the 1000-hertz audio. The three individual signals are shown in the illustration. In this particular case there is only a single audio frequency. Therefore there are only two sideband signals [1399 kilohertz, the lower sideband (LSB), and 1401 kilohertz, the upper sideband (USB)]. If the music from a symphony orchestra were being used to modulate the 1400 kilohertz, there would be a wide variety of audio frequencies. So instead of having two individual distinct sidebands, as shown in Fig. 12-7*a*, you would have a large number of sidebands on both sides of the carrier.

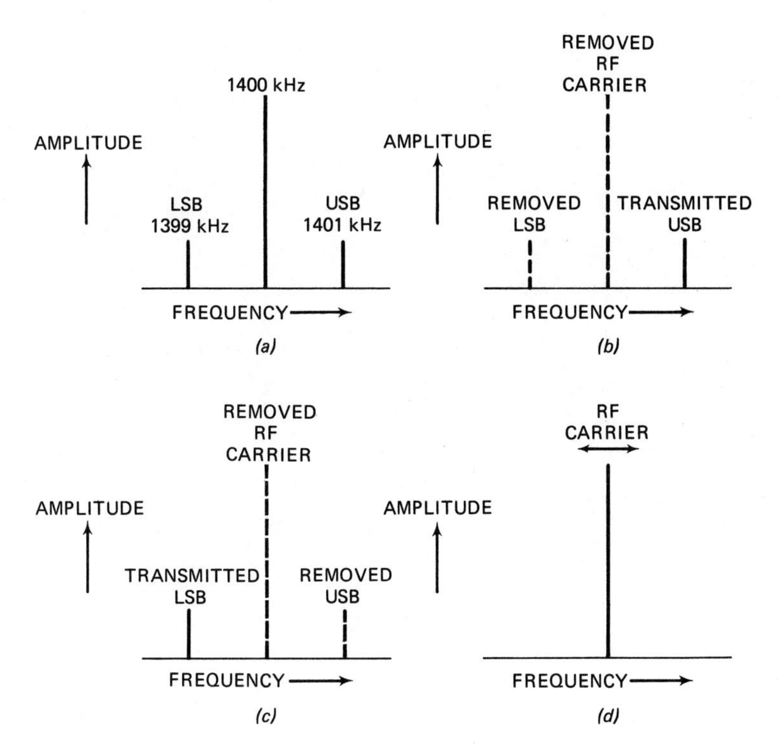

Fig. 12-7 Types of modulated signals. (*a*) AM (*b*) AM single upper sideband. (*c*) AM single lower sideband. (*d*) FM.

The type of transmission illustrated in Fig. 12-7*a* is called *amplitude-modulated* (*AM*). This is the type of signal used for AM broadcast radio. One additional important point must be made regarding Fig. 12-7*a*. The height of the sideband signals depends upon loudness of the sound being transmitted. The height shown is maximum, and it represents only one-fourth of the total height of the carrier. In most cases the height of the sidebands will be much less, so for any AM transmitted signal, most of the energy is used for sending the carrier.

When you listen to a radio, you do not listen to the carrier. Obviously all the useful information that you want to hear is in the sidebands. Therefore you should be able to eliminate the carrier and transmit only the sidebands. When this is done the result is called *single-sideband* (*SSB*) *transmission*. It is illustrated in Fig. 12-7*b*. The carrier is removed, and the lower sideband is also removed. Both of these are shown in dotted lines. Only the upper sideband is transmitted. Thus this is *upper-sideband* (*USB*) *transmission*. It is also possible to send only the lower sideband and eliminate the carrier and upper sideband. That is called *lower-sideband* (*LSB*) *transmission* (Fig. 12-7*c*). These are both examples of single-sideband transmission.

In all the systems discussed so far, the carrier frequency was one single fixed frequency value. Another way to transmit intelligence is simply to shift the carrier back and forth in frequency, in accordance with the audio signal. This is called *frequency modulation* (*FM*). It is illustrated in Fig. 12-7*d*. The carrier frequency is moved rapidly back and forth for high-frequency sounds and slowly back and forth for low-frequency sounds. The amount of change that the carrier frequency is moved from the center frequency is determined by the loudness of the audio being transmitted. FM signals must be

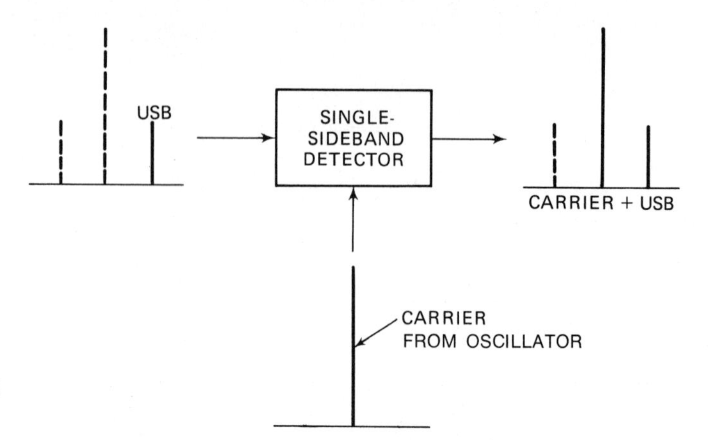

Fig. 12-8 The carrier is reinserted in the SSB receiver.

received by radios that are designed to detect or demodulate this kind of signal. It turns out that if the receiver is capable of demodulating an FM signal, it is not capable of demodulating amplitude-modulation (AM) signals. Static caused by lightning and machinery is mostly amplitude-modulated, so FM receivers have little or no static from their speakers.

Eliminating the carrier in single-sideband transmission makes the transmission more efficient, but you do run into problems in the receiver. It is not possible for the receiver to utilize only one sideband and reproduce the original audio. In order for the receiver to reproduce the audio signal it is necessary to first reinsert the carrier in the receiver. To do this, an oscillator that produces a constant RF signal at the carrier frequency is used to inject a signal that is mixed with the single sideband. This is shown in Fig. 12-8. The single sideband (SSB) and the carrier produced by the oscillator are injected into a *single-sideband detector*. The output of the detector contains the sideband and the carrier. Receivers are able to convert this type of a signal into a useful audio signal. Single sideband is not used in regular AM broadcast transmission. The tuning necessary to reinsert the carrier must be done very carefully. The tuning procedure has been judged too difficult for the average AM radio listener.

What Are the Sections of an AM Radio Transmitter?

Figure 12-9 shows the block diagram of an AM transmitter. The radio frequency is produced by the oscillator. It produces a very exact frequency which does not vary more than a few hertz throughout the day. Often a crystal-oscillator circuit is used. The component that determines the frequency is a piezoelectric crystal which is designed to flex at a very specific rate.

In many cases the crystal-oscillator output frequency is lower than what is needed for an RF carrier signal. *Frequency multipliers* are used to increase the frequency of the oscillator signal to the carrier frequency. The *buffer* presents a constant load to the oscillator so that it will not be pulled off frequency by changes in the load in later stages.

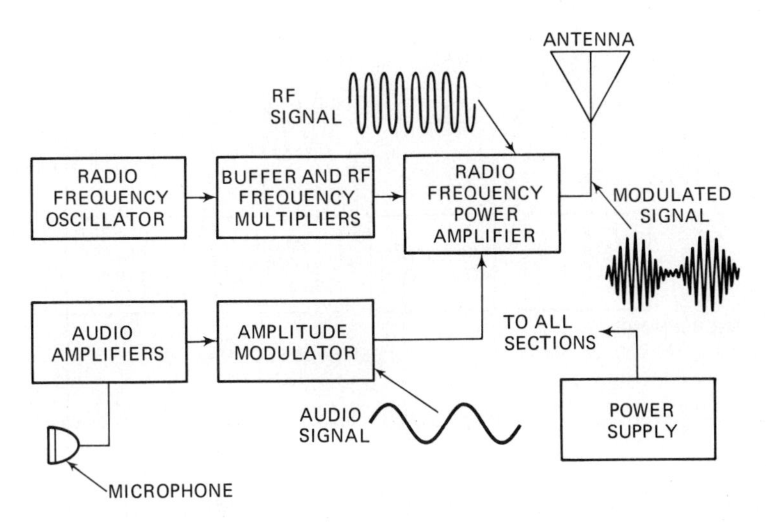

Fig. 12-9 AM transmitter block diagram.

Once the multipliers have brought the signal up to the carrier frequency, it is necessary to have a stage of radio-frequency power amplification.

So far we have shown how the carrier is generated when there is no audio. The audio signal to be transmitted is picked up by a microphone and amplified by a number of audio voltage amplifiers. The amplified audio signal is delivered to an audio *modulator,* which has a power output that is high enough to vary the amplitude of the radio-frequency power-amplifier signal. The result is the amplitude-modulated signal that is delivered to an antenna, where it is radiated. As with all electronic systems, the power supply is needed to provide the voltages and currents for operating the electronic components.

To make the transmitter in Fig. 12-9 into a single sideband type, you could add filters after the RF power amplifier. Filters are electrical circuits that will pass some frequencies and eliminate all others. By properly designing the filter it would be possible to pass only one sideband. Or you could eliminate only the carrier. This is not the only way to do it, but it is one way.

What Are the Sections of an FM Radio Transmitter?

Figure 12-10 shows the block diagram of an FM transmitter. Notice that this diagram looks a lot like the one of the AM transmitter of Fig. 12-9. In the FM transmitter we also have audio amplifiers, a radio-frequency oscillator, buffer and RF-frequency multipliers, a radio-frequency power amplifier, and a power supply.

The main difference in Fig. 12-10 is in the method of modulation. Here, as you see, the output of the audio amplifiers is fed to a *frequency modulator.* (In Fig. 12-9 note that the output of the audio amplifiers is fed to an amplitude modulator, which then modulates the power amplifier.) The frequency modulator, in turn, frequency-modulates the radio-frequency oscillator. The type of output of the oscillator after modulation is shown by the waveform above the oscillator block in Fig. 12-10. Note that the frequency of the oscillator is

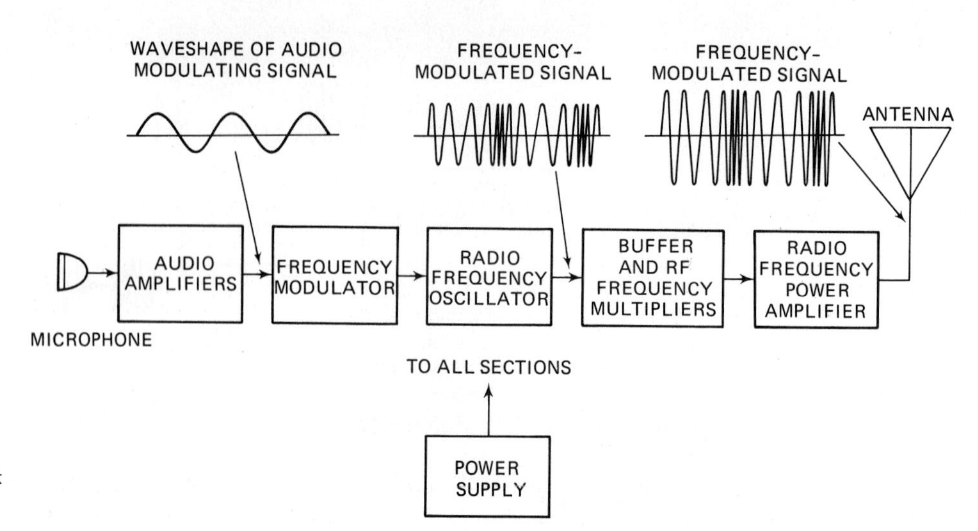

Fig. 12-10 FM transmitter block diagram.

caused to change in accordance with the amplitude of the audio wave. On the other hand, the rate at which the frequency is caused to change is the rate of the audio frequency itself.

For example, let us take an audio frequency of 400 hertz, at some given amplitude. When the audio signal reaches maximum positive amplitude, the oscillator frequency will become maximum. When the audio signal is at zero amplitude, the oscillator is at its center frequency. And when the audio signal reaches maximum negative amplitude, the oscillator frequency will become minimum. Since the audio frequency is 400 hertz, this is also the rate at which the oscillator frequency will go through its variations.

Summary

1. Two ways of showing waveforms are with frequency-domain and time-domain graphs.
2. Frequency-domain graphs show amplitude versus frequency.
3. Time-domain graphs show amplitude versus time.
4. Transmitted signals can often be shown better with frequency-domain graphs.
5. A broadcast AM signal has a carrier, upper-sideband frequencies, and lower-sideband frequencies.
6. Since all the useful signal is in the sidebands, the carrier and one sideband can be eliminated. This is called *single-sideband transmission.*
7. Single sideband takes up less of the radio-frequency spectrum, but it is harder to tune on a receiver.
8. With double-sideband carrier-suppressed transmission, only the upper and lower sidebands are transmitted.
9. When the carrier frequency is varied by the audio signal, we have a frequency-modulated (FM) wave.
10. The radio frequency is generated in a transmitter in the oscillator section. Frequency multipliers may be used to bring the oscillator frequency up to the desired carrier frequency.

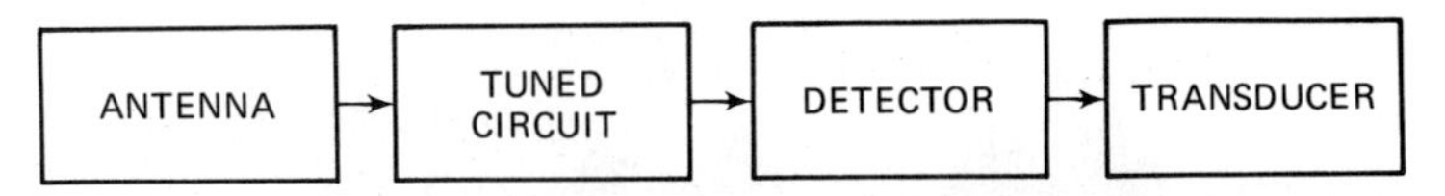

Fig. 12-11 **All receivers have these basic sections.**

11. The AM modulator causes the carrier to vary in amplitude at the audio rate.

12. In FM the amplitude of the audio signal causes the changing of the oscillator frequency. On the other hand, the frequency of the audio signal determines the rate at which the oscillator frequency changes.

What Are the Basic Sections of a Receiver?

The study of radio receivers is simplified by the fact that all receivers consist of at least four basic sections. The different kinds of receivers are obtained by adding more circuits to these basic four sections. However, in no case will you find a radio receiver that has less than the basic four. Figure 12-11 shows the basic four sections.

An antenna is necessary to convert the electromagnetic radio waves into electrical signals for operating the receiver. You may know of a case where there is a radio or television set that appears to be operating without an antenna. Technically this is impossible. If no external antenna is connected, the radio uses the wires and the connections within the receiver to serve as an antenna. It is *not* possible for a receiver to operate unless it has some type of an antenna system.

The antenna can be mounted outdoors, as in the case of many television receivers. Such antennas are, of course, located at some distance from the television transmitters. The high frequencies related to television signals do not follow the curvature of the earth, and they are not reflected back to the earth from the ion layer. Therefore it is usually possible to receive only the signals from what is known as the *line-of-sight distance.* (This is about 50 miles.)

The antenna delivers a signal to the tuned circuit in the receiver. It would not be possible to have radio communication as we know it unless you could select one station at a time and reject all others. The usual method of rejecting undesired stations is to use *LC* (inductance-capacitance) tuned circuits. Either a capacitance or an inductance is varied to enable the listener to select the desired station.

After the signal has been selected, it is delivered to a section called a *detector.* Basically the detector is the stage that changes the modulated RF carrier back to the original audio signals.

The audio signals are delivered to a transducer. This may be a set of earphones or a loudspeaker.

How Does a Crystal Radio Work?

The simplest possible receiver, called a *crystal* receiver, is shown in Fig. 12-12. It uses only the four sections shown in Fig. 12-11.

The antenna delivers a signal to a transformer *T*. This transformer couples

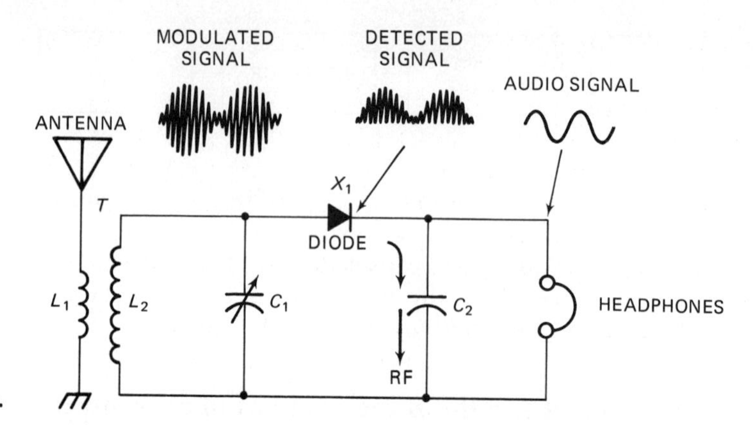

Fig. 12-12 Crystal receiver.

the signal from the primary L_1 to the secondary L_2. The secondary of the transformer is an inductor which works with variable capacitor C_1 to form the tuned circuit of the receiver. This section selects the desired station.

The diode X_1 is usually referred to as the *detector*. In reality its purpose is to permit the signal current to pass in one direction but not in the other. In other words, the modulated signal is converted to a pulsating dc (rectified). This is necessary because the amplitude-modulated signal contains two audio signals that are 180° out of phase. If both of these audio signals were passed to the headphones, they would cancel. Therefore one of them is eliminated by using the diode.

The signal at the output of the diode detector shown in Fig. 12-12 is the signal that would be present if it were not for capacitor C_2. Capacitor C_2 bypasses the RF signal around the headphones, and only the audio signal is delivered to the headphones.

What Is a TRF Receiver?

One way to improve the sensitivity of a receiver—that is, improve its ability to receive weak signals—is to use a *TRF receiver* like the one shown in Fig. 12-13. The letters TRF stand for *T*uned *R*adio *F*requency. It has one or more tuned RF amplifiers. These amplifiers increase the very weak RF

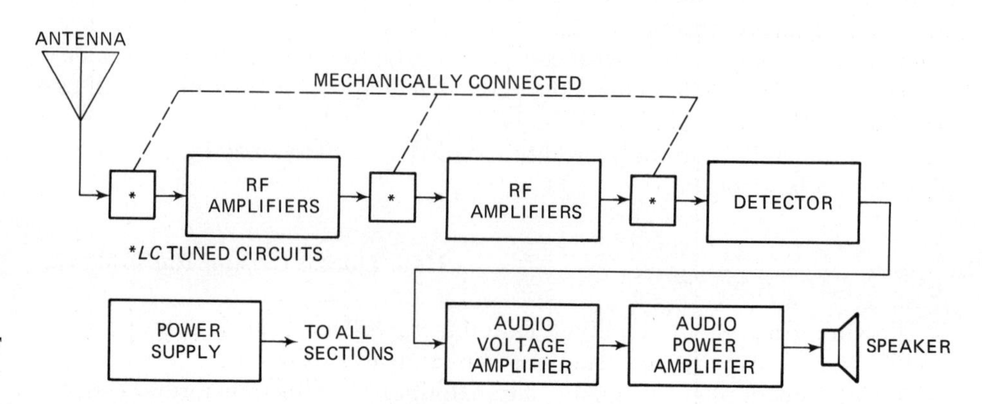

Fig. 12-13 Block diagram of a TRF receiver.

signal from the antenna to a strong RF signal that can be used to operate the detector.

Tuned circuits are employed as part of each RF amplifier so that the receiver will receive only on the one desired radio frequency at any given time.

The output of the last RF amplifier goes to the detector. After the detector the audio signal is amplified first by a audio voltage amplifier and then by a audio power amplifier. The output of the audio power amplifier drives the loudspeaker. A dc power supply is necessary for operating the RF amplifiers and the audio amplifiers.

Early models of the TRF receiver were difficult to tune because three separately tuned circuits were necessary. In later designs the tuned circuits were combined in such a way that a single dial controlled all three. This made it much easier to tune. The TRF receiver is used even today in some special applications, such as garage-door-opener receivers.

What Is an AM Superheterodyne Receiver?

A great improvement over the TRF receiver is the superheterodyne receiver, shown in Fig. 12-14. This receiver can tune weaker signals. In other words, it is more sensitive than the TRF receiver. It is also more *selective.* That is, it can better separate stations that are close together in their transmitting frequencies.

The block diagram shows you that the signal is delivered to an RF amplifier through a tuned circuit, and then to another tuned circuit, to a mixer stage. The RF signal from the radio station is combined in the mixer with the oscillator signal. When you combine two signals in a nonlinear circuit, like the mixer, there are four main output frequencies: the RF signal, the oscillator signal, the RF-signal frequency plus the oscillator-signal frequency, and the RF-signal frequency minus the oscillator-signal frequency. In a superheterodyne receiver it is the difference frequency that is selected by the tuned IF amplifier(s). This difference frequency, which is called the *intermediate fre-*

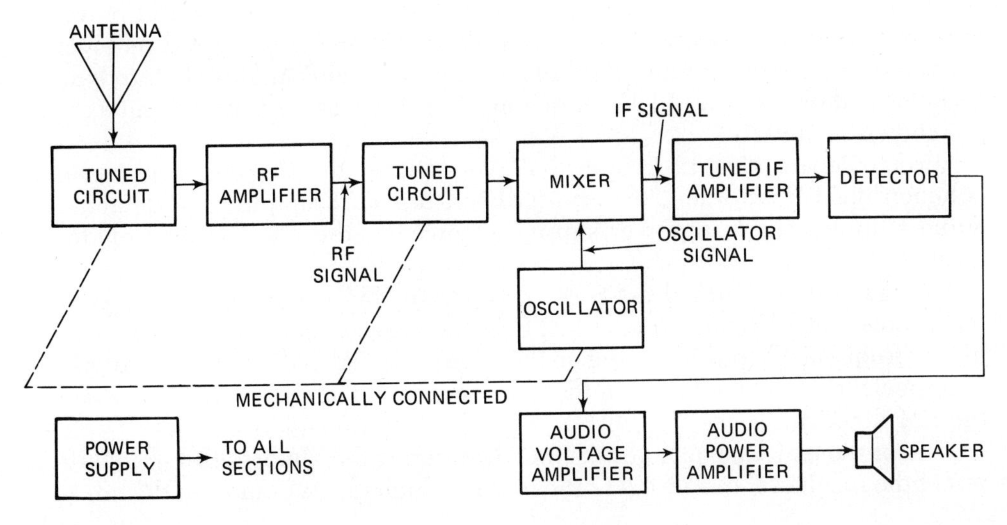

Fig. 12-14 Block diagram of a superheterodyne receiver.

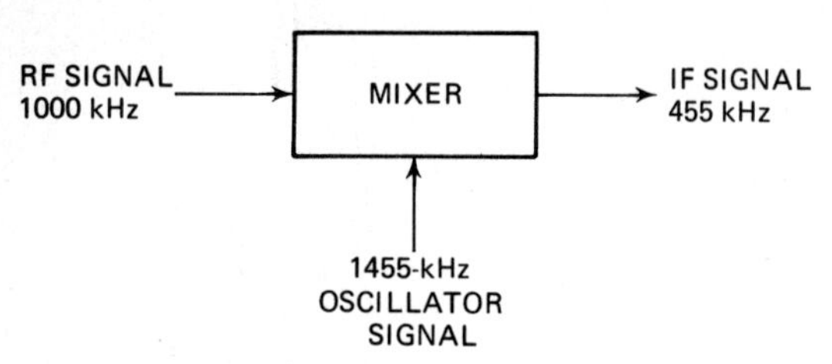

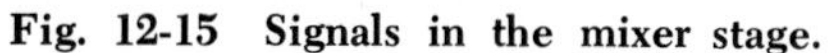
Fig. 12-15 Signals in the mixer stage.

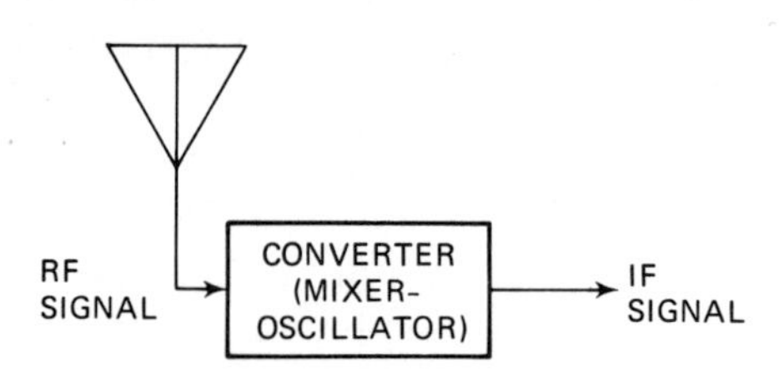

Fig. 12-16 Signals at the converter stage.

quency (*IF*), is lower than the radio frequency. The lower frequency makes it possible to get greater gain and selectivity in the amplifier stages. The strong output signal from the IF amplifier is delivered to the detector, and from that point on the receiver is identical to the TRF receiver.

Figure 12-15 shows you how signals are combined in the mixer to produce the intermediate frequency. The RF signal delivered to the mixer has a frequency of 1000 kilohertz in this example. The oscillator frequency is at 1455 kilohertz. The difference frequency between these two is 455 kilohertz, and that is the intermediate frequency of the receiver. As a matter of fact, most broadcast receivers use a 455-kilohertz intermediate frequency.

No matter what frequency is being tuned by the receiver, the oscillator is always adjusted in such a way that when the two signals are combined in the mixer the output is always 455 kilohertz. This simplifies the tuning of the IF amplifier stages because there is only one frequency that is ever delivered to these stages.

Instead of an RF amplifier, mixer, and oscillator, as shown in Fig. 12-14, some receivers have the received signal delivered directly to a stage called a *converter*. (See Fig. 12-16.) This converter is a combination mixer and oscillator.

What Is an FM Superheterodyne Receiver?

FM superheterodyne receivers, similar to the ones that receive FM entertainment broadcasts, operate on the same basic principle as the AM receiver shown in Fig. 12-14. A simplified block diagram of an FM superheterodyne receiver is shown in Fig. 12-17. For simplicity the various tuned circuits are not shown separately.

If you compare Figs. 12-14 and 12-17, you will see that each receiver has in common the RF amplifier, mixer, oscillator, tuned IF amplifier, a detector, audio amplifiers, and a power supply. However, there are important differences.

Whereas the AM broadcast receiver operates on a frequency band centering on about 1000 kilohertz, the FM broadcast receiver operates on a much higher frequency band, centering on 98 megahertz. Also the FM receiver has two special circuits not found in the AM receiver. These are the limiter and the FM detector.

One of the major advantages of FM reception is the almost noise-free output of the receiver. In Fig. 12-17 the circuit which removes most of the noise

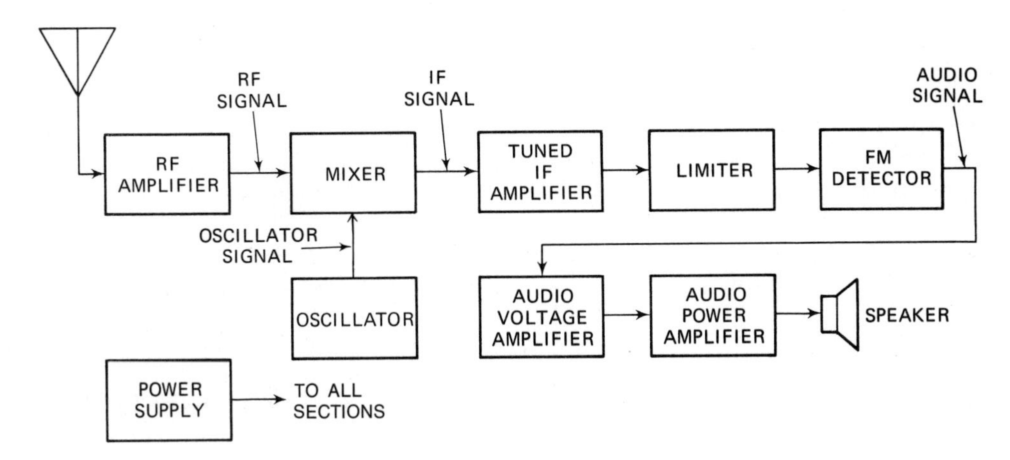

Fig. 12-17 Simplified block diagram of an FM superheterodyne receiver.

from the signal is the limiter. Amplitude-modulated noise, which is most of the received noise, is rejected by the limiter.

The FM detector must respond to a signal which is varying in frequency (rather than amplitude, as is the case with the AM detector). Thus the FM detector has specially tuned circuits which respond to the changing frequencies of the FM signal. The output of the FM detector is an audio signal, just as the output of the AM detector is an audio signal. As you see in Fig. 12-17, the detected audio signal is amplified and then fed to the speaker.

An important feature of FM broadcast reception is the wider band of audio frequencies, the so-called high-fidelity reception. You should remember however, that this feature is not restricted to the FM system. An AM system is capable of providing the same high-fidelity audio. However, at the frequency band used by AM there is not enough room for each AM station to transmit high-fidelity audio. By room we mean bandwidth.

Another feature of many FM broadcast receivers (not shown in Fig. 12-17) is so-called stereophonic reception. In this case the FM transmitter broadcasts a special kind of FM signal. This signal contains the information of sound coming from two physical directions, left and right of the listener. At the FM receiver the right and left audio signals are separated and fed to two speakers, one to the right and the other to the left of the listener. This provides a much more realistic sound.

Summary

1. All receivers have at least four basic sections: antenna, tuner, detector, and transducer.
2. A crystal radio has only the four basic receiver sections. The crystal is really a diode detector.
3. Tuned radio-frequency (TRF) receivers may have several tuned RF sections.
4. Superheterodyne receivers have a local oscillator. The local-oscillator signal is mixed with the RF signal to produce an intermediate frequency.

5. Not all superheterodyne receivers use a separate RF amplifier. In these sets a converter stage may be used. This is a combination mixer-oscillator.
6. In most broadcast AM superheterodyne receivers, the difference between the RF and oscillator frequencies is 455 kilohertz regardless of what station is being tuned. This means that the intermediate frequency is always 455 kilohertz.
7. An FM superheterodyne receiver is similar to an AM set. However, the FM set uses a limiter stage to reduce noise and a special type of FM detector.

PROGRAMMED REVIEW QUESTIONS

(Instructions for using this programmed section are given in Chapter 1.)
We will now review the important concepts of this chapter. If you have understood the material, you will progress easily through this section. Do not skip this material because some additional theory is presented.

1. A detector circuit in a radio receiver must use a nonlinear component. Which of the following components is used as a detector in a radio receiver?
   A. A diode. (Proceed to block 17.)
   B. A resistor. (Proceed to block 9.)

2. *The correct answer to the question in block 24 is* **C**. *Either the capacitor or the inductor in the tuned LC circuit can be varied. In some receivers a* ***varactor*** *diode may be used. This is a* ***voltage-variable capacitor.*** *It is a diode that acts like a capacitor, so the circuit is considered to be capacitor-tuned.* Here is your next question.
   A broadcast AM receiver is tuned to a frequency of 1200 kilohertz. If the receiver is a superheterodyne type with an intermediate frequency of 455 kilohertz, what is the oscillator frequency?
   A. 1655 kilohertz. (Proceed to block 8.)
   B. 1200 kilohertz. (Proceed to block 15.)

3. *The correct answer to the question in block 19 is* **A**. *The detector is sometimes called a* ***demodulator.*** Here is your next question.
   An advantage of single-sideband transmission compared with regular AM transmission is that the single-sideband-transmission signal
   A. is easier to tune. (Proceed to block 10.)
   B. does not require as much of the radio-frequency spectrum. (Proceed to block 16.)

4. *The correct answer to the question in block 17 is* **A**. *When the radio is tuned from one station to another the oscillator frequency is automatically changed. The oscillator frequency minus the frequency of the station being tuned always equals the intermediate frequency.*

Here is your next question.
Which of the following is true?

A. Longer transmitting antennas are usually used for lower frequencies. (Proceed to block 24.)
B. Longer transmitting antennas are usually used for higher frequencies. (Proceed to block 7.)

5. *Your answer to the question in block 20 is* **B**. *This answer is wrong. The carrier does not contain any audio information, which is actually in the sidebands.* Proceed to block 26.

6. *The correct answer to the question in block 8 is* **B**. *There are some transmitters made in which a variable-frequency oscillator (VFO) is used instead of a crystal oscillator. This permits the operator to tune to different carrier frequencies.* Here is your next question.
The reason for outlawing radio communications with damped waves is that

A. they are very dangerous. (Proceed to block 13.)
B. they used too much of the radio spectrum. (Proceed to block 20.)

7. *Your answer to the question in block 4 is* **B**. *This answer is wrong. The lower the transmitting frequency, the longer the antenna.* Proceed to block 24.

8. *The correct answer to the question in block 2 is* **A**. *Here is the calculation:*

$$\text{Oscillator frequency} - \text{station frequency} = \text{intermediate frequency}$$

Therefore

$$\begin{aligned}\text{Oscillator frequency} &= \text{station frequency} + \text{intermediate frequency}\\ &= 1200 \text{ kilohertz} + 455 \text{ kilohertz}\\ &= 1655 \text{ kilohertz}\end{aligned}$$

Here is your next question.
In an AM transmitter the carrier frequency is produced by

A. an audio oscillator and a modulator. (Proceed to block 22.)
B. a crystal oscillator and frequency multipliers. (Proceed to block 6.)

9. *Your answer to the question in block 1 is* **B**. *This answer is wrong. A resistor is a linear component. It cannot be used as a detector.* Proceed to block 17.

10. *Your answer to the question in block 3 is* **A**. *This answer is wrong. The fact that single-sideband receivers are more difficult to tune explains why this method of transmission is not used for AM broadcast sets.* Proceed to block 16.

11. *Your answer to the question in block 17 is* **B**. *This answer is wrong. The intermediate frequency (IF) does not change when the radio is tuned.* Proceed to block number 4.

12. *Your answer to the question in block 26 is* **A**. *This answer is wrong. A varactor is a diode that acts like a capacitor.* Proceed to block 19.

13. *Your answer to the question in block 6 is* **A**. *This answer is wrong. There has been no discussion of danger regarding damped waves.* Proceed to block 20.

14. *Your answer to the question in block 24 is* **A**. *This answer is wrong. In many car radios the tuning circuit uses variable inductors. These are preferred here over variable capacitors because car radios must be exceptionally sensitive. Using variable inductors enables car radios to be made more sensitive.* Proceed to block 2.

15. *Your answer to the question in block 2 is* **B**. *This answer is wrong. Add the radio frequency to the intermediate frequency to get the oscillator frequency.* Proceed to block 8.

16. *The correct answer to the question in block 3 is* **B**. *The radio spectrum is becoming more and more crowded. So the fact that single-sideband transmission takes less space in the spectrum is becoming more and more important.* Here is your next question.
A 1500-kilohertz signal is modulated by a 500-hertz (0.5 kilohertz) audio sine wave. The upper sideband is at
A. 1500.5 kilohertz. (Proceed to block 23.)
B. 1505 kilohertz. (Proceed to block 27.)

17. *The correct answer to the question in block 1 is* **A**. *The diode must have a very low forward-voltage drop. In other words, when current flows from cathode to anode, there should be very little voltage drop.* Here is your next question.
Which of the following statements is true?
A. The intermediate frequency in a radio does not change when the radio is tuned from one station to another. (Proceed to block 4.)
B. The intermediate frequency in a radio becomes higher when the radio is tuned to a station with a higher frequency. (Proceed to block 11.)
C. The intermediate frequency in a radio becomes lower when the radio is tuned to a station with a lower frequency. (Proceed to block 18.)

18. *Your answer to the question in block 17 is* **C**. *This answer is wrong. The intermediate frequency does not change when the radio is tuned.* Proceed to block 4.

19. *The correct answer to the question in block 26 is* ***B****. In tube-type receivers a special tube with five grids may be used for a converter. The resulting circuit is called a* **pentagrid converter.** Here is your next question.
There are four basic sections common to all receivers. They are the output transducer, the tuned circuit, the antenna, and the
A. detector. (Proceed to block 3.)
B. power supply. (Proceed to block 25.)

20. *The correct answer to the question in block 6 is* ***B****. The damped waves cover such a wide range of frequencies that only a few stations could operate with that type of transmission. Also, it would not be possible to amplitude-modulate or frequency-modulate the damped waves.*
Here is your next question.
The useful audio information in a broadcast signal is
A. in the sidebands. (Proceed to block 26.)
B. in the carrier. (Proceed to block 5.)

21. *Your answer to the question in block 24 is* ***B****. This answer is wrong. Most table-model radios are tuned with variable capacitors.* Proceed to block 2.

22. *Your answer to the question in block 8 is* **A***. This answer is wrong. Study the block diagram of Fig. 12-9, then* proceed to block 6.

23. *The correct answer to the question in block 16 is* **A***. The upper-sideband frequency is equal to the carrier frequency plus the audio frequency:*

$$\text{USB} = \text{carrier frequency} + \text{audio frequency}$$

The carrier frequency is 1500 kilohertz, and the audio frequency is 500 hertz, or 0.5 kilohertz. Therefore

$$\begin{aligned}\text{USB} &= 1500 \text{ kilohertz} + 0.5 \text{ kilohertz}\\ &= 1500.5 \text{ kilohertz}\end{aligned}$$

Here is your next question.
The process of combining an audio signal with an RF carrier is called ________________. (Proceed to block 28.)

24. *The correct answer to the question in block 4 is* **A***. The length of the transmitting antenna is related to the wavelength of the frequency being transmitted. Mathematically,*

$$\text{Wavelength in meters} = \frac{300{,}000{,}000}{\text{frequency in hertz}}$$

This equation shows that higher frequencies have shorter wavelengths. Here is your next question.
An *LC* circuit is used in a radio tuner. This permits the listener to select one station and reject all others. Tuning a station involves

A. varying the capacitance but never the inductance of the *LC* circuit. (Proceed to block 14.)
B. varying the inductance but never the capacitance of the *LC* circuit. (Proceed to block 21.)
C. varying either the inductance or the capacitance of the *LC* circuit. (Proceed to block 2.)

25. *Your answer to the question in block 19 is **B**. This answer is wrong. Radios can be made without a power supply. The crystal receiver is an example.* Proceed to block 3.

26. *The correct answer to the question in block 20 is* **A**. *The carrier is needed for demodulating the signal in the receiver, but the useful audio information is in the sidebands.* Here is your next question.
The circuit that is a combination of an RF amplifier and a mixer in a superheterodyne receiver is called

A. a varactor. (Proceed to block 12.)
B. a converter. (Proceed to block 19.)

27. *Your answer to the question in block 16 is **B**. This answer is wrong. The audio frequency is 500 hertz, which is equal to 0.5 kilohertz. The upper-sideband frequency is obtained by adding the audio frequency to the carrier frequency.* Proceed to block 23.

28. *Modulation.*
You have now completed the programmed questions. The next step is to put some of these ideas to work in laboratory experiments. Proceed to the Experiment section of this chapter.

EXPERIMENT

(The experiment described in this section may be performed on the circuit board described in Appendix C or on a similar laboratory setup.)

Purpose To demonstrate the sound of the 60-hertz and 120-hertz ripple frequencies, and to show how an audio oscillator can be made with feedback to the inverting input terminal of an operational amplifier.

Theory In Chapter 6 you studied the theory of operation for power-supply circuits. In that chapter you learned that a rectifier converts the ac power into pulsating dc. A capacitor filters the pulsating dc, and the output of the supply is a nearly pure dc.

Figure 12-18 shows the waveforms in a half-wave supply. In Fig. 12-18*a* the output waveform is not filtered. It is a 60-hertz pulsating wave. In Fig.

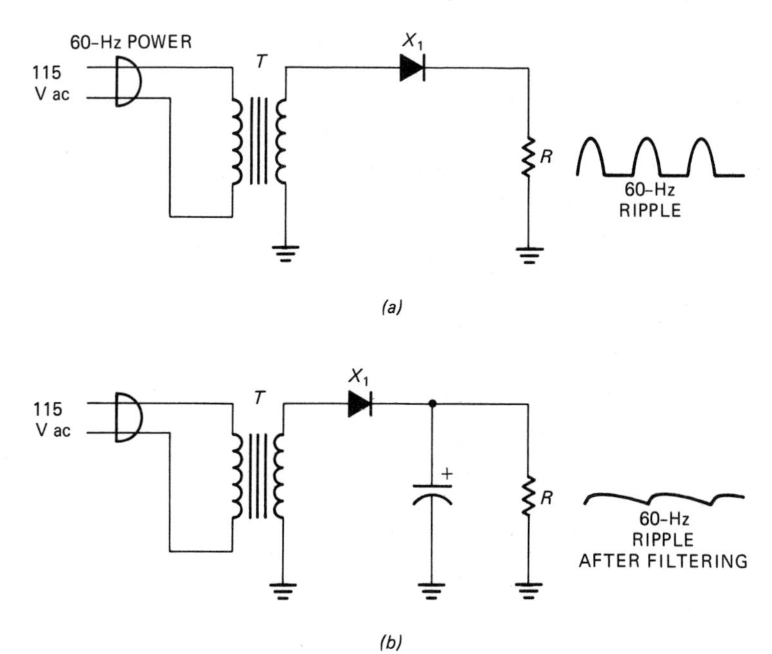

Fig. 12-18 Half-wave rectifier. (*a*) Without filtering. (*b*) With filtering.

12-18*b* a filter capacitor has been added, and the output waveform is a nearly pure dc.

As the power-supply electrolytic filter capacitors age, they become less effective in removing ripple. One of the symptoms of a defective filter capacitor in consumer products is a loud hum in the speaker. This occurs because the current for the power amplifier is no longer pure dc, but is pulsating at the ripple frequency.

The waveforms for a full-wave rectifier circuit are shown in Fig. 12-19. A two-diode full-wave supply is shown. However, the waveforms would be the same in any other full-wave supply, such as a bridge rectifier. In Fig. 12-19*a* the unfiltered waveform is seen to be 120 hertz. The filtered waveform is shown in Fig. 12-19*b*.

With a little practice you can learn to recognize 60-hertz and 120-hertz sounds. This is important because these sounds are symptoms of defective filtering. In this experiment you will listen to the sounds of power-supply ripple.

There are two ways to get the desired feedback for op amp oscillators, and they are shown in Fig. 12-20. In Fig. 12-20*a* the amplifier output is fed back to the noninverting input. The input and output signals are in phase, so oscillation takes place. This is the type of oscillator circuit used in the experiment for Chapter 10.

In Fig. 12-20*b* the amplifier output is fed back to the inverting input. The inverting input signal and the output signal are normally out of phase. This means that if the signal is fed back directly, as it would be with only a resistor in the feedback path, the feedback would be degenerative. That is why a phase inverter is used here to make the feedback positive.

Although this discussion has been related to the use of operational amplifi-

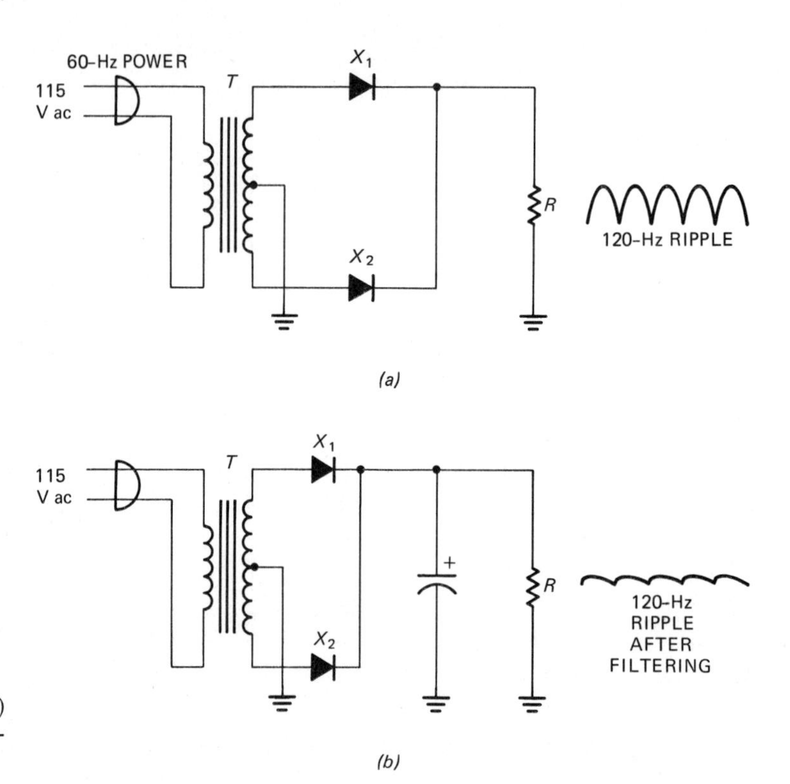

Fig. 12-19 Full-wave rectifier. (*a*) **Without filtering.** (*b*) **With filtering.**

ers as oscillators, it is also true for oscillators that use tubes, or bipolar transistors, or FETs. The conventional amplifier connection is used most often for the amplifier portion of the oscillators. Figure 12-21 shows how the feedback is usually obtained. Note that a phase-inverting circuit is needed between the two points in order to make the feedback positive.

The subjects covered in this experiment are related to all electronic theory, not just transmitters and receivers. When you are troubleshooting a radio receiver you must keep in mind that it has an audio section. Furthermore, receivers have audio playback systems and audio amplifiers in their disc and tape recorders. In this experiment you will learn to recognize audio sounds that are symptoms of bad filters in power supplies. You may hear these sounds in any equipment that has a power supply.

In an earlier chapter you learned that oscillators may employ positive feedback. In this chapter you will use an operational amplifier to make an oscil-

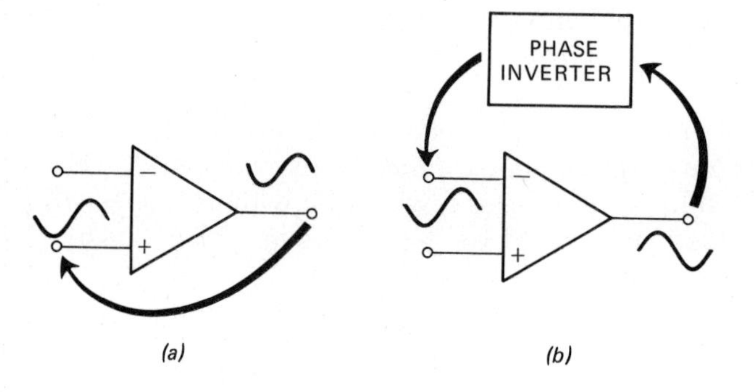

Fig. 12-20 Two feedback paths for oscillators. (*a*) **To the noninverting input.** (*b*) **To the inverting input.**

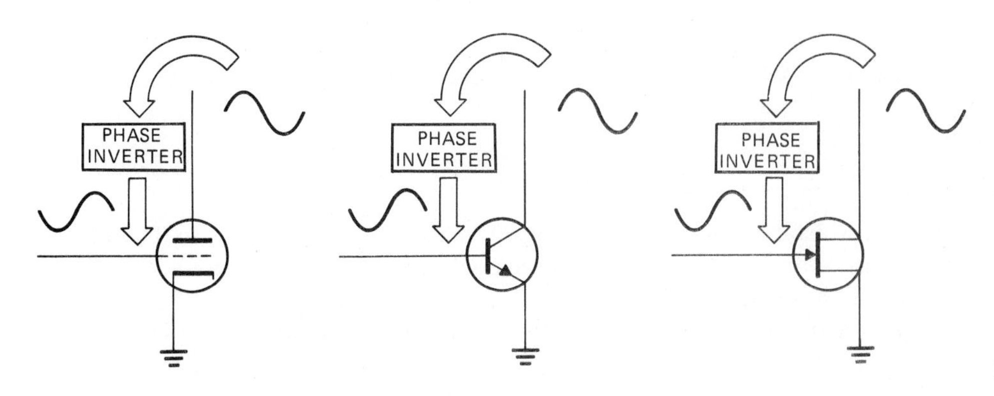

Fig. 12-21 Oscillator feedback.

lator. The feedback will be to the positive (noninverting) terminal. The oscillator that you are constructing in the lab is similar to the type of broadband audio oscillator used by technicians for troubleshooting audio equipment.

PART I

Test Setup Figure 12-22 shows the half-wave rectifier circuit. The schematic is shown in Fig. 12-22*a*, and the wiring diagram is shown in Fig. 12-22*b*. Do not connect the capacitor at this time, but wire the circuit.

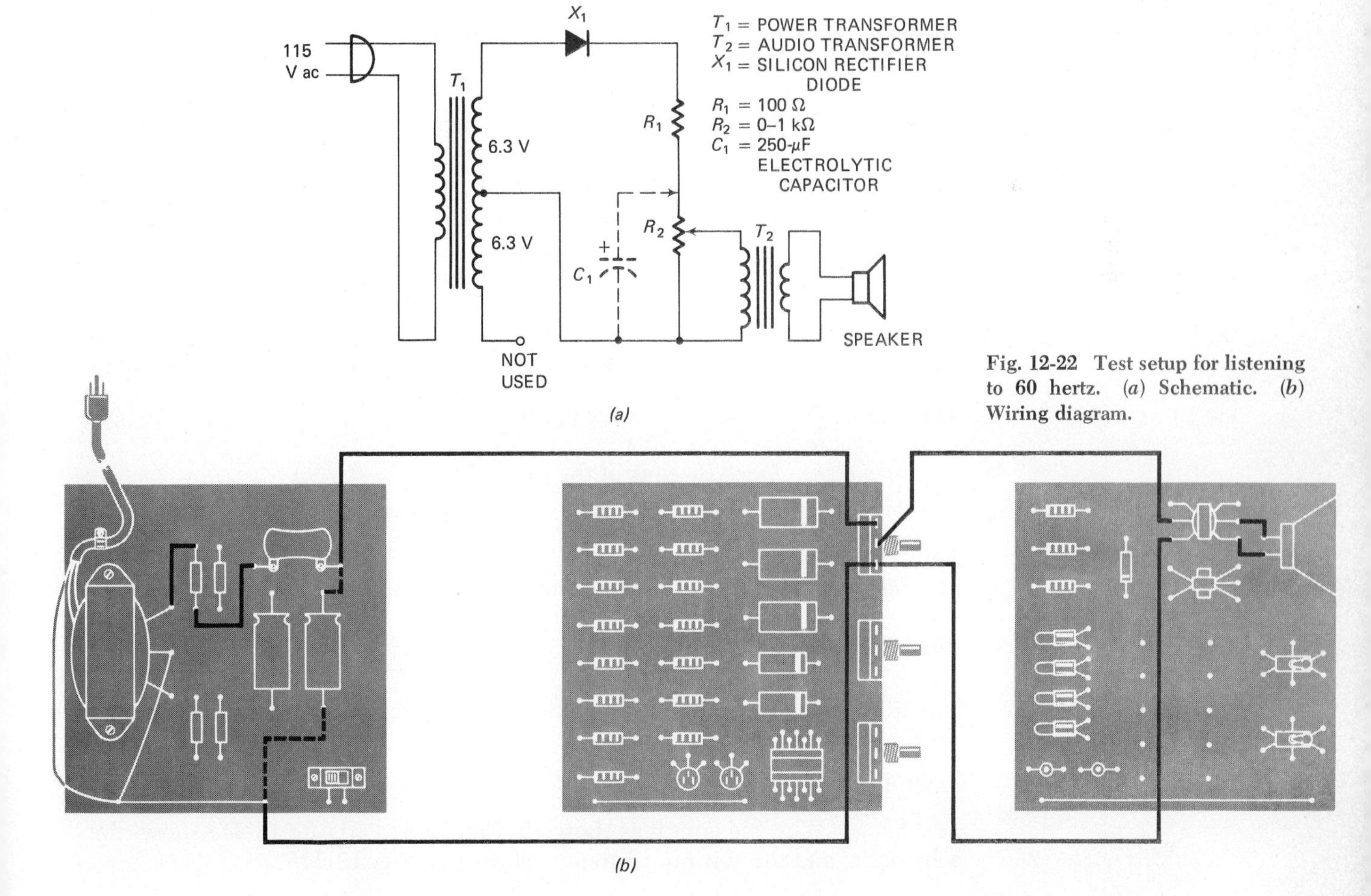

Fig. 12-22 Test setup for listening to 60 hertz. (*a*) Schematic. (*b*) Wiring diagram.

Procedure

step one Adjust R_2 to the center of its position. Do you hear a sound in the speaker? ________________ Yes or No

You should hear a low-frequency sound in the speaker. Listen to it carefully. This is the sound you will hear in the speaker of a system in which the power-supply filter is defective.

step two Adjust R_2 and note the effect on the speaker sound. ________________

Adjusting R_2 should change the volume of sound. This connection of R_2 is the same as the connection of a volume control in a radio or television receiver.

step three Readjust R_2 to the center of its range.

step four Connect the filter capacitor across the output. What happens to the speaker sound? ________________

The filter capacitor should remove the sound from the speaker.

step five Wire the circuit shown in Fig. 12-23. Do not connect the filter capacitor at this time.

step six Adjust R_2 to the center of its range. Do you hear a sound coming from the speaker? ________________ Yes or No

step seven In Step 6 you should hear a ripple sound. Is the sound higher or lower in frequency than the 60-hertz sound? ________________ Higher or Lower

The sound should have a higher frequency.

step eight Does R_2 control the volume? ________________ Yes or No

Your answer should be yes.

step nine Connect the capacitor across the rectifier output. How does this affect the sound? ________________

The filter capacitor should remove the ripple, and therefore it should remove the sound.

PART II

Test Setup Wire the circuit as shown in Fig. 12-24. Figure 12-24*a* shows the schematic, and the wiring diagram is shown in Fig. 12-24*b*.

115 V ac

T_1 X_1 X_2 C_1 R_1 R_2 T_2

R_1 = 100 Ω
R_2 = 0–1 kΩ
X_1 = X_2 = SILICON RECTIFIER DIODES
C_1 = 250 μF ELECTROLYTIC CAPACITOR
T_1 = POWER TRANSFORMER
T_2 = AUDIO TRANSFORMER

SPEAKER

(a)

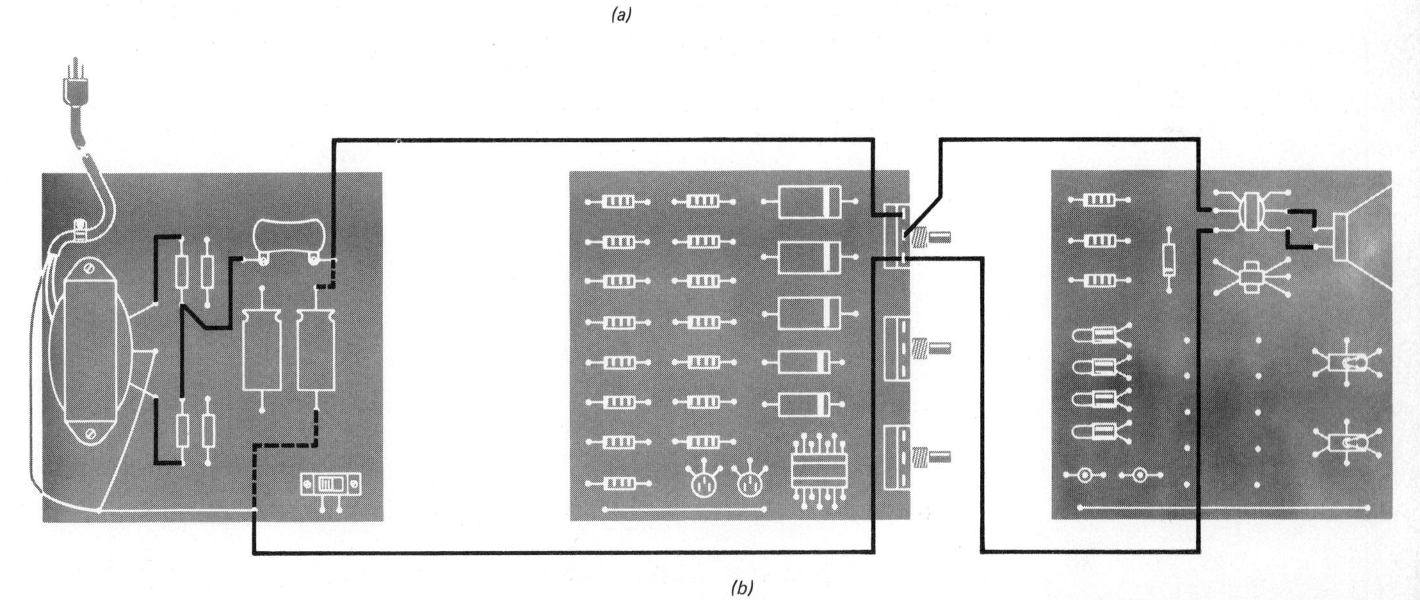

(b)

Fig. 12-23 Test setup for listening to 120 hertz. (*a*) Schematic. (*b*) Pictorial wiring diagram.

Procedure

step one What type of power-supply rectifier is used in this circuit?

____

A bridge rectifier is used.

step two Measure the positive power-supply voltage between the busbar and point *a*. Record the voltage. ____________________ volts

If you do not measure *about* +9 volts (usually slightly less), then remove power from the circuit and carefully check the wiring. The circuit will not operate properly unless you have the proper positive supply voltage.

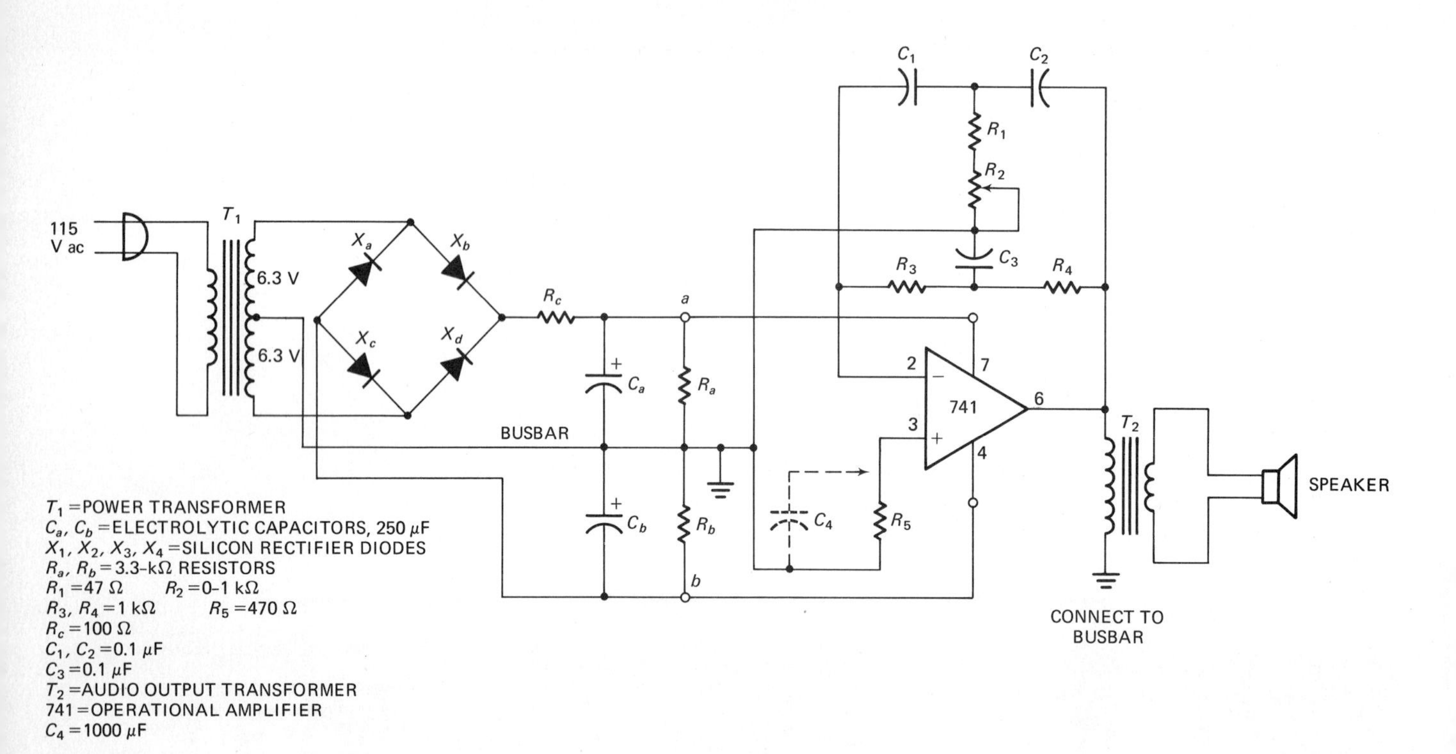

(a)

a
b

(b)

Fig. 12-24 Test setup for oscillator experiment. (*a*) **Schematic.** (*b*) **Wiring diagram.**

step three Measure the negative power-supply voltage between the busbar and point *b*. Record the voltage. ______________ volts

If you do not measure *about* −9 volts (usually slightly less), then remove power from the circuit and carefully check the wiring. The circuit will not operate properly.

step four Adjust R_2 and determine its effect on the output sound. Does it control the volume or the frequency? ______________

Resistor R_2 may affect both the volume and the frequency, but the greatest effect is on the frequency.

step five Place a 1000-microfarad capacitor across the resistor that is connected between the noninverting terminal (pin 3) and common. This capacitor is shown with dotted lines in Fig. 12-24. Does the oscillator stop working? ______________

Your answer should be no. This proves that there is no input signal to the noninverting terminal.

Conclusion The ripple frequency of a half-wave rectifier is 60 hertz, and the ripple frequency of a full-wave rectifier is 120 hertz when the power-line frequency is 60 hertz. A technician should learn these sounds. When an electrolytic filter capacitor in the power supply of an audio system or receiver ages, the ripple voltage increases. The ripple frequency sound will then be heard in the speaker.

Oscillators require positive (regenerative) feedback for their operation. The feedback signal may go directly to a noninverting input. If it goes to an inverting input, then some form of phase-shifting network must be used to make the feedback signal regenerative.

SELF-TEST WITH ANSWERS

(Answers with discussions are given at the end of the chapter.)

1. Damped waves are not used in communications because (*a*) they cannot transmit a single frequency; (*b*) they require too much power; (*c*) they require too much voltage; (*d*) expensive equipment is needed for making them.
2. One way to communicate is to turn the RF carrier ON and OFF to make Morse code characters. This is called (*a*) AM; (*b*) FM; (*c*) CW; (*d*) RF.
3. Which of the following is *not* a section found in all receivers? (*a*) Antenna; (*b*) Tuned circuit; (*c*) Detector; (*d*) Power supply.
4. A certain circuit separates the audio intelligence from the RF signal in a

receiver. This circuit is called a (*a*) detector; (*b*) mixer; (*c*) converter; (*d*) modulator.

5. A 1000-kilohertz RF carrier signal is combined in an AM transmitter with an audio signal of 1 kilohertz. Which of the following is correct? (*a*) There are no sidebands; (*b*) There is one sideband at 1100 kilohertz; (*c*) There are two sidebands, one at 1001 kilohertz and one at 1002 kilohertz; (*d*) There are two sidebands, one at 999 kilohertz and one at 1001 kilohertz.
6. For a frequency-domain display the horizontal axis shows frequency and the vertical axis shows (*a*) time; (*b*) amplitude; (*c*) delay; (*d*) none of these.
7. In a receiver the speaker converts electrical signals into sound signals. The speaker is also called a (*a*) modulator; (*b*) transducer; (*c*) detector; (*d*) reactor.
8. In an FM transmitter modulation is often applied to which stage? (*a*) Detector; (*b*) Oscillator; (*c*) Amplifier; (*d*) Multiplier.
9. In a certain receiver the radio frequency is 1000 kilohertz and the local-oscillator frequency is 1250 kilohertz. What is the intermediate frequency in this receiver? (*a*) 750 kilohertz; (*b*) 2250 kilohertz; (*c*) 250 kilohertz; (*d*) 455 kilohertz.
10. Assuming that the line-power frequency is 50 hertz, the ripple frequency of a full-wave rectifier is (*a*) 50 hertz; (*b*) 100 hertz; (*c*) 60 hertz; (*d*) 120 hertz.

ANSWERS TO SELF-TEST

1. (*a*)—In the early days of radio communications the damped waves were sent in the form of dots and dashes. These dots and dashes formed letters by Morse code. However, the damped waves used up so much of the radio spectrum that it was not possible to tune one station and reject all others.
2. (*c*)—CW stands for continuous waves.
3. (*d*)—Not all receivers require a power supply. For example, a crystal radio does not have a power supply.
4. (*a*)—In the superheterodyne receiver the detector separates the intermediate frequency and audio signals. The intermediate frequency is discarded.
5. (*d*)—The sideband frequencies are *the carrier frequency plus the audio frequency* and *the carrier frequency minus the audio frequency:*

$$1000 \text{ kilohertz} + 1 \text{ kilohertz} = 1001 \text{ kilohertz (USB)}$$
$$1000 \text{ kilohertz} - 1 \text{ kilohertz} = 999 \text{ kilohertz (LSB)}$$

6. (*b*)—The vertical axis shows amplitude in both time- and frequency-domain displays.
7. (*b*)

8. (*b*)—In many FM transmitters the audio modulation is applied directly to the oscillator stage.
9. (*c*)—Automobile radios may have intermediate frequencies of 250 kilohertz.
10. (*b*)—The ripple frequency of a full-wave rectifier is twice the power-line frequency.

how do audio systems work? 13

INTRODUCTION

When you think of the word *audio*, you probably think of the high-fidelity systems. Many of these home entertainment systems are the most elaborate of all audio systems. However, technicians are working on other types of audio equipment too.

One application of audio electronics is the public address system. These systems may be simple power amplifiers or very elaborate concert hall layouts. A technician who works with public address must have a very good knowledge of the principles of sound waves and hearing.

Another field of audio electronics is *intercoms* (intercommunications systems). Again, these range from relatively simple systems between two rooms or between two offices to very elaborate systems with telephone switchboards and dialing provisions.

Electronic musical instruments are another important field of audio. Instruments range from relatively simple guitar amplifiers to electronic organs that make the sounds of many musical instruments.

Each of these fields of audio has its own special equipment for testing and troubleshooting. In this chapter we will concentrate on the subject of high-fidelity systems. They have the same electronic components that are used in the other audio systems, but the power requirements may differ. For example, a loudspeaker in a high-fidelity system is not likely to be able to deliver as much power as an outdoor speaker designed for use in a stadium. Even so, the basic principles of operation are the same.

You will be able to answer these questions after studying this chapter:

- ☐ **What are the parts of a high-fidelity audio system?**
- ☐ **How are transducers used in audio systems?**
- ☐ **How is the quality of speaker sound controlled?**
- ☐ **What does the tuner do in an audio system?**
- ☐ **What is special about audio voltage and power amplifiers?**

☐ **Why was quadraphonic sound developed?**

☐ **What are the variations in tape and disc recording?**

INSTRUCTION

What Are the Parts of a High-Fidelity Audio System?

Figure 13-1 shows a block diagram of a simple hi-fi system. There are four possible signal inputs here: *tape recorder, AM-FM tuner, disc player (phonograph)*, and *TV receiver.* Switch SW_1 selects the desired input and delivers the signal to the voltage amplifiers.

The voltage amplifiers are not greatly different from the voltage amplifiers used in other systems. One difference here is that there is more emphasis on bandwidth and less emphasis on gain than in some other fields of electronics.

The output of the voltage amplifiers goes to frequency-compensation circuits. In some systems these circuits may be part of the amplifiers. The frequency-compensation circuits make it possible to adjust the system for the most pleasing sound. The tone control, or bass and treble controls, are examples of the circuits in this section.

The power amplifiers are the next stage. These amplifiers are capable of delivering a large amount of audio current to the output transducers (speakers).

In a high-fidelity system the output transducers are the *loudspeakers.* A large part of the total cost of a good audio system is likely to be in the loudspeakers.

The enclosure for the speakers is very important to the output sound. An *enclosure* is the box the speaker is located in. It gives the output sound its sound quality. Regardless of how much money you spend on the speaker, the system will not sound right unless it is in a well-designed enclosure.

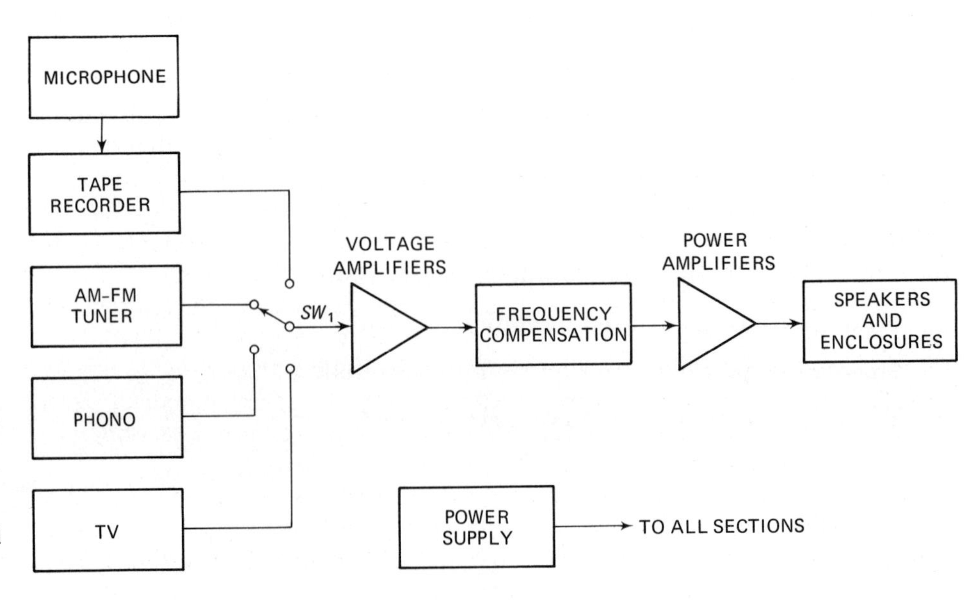

Fig. 13-1 Block diagram of hi-fi audio system.

All electronic systems have power supplies. In a very high quality audio system, it is possible that separate power supplies would be used for the power amplifier and voltage amplifier. Also each of the units that supplies a signal to the system may have its own individual power supply. This is certainly true of the television set and the AM-FM tuner. As with any electronic system, the audio system cannot function properly unless every power supply is delivering the required rated amount of voltage and current.

How Are Transducers Used in Audio Systems?

A transducer is a component that permits the energy of one system to control the energy in another system. A microphone is an example of a transducer because sound energy controls the amount of electrical energy from the microphone. Likewise, a loudspeaker is a transducer because the amount of electrical energy delivered to the speaker controls the amount of sound energy that leaves it.

The definition of *transducer* is often simplified to mean a device that converts energy from one form to another. Physically this is not correct, but it is a popular definition. In the audio system the important transducers are the microphone for the tape recorder, the tape heads for recording and playing tapes, the cartridge for playing records, and the loudspeaker. We will now discuss the principles of operation of these transducers.

How Does a Microphone Work?

Figure 13-2 shows five basic types of microphones used in audio systems. There are advantages and disadvantages to each of these, so no one microphone is used for all purposes.

The *carbon microphone* is one of the older types. It is shown in Fig. 13-2*a*. It consists of a package or button of loosely packed carbon granules. Battery E causes current to flow through the carbon and through the load resistor R. The amount of current depends upon the amount of voltage used, the resistance of the carbon button, and the load resistance.

When the varying air pressures of sound strike the microphone the carbon granules become tightly packed (under high pressure) and then loosely packed (under low pressure). When they are tightly packed, their resistance is low, and when they are loosely packed, their resistance is high. This means that the sound varies the resistance of the circuit by varying the packing of the carbon granules. This in turn varies the amount of current flowing in the circuit and the amount of voltage across the load resistor. The output signal from the carbon-microphone system is an audio voltage.

One disadvantage of the carbon microphone is that a power supply must be used to provide the current flow. Most other types of microphones generate their own electrical power. They also have limited frequency response, but high output-signal voltage.

Another kind of microphone is the *ceramic crystal* type like the one shown in Fig. 13-2*b*. Originally these were called crystal microphones, but in modern systems the crystals have been replaced with a ceramic material which is

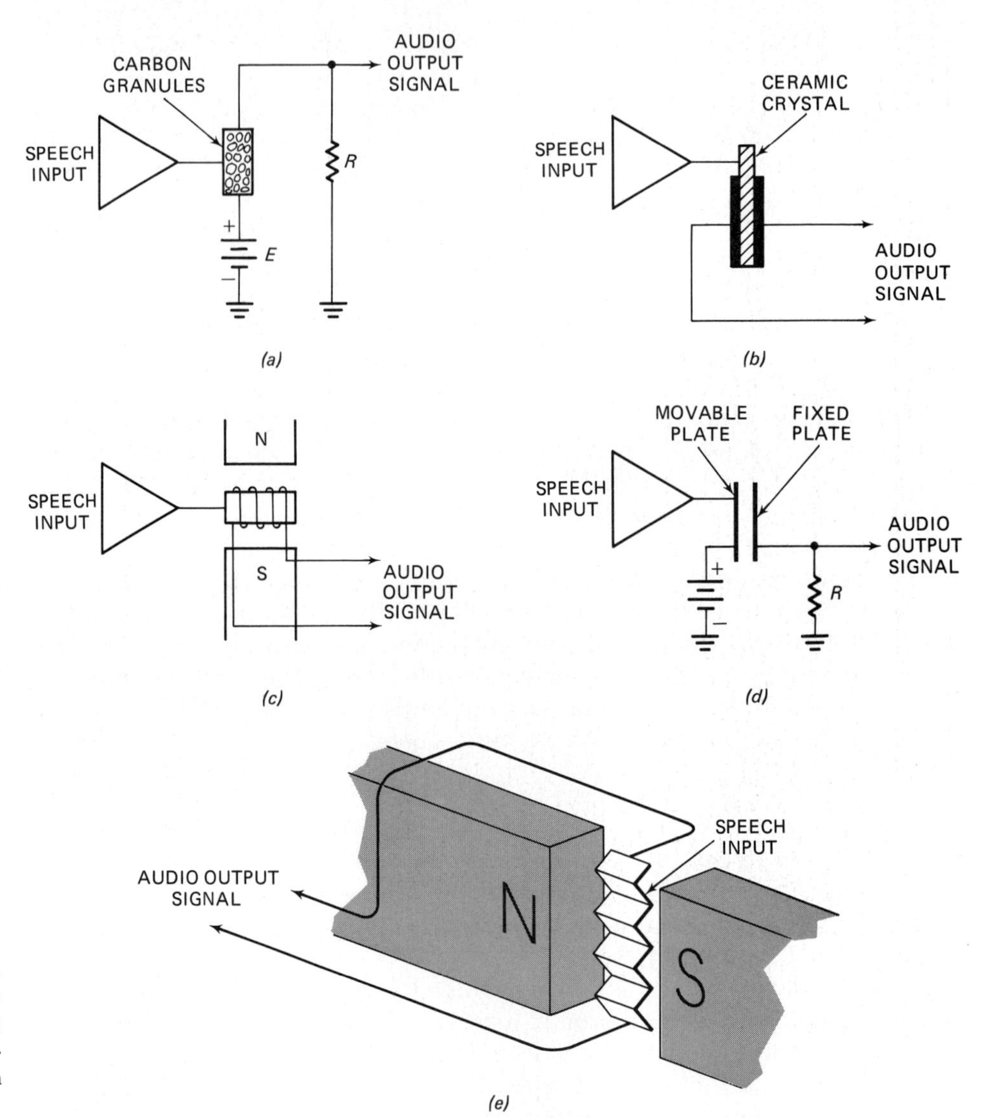

Fig. 13-2 Types of microphones. (*a*) Carbon microphone. (*b*) Ceramic crystal microphone. (*c*) Dynamic microphone. (*d*) Condenser microphone. (*e*) Ribbon microphone.

more rugged. The principle of operation is the piezoelectric effect. Sound striking the microphone causes the crystal to move back and forth. As you know, when you flex a crystal, it causes a voltage to be generated across its surface. The amount of voltage is related directly to the amount of sound striking the microphone at any time. The output is an audio voltage which is directly related to the input sound.

Ceramic crystal microphones are very rugged, and it is easy to design a cricuit for use with a crystal microphone. Therefore you will see them used extensively. They have good frequency response and high output.

The *dynamic microphone* in Fig. 13-2*c* works on the principle of Faraday's law. This law states that any time a conductor moves in a magnetic field, a voltage is generated. The microphone causes a coil to move in a permanent-magnet field. As the coil moves, the voltage is induced which is directly related to the sound energy.

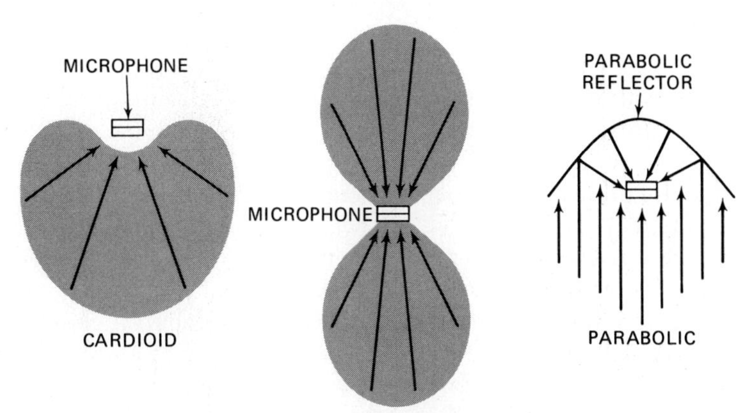

Fig. 13-3 Three microphones showing directional characteristics.

Whenever you see the word *dynamic* applied to a transducer, it is generally some type in which a conductor is caused to move in a magnetic field. This type has excellent frequency response but low output.

Figure 13-2*d* shows the principle of operation for a *condenser microphone.* Quite a while ago capacitors were called *condensers,* and that is the reason for the name. Even though the word *condenser* is obsolete, the name *condenser microphone* still persists. One of the plates of the capacitor in the microphone is movable, and the other is in a fixed position. As the sound strikes the microphone it causes the movable plate to move closer with high sound pressure and farther away with low sound pressure. Since the capacitance of a capacitor depends upon the distance between the plates, the capacitance varies with the sound.

When the capacitance increases, it can hold a greater charge, and a charging current flows in one direction through *R*. When the capacitance is smaller, it cannot hold its charge and it must discharge through *R*, causing current to flow in the opposite direction. The result is a varying current through *R* that produces an audio voltage. These microphones have good frequency response but low output.

Figure 13-2*e* shows the basic parts of a *ribbon microphone.* The ribbon is actually a piece of corrugated aluminum that is highly flexible. When a sound strikes this aluminum it vibrates. Aluminum is a conductor, so when it moves in the magnetic field a voltage is produced across it. This is an audio voltage. The ribbon microphone, like the dynamic microphone, works on the principle of Faraday's law. These microphones have excellent frequency response but low output.

Microphones are also identified by their directional characteristics. Figure 13-3 shows three examples. The *cardioid* type can receive audio energy from directly in front and to a smaller extent from the sides. The response is heart-shaped.

Bidirectional microphones can receive from either of two directions. A *parabolic microphone* is a highly directional one which can receive energy only from a narrow area from a single direction—that is, in front. Users choose the microphone with the directional characteristic that best meets their needs.

Summary

1. An audio technician may work on high-fidelity (hi-fi) systems, public address systems, intercommunications systems, or electronic musical instruments.
2. The high-fidelity system has the largest number of individual components.
3. Examples of inputs to a hi-fi system are tape players, record players, AM-FM tuners, and the TV audio signal.
4. This is the correct definition of a transducer: a component in which the energy of one system controls the energy of another system.
5. This is a simpler (but not as accurate) definition of a transducer: a component that converts energy from one form to another.
6. Examples of input transducers in a hi-fi system are microphones, tape heads, and phono cartridges.
7. A loudspeaker is an example of an output transducer in a hi-fi system.
8. Some of the more common types of microphones are dynamic and condenser microphones.

How Does a Tape Recorder Work?

The first attempt at magnetic recording used a wire made of material that could be magnetized by an audio-signal current. There were many problems with the wire recorder. One serious problem was that the wire would break and tangle. Another problem was the poor frequency response of the magnetic wire.

To get around these problems, the tape recorder was developed. The basic principle of the tape recorder is shown in Fig. 13-4. The tape consists of a coating of magnetic material on a plastic-tape backing. To make a recording the tape is passed across a head gap. At the same time the audio signal magnetizes the head first in one direction (for one half cycle) and then in the other direction (for the next half cycle). The magnetic field at the gap passes readily through the oxide coating and causes it to become permanently magnetized.

The playback transducer is similar to the one shown in Fig. 13-4. What happens is that the magnetized tape is passed across the gap in the head. The permanent magnetic field on the tape causes flux lines to flow in the tape

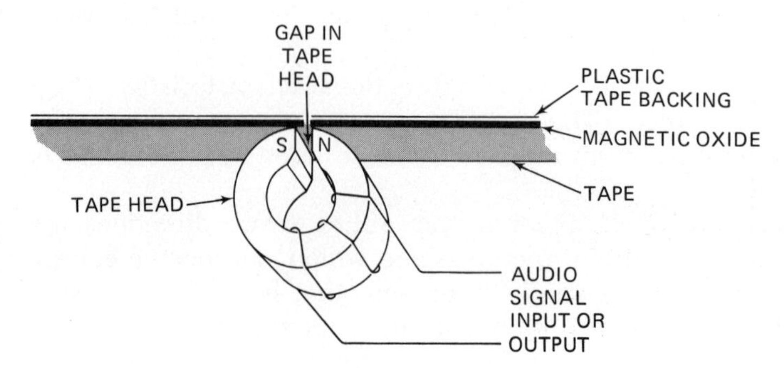

Fig. 13-4 Transducer for tape recording and playback.

head. As the magnetic field varies, an audio voltage is induced in the coil. This is the desired audio-voltage output.

Theoretically the same head could be used for recording and playback, but for the better tape recorders a separate head is used for each function. This makes it possible to design the heads for maximum efficiency for playback and for recording.

One of the problems with magnetic recorders is that the magnetic materials may produce distortion of the recorded signals. This problem is illustrated in Fig. 13-5.

Figure 13-5*a* shows an ideal response for magnetizing force versus the amount of permanent magnetism established by that force. The magnetizing force comes from the magnetic field of the coil on the tape head. Note that as the audio signal varies the magnetizing force, the amount of magnetism established on the tape follows the audio signal exactly. Unfortunately this is an ideal condition and can never be reached.

The actual magnetization curve is shown in Fig. 13-5*b*. Note that the curvature in this magnetization curve causes the audio signal to become distorted, especially at the crossover point. The crossover point is the point

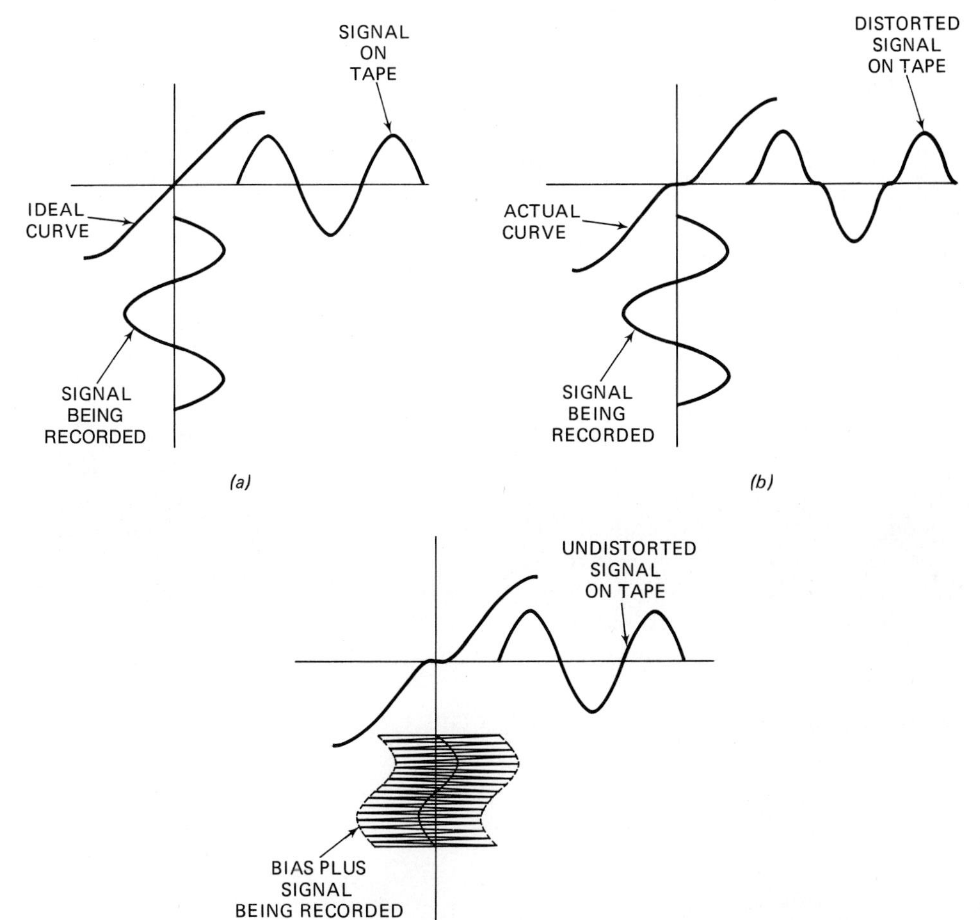

Fig. 13-5 How bias eliminates distortion. (*a*) Linear magnetizing curve. (*b*) Distortion due to nonlinear curve. (*c*) Use of bias to eliminate distortion.

where the magnetizing force is nearly zero. The characteristic curve, as shown in Fig. 13-5*b*, causes the recorded signal to be distorted.

To get around the problem of distortion a high-frequency, ac bias signal is applied constantly to the tape head when a recording is made. The audio signal being recorded causes the bias signal to increase and decrease, as shown in Fig. 13-5*c*. This illustration shows the maximum amplitude of audio that can be recorded. You will note that the recorded signal never is permitted to go into the nonlinear region of the curve.

At first you might think that the bias signal would cause a problem of interference because it is also recorded on the tape. However, the frequency of the bias signal is well above the frequency that the human ear can hear. Therefore when the tape is replayed you cannot tell there is a bias signal present.

To summarize, the bias signal is an important part of all tape recorder systems. It is necessary because the nonlinear magnetizing curve causes distortion.

What Are the Parts of a Tape Head Assembly?

Figure 13-6 shows a tape-head assembly. This could be an assembly for either a recording or playback head. You will note that the assembly is designed in such a way that the tape presses against the surface as it passes by. A pressure pad (not shown) holds the tape tight against the slot in the tape-head assembly. This pressure pad can be seen in the board assembly shown in Fig. 13-7.

The recording surface on the tape is coated with a magnetic material that can be partially rubbed off over a long period of time. This leaves particles of magnetic material on the tape head. When these particles begin to gather on the gap, the tape head can no longer do its job. Therefore part of the maintenance procedure for tape recorders is to clean the head. There are special solutions made for this purpose.

The speed of the tape as it passes through the tape assembly should be held constant in order to make it possible to play the tape on any recorder. If you allow the take-up reel to pull the tape through the head assembly, then the speed of the tape will change as the tape builds up on the reel. You will remember from basic science that the linear velocity (that is, the speed of the tape) is equal to the angular velocity [the revolutions per minute (r/min) of the tape reel] times the radius (of the tape on the take-up reel). For a small radius the speed of the tape is slow, and as the radius gets larger the speed of the tape increases.

To get around this problem the better tape recorders do not permit the tape

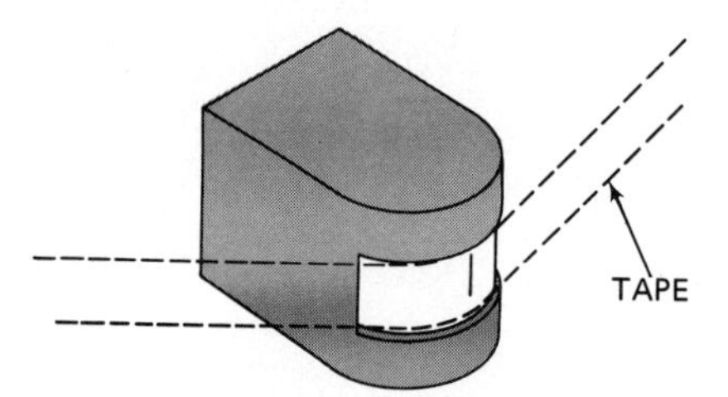

Fig. 13-6 The tape-head assembly.

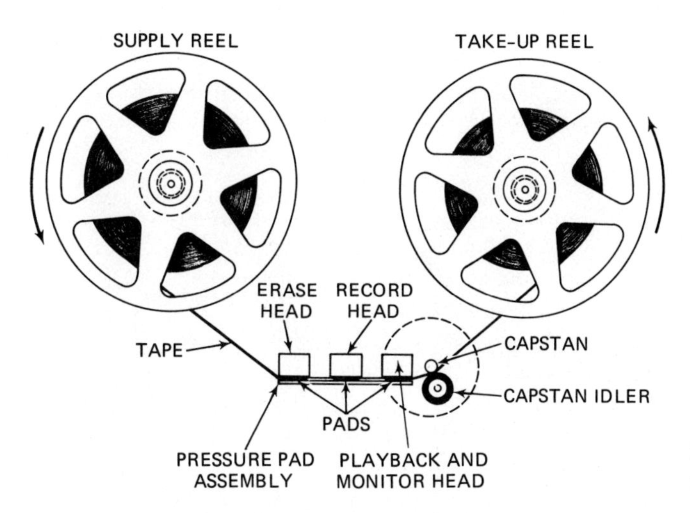

Fig. 13-7 Board assembly for tape recorder.

reel to pull the tape ahead. Instead, a *capstan,* which is nothing more than a small metal cylinder, pulls the tape. The tape is pinched between the turning capstan and the idler. The capstan-idler assembly is shown in Fig. 13-7. With this kind of arrangement the tape speed is not dependent in any way upon the amount of tape on the take-up reel. The take-up reel simply turns through a clutching mechanism to wind up the tape after it has passed through the capstan.

Inexpensive tape recorders sometimes do not have this capstan drive. The tape recorded on such a recorder cannot be played through a constant-speed capstan-drive tape recorder.

Not only must the tape be passed across the head at a constant speed, but the alignment of the head must also be correct in order for the system to work without distortion.

Figure 13-8 shows the problems of head-gap alignment. The gap must be

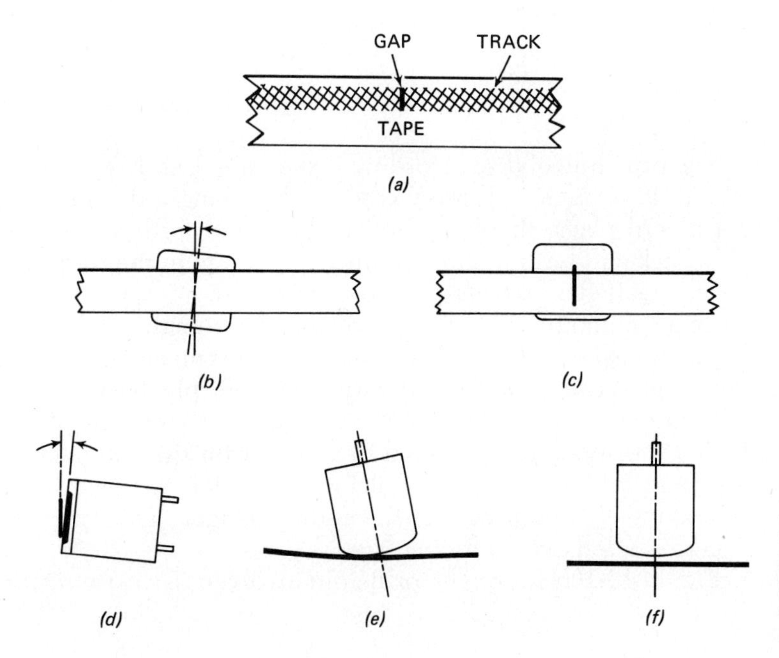

Fig. 13-8 Alignment problems with tape heads. (*a*) Correct gap alignment. (*b*) Incorrect azimuth. (*c*) Incorrect height. (*d*) Tilt. (*e*) Tangency. (*f*) Contact.

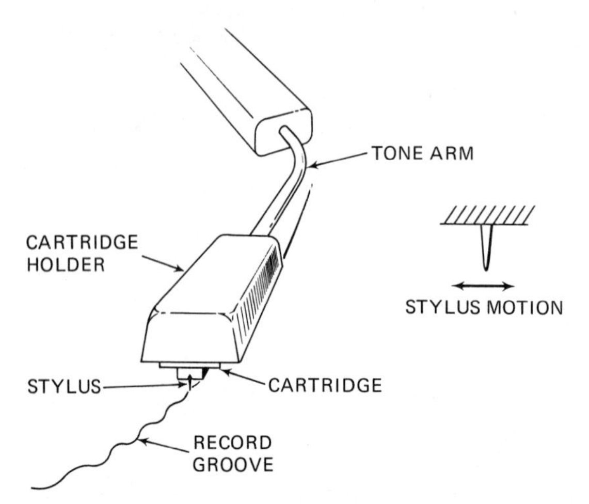

Fig. 13-9 Principle of disc recording.

perfectly vertical and in correct alignment, as shown in Fig. 13-8*a*. The gap, of course, is a fixed part of the tape-head assembly.

There are several ways that the gap can get out of adjustment. The one shown in Fig. 13-8*b* is called an incorrect *azimuth* adjustment. As you can see, the gap is at an angle with the track. If you assume that the gap in the recording head is perfectly vertical, then a playback head with an incorrect azimuth adjustment will not be able to reproduce audio signals properly. In this case the head is not at right angles to the tape.

Figure 13-8*c* shows another way that the gap can be out of alignment. Here the gap is not at the center on the recorded track, so the amount of magnetic field induced in the gap is not going to be large enough to produce an undistorted signal. Alignment errors in tilt, tangency, and contact are shown in Fig. 13-8*d*, *e*, and *f*.

Most tape recorder systems and tape playback systems have their own amplifier, but in the audio system of Fig. 13-1 the output of the tape playback is passed through the hi-fi system.

How Does a Record Player Work?

The principle of disc recording is shown in Fig. 13-9. On the surface of the record there are small, wavy grooves that contain the audio information. As the record plays, the stylus moves back and forth as it follows the groove. This back-and-forth motion produces a voltage in the transducer. This transducer is called a *cartridge*. The cartridge, then, is a transducer that converts mechanical motion into an electrical audio signal.

Any transducer that will accurately convert mechanical motion into an electrical signal can be used in the disc-recorder playback system.

Figure 13-10 shows the two most popular methods of doing it. In Fig. 13-10*a* the stylus is connected to a holder for the coil. As the stylus moves back and forth in the groove, it forces the coil to move with relation to the magnetic field and produces the audio voltage. This type of cartridge transducer is called a *dynamic pickup*.

Figure 13-10*b* shows the principle involved in a *ceramic pickup*. The pie-

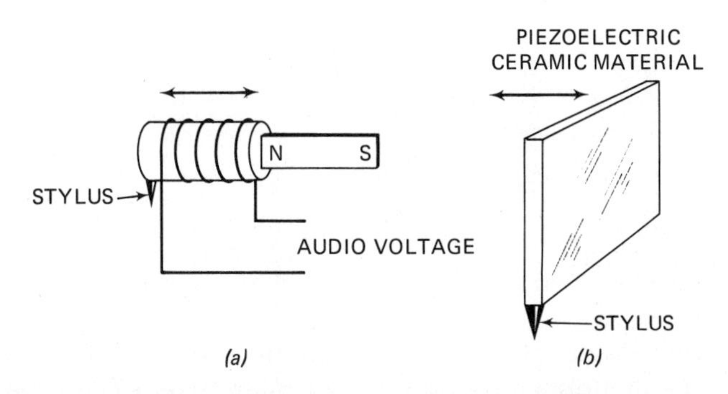

Fig. 13-10 Transducers for disc players. (*a*) Electromagnetic. (*b*) Ceramic.

zoelectric effect is used here. The ceramic material generates a voltage as the moving stylus bends it back and forth.

In order to get the maximum *fidelity* (faithfulness of reproduction of recorded sound), the stylus must sit properly in the groove. A worn stylus tip will destroy the record, so an important part of the audio technician's job is to look at the stylus through a microscope or very strong magnifying glass.

Summary

1. The earliest magnetic recorders used magnet wire instead of tape. They had a serious problem in that their frequency response was poor.
2. The tape used in modern systems has a coating of magnetic material on a plastic tape.
3. A magnetic recording is made by passing the tape across a gap. A varying field at the gap is produced by an audio current in the coil around the head.
4. During playback the magnetic fields on the tape induce a flux into the head. This, in turn, induces an audio voltage in the coil around the head.
5. The magnetization curve of the head material is not linear. This would cause distortion of the recorded audio if an ac bias signal were not used.
6. The ac bias signal causes the recorded audio to be in the linear part of the magnetization curve. The bias frequency is so high that it does not produce audible signals during playback.
7. A capstan drive is usually used to pull the tape across the head at a constant speed. [Typical speeds are $1\frac{7}{8}$, $3\frac{3}{4}$, and $7\frac{1}{2}$ inches per second (in/s).]
8. The transducer for disc playback is a cartridge. It converts the mechanical motion of a stylus into audio electrical signals.
9. The stylus moves back and forth as it follows the record groove. This motion is related to the audio that has been recorded on the disc.
10. Two popular types of cartridges are the dynamic cartridge and the ceramic crystal.

How Does a Loudspeaker Work?

The transducers discussed so far have been used to deliver signals to the input of the audio system. There is, of course, also a transducer that converts

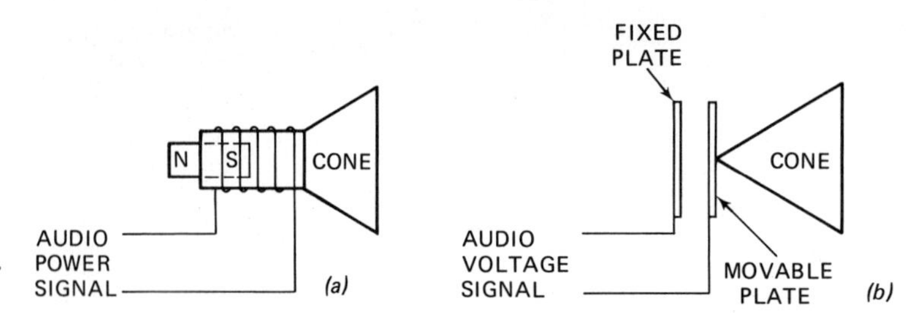

Fig. 13-11 **Two kinds of speakers. (*a*) Dynamic. (*b*) Electrostatic.**

the electrical signals into sound energy. The transducer may be a loudspeaker or headphones. Two main types of speakers are shown in Fig. 13-11.

In the *dynamic speaker* of Fig. 13-11*a* the audio signal causes audio currents to flow in the coil. This coil is connected to the cone. The audio current causes a varying magnetic field around the coil. This field reacts with the field of the permanent magnet and causes the coil (and the cone) to move back and forth. The cone causes the air molecules to vibrate and produces the sound that you hear. Almost all speakers in use operate on this principle.

An *electrostatic speaker* is shown in Fig. 13-11*b*. It works on the principle that unlike charges attract and like charges repel. An audio voltage is applied between the fixed plate and the movable plate. You can see that the two plates and dielectric make a capacitor. As this capacitor charges and discharges, the attraction between the plates varies in step with the audio signal. Since the cone is attached to the movable plate, it moves back and forth with the audio signal. This causes the air molecules to vibrate and produces the sound that you hear.

How Is the Quality of Speaker Sound Controlled?

In general, no single speaker produces the total range of audio sounds that the human ear can hear. Speakers with large-diameter cones (10 inches or more) can reproduce the lower-frequency tones better, and speakers with small-diameter cones (2 to 4 inches) can reproduce the higher frequencies better. A speaker that is especially designed for reproducing low-frequency sounds is called a *woofer*. Speakers especially designed for reproducing high-frequency sounds are called *tweeters*. In some audio systems there is a third speaker designed for reproducing sound in the center range, and these speakers are sometimes called *squawkers*.

Figure 13-12 shows a *coaxial speaker*. This is simply a woofer and tweeter mounted in the same frame. This speaker can produce a wide range of audio frequencies, but it is necessary for the audio delivered to the speaker to be

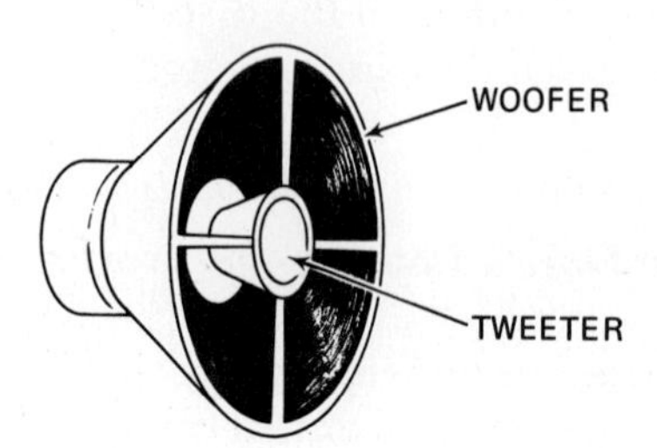

Fig. 13-12 **Coaxial speaker.**

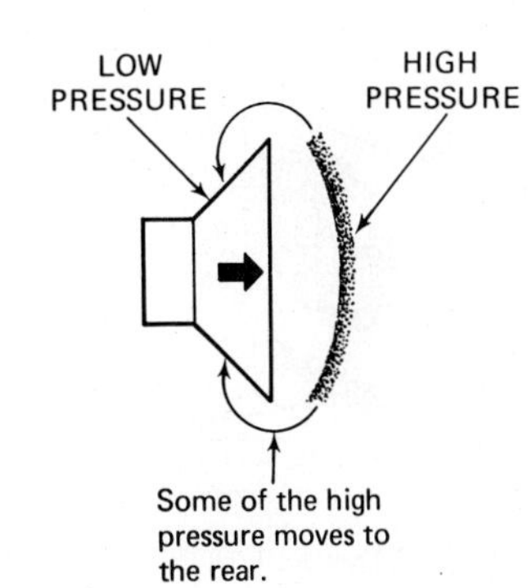

Fig. 13-13 Cause of low volume from unmounted speaker.

separated into two distinct components. The circuit used for separating them is called a *crossover network.* This is simply a combination of filters for passing low and middle frequencies to the woofer and high frequencies to the tweeter.

Instead of using a coaxial speaker like the one shown in Fig. 13-12, some designers prefer to mount several different sizes of speakers on the same board. When you do this, you still have the same problem of delivering the lower and middle frequencies to the woofers and the high frequencies to the tweeters.

The speaker by itself does not produce an acceptable sound for high fidelity. You can compare the speaker to the voice of a human. It is not only the sound from the voice box that gives the human voice its individual quality, but it is also the shape of the mouth cavity which helps to form the vowels and consonant sounds. Likewise, the characteristic sound of the speaker is dependent not only upon the speaker quality but also upon the way the speaker is mounted.

Unequal air pressures always try to equalize, and this tends to reduce the amount of output sound from the speaker. Figure 13-13 shows how this occurs. When the speaker is moving in the forward direction, it produces high air pressure in front of the speaker. Instead of all the air pressure simply moving out ahead of the speaker, some of it moves around behind to partially equalize the low pressure created by the cone moving outward. The result is reduced sound, particularly of the lower frequencies.

The simple speaker mounting shown in Fig. 13-14 is called a *baffle.* The baffle makes a longer distance for the air to travel between the front and back of the speaker. Therefore it greatly reduces the loss of sound pressure that is shown in Fig. 13-13. When you mount a speaker on a baffle you can usually hear the increase in loudness.

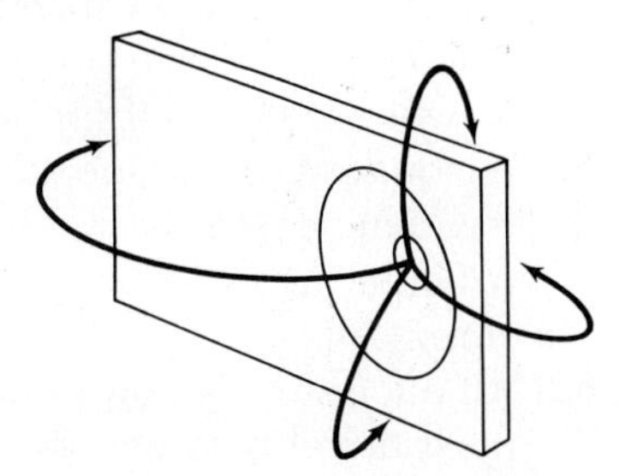

Fig. 13-14 Baffle reduces problem of pressure loss.

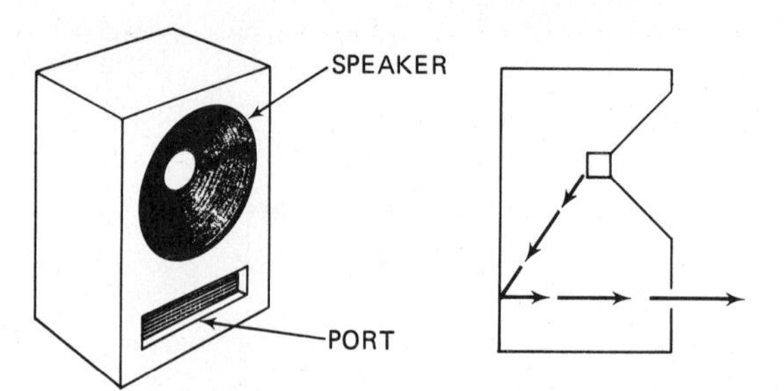

Fig. 13-15 Bass reflex enclosure.

If a speaker is mounted in the center of the baffle, it may tend to have maximum volume for one particular frequency. By mounting it off center, as shown in Fig. 13-14, the speaker response is flatter over a wide range of frequencies.

The baffle does a good job, but the air pressure can still get around behind the speaker to some extent. An even better way to improve sound is to use a speaker *enclosure.* An example is shown in Fig. 13-15. Since the speaker is mounted in a box, it is more difficult for the pressures to equalize.

The box shown in Fig. 13-15 is called a *bass reflex enclosure.* It has a port for improving low-frequency response. The low-frequency sounds bounce off the inner back of the enclosure, pass through the port, and reinforce the low-frequency sound. The path is shown with arrows in the side view. Only the low frequencies are reinforced. The high frequencies do not have the correct wavelength to reinforce the signal.

The bass reflex is not the only kind of enclosure that is used. In fact, many books have been written on enclosures, and audio experts make extensive claims about the advantages of one enclosure over another. However, you must remember this about enclosures: There is a certain amount of personal opinion involved in making the decision of which speaker sound system is best. Actually there is no one single enclosure-speaker combination that will satisfy everyone.

Summary

1. Two types of speakers are the dynamic and the electrostatic.
2. In a dynamic speaker the magnetic field produced by audio current in the voice coil reacts with a permanent magnetic field. This causes the speaker cone to move back and forth and produce sound.
3. In an electrostatic speaker the plates of a capacitor are charged and discharged by the audio-signal voltage. One of the plates moves the cone as the charge on the capacitor changes with the audio signal.
4. A woofer is a large speaker used mainly for low-frequency audio sounds.
5. A tweeter is a small speaker used mainly for high-frequency audio sounds.
6. A squawker is a speaker used mainly for sound in the middle audio-frequency range.
7. A baffle increases speaker volume by reducing the equalization of pressure.
8. An enclosure is more effective than a baffle because the distance the air must travel between the front and back of the speaker is greater.
9. Some enclosures, such as the bass reflex, change the quality of sound.

What Does the Tuner Do in an Audio System?

We have talked about the input and output transducers for the audio system. There is also an input to an audio voltage amplifier that comes from an AM-FM tuner. The tuner is actually a radio that has all sections except the audio. This radio can receive either AM or FM broadcasts, as selected.

Most hi-fi listeners prefer FM over AM. There are three major reasons for this. First, the FM system makes it possible to transmit and receive a greater range of audio frequencies. Second, the FM system greatly reduces all noise in the audio output. And third, many FM stations transmit their programs in "stereophonic" (two dimensions of sound) or "quadraphonic" (four dimensions of sound). The AM standard broadcast system does not provide these three advantages.

What Is Special about Audio Voltage and Power Amplifiers?

You have already studied the basic principles of voltage amplifiers and power amplifiers. In any amplifier there is a tradeoff between the range of frequencies that it can amplify—called the *bandwidth*—and the ratio of output-signal voltage to input-signal voltage—called the *voltage gain*. This means that, generally, anything you do to reduce gain will increase bandwidth.

In high-fidelity audio systems the gain per stage is usually moderate. Negative feedback is used to improve the bandwidth, and it also incidentally reduces the gain. With negative feedback a small portion of the output signal is mixed with the input in such a way that it reduces the amount of input signal.

Figure 13-16 shows three methods of getting negative feedback in audio amplifiers. In Fig. 13-16*a*, resistors at the dc input electrode (emitter, cathode, or source) are used for *current feedback*. Signal currents flowing through these resistors produce voltage drops that reduce gain by reducing the effect of the input signal.

In Fig. 13-16*b* the bias resistor R_1 is connected to the collector instead of to the positive power-supply terminal. Audio-voltage variations occur across the load resistor R_3. The audio voltage at the collector is opposite in phase to the input signal. A small part of this output is returned to the transistor base through the bias voltage divider, consisting of R_1 and R_2, and so it subtracts from the input signal.

Figure 13-16*c* shows a third feedback circuit. With this method a feedback voltage is taken from the secondary of a coupling transformer. This signal is returned to the base of the transistor through C_1. As with the other circuits, the feedback signal partly subtracts from the input signal to reduce the gain and increase the bandwidth of the stage.

Negative feedback in an amplifier has two other important advantages. It reduces audio distortion and also improves the stability of the amplifier. That is, it reduces the tendency of the amplifier to oscillate.

The feedback circuit of Fig. 13-16*b* can be used only with bipolar transistors or enhancement MOSFETs. The circuits of Fig. 13-16*a* and *c* can be used with all amplifying components.

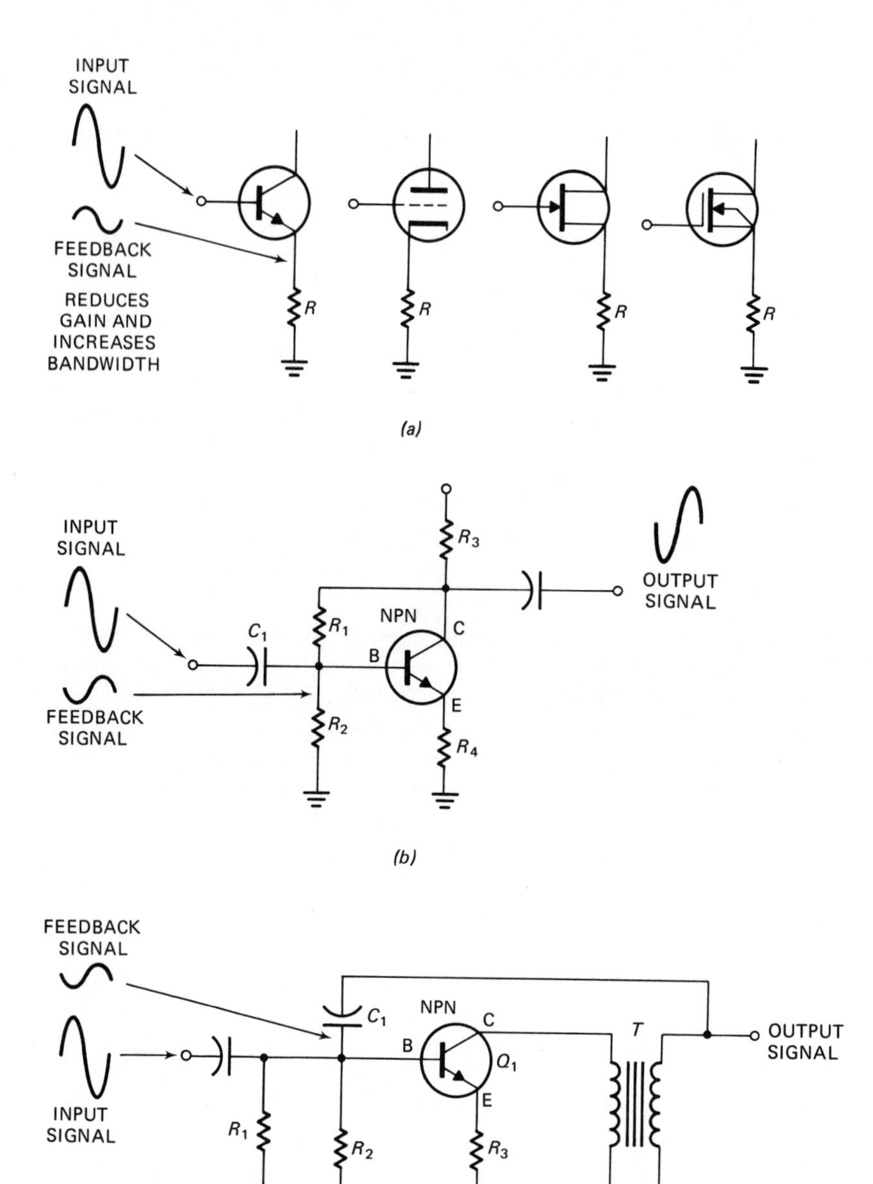

Fig. 13-16 Feedback circuits. (*a*) **Dc input circuit.** (*b*) **Collector-base.** (*c*) **Transformer output.**

Figure 13-17 shows two important types of power amplifiers used in audio.

The *push-pull amplifier* of Fig. 13-17*a* uses two NPN power transistors. The circuit will also work with PNP transistors, vacuum tubes, and FETs. In this circuit the audio input signal goes to the primary of T_1. The secondary of T_1 is center-tapped so that the two output signals from the secondary (*a* and *b*) are 180° out of phase. In other words, *a* is positive when *b* is negative, and *a* is negative when *b* is positive.

Since Q_1 and Q_2 are NPN transistors, they will conduct when the voltage on

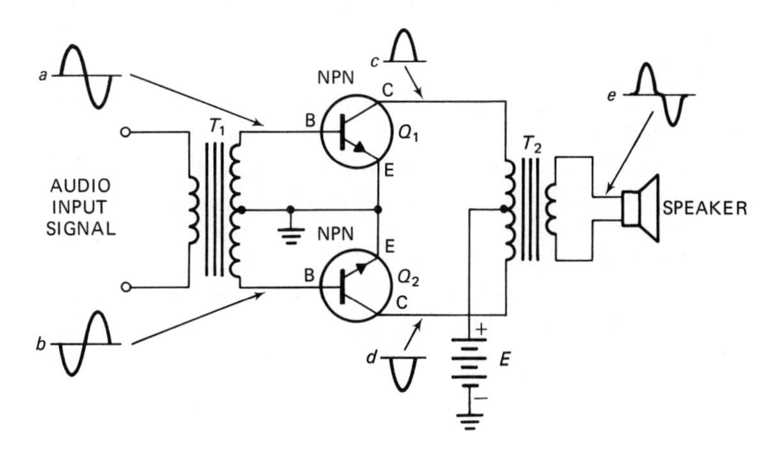

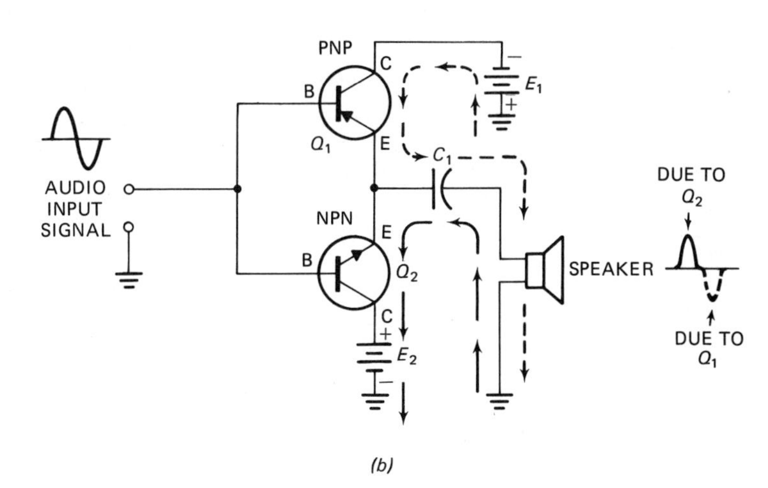

Fig. 13-17 Two important power-amplifier circuits. (*a*) Push-pull. (*b*) Complementary symmetry.

their base is positive. On one half cycle Q_1 conducts because a is positive. At the same time, b is negative and Q_2 is cut off. On the next half cycle a is cut off with a negative half cycle on its base, but Q_2 conducts because its base is positive.

The currents c and d flowing in the primary of output transformer T_2 flow in opposite directions. They combine to make the full cycle e.

You will note that the output waveform e is distorted near the zero amplitude points. This is called *crossover distortion*. It is caused by the fact that silicon bipolar transistors do not conduct when the input voltage to their base is less than about 0.7 volt. (Germanium transistors will not conduct when the input to their base is less than about 0.2 volt.)

Crossover distortion can be avoided by applying a small amount of forward bias to the bases of the transistors.

Figure 13-17*b* shows a *complementary symmetry* power amplifier. It uses a PNP and an NPN transistor. An obvious advantage of this circuit is that it does not require a device to split the phase of the input signal. (Transformer T_1 does this job in the push-pull circuit.) The same signal is applied to both bases.

On the positive half cycle of input signal, PNP transistor Q_1 is cut off and Q_2 conducts. The conduction path is shown with solid arrows. The positive half cycle of signal passes through C_1 and activates the speaker voice coil.

On the negative half cycle of input signal, NPN transistor Q_2 is cut off and Q_1 conducts. This conduction path is shown with dotted arrows. Now the negative half cycle of the audio signal passes through C_1 and activates the speaker voice coil.

The result is that an alternating audio-frequency current flows through the speaker voice coil. The output-current waveform is shown as a solid line for the half cycle that Q_2 conducts and as a dotted line for the half cycle that Q_1 conducts. As with push-pull amplifiers, crossover distortion may occur when bipolar transistors are permitted to operate with no forward bias, as in this case.

Why Was Quadraphonic Sound Developed?

Figure 13-18 shows how audio systems have changed. In the earliest system all the sound came from one location. This is called *monaural sound*, and it is illustrated in Fig. 13-18*a*.

Audio experts complained about monaural sound. They felt that the sound coming from one single location was unnatural. They said it was like listening to a symphony orchestra in a tunnel.

In real life, sounds do not all come from one single direction. In an auditorium, for example, you hear sounds that are reflected from the walls as well as the original sounds. The reflected sounds are an important part of the total sound that you hear.

The next step in the audio system was to develop *stereophonic sound*.

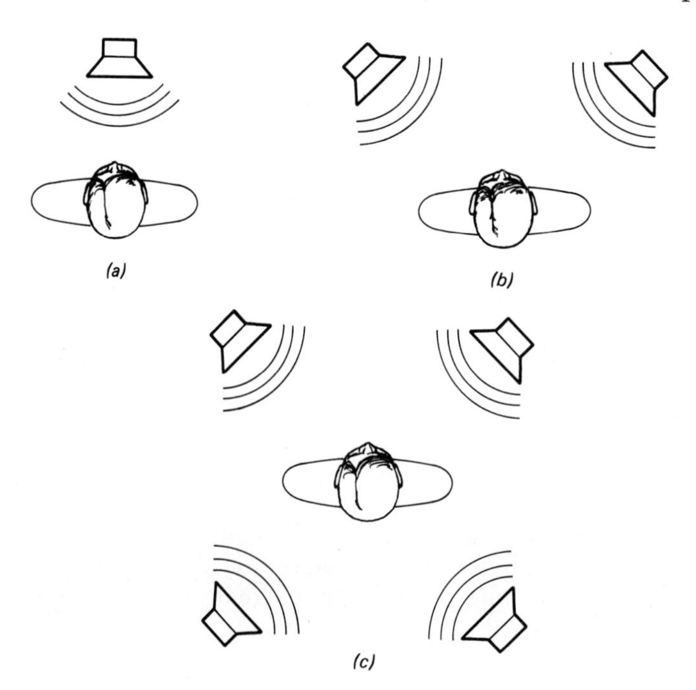

Fig. 13-18 Position of a listener for different types of sound systems. (*a*) Monaural. (*b*) Stereo. (*c*) Quad.

This is shown in Fig. 13-18*b*. The source of sound is divided into two parts, left and right. According to the audio experts this was much more realistic, but it was not without problems also. One problem is called the *Ping-Pong effect*. When you hear a Ping-Pong game the sound comes from the left or the right, but nothing happens in the middle. The Ping-Pong effect, or *hole-in-the-middle* effect, can also occur in stereo. This is especially true when the speakers are too far apart. It can be partly eliminated by putting a third speaker between the two, and this is often done. Another problem of stereo is that it does not produce *ambience* or *surround sound*. This is the sound that is bounced from the walls. It produces a three-dimensional sound which the audio expert likes to call ambience.

To reproduce ambience, quadraphonic sound was developed. As shown in Fig. 13-18*c*, it has basically four speakers. Two speakers reproduce the sounds coming from the front, and the other two reproduce the sounds that bounce off the walls in the symphony hall.

What Are the Kinds of Quadraphonic Sound?

There are three ways to obtain quad sound. They are illustrated in Fig. 13-19.

The *imitation system* of Fig. 13-19*a* is also called the *simulated system*. It works with any stereo recording. The two tracks of information into the stereo playback can come from tape, disc, or stereo-multiplex FM. The left front (LF) and right front (RF) speakers reproduce the stereo sound. There is a delay between the front speakers and rear speakers (left rear and right rear). The delay causes the sound from the rear speakers to sound like echoes from the walls of an auditorium. This delay time must be very short because sound travels at a speed of about 1100 feet per second. Therefore sound that is reflected from the walls of an auditorium does not arrive much later than sound from the stage.

Imitation quad sound is the least expensive of the methods used. It has the added advantage that it can be used with existing stereo equipment. However, some audio experts claim that it is not as realistic as other methods.

The *matrix method* of Fig. 13-19*b* is also known as *encoded quadraphonics*. It takes the electrical signals from four microphones and combines them electrically into two signals. These two signals are recorded as two tracks. The playback equipment takes the signals from the two tracks and converts them into four signals for the speakers. The recording matrix is also called the *encoder*, and the playback matrix is also called the *decoder*.

The electrical circuitry for the matrixing can be rather complicated if great realism is desired. Unfortunately, several different companies designed their own matrix circuits, and one recording system would not play back on another. This led to confusion and lack of progress in quadraphonics.

The third system is called *discrete*, and it is illustrated in Fig. 13-19*c*. The outputs of four microphones are recorded on four separate tracks. At the playback equipment each track is treated as a separate recording to a four-speaker system.

There is only one type of recording system that can produce discrete qua-

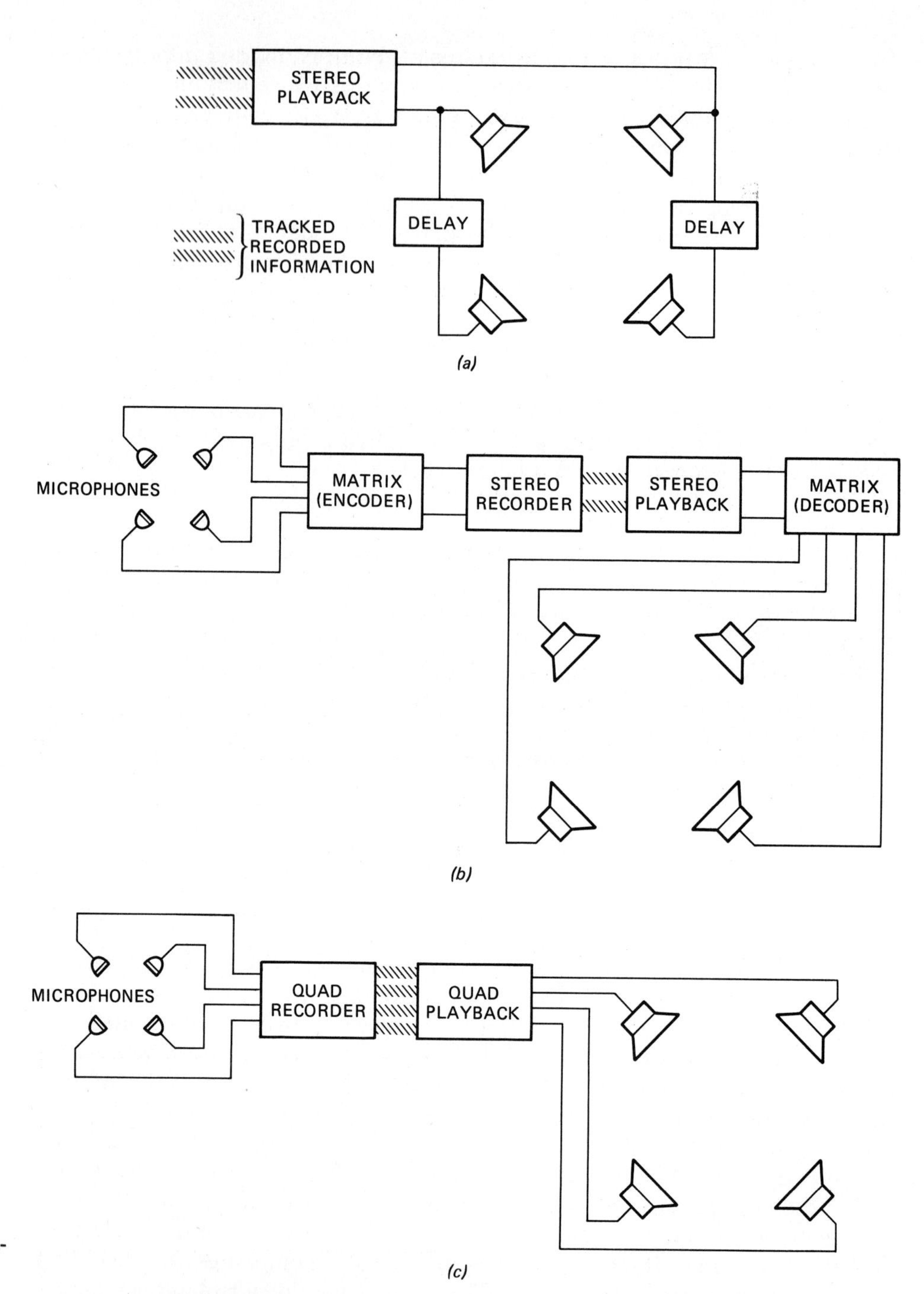

Fig. 13-19 Quad systems. (*a*) Imitation. (*b*) Matrix. (*c*) Discrete.

draphonics, and that is the tape recording system. The so-called discrete disc (record) recording system does not record four individual audio channels, as shown in Fig. 13-19*c*. Instead, it converts two of the microphone inputs to higher-frequency signals. At the playback system the higher-frequency signals are converted back to audio.

What Are the Variations in Tape and Disc Recording?

There are a few basic ideas related to tape recording that you should understand.

What Are Some Basic Principles of Tape Recording?

The faster the tape moves, the higher the audio frequencies that can be reproduced. In other words, fast tapes give higher fidelity than slow tapes. However, when faster tape speeds are used, the amount of playing time is reduced.

The width of the gap in the head can affect the fidelity. As a general rule the narrower the gap, the higher the frequency that can be recorded and replayed. This is one reason why the quality of playback is reduced by a worn head. As the head wears, the gap size increases and the high-frequency response is somewhat lost.

The basic principles of tape recording are the same for reel-to-reel tape recorders, cassettes, and cartridges. The most important difference between these systems is the method of storing the tape.

Figure 13-20 shows several popular recording formats for tape recording.

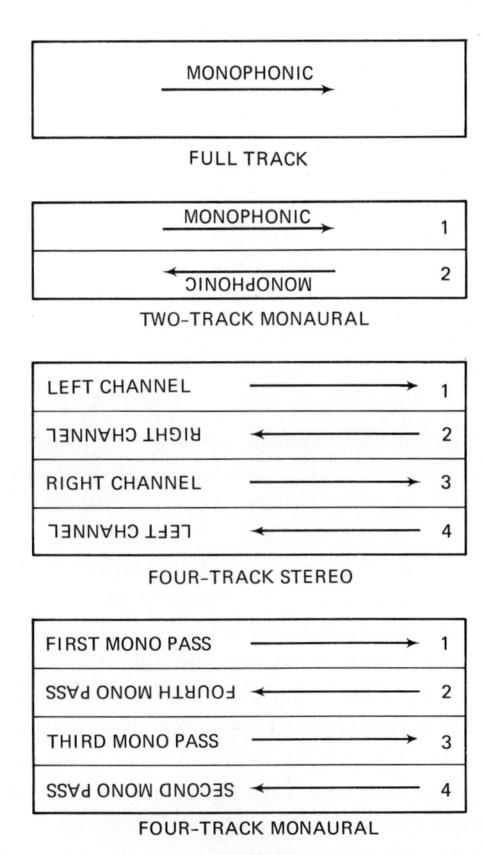

Fig. 13-20 Several popular recording formats.

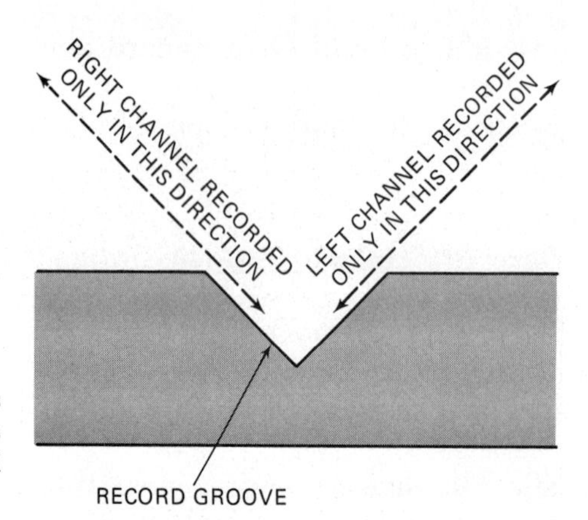

Fig. 13-21 How the stylus moves to produce stereo. The stylus moves in directions of dotted arrows to produce stereo.

What Are Some Basic Principles of Disc Recording?

For monaural disc recording the groove moves the stylus back and forth. Figure 13-9 shows the principle involved.

For stereophonic sound the information for the left speaker is recorded on one side of the groove and the information for the other speaker is recorded on the other side of the groove. Two speakers are used—one for each *channel of sound.* (For stereo there are two channels, left and right.) Figure 13-21 shows how the pickup stylus moves to reproduce the two channels.

For four-channel recording there are, in addition, very small bumps on each side of the groove. They represent the high-frequency signals that will be used for the two rear speakers. The information for the rear speakers is changed to the high frequencies for recording on the disc. For playback, the high-frequency signals are taken from the disc and converted back to audio frequencies for the rear speakers. Figure 13-22 shows a quadraphonic disc groove (one side only is shown for simplicity).

Summary

1. The tuner makes it possible to play AM and FM stations through the audio system.
2. The FM standard broadcast system has the advantages of a greater frequency range, greatly reduced noise in the audio, and ability to transmit stereophonic or quadraphonic sound.
3. Voltage amplifiers in audio systems usually have negative feedback to increase their bandwidth.
4. Push-pull amplifiers and complementary symmetry are two popular types of power amplifiers used in audio systems. When bipolar transistors are used in these circuits they must have a small amount of forward bias if crossover distortion is to be prevented.
5. Monaural sound is sound coming from one direction. Audio experts feel that monaural sound is not realistic. We are accustomed to sounds coming from many directions, including reflections from walls.
6. Stereophonic sound is more realistic than monaural sound. However, it

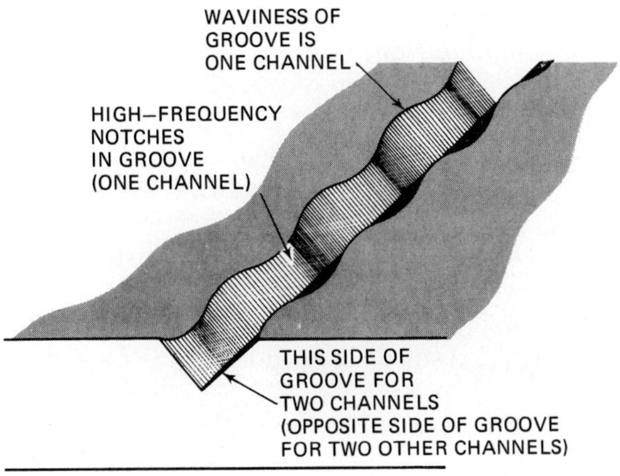

WAVINESS OF GROOVE IS ONE CHANNEL
HIGH–FREQUENCY NOTCHES IN GROOVE (ONE CHANNEL)
THIS SIDE OF GROOVE FOR TWO CHANNELS (OPPOSITE SIDE OF GROOVE FOR TWO OTHER CHANNELS)

Fig. 13-22 A quadraphonic disc groove.

can produce an unrealistic Ping-Pong effect if the speakers are not properly located.

7. Quadraphonic sound provides ambience. This means that listeners get the same type of sound as they hear in an auditorium. Another name for ambience is *surround* sound.
8. There are three kinds of quadraphonic sound: imitation, matrix, and discrete.
9. Imitation quad sound can be used with any existing stereo system. It simply delays sound signals to the rear speakers to imitate wall reflections.
10. Matrixed quad sound is recorded with signals from four microphones, but these are converted into two signals. During playback the two signals are converted back into four signals for four speakers.
11. Discrete quad sound records the four individual signals from four microphones. During playback the four individual signals are used for four speakers.

PROGRAMMED REVIEW QUESTIONS

(Instructions for using this programmed section are given in Chapter 1.)

We will now review the important concepts of this chapter. If you have understood the material, you will progress easily through this section. Do not skip this material because some additional theory is presented.

1. Which of the following operates on the principle of Faraday's law?
   A. A dynamic microphone. (Proceed to block 17.)
   B. A ceramic microphone. (Proceed to block 9.)

2. *Your answer to the question in block 18 is* ***B****. This answer is wrong. The bass reflex has a port, or opening, that permits the bass sounds from behind the speaker to reinforce the bass sounds coming from the front of the speaker. This increases the volume of the bass.* Proceed to block 12.

3. *The correct answer to the question in block 14 is* **B**. *PM, or phase modulation, is the same as FM, or frequency modulation. Any receiver that can receive PM can also receive FM. The output of these receivers does not respond to changes in signal amplitude. They are sensitive only to signals with changes of frequency. Since the signals that produce static are AM, the FM (and PM) receivers do not reproduce them.* Here is your next question.
For a given tape recorder a faster recording and playback tape speed will
A. give better fidelity. (Proceed to block 10.)
B. result in poor fidelity. (Proceed to block 16.)

4. *Your answer to the question in block 10 is* **A**. *This answer is wrong. If the reel turns at a constant speed, then the tape speed will gradually change. This is because the tape speed is directly related to the radius of the tape on the reel. In other words, as the tape builds up on the reel the tape speed changes. You cannot get a constant tape speed by using a constant reel speed.* Proceed to block 15.

5. *Your answer to the question in block 7 is* **A**. *This answer is wrong. Class A amplifiers never stop conducting. In other words, in class A amplifiers, with a vacuum tube there is always a plate current, with a transistor there is always a collector current, and with an FET there is always a drain current. When a signal is applied to a class A amplifier, the amplifier is operated in the linear range. This prevents the problem of crossover distortion.* Proceed to block 11.

6. *Your answer to the question in block 11 is* **A**. *This answer is wrong. Monaural sound is sound coming from a single direction. Ping-Pong effect occurs when sound comes from two directions. It is as if the listener were listening to a Ping-Pong game, where sound comes from one side of the table or the other, but not from the middle.* Proceed to block 18.

7. *The correct answer to the question in block 17 is* **A**. *Magnetic materials are nonlinear. This means that the amount of magnetism on the tape is not directly related to the amount of magnetizing force. In other words, if you double the magnetizing force, you will not necessarily obtain twice the amount of permanent magnetism on the tape.*
The most nonlinear portion of the magnetism curve is the point at which there is a very low magnetizing force, and also the point at which there is a very high magnetizing force.
The bias oscillator provides an ac signal, and the audio to be recorded is added to this ac. This causes the audio to produce a magnetizing force that is in the linear part of the magnetizing curve. Here is your next question.
Crossover distortion is more likely to occur in
A. class A voltage amplifiers. (Proceed to block 5.)
B. class B power amplifiers. (Proceed to block 11.)

8. *Your answer to the question in block 12 is* **A**. *This answer is wrong. The word* **ambience** *is used to mean surround sound.* Proceed to block 14.

9. *Your answer to the question in block 1 is* **B**. *This answer is wrong. A ceramic microphone operates on the principle of the piezoelectric effect. This refers to the fact that certain materials generate a voltage across their surface when they are bent, twisted, or otherwise deformed.* Proceed to block 17.

10. *The correct answer to the question in block 3 is* **A**. *By moving the tape rapidly across the gap, the higher-frequency audio signals can be more accurately recorded.*

If the tape is moved rapidly past the gap, the length of tape that records a high frequency is longer. Thus the magnetic signal on the tape is in effect "stretched out" and can be more faithfully reproduced as it passes over the gap. Here is your next question.

To get a constant tape speed across the tape head,

A. the reels must turn at a constant speed. (Proceed to block 4.)
B. a capstan drive is used. (Proceed to block 15.)

11. *The correct answer to the question in block 7 is* **B**. *Crossover distortion is a problem with bipolar transistor amplifiers not using any forward bias, but it can also occur with tubes and FETs. It occurs in push-pull class B and complementary class B circuits. The problem occurs when the signal is at a point that causes one amplifier to stop conduction and just before the second amplifier starts into conduction. When the signal voltage drops below 0.7 volt (for silicon transistors) or below about 0.2 volt (for germanium transistors), the collector current stops flowing for a short period. That is what produces the crossover distortion.* Here is your next question.

Ping-Pong effect is a possible problem in

A. monaural sound. (Proceed to block 6.)
B. stereo sound. (Proceed to block 18.)

12. *The correct answer to the question in block 18 is* **A**. *The bass-reflex enclosure permits sound reflected from behind the speaker to reinforce sound in front of the speaker at the lower, or bass, frequencies. This is an important feature of the bass-reflex enclosure.* Here is your next question.

Which of the following gives more ambience?

A. Stereo sound. (Proceed to block 8.)
B. Quad sound. (Proceed to block 14.)

13. *Your answer to the question in block 17 is* **B**. *This answer is wrong. If an amplifying component has the wrong dc bias, it can produce a large amount of distortion. In other words, the base bias on a transistor, the grid bias on a vacuum tube, or the gate bias on an FET can be directly responsible for distortion in an amplifier when the bias* *value is not correct. Remember, though, that these are dc biases on the amplifying component. A bias oscillator in a tape recorder produces ac bias.* Proceed to block 7.

14. *The correct answer to the question in block 12 is* **B**. *The purpose of recording quad sound is to give realism to the listener. It permits the listener to hear sounds coming from the front and sounds coming from behind. The sounds coming from behind imitate sounds that are reflected from the walls of an auditorium. The word* **ambience** *means surround sound, and this is best achieved with the quad system.* Here is your next question.
Which of the following gives "static-free reception"?
A. AM. (Proceed to block 19.)
B. PM. (Proceed to block 3.)

15. *The correct answer to the question in block 10 is* **B**. *A capstan is a small rotating cylinder. The tape is pinched between the capstan and an idler. Details of a capstan drive are shown in Fig. 13-23.* Here is your next question.
One way to increase the bandwidth of an amplifier is to ________________ (increase or decrease) its gain. (Proceed to block 20.)

16. *Your answer to the question in block 3 is* **B**. *This answer is wrong. If you move the tape very slowly across the gap, the amount of highest frequency signal that can be recorded is limited.* Proceed to block 10.

17. *The correct answer to the question in block 1 is* **A**. *A dynamic microphone consists of a coil in a magnetic field. Sound causes the coil to move in the magnetic field, and the motion is directly related to the sound. The voltage generated in the coil is the audio voltage.* Here is your next question.
A bias oscillator is used in tape recorders to eliminate distortion caused by
A. nonlinear magnetism curves. (Proceed to block 7.)
B. poor amplifier response. (Proceed to block 13.)

18. *The correct answer to the question in block 11 is* **B**. *The Ping-Pong effect occurs in stereo when the speakers are located too far apart.*

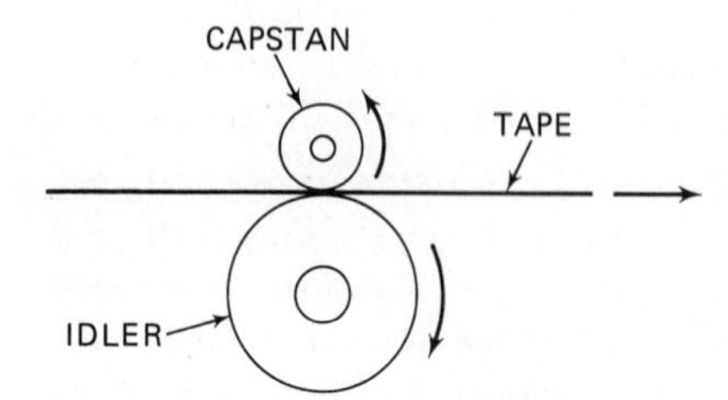

Fig. 13-23 Capstan drive.

This causes the sound to come from one point at the left and one point at the right, with no sound coming from the middle. It gets its name from the fact that you would get the same effect from listening to a Ping-Pong game if you were standing at the net. Here is your next question.

A bass-reflex enclosure is used to

A. increase the volume of bass sounds. (Proceed to block 12.)

B. decrease the volume of bass sounds. (Proceed to block 2.)

19. *Your answer to the question in block 14 is* A. *This answer is wrong. AM stands for amplitude modulation. The signals that produce noise and static are usually amplitude-modulated. Any receiver that reproduces AM signals is subject to noise interference.* Proceed to block 3.

20. *The answer to the question in block 15 is "decrease." There is a tradeoff between amplifier gain and bandwidth. Almost anything you do to decrease the gain will automatically increase the bandwidth. Negative feedback—also called* **degenerative** *feedback—is used to reduce the gain and increase the bandwidth of audio amplifiers. As a result, the amplifier can amplify a wider range of audio frequencies.* You have now completed the programmed questions. The next step is to put some of these ideas to work in laboratory experiments. Proceed to the Experiment section of this chapter.

EXPERIMENT

(The experiment described in this section may be performed on the circuit board described in Appendix C or on a similar laboratory setup.)

Purpose To demonstrate the complementary power-amplifier circuit used in audio systems.

Theory In many audio power-amplifier systems more than one amplifying component is used to achieve an output. This is done for the same reason that there is more than one engine on many airplanes. Two engines give more power than one, and two amplifying devices permit a greater audio output power than one.

There are a number of different ways to connect the amplifying components to get a greater output power. A very popular way is the push-pull circuit in Figure 13-24. In order to simplify this illustration, the base bias is not shown. To get this circuit to operate it is necessary to modify the input signal. The modification required is *phase splitting.* This involves splitting it into two identical signals that are 180° out of phase. The two signals are used for the inputs of Q_1 and Q_2 in Fig. 13-24.

The signals are sine waves in this application, but any audio waveform could be used. During period *a* the base of Q_1 becomes positive, and this transistor conducts through one-half of the output transformer T_1. During this

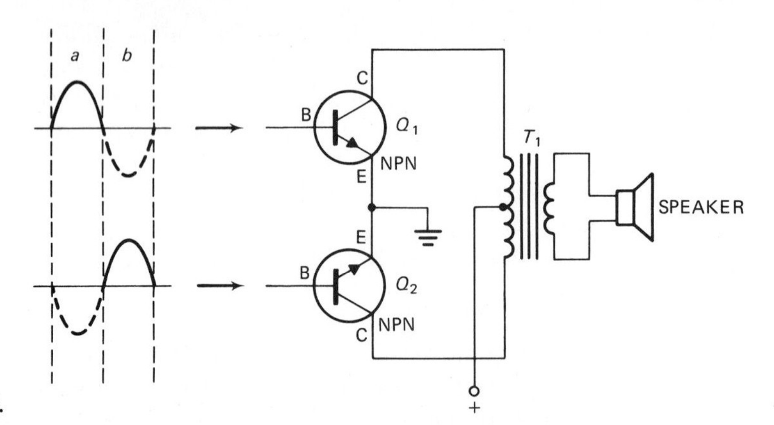

Fig. 13-24 Push-pull circuit.

same period of time the base of Q_2 is negative, and that transistor cannot conduct.

On the next half cycle (period b) the base of Q_1 is negative, and therefore it is cut off. At the same time the base of transistor Q_2 is positive, so it can conduct.

The overall result is that Q_1 conducts for one-half cycle, and Q_2 conducts for the other half cycle. This permits a greater output power with less distortion than can be obtained with a single transistor in a power-amplifier circuit. (When a single transistor is used in a power-amplifier circuit, it is called a *single-ended amplifier.*)

The push-pull, out-of-phase signals can be obtained in several different ways. Two methods are shown in Fig. 13-25. In Figure 13-25*a*, a transformer with a center-tapped secondary is used. This is similar to the power transformers used in full-wave rectifiers. When the center tap is grounded, the waveforms at the ends of the transformer are 180° out of phase, as required for operating the push-pull circuit.

Another method of getting the required out-of-phase signals is shown in Fig. 13-25*b*. As you know, the voltage at the collector of an amplifier is 180° out of phase with the voltage at the emitter. Here one signal output is taken from the collector and one is taken from the emitter, providing the required two 180° out-of-phase signals.

Although bipolar transistors are shown for the push-pull amplifier in Fig.

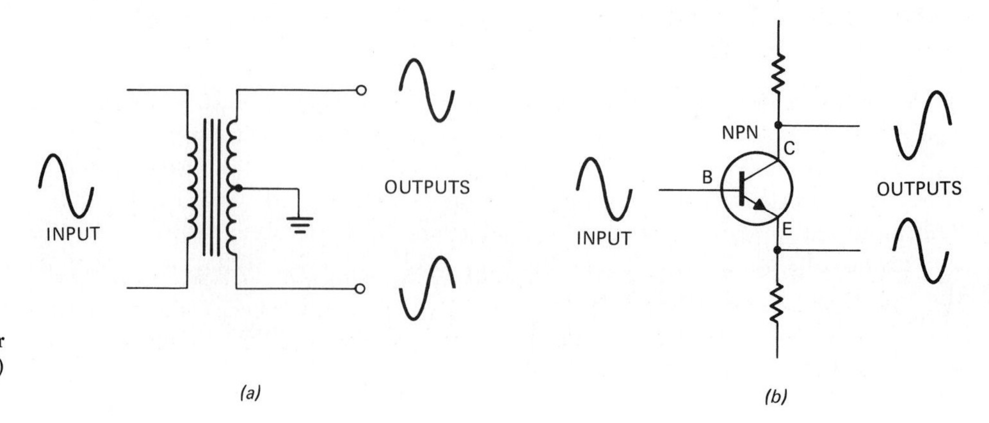

Fig. 13-25 Phase inversion for push-pull. (*a*) A transformer. (*b*) An amplifier.

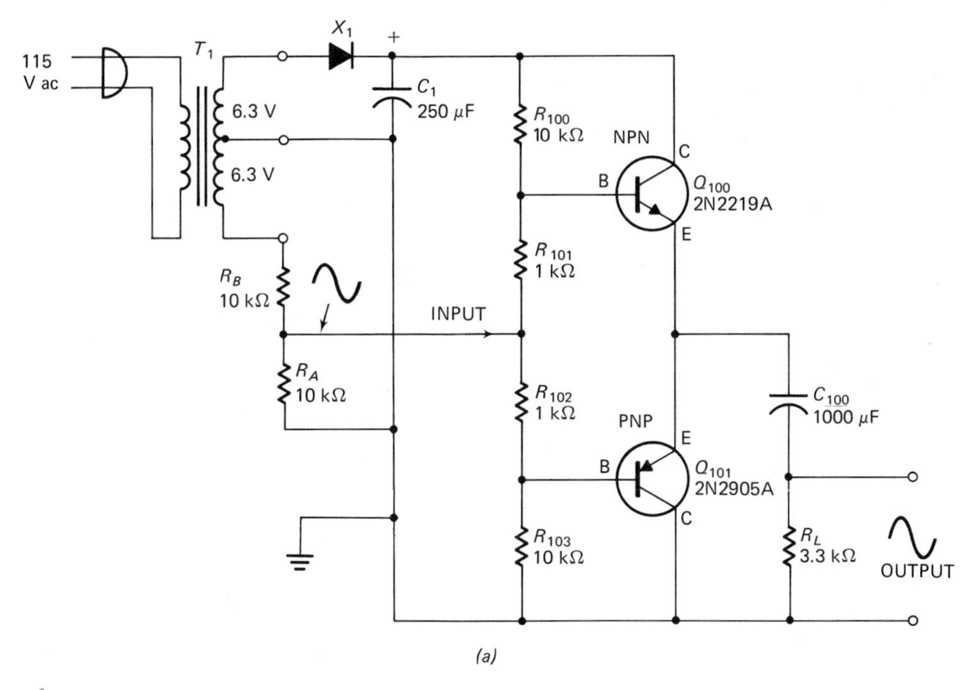

(a)

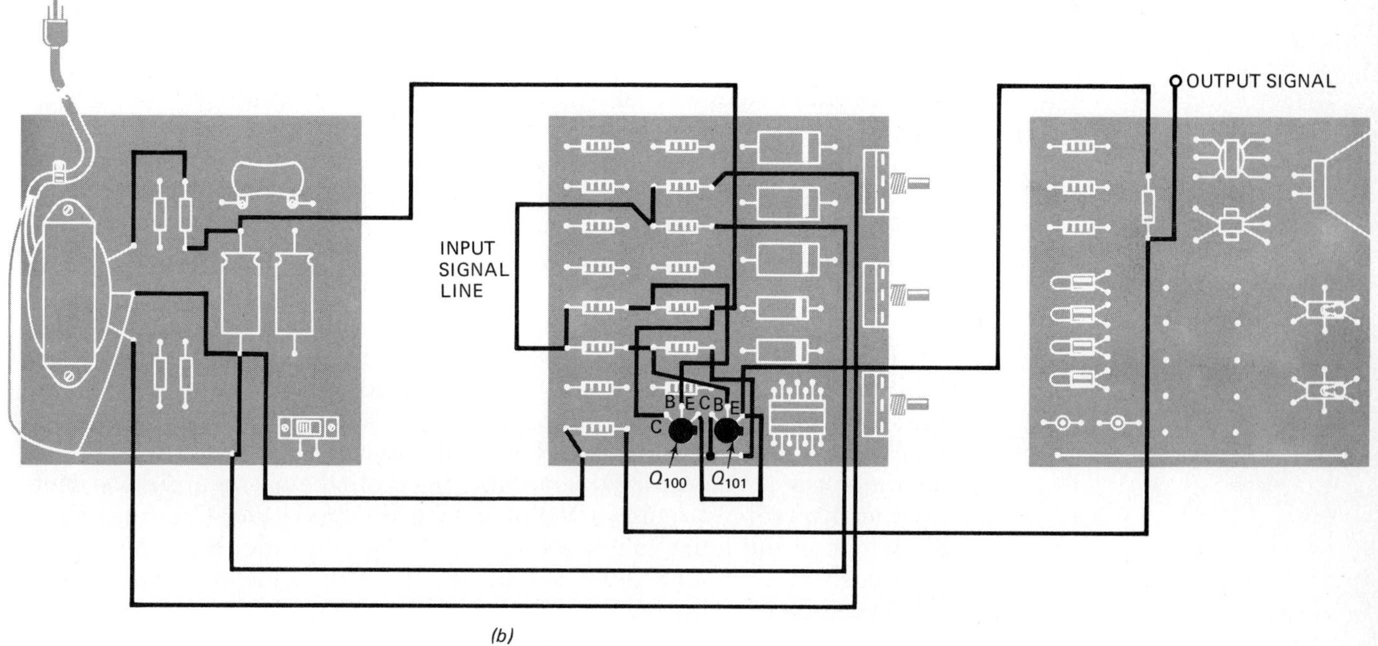

(b)

Fig. 13-29 Complementary circuit with signal at input terminal. (*a*) Schematic drawing. (*b*) Wiring diagram.

sine wave over one complete cycle is 0 volts. Using an ac meter, measure the voltage across R_L and record the value.

ac voltage across R_L _______________ volts

(You should measure several volts. This voltage should be greater than the input ac voltage to the amplifier.)

You have demonstrated that a complementary amplifier can produce an amplified ac voltage across the load and that a phase inverter is not required. You have also shown that there is no idling current in the complementary circuit.

Since this is a power amplifier, it would be meaningless to measure the voltage gain. Most power amplifiers have a low voltage gain. What you want in a power amplifier is a large output current, not voltage.

Conclusion Complementary amplifiers make use of two transistors in a push-pull circuit, for large power gain. They do not require additional circuitry for phase splitting as required by other push-pull amplifiers. They do not have an idling current.

SELF-TEST WITH ANSWERS

(Answers with discussions are given at the end of the chapter.)

1. The output transducer for an audio system is the (*a*) enclosure; (*b*) speaker; (*c*) power amplifier; (*d*) power supply.
2. Which one of the following types of microphones requires a direct current for its operation? (*a*) Condenser microphone; (*b*) Dynamic microphone; (*c*) Ribbon microphone; (*d*) Carbon microphone.
3. The response of the magnetic material on recording tape is not linear. This could cause distortion, but distortion is prevented by using (*a*) ac bias; (*b*) double-sided tape; (*c*) smaller recording currents; (*d*) faster tape speeds.
4. Which of the following is used to assure that the speed of the tape past the heads is constant? (*a*) Constant speed of take-up reel; (*b*) Constant speed of supply reel; (*c*) Variable tape thickness; (*d*) Capstan drive.
5. Which of the following is not a factor in tape-recorder head-gap alignment? (*a*) Azimuth adjustment; (*b*) Height adjustment; (*c*) Tape thickness; (*d*) Tilt.
6. The transducer for disc playback is the (*a*) stylus; (*b*) head; (*c*) cartridge; (*d*) tone arm.
7. Three types of speakers are used for three different frequency ranges. They are (*a*) tooters, barkers, and squawkers; (*b*) squawkers, hawkers, and rooters; (*c*) rooters, tooters, and hooters; (*d*) woofers, squawkers, and tweeters.
8. Adding a baffle to a speaker will increase its (*a*) low-frequency volume; (*b*) frequency range; (*c*) power consumption; (*d*) current requirement.

9. For a 60-hertz power supply, the ripple frequency of a full-wave rectifier is (*a*) always 60 hertz; (*b*) always 120 hertz; (*c*) equal to the line frequency; (*d*) equal to three times the line frequency.
10. Which of the following is a type of distortion in push-pull and complementary amplifiers? (*a*) Loudness; (*b*) Reverse; (*c*) Crossover; (*d*) Forward.

ANSWERS TO SELF-TEST

1. (*b*)—A speaker is the only transducer in the choices listed.
2. (*d*)—The carbon-element resistance (and current) varies with sound pressures.
3. (*a*)—The high-frequency bias causes recording to occur only within the linear response of the tape.
4. (*d*)—A capstan tape drive is shown in Fig. 13-23.
5. (*c*)—Tape thickness is not involved in head-gap alignment.
6. (*c*)—Choices (*a*) and (*d*) are other parts of the disc play system, and choice (*b*) is used in tape playback. Only the cartridge is the actual transducer.
7. (*d*)—Woofers for low frequencies, squawkers for middle frequencies, and tweeters for high frequencies.
8. (*a*)—The volume is increased because there is less low-frequency canceling of pressures in front and back of the speaker. Figure 13-14 shows the longer paths needed for canceling pressures when the baffle is present.
9. (*b*)—This question was not directly answered in the text. In the previous chapter the experiment demonstrated that half-wave rectifiers have a 60-hertz ripple and full-wave rectifiers have a 120-hertz ripple when the input ac power frequency is 60 hertz.
10. (*c*)—Of the choices listed, only crossover is a type of distortion.

how do monochrome TV receivers work? 14

INTRODUCTION

In this chapter you will study the *monochrome* (black-and-white) television system. This study will be in block-diagram form to acquaint you with how the TV picture is transmitted and received.

Television pictures are sent as halftones. This means that the *video* or picture information must be able to convey all tones of light and dark from white to gray to black in order to represent a complete picture. The video signal determines the amount of blackness or darkness of the scene at various points of the picture.

Two transmitters are used in the TV system. One is the *aural* transmitter, which converts the sound into a frequency-modulation signal. This is very much like the basic FM signal that you studied in the previous chapter. The other transmitter converts the video into an AM signal.

Transducers are very important in both the transmitter and receiver systems. They convert the sound and picture into electrical signals in the transmitter, and they convert electrical signals into sound and picture in the receiver. At the transmitter it is necessary to use a transducer to convert the picture into a video signal. A microphone is used to convert the sound into electrical signals. At the receiver the sound signals are converted back to sound in the loudspeaker. The video signals are converted back into a lighted scene by the picture tube.

To summarize, the transducers convert light and sound into electrical signals at the transmitter. They convert electrical signals into picture and sound at the receiver.

A good place to start your study of the television system is to learn how the basic transducers work. You already know how a microphone and a loudspeaker work from your previous studies, so we will concentrate on the television camera tube and the television picture tube.

You will be able to answer these questions after studying this chapter.

☐ **How are black-and-white pictures converted into electrical signals?**

☐ **How are electrical signals converted to lighted scenes?**

☐ **What is scanning and what is synchronizing?**

☐ **What is the makeup of the monochrome signal?**

☐ **What are the circuits in a monochrome transmitter?**

☐ **What are the circuits in a monochrome television receiver?**

☐ **What are the television channel frequencies?**

INSTRUCTION

How Are Black-and-White Pictures Converted into Electrical Signals?

There are certain materials that change their electric charge or generate a voltage when light falls on them. There are also some materials that emit electrons when exposed to light. Selenium is an example of a material that will emit electrons when exposed to light. Most semiconductor materials, like the ones used in making transistors, change their conduction when exposed to light.

The camera tube in a TV transmitter makes use of these principles in converting a black-and-white scene into electrical signals. Figure 14-1 shows two examples of how it is done. In Fig. 14-1*a* the object in the scene to be televised is focused through a lens system onto a photosensitive material. The image on the photosensitive material causes that material to become charged highly positive in some (white) areas and have very little charge in other (dark) areas. The electron gun sends a beam of electrons across the scene one line at a time. Each line is called a *trace*. On the line shown in Fig. 14-1*a* the beam starts at the left (viewed from front), moves through a white area, then into a dark area, and then into a light area again.

When the electron beam strikes the photosensitive board, secondary (or return) electrons are emitted. The number of secondary electrons depends upon how much positive charge the board has at each point. The amount of positive charge depends upon the light or dark area at each point in the scene. Thus the return electrons vary in number in accordance with the light and dark areas on the scene as it is traced by the electron beam.

The secondary electrons return to the positive power supply through resistor R. The varying electron current through R produces a voltage across R, which is dependent at each instant of time upon the number of secondary electrons. This voltage is called the *video-signal voltage*.

Figure 14-1*b* shows another way of converting a light scene into an electrical signal. The object of the scene is focused through a lens system onto a board which is made of a different kind of material than used in the camera

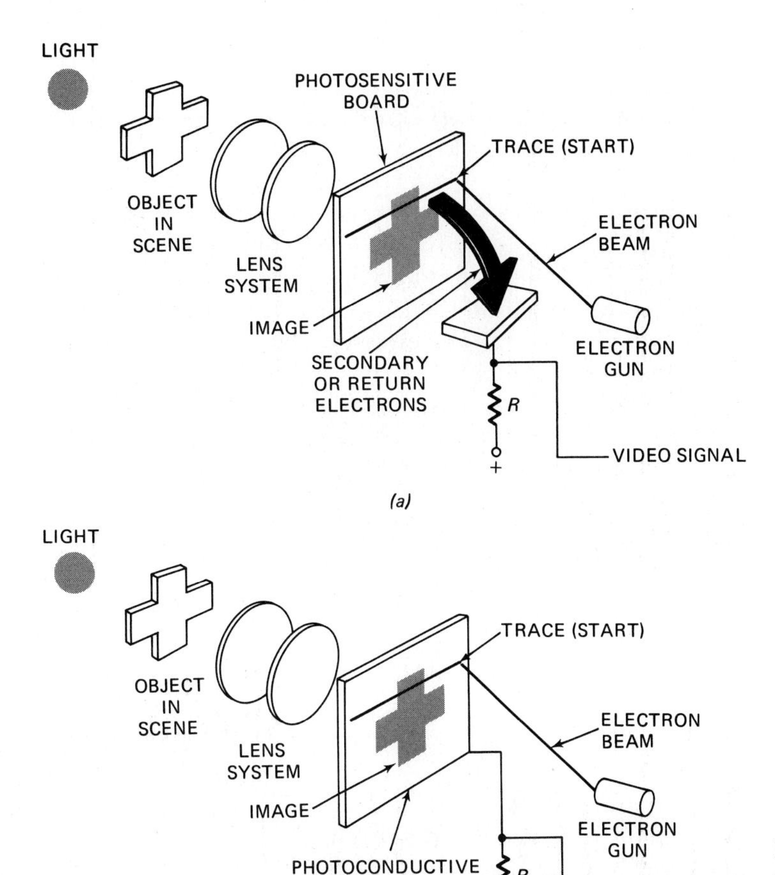

Fig. 14-1 Two ways of converting a scene to an electrical signal. (*a*) Video signal produced by secondary or return electrons. (*b*) Video signal produced by a photoconductive plate.

tube of Fig. 14-1*a*. In this case the board is a conductor. The electrons in the beam that strike the board flow out of the corner of the board through resistor R.

The resistance of the board at any point depends upon the amount of light striking that point. Therefore, as the electron beam sweeps across the image, the amount of conduction through R changes with the light and dark areas of the scene. The return electrons flowing through R produce a variable voltage which is called the video-signal voltage.

Regardless of which of the methods of Fig. 14-1 is used for producing the camera tube, it should be apparent that it is necessary to have an electron gun that sweeps an electron beam back and forth across the light-sensitive material.

The technique of sweeping this beam back and forth across the material is called *tracing* and *retracing*. There are many ways it can be done, but for commercial television the method that is used is to sweep one line at a time (from left to right) and return the beam to the left side quickly for the next line. More will be said about the tracing and retracing of the beam later in this chapter.

Summary

1. The aural signal (FM) carries the sound information in a TV system.
2. The video signal (AM) carries the black-and-white picture information in a TV system.
3. Transducers convert picture and sound information into electrical signals at the transmitter. In the receiver, transducers convert the electrical signals back into sound and picture.
4. In the camera tube the image is focused onto a photosensitive or photoconductive board. An electron beam traces across the board, and a video signal is produced.
5. Camera tubes use materials that either change their conductance or emit electrons when exposed to light. The amount of change in conductance or in the number of electrons emitted depends upon the amount of light.

How Are Electrical Signals Converted to Lighted Scenes?

In the complete television system it is not only necessary to change the scenes into electrical signals, but it is also necessary to change the electrical signals into scenes. Cathode-ray tubes (CRTs) are used for changing the electrical signals into scenes. More exactly, CRTs change the electrical impulses into light.

There are two basic kinds of cathode-ray (or picture) tubes: the *electrostatic deflection type* and the *electromagnetic deflection type*. The two types are very similar except for the way the beam is moved back and forth across the face or screen of the tube.

Figure 14-2 shows the principle of operation for an electrostatic deflection cathode-ray tube. An electron gun shoots the beam of electrons toward the screen. This screen is coated with a material that emits light when struck by an electron beam. The video signal is used to control the number of electrons in the electron beam, and therefore it controls the amount of light on the screen.

As in the case of the transmitter tube, the beam of the cathode-ray tube must be swept back and forth across the screen. In the CRT of Fig. 14-2 two sets of metal plates—called *deflection plates*—are used to deflect the beam on the face of the screen. Voltages on the vertical deflection plates cause the

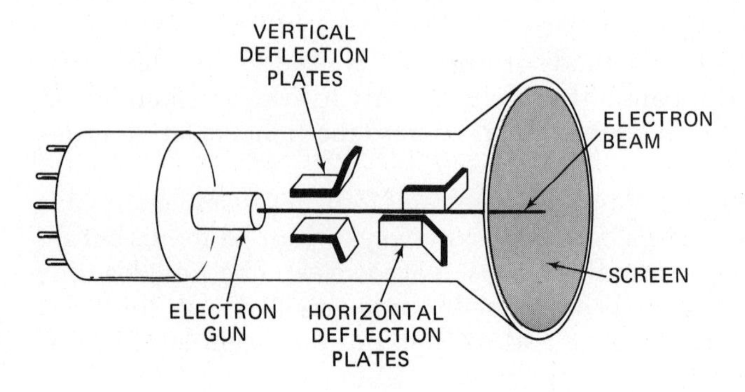

Fig. 14-2 Illustration of basic CRT principle of operation.

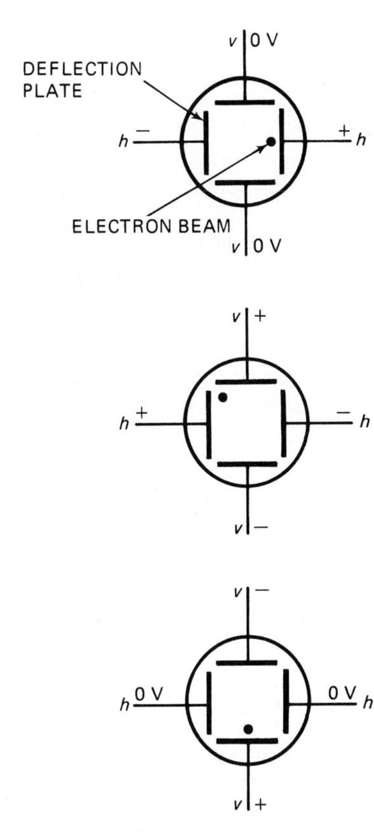

Fig. 14-3 Examples of deflection in a CRT.

beam to move up and down. Voltages on the horizontal deflection plates move the beam from left to right or right to left.

The deflection plates work on the principle that like charges repel and unlike charges attract. Suppose one deflection plate is made positive and the other is made negative; the positive plate will tend to attract the negative electrons while the negative plate repels the negative electrons. Figure 14-3 illustrates the cathode-ray tube looking from the front. The position of the beam is shown for various voltages on the deflection plates. In each case you will see that the beam is attracted toward the positive plates and away from the negative plates.

At first you might think that the beam would just go all the way over to the positive plate and never reach the screen. If you refer again to Fig. 14-2, you will see that these plates are a distance from the electron gun. The speed of the electrons at this point is so extremely high that they cannot make a right-angle turn. As shown in Fig. 14-4, they strike the screen at some point off-center, but they do not strike the deflection plates.

The fact that you can use a positive or negative voltage to move the beam to any point on the screen is useful in a number of applications. An oscilloscope, for example, works with this kind of tube. The beam is moved so rapidly that it appears to produce a continuous line on the screen. In a television-receiver CRT the beam is caused to move back and forth across the screen from top to bottom one line at a time to produce a lighted rectangular area. This area is referred to as the *raster*.

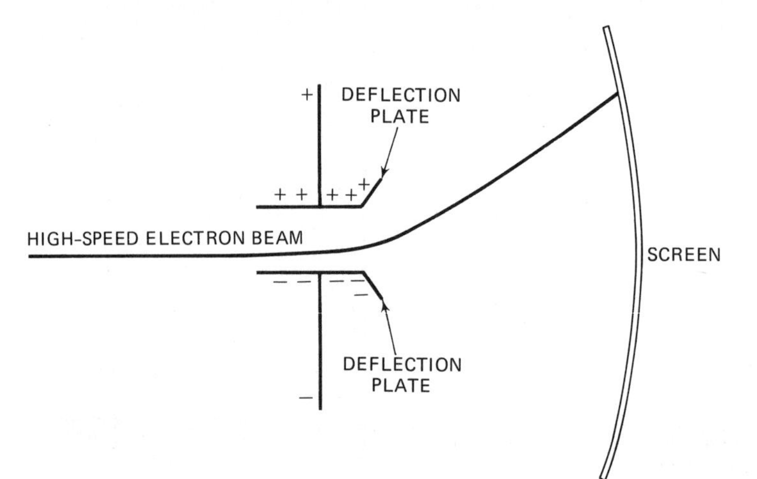

Fig. 14-4 Deflection-plate voltages turn the electron beam.

There are two ways to move an electron beam as it moves through a picture tube. The one just described is called *electrostatic deflection* because electric voltages cause the beam to move in some desired way. It is also possible to deflect the beam with an electromagnetic field. When an electron beam passes through a magnetic field it is deflected by the beam. The amount of deflection can be controlled by changing the magnetic field.

Figure 14-5 shows a cathode-ray tube with *electromagnetic deflection.* There is a pair of deflection coils for both vertical deflection and horizontal deflection. The coils are located around the neck of the cathode-ray tube. The electron gun shoots the electrons toward the screen, and these electrons pass through the magnetic field from the deflection coils.

Varying electric currents flowing through the deflection coils produce magnetic fields that move the beam back and forth and up and down as desired. This is how the deflection is controlled.

In modern television receivers electromagnetic deflection is more popular because of the extremely high voltages that would be required for electrostatic deflection. When a picture tube is changed in a television receiver, the deflection coils need not be changed.

The two types of deflection used with CRTs are summarized in Fig. 14-6. The deflection plates that are marked *V* in Fig. 14-6*a* move the beam up and down, and those marked *H* move the beam horizontally. In Fig. 14-6*b* two pairs of deflection coils are used. Note that horizontal deflection is obtained by the top and bottom coils, and vertical deflection is obtained from the coils at the sides of the neck. The reason for this is that electrons are deflected at

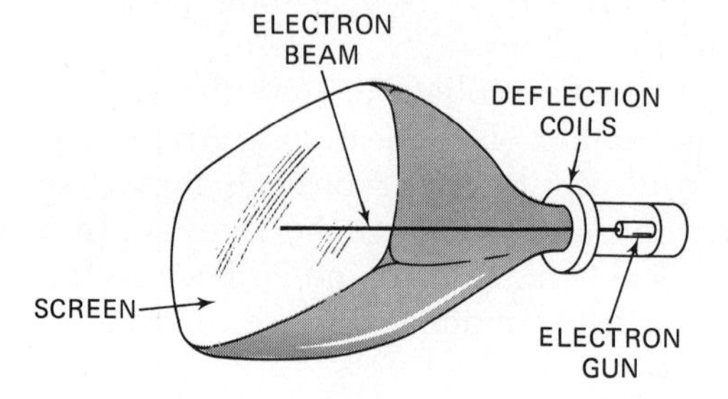

Fig. 14-5 Cathode-ray tube with electromagnetic deflection.

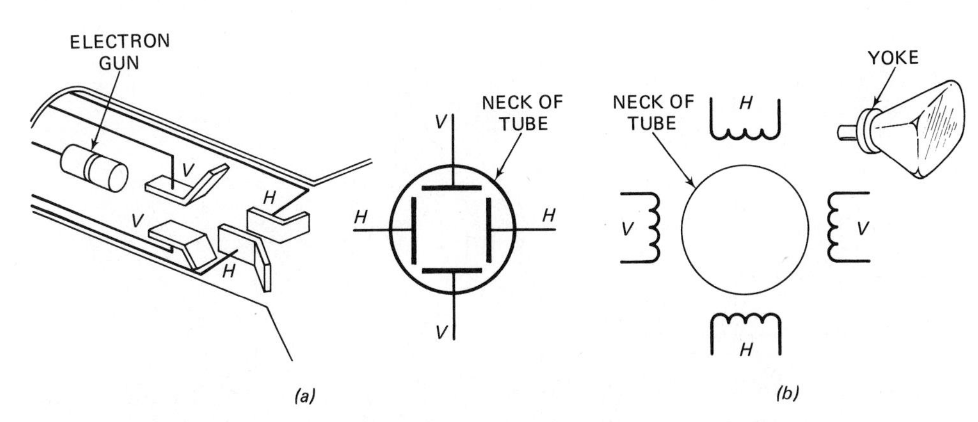

Fig. 14-6 Two ways to deflect the beam in a cathode-ray tube. (*a*) With deflection plates. (*b*) With deflection coils.

right angles to a magnetic field. The deflection coils are located in a container called the *deflection yoke.*

Summary

1. TV picture tubes are also called *cathode-ray tubes* or *CRTs.* They are used to change electrical signals into television scenes.
2. Two kinds of picture tubes are in use. They are electrostatic deflection tubes and electromagnetic deflection tubes. The electromagnetic deflection type is the most popular one for television use.
3. The lighted rectangular area on the face of the picture is made by scanning the electron beam back and forth. This lighted area is called the *raster.*

What Is Scanning and What Is Synchronizing?

Figure 14-7 shows an image on the screen of a camera tube. The path of the electron beam is shown in Fig. 14-7*a* as it traces one line across the screen. The video signal for this line is shown in Fig. 14-7*b*. In this example, when the beam is in the light area, the video voltage is relatively low. When the line is in the dark area, the video voltage is high.

To reproduce a picture at the receiver it is necessary for the receiver CRT to produce a line of trace exactly like the one at the transmitter.

Figure 14-8 shows a diagram of how the beams in the camera tube and in the receiver CRT are moved across the screens. When the two beams are in

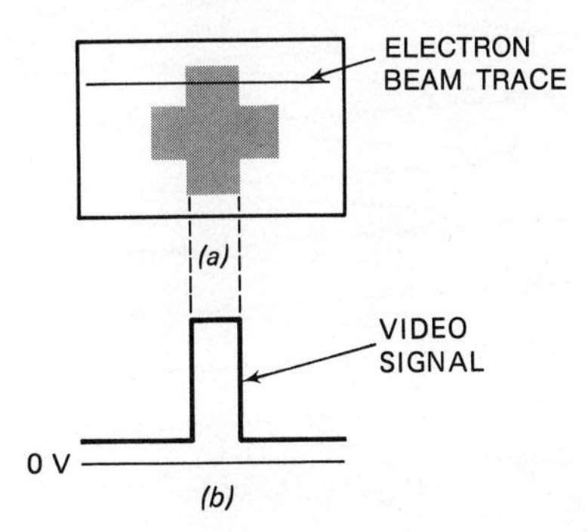

Fig. 14-7 Generation of the video signal. (*a*) One line of trace produces the video signal shown in (*b*).

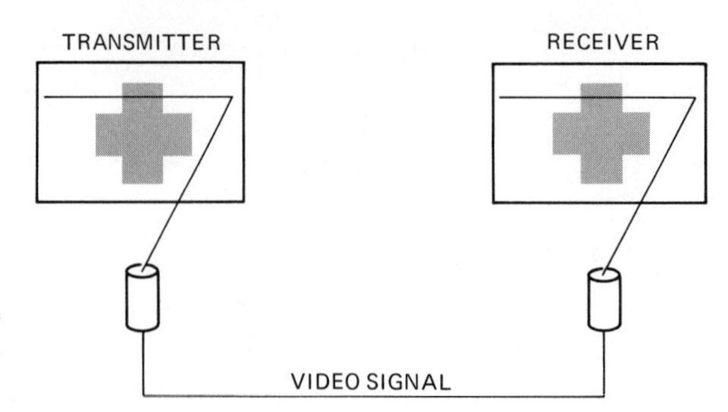

Fig. 14-8 The beam in the receiver tube must follow the beam in the transmitter tube.

step, as shown in Fig. 14-8, they are said to be *synchronized.* In order to do this it is necessary to send a special signal from the transmitter to tell the receiver exactly the precise moment to start each line of trace.

The method of tracing lines in both the transmitter and receiver is called *interlaced scanning.* The idea is shown in Fig. 14-9.

The solid lines show the trace (from left to right). The dotted lines show the retrace (from right to left). The trace starts at 0 in the center of the screen and traces line 1. Then it retraces and traces line 3. This continues until all the odd lines are traced.

There are $262\frac{1}{2}$ odd lines. After the last odd line the beam is sent to the top left side of the picture, where it starts tracing the even-numbered lines.

To summarize, the beam has to completely trace the odd-numbered lines and then completely trace the even-numbered lines. To your eye it looks like the scene is being presented continuously because the beam is moving so fast.

What Are Fields and Frames?

All the odd lines together are called a *field,* and all the even lines put together are called a *field.* When the odd and even lines are both combined, the picture is complete. The complete picture is referred to as a *frame.* So it takes two fields to make a frame.

You might wonder how this intricate interlacing is accomplished. In real-

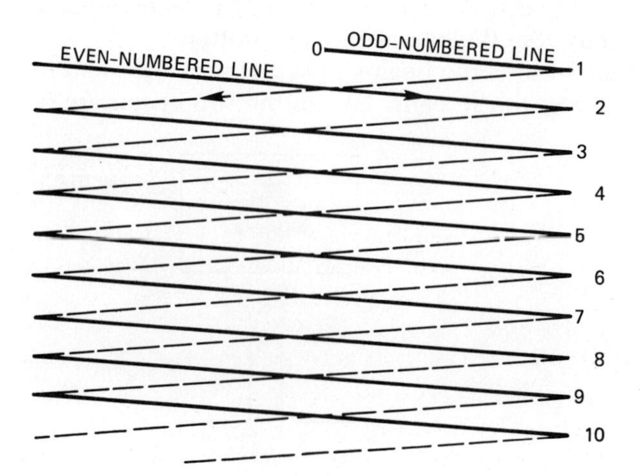

Fig. 14-9 Interlaced scanning.

ity the interlacing is controlled by the signal from the transmitter which tells the receiver when to trace the even lines and when to trace the odd lines.

There are 60 fields traced each second. Since two complete fields are needed for one frame, there are only 30 frames sent per second. Your eye sees only 30 complete pictures per second.

Since 60 half-pictures (fields) are flashed on the screen, your eye is fooled into thinking that it is looking at a continuous picture. You cannot see 60 flashes per second.

Suppose they did not divide the picture into two halves, but sent only 30 complete pictures each second. Your eye would see a flicker—that is, a rapidly changing brightness. Interlaced scanning is used so you cannot see this flicker, since the eye thinks it is seeing 60 pictures per second.

Summary

1. The trace in the picture tube is moved in step with the trace in the camera tube. This is called *synchronization.*
2. Interlaced scanning is used for producing the raster.
3. With interlaced scanning, all the odd lines make up one field and all the even lines make up the next field.
4. Two fields are required to make a frame. A frame is one complete television picture.
5. Interlaced scanning is used so you will not see the flickering that would occur with only 30 pictures per second. The eye thinks it is seeing 60 pictures per second.

What Is the Makeup of the Monochrome Signal?

The video signal consists of voltage variations which correspond to light and dark areas on the camera-tube screen. As the beam strikes the receiver CRT, light is emitted. The amount of light depends upon the number of electrons in the beam at that instant. The number of electrons, in turn, is controlled by the video signal delivered to the CRT electron gun. It is now possible to understand the television signal that is used to transmit the scene from the transmitter to the receiver.

We will start by looking at the video signal for one line of scan. This is shown in Fig. 14-10. Figure 14-10*a* shows a line of video-information waveform and identifies each part of the waveform. (The next line is shown with a dotted line.) The horizontal *blanking pedestal* is used to shut the beam off during the time that it is retraced from the right side of the picture back to the left side of the picture. The *sync pulse* sitting on top of the blanking pedestal is the part of the signal that tells the TV receiver to start a new line synchronized with the camera-tube scanning.

After the new line is started, it is blanked out for a short time by the *back porch* of the blanking pedestal; then the video signal begins. The video signal varies the number of electrons in the beam as it traces from left to right across the screen. When it gets to the right of the screen, the blanking pedes-

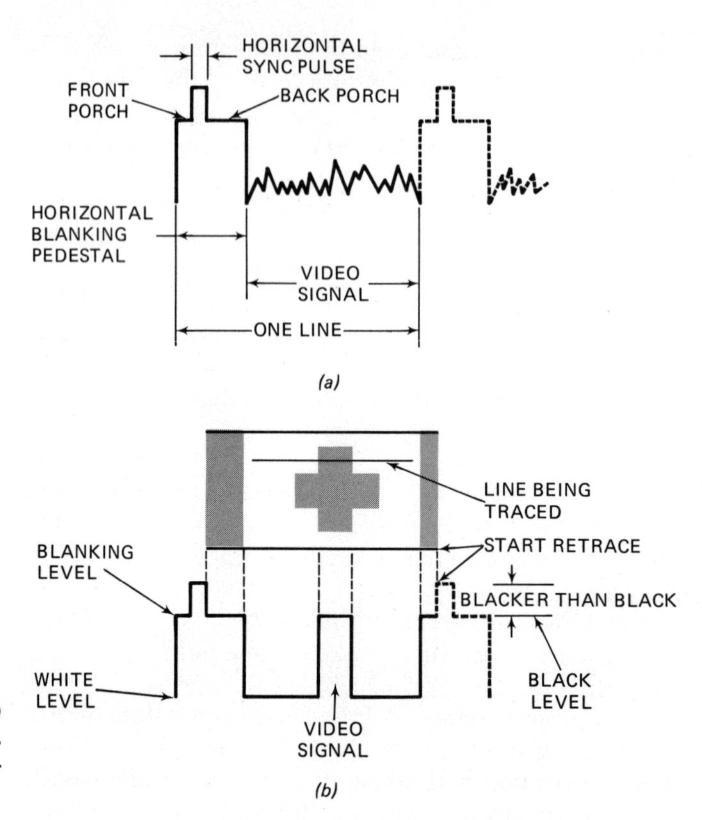

Fig. 14-10 One line of video. (*a*) Typical signal for one line of video. (*b*) Relationship between the picture and the video signal.

tal again shuts the beam OFF. Then another horizontal sync pulse tells the TV receiver to start a new line.

Figure 14-10*b* shows how the video signal for one horizontal line affects the intensity of the scan on the picture-tube screen. The top of the blanking pedestal shuts the beam OFF. The sync pulse is more positive than the blanking signal required to turn the screen black. The sync pulse is said to be in the *blacker-than-black* region. The line of video-signal level is white until the vertical portion of the cross is encountered. Then the video signal goes quickly positive and shuts the electron beam OFF.

After the electron beam has passed this black portion of the picture, it returns to the white-voltage level and remains at this level until the next blanking pedestal is encountered.

Always keep in mind that although it takes a little while to tell about this, in reality the events occur very rapidly. They are so rapid that it looks to the eye as though the screen were lighted at all times continuously. There are $262\frac{1}{2}$ lines for each field, and two fields are required to make a frame. Thus there are 525 lines per frame and 30 frames every second. In other words, your television set displays 30 complete pictures every second. Since there are 525 lines for each picture, and there are 30 pictures per second, it follows that there are

$$30 \times 525 = 15{,}750 \text{ individual lines}$$

traced on your screen every second.

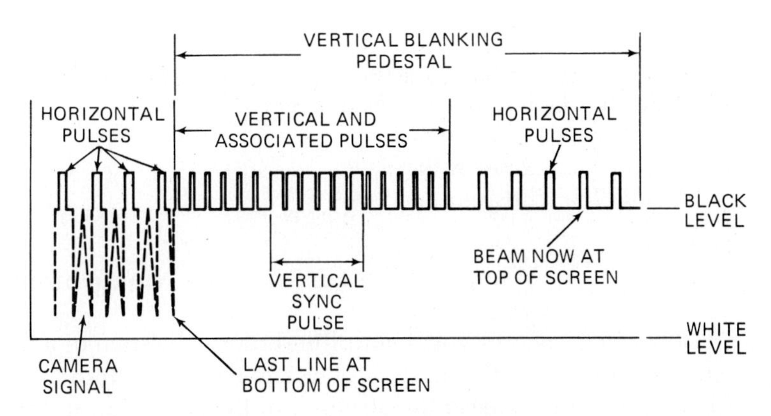

Fig. 14-11 The vertical blanking pedestal.

It is necessary for the transmitter to send a signal to tell the receiver when to start a new line. It is also necessary to send a special signal to tell the receiver to start a new field. After $262\frac{1}{2}$ lines have been traced, the beam must be brought from the bottom of the picture up to the top to start the new field. Therefore a vertical blanking pedestal and a vertical sync pulse must be added to the composite signal. Figure 14-11 shows how the vertical blanking pedestal looks.

It takes several lines to get the beam from the bottom of the picture back up to the top. It is necessary to add horizontal sync pulses on top of the vertical blanking pedestal in order to keep the horizontal scan in the receiver in step with the transmitter. If you did not do this, when the beam got to the top left side of the picture it would take a short instant of time to get the receiver back into sync with the horizontal lines of the transmitter. Therefore during the vertical blanking period there are still horizontal lines being produced. You cannot see them because the picture-tube beam is turned OFF by the vertical blanking pedestal. You can see these in Fig. 14-11.

The vertical sync pulse is slotted, or serrated. This is necessary to keep the horizontal sync in operation during the retrace period. Following the vertical sync pulse there are some additional pulses used to keep the receiver in horizontal synchronization until the beginning of the next field.

You have now seen how the camera tube at the transmitter produces a video signal and also that this video signal is added to blanking pedestals and sync pulses.

Summary

1. The complete television signal has these basic parts:
   (*a*) the video signal
   (*b*) the horizontal blanking pedestal
   (*c*) the horizontal sync pulse
   (*d*) the vertical blanking pedestal
   (*e*) the vertical sync pulse
2. The video signal varies the brightness of the scanned line as the electron beam moves across the screen.
3. The blanking pedestals turn the electron beam OFF during the retrace time.

4. The horizontal sync pulse tells the receiver to start a new line. The vertical sync pulse tells the receiver to start a new field.
5. In the television system there are
   (*a*) 60 fields per second;
   (*b*) 30 frames per second;
   (*c*) 15,750 lines per second.

What Are the Circuits in a Monochrome Transmitter?

Figure 14-12 shows you a simplified block diagram of a monochrome television transmitter. The camera converts the scene into a video signal. The video amplifiers increase the strength of these signals and then deliver them to an amplitude-modulation (AM) transmitter.

A separate section of the (AM) transmitter (pulse generator) produces the *pulse signals*. The pulse signals include the blanking pedestals and the sync pulses. Note that the pulse signals are delivered to the camera for sweeping the camera beam and also to the AM transmitter so that they can be sent to the receiver.

The output of the AM transmitter is delivered to a *diplexer*. A diplexer is simply a circuit that permits an antenna to be used for transmitting two different signals at the same time.

The sound section of the transmitter is completely separate from the video section. A microphone picks up the sound signal and converts it to audio-voltage variations. These variations are increased in strength by the audio amplifiers and then delivered to an FM transmitter. The FM transmitter also delivers its signal to the diplexer. The diplexer delivers the audio and video signals to the antenna, where they are radiated.

It is very important to note that the video signal and the pulse signals are sent as amplitude-modulated signals, but the sound portion of the signal is sent as an FM signal from the FM transmitter. Thus a television receiver must be able to detect both AM and FM signals.

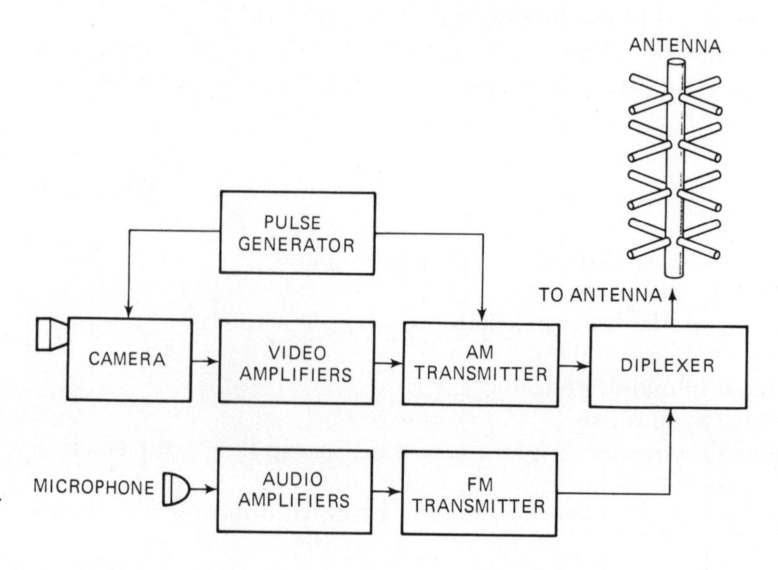

Fig. 14-12 Simplified block diagram of a TV transmitter.

What Are the Circuits in a Monochrome Television Receiver?

Figure 14-13 shows a block diagram of a monochrome television receiver. The signal is picked up by the antenna and delivered to the tuner through a transmission line. The tuner accomplishes two purposes: first, it enables you to select one station and reject all others; and second, it converts the high-frequency signals to an intermediate frequency as required for superheterodyne operation. All television receivers are superheterodyne types.

The output of the tuner is an intermediate frequency signal, usually in the range of about 40 megahertz. Amplifiers increase the strength of the IF signal. This is usually done in several stages.

The output of the IF amplifier goes to a video detector stage. It is used to demodulate only the video, sync, and blanking signals. The FM signal cannot be detected by this stage.

At the output of the video detector you will note that there is a takeoff point for the FM signal. The FM signal is delivered to a separate FM section.

The video signal and the sync and blanking signals go to separate sections in the receiver. The video amplifier converts the weak video signals from the detector to signals that are strong enough to control the electron beam on the picture tube. The sync signals, both horizontal and vertical, are delivered to a stage in the receiver called the *sync separator*. The sync separator permits the vertical and horizontal sync pulses to be separated into two individual signals—one for operating the vertical and one for horizontal scan

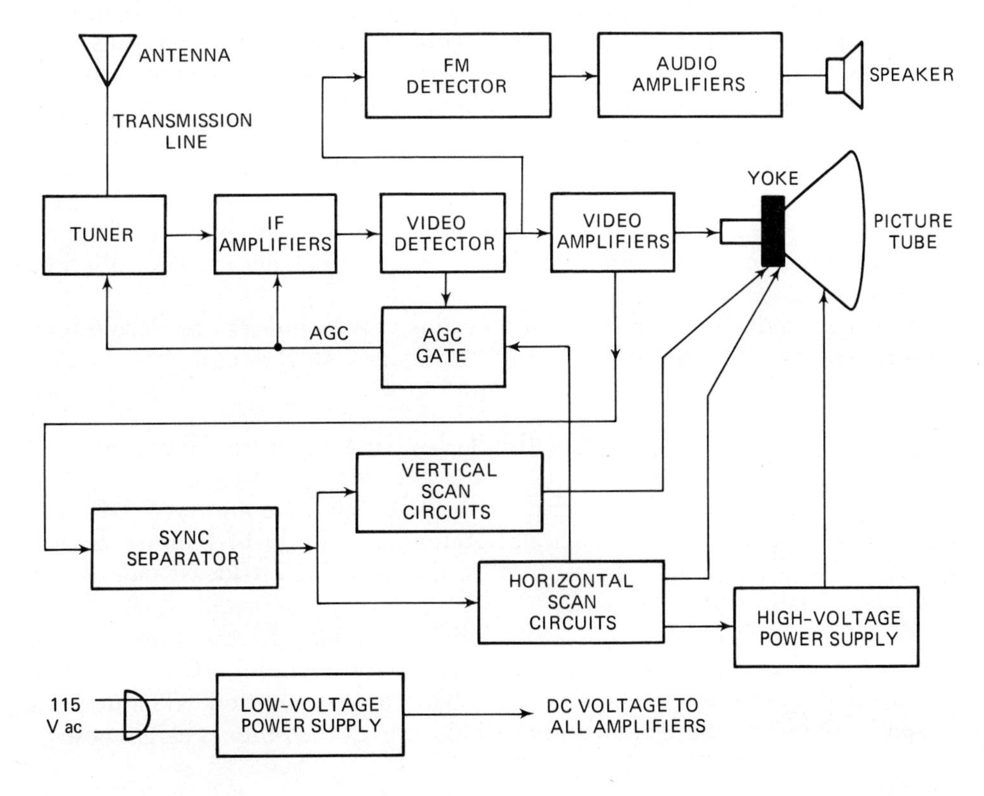

Fig. 14-13 Block diagram of a TV receiver.

currents. Picture-tube deflection coils are located in the yoke. The scan currents move the beam back and forth and up and down on the screen to produce the raster.

What Does the Flyback High-Voltage Circuit Do?

The horizontal scan circuits also operate a separate *high-voltage power supply.* The cathode-ray tube needs a very high positive voltage to move the electrons from the gun to the face or screen of the picture tube. This high voltage may range from 12,000 volts for smaller tubes up to 30,000 volts for a large picture tube. The high voltage is generated by a special horizontal circuit transformer, called the "flyback" transformer.

What Does the AGC Circuit Do?

The purpose of the automatic-gain-control (AGC) circuit is to set the gain of the receiver in accordance with the strength of the signal being received. For weak signals the gain of the receiver is automatically set to a high value, and for strong signals the gain is set to a low value. This is automatically accomplished by the AGC circuit. However, it is not possible to use the composite video signal to produce the AGC voltage. If it were used, the amount of AGC voltage would depend upon whether a light scene or a dark scene were being transmitted.

To get around this problem only the sync pulses in the blacker-than-black region are used for producing the AGC voltage. This voltage has an amplitude that depends only upon the strength of the signal, but does not change in accordance with light and dark scenes being transmitted.

A pulse output from the horizontal scan circuits turns the AGC gate ON at precisely the instant the sync pulse arrives at it. The AGC gate is OFF for all other periods of time. Thus only the sync pulses can affect the AGC gate. These pulses are converted to a dc voltage for controlling the gain of the receiver. The gains of the tuner and the IF amplifier stages are controlled by the AGC voltage.

The type of AGC circuit described here is very popular in modern television receivers. It is called a *keyed AGC* or *gated AGC* circuit.

What Are the Television Channel Frequencies?

Television signals are broadcast in the very high frequency (VHF) and ultra-high frequency (UHF) range of frequencies. Table 14-1 shows the frequencies for the channels in use. Note that the FM broadcast band is located between channels 6 and 7.

Each television channel has a bandwidth of 6 megahertz. Figure 14-14 shows how the frequencies are arranged in each channel. One complete sideband and a portion of the lower sideband are transmitted. The method of sending one complete sideband and a portion of the other is called *vestig-*

Table 14-1 Frequencies for TV Channels

P = PICTURE CARRIER FREQ. (MHz) S = SOUND CARRIER FREQ. (MHz)

CHANNEL NUMBER	P	S	FREQ. LIMITS OF CHANNEL (MHz)
2	55.25	59.75	54 – 60
3	61.25	65.75	60 – 66
4	67.25	71.75	66 – 72
5	77.25	81.75	76 – 82
6	83.25	87.75	82 – 88
FM BROADCAST 88 – 108 MHz			
7	175.25	179.75	174 – 180
8	181.25	185.75	180 – 186
9	187.25	191.75	186 – 192
10	193.25	197.75	192 – 198
11	199.25	203.75	198 – 204
12	205.25	209.75	204 – 210
13	211.25	215.75	210 – 216
14	471.25	475.75	470 – 476
15	477.25	481.75	476 – 482
16	483.25	487.75	482 – 488
17	489.25	493.75	488 – 494
18	495.25	499.75	494 – 500
19	501.25	505.75	500 – 506
20	507.25	511.75	506 – 512
21	513.25	517.75	512 – 518
22	519.25	523.75	518 – 524
23	525.25	529.75	524 – 530
24	531.25	535.75	530 – 536
25	537.25	541.75	536 – 542
26	543.25	547.75	542 – 548
27	549.25	553.75	548 – 554
28	555.25	559.75	554 – 560
29	561.25	565.75	560 – 566
30	567.25	571.75	566 – 572
31	573.25	577.75	572 – 578
32	579.25	583.75	578 – 584
33	585.25	589.75	584 – 590
34	591.25	595.75	590 – 596
35	597.25	601.75	596 – 602
36	603.25	607.75	602 – 608
37	609.25	613.75	608 – 614
38	615.25	619.75	614 – 620
39	621.25	625.75	620 – 626
40	627.25	631.75	626 – 632
41	633.25	637.75	632 – 638
42	639.25	643.75	638 – 644
43	645.25	649.75	644 – 650
44	651.25	655.75	650 – 656
45	657.25	661.75	656 – 662
46	663.25	667.75	662 – 668
47	669.25	673.75	668 – 674
48	675.25	679.75	674 – 680
49	681.25	685.75	680 – 686
50	687.25	691.75	686 – 692
51	693.25	697.75	692 – 698
52	699.25	703.75	698 – 704
53	705.25	709.75	704 – 710
54	711.25	715.75	710 – 716
55	717.25	721.75	716 – 722
56	723.25	727.75	722 – 728
57	729.25	733.75	728 – 734
58	735.25	739.75	734 – 740
59	741.25	745.75	740 – 746
60	747.25	751.75	746 – 752
61	753.25	757.75	752 – 758
62	759.25	763.75	758 – 764
63	765.25	769.75	764 – 770
64	771.25	775.75	770 – 776
65	777.25	781.75	776 – 782
66	783.25	787.75	782 – 788
67	789.25	793.75	788 – 794
68	795.25	799.75	794 – 800
69	801.25	805.75	800 – 806
70	807.25	811.75	806 – 812
71	813.25	817.75	812 – 818
72	819.25	823.75	818 – 824
73	825.25	829.75	824 – 830
74	831.25	835.75	830 – 836
75	837.25	841.75	836 – 842
76	843.25	847.75	842 – 848
77	849.25	853.75	848 – 854
78	855.25	859.75	854 – 860
79	861.25	865.75	860 – 866
80	867.25	871.75	866 – 872
81	873.25	877.75	872 – 878
82	879.25	883.75	878 – 884
83	885.25	889.75	884 – 890

ial sideband. One obvious advantage is that it permits a greater number of television stations in a given range of frequencies.

The video and sound carrier frequencies are exactly 4.5 megahertz apart in each channel. Receivers that amplify both the sound and video frequencies in their IF stage and separate the sound and video signals at or after the detector (see Fig. 14-13) are called *intercarrier* sets. Almost all TV sets are made this way.

In intercarrier receivers the video carrier is mixed with the sound carrier to produce a sound intermediate frequency. Since these frequencies are transmitted 4.5 megahertz apart, it follows that the sound intermediate frequency is always 4.5 megahertz.

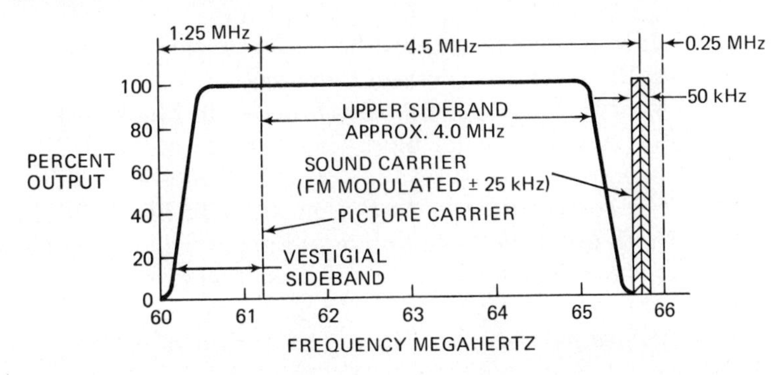

Fig. 14-14 Arrangement of frequencies in each TV channel.

Summary

1. In the television system the video and the pulse signals are amplitude-modulated.
2. The sound signal in the television system is frequency-modulated.
3. Both signals are radiated from a single antenna.
4. The tuner in a TV receiver is used for selecting the desired station. It also converts the station carrier signal to an intermediate frequency.
5. Keyed automatic-gain-control (AGC) circuits are used in most television receivers. This circuit produces a dc voltage that is directly related to the strength of the signal coming from the TV station.
6. The horizontal scan circuit operates a power supply for producing the high voltage to operate the picture tube.

PROGRAMMED REVIEW QUESTIONS

(Instructions for using this programmed section are given in Chapter 1.)
We will review the important concepts of this chapter. If you have understood the material, you will progress easily through this section. Do not skip this material because some additional theory is presented.

1. The FM broadcast band is between television channels
   A. 4 and 5. (Proceed to block 17.)
   B. 6 and 7. (Proceed to block 9.)

2. *The correct answer to the question in block 10 is **B**. Since there are $262\frac{1}{2}$ lines per field and there are 60 fields per second, it follows that there are*

$$262.5 \text{ lines} \times 60 = 15{,}750 \text{ lines per second}$$

Another way to do this is to mulitply the number of lines per frame (525) by the number of frames per second (30):

$$525 \text{ lines} \times 30 = 15{,}750 \text{ lines per second}$$

Here is your next question.
In the television system there are
   A. two fields per frame. (Proceed to block 14.)
   B. two frames per field. (Proceed to block 20.)

3. *The correct answer to the question in block 14 is **B**. Even though there are only 30 complete pictures (frames) sent each second, your eye does not see a flicker because there are two half-pictures (fields) sent for each frame. So 60 fields per second are needed to produce 30 frames per second.* Here is your next question.

At the transmitter the sound (aural) and video signals are transmitted with a single antenna. This is made possible by using a
A. diplexer. (Proceed to block 12.)
B. duplexer. (Proceed to block 19.)

4. *The correct answer to the question in block 15 is **B**. The 4.5-megahertz separation between the sound and video carrier frequencies is important. Intercarrier sets mix the video carrier frequency with the sound carrier frequency to produce the sound intermediate frequency of 4.5 megahertz.* Here is your next question.
With a keyed AGC system the amount of AGC voltage depends upon
A. the sync-pulse amplitude. (Proceed to block 10.)
B. the amount of video signal. (Proceed to block 25.)

5. *Your answer to the question in block 11 is **B**. This answer is wrong. It is not correct to call the high-voltage supply a "horizontal supply."* Proceed to block 21.

6. *The correct answer to the question in block 24 is **B**. One of the clues that television technicians use for finding trouble in a TV set is the presence of a raster. They know that when there is no raster the high voltage **may** be faulty. Of course, there are many other clues that they will also look for.* Here is your next question.
A transducer converts energy from one system to another. Name two transducers used in television receivers. (Proceed to block 28.)

7. *Your answer to the question in block 23 is **A**. This answer is wrong. A TRF receiver **could** be made for television, but it would be difficult to amplify the 6-megahertz channel bandwidth through several RF stages.* Proceed to block 11.

8. *Your answer to the question in block 15 is **A**. This answer is wrong. Review the frequency distribution shown in Fig. 14-14, then* proceed to block 4.

9. *The correct answer to the question in block 1 is **B**. There is a 4-megahertz space between channels 4 and 5, but the FM band is not located in this space. The FM band is in the 88- to 108-megahertz range, which is between channels 6 and 7.* Here is your next question.
The picture signal, or video signal, is
A. amplitude-modulated. (Proceed to block 23.)
B. frequency-modulated. (Proceed to block 18.)

10. *The correct answer to the question in block 4 is **A**. If you used the video signal to obtain the AGC voltage, the voltage would be different when bright scenes were being shown from the voltage when dark scenes were shown. The AGC voltage is a dc voltage that de-*

pends only upon the strength of the TV signal that is being received. The sync pulses are the best indication of signal strength. Here is your next question.
In the television system there are
A. $262\frac{1}{2}$ lines per second. (Proceed to block 16.)
B. 15,750 lines per second. (Proceed to block 2.)

11. *The correct answer to the question in block 23 is* ***B****. The superheterodyne-receiver design is a logical choice for television. It is more efficient to pass the required range of frequencies through IF stages.* Here is your next question.
The high voltage for operating the picture tube may be obtained from a signal in the horizontal sweep circuit. This type of high-voltage power supply is called a
A. flyback supply. (Proceed to block 21.)
B. horizontal supply. (Proceed to block 5.)

12. *The correct answer to the question in block 3 is* **A**. *For your information, a* **duplexer** *permits one antenna to be used for both transmitting and receiving. A* **diplexer** *permits one antenna to be used for radiating two different signals.* Here is your next question.
Which of the following describes the method of transmitting television video signals?
A. Vestigial sideband. (Proceed to block 24.)
B. Single sideband. (Proceed to block 22.)

13. *Your answer to the question in block 21 is* **A**. *This answer is wrong. A field is only half of a picture.* Proceed to block 15.

14. *The correct answer to the question in block 2 is* **A**. *Two fields are required to make a complete frame. One field will have all the odd-numbered lines, and the alternate field will have all the even-numbered lines. (You can remember it this way: You would not frame a field; you would frame a complete picture.)* Here is your next question.
The number of TV frames per second is
A. 60. (Proceed to block 26.)
B. 30. (Proceed to block 3.)

15. *The correct answer to the question in block 21 is* ***B****. Since a field contains only the even lines or only the odd lines, it is only half of a picture. A full picture has both the even and the odd lines and is called a frame.* Here is your next question.
The sound and video carrier frequencies are
A. 3.58 megahertz apart. (Proceed to block 8.)
B. 4.5 megahertz apart. (Proceed to block 4.)

16. *Your answer to the question in block 10 is* **A**. *This answer is wrong.*

There are 262½ lines for each field, and there are 60 fields per second. Proceed to block 2.

17. *Your answer to the question in block 1 is* **A**. *This answer is wrong. Refer to Table 14-1,* then proceed to block 9.

18. *Your answer to the question in block 9 is* **B**. *This answer is wrong. The sound signal, or aural signal, is frequency-modulated. The video signal is amplitude-modulated.* Proceed to block 23.

19. *Your answer to the question in block 3 is* **B**. *This answer is wrong. Study the block diagram of a simple television transmitter in Fig. 14-12, then* proceed to block 12.

20. *Your answer to the question in block 2 is* **B**. *This answer is wrong. Always remember that it takes two fields to make a frame.* Proceed to block 14.

21. *The correct answer to the question in block 11 is* **A**. *The flyback supply gets its name from the fact that the high voltage is generated during the flyback time—that is, during retrace time. This is a very short period of time when the electron beam is moved from the right side to the left side of the screen. A high rate of change of current is required for moving the beam so rapidly. The voltage generated is proportional to the rate of change of current and the number of turns in the flyback transformer.* Here is your next question.
A complete television picture is called a
A. field. (Proceed to block 13.)
B. frame. (Proceed to block 15.)

22. *Your answer to the question in block 12 is* **B**. *This answer is wrong. You studied single sideband in an earlier chapter. Do you remember the disadvantages of single sideband?* Proceed to block 24.

23. *The correct answer to the question in block 9 is* **A**. *The sound signal is FM, but the video signal, blanking pedestals, and sync pulses are all transmitted as amplitude-modulated signals.* Here is your next question.
Television receivers are
A. tuned-radio-frequency (TRF) receivers. (Proceed to block 7.)
B. superheterodyne receivers. (Proceed to block 11.)

24. *The correct answer to the question in block 12 is* **A**. Here is your next question.
The white rectangular display obtained on the picture tube when it is scanned in a horizontal vertical direction is called the
A. tracer. (Proceed to block 27.)
B. raster. (Proceed to block 6.)

25. *Your answer to the question in block 4 is* ***B****. This answer is wrong. The video signal changes with the amount of brightness in the scene. It would not be a good indication of signal strength.* Proceed to block 10.

26. *Your answer to the question in block 14 is* A. *This answer is wrong. The number of fields per second (60) is greater than the number of frames (30).* Proceed to block 3.

27. *Your answer to the question in block 24 is* A. *This answer is wrong. The word* **tracer** *has not been mentioned in relation to TV.* Proceed to block 6.

28. *The speaker and the picture tube are both transducers.*
You have now completed the programmed questions. The next step is to put some of these ideas to work in laboratory experiments. Proceed to the Experiment section of this chapter.

EXPERIMENT

(The experiment described in this section may be performed on the circuit board described in Appendix C or on a similar laboratory setup.)

Purpose To demonstrate the operation of basic *RC* circuits and relaxation oscillators.

Theory Many of the circuits in electronic systems (including TV sets) depend upon the relationship between resistance and capacitance for their operation. These *resistance-capacitance* (*RC*) circuits are very often referred to as *time-constant circuits. Resistance-inductance* (*RL*) circuits, also used in electronic systems, are also called time-constant circuits.

Figure 14-15 shows the basic *RC* and *RL* circuits. In Fig. 14-15*a* a switch, a capacitor, and a resistor are connected across a battery. Assume that the capacitor is not charged.

When the switch is closed, charging current will flow. This is shown by the arrows. This current charges the capacitor. The voltage across the capacitor increases with time. The greater the resistance (or capacitance), the longer it takes the capacitor to charge.

The waveform to the right of *C* shows how the voltage across the capacitor changes with time. At the instant the switch is closed, there is no voltage across the capacitor because it is not charged. As the charging current flows, the voltage across the capacitor increases.

The voltage across the resistor is maximum at the instant the switch is closed. As the capacitor becomes more fully charged, the rate of current flow decreases. This causes the voltage across *R* to decrease as the voltage across the capacitor increases.

When the capacitor is fully charged, all the applied voltage is across

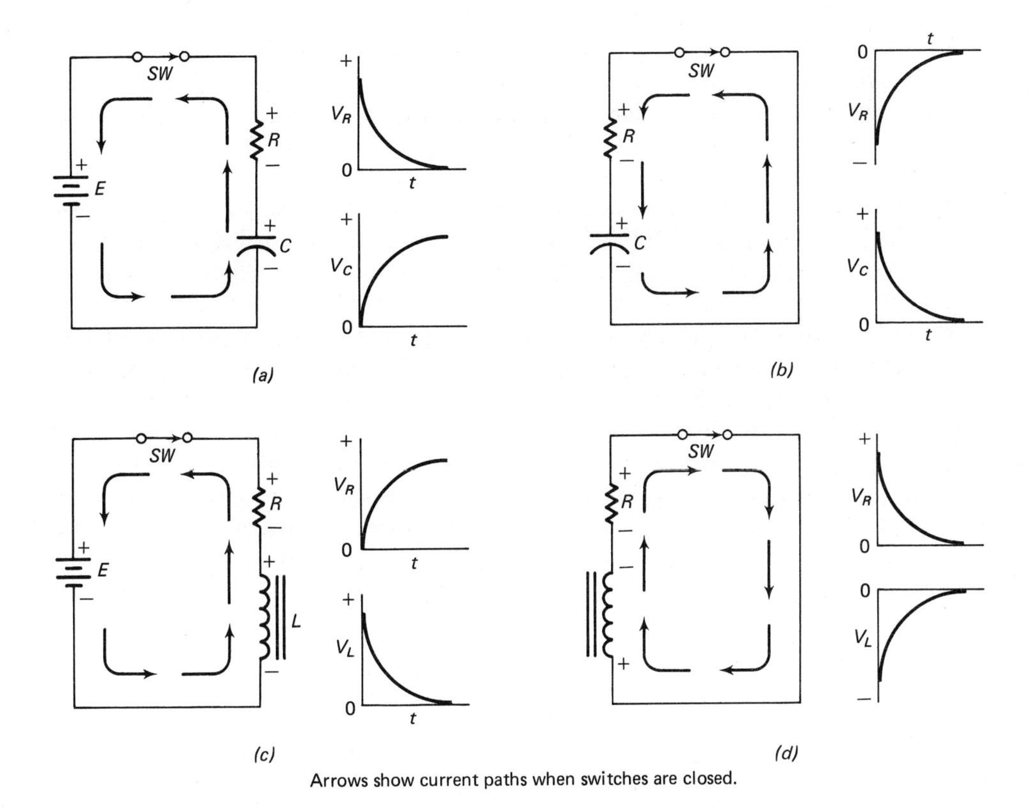

Arrows show current paths when switches are closed.

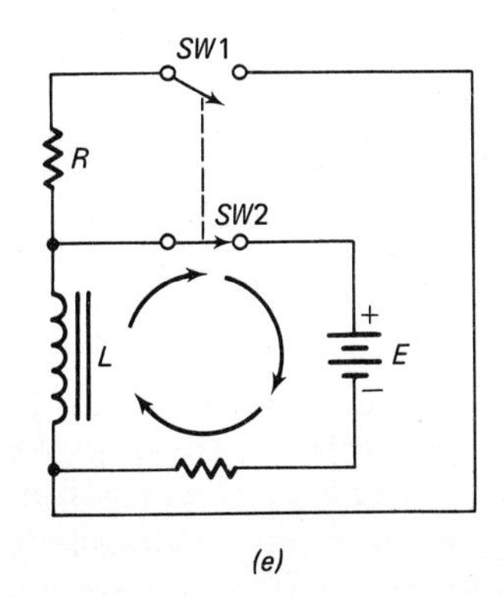

Fig. 14-15 Time constants in *RC* and *RL* circuits. (*a*) Capacitor charging. (*b*) Capacitor discharging. (*c*) Inductor current increasing. (*d*) Inductor current decreasing. (*e*) Induced voltage causes arc across SW_2 (see text).

C. There is no longer any charging current flowing, so there is no voltage across *R*.

Suppose now that the capacitor is fully charged and you wish to discharge it through a resistor. This condition is shown in Fig. 14-15*b*. Closing the switch in this circuit permits the capacitor to discharge as shown by the arrows. As the capacitor discharges, the rate of current flow decreases. This means that the negative-going voltage across the resistor decreases.

Resistors and inductors also have a time-constant relationship. However, it is not as easily expressed as in the case of *RC* circuits.

Figure 14-15*c* shows an inductor and a resistor in series with a battery *E*.

At the instant the switch is closed, the current in the circuit tries to rise very rapidly. This causes a countervoltage across the inductor. (The countervoltage is maximum when the rate of change of current through the inductor is maximum.) The overall result is that the countervoltage prevents the current from rising to its maximum value quickly. Instead, the current increases gradually. The increase in current is similar to the increase in voltage across the capacitor. (See Fig. 14-15*a*.) As the inductor countervoltage decreases, the circuit current increases and the voltage across the resistor increases.

Eventually the dc voltage across the inductor will drop to 0 volts. This is because the only resistance in the inductor is due to the resistance of the windings. Therefore the dc voltage drop ($I \times R$) is very small.

Assume now that maximum current is flowing through the inductor and resistor and that we wish to examine the circuit conditions if we place a short across the resistor and inductor instantaneously. This condition is shown in Fig. 14-15*d*. Note that current continues to flow through the inductor and resistor in the same direction, but at a continuously decreasing rate. After a certain time, therefore, the current through the inductor (and the voltage across it) decreases to zero. Also, the current through the resistor (and the voltage across it) decreases to zero.

To summarize, *RC* and *RL* circuits both produce a delay in the time required for the circuit current to reach its maximum or minimum value.

In Fig. 14-15*e* we have shown a circuit with an inductor and a resistor. Two switches are used. In the inductive circuit the energy is stored in the form of an electromagnetic field around the inductor. When SW_1 is open and SW_2 is closed, as shown in the illustration, there is current through the coil. Therefore there is a magnetic field across it.

The dotted line between the switches means that they are mechanically connected together. When SW_2 opens, SW_1 closes. When SW_1 closes, SW_2 opens.

If you open SW_2, the magnetic field around the inductor collapses. This induces a voltage across the inductor and causes current to flow through the resistor and SW_1. This current will decrease as the inductor-induced voltage drops to zero.

The circuit of Fig. 14-15*e* is shown for reference only. You will note that no curves have been included. There is a very important reason for this that you should understand as an electronic technician. When SW_2 is open, the current through the inductor is rapidly changing. The field around the inductor begins to collapse very rapidly and induces a very high voltage across the coil. The induced voltage is so high that there will be an arc across SW_2. The arc is actually a current flowing between the open contacts. In other words, you cannot instantly stop current flow in an inductive circuit. For this reason special circuitry is often used to avoid the arcing from the opening and closing of switches in inductive circuits.

By definition, the *time constant* of an *RC* circuit *T* is the amount of time that it takes a capacitor to charge to 63 percent of the applied voltage or to discharge to 37 percent of its initial voltage. These two points are clearly marked on the curves of Fig. 14-16. Note that the curve in Fig. 14-16*a* represents the voltage across a capacitor during charge. You can see that at the end of one time constant the voltage has reached 63 percent of the total voltage.

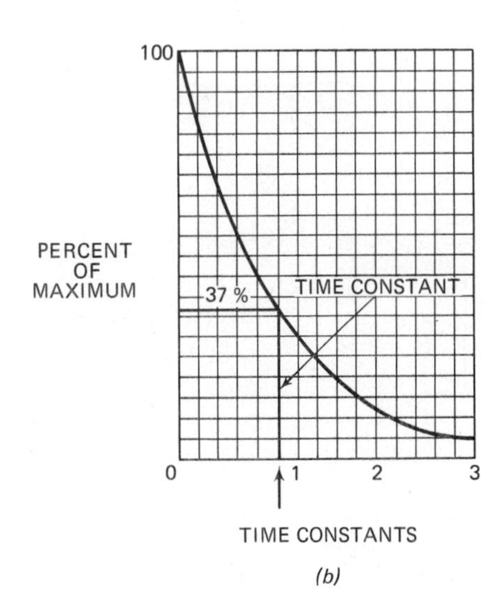

Fig. 14-16 Time-constant curves. (*a*) Charging curve. (*b*) Discharging curve.

Figure 14-16*b* shows a graph of the voltage across a capacitor as it is discharging. The voltage across the capacitor drops to 37 percent of the full value in one time constant.

It is fortunate that the calculation of time constant is very simple. For an *RC* circuit the time constant in seconds is simply the product of the resistance in ohms and the capacitance in farads:

$$T = RC$$

For inductive circuits the time constant can also be expressed:

$$T = \frac{L}{R}$$

where L = inductance, in henrys
R = resistance, in ohms
T = the time constant, in seconds

At the end of five time constants a capacitor is considered to be either fully charged or fully discharged.

Figure 14-17 shows an *RC* time-constant circuit to be used in this experiment. (A power supply will be used instead of a battery.) Here the resistance is 50 kilohms and the capacitance is 250 microfarads. The time constant of this circuit is

$$\begin{aligned} T &= RC \\ &= 50{,}000 \times 0.000\,250 \\ &= 12.5 \text{ seconds} \end{aligned}$$

This means that it would take 12.5 seconds for the capacitance in this circuit to charge to 63 percent of the applied voltage. In the same amount of time the

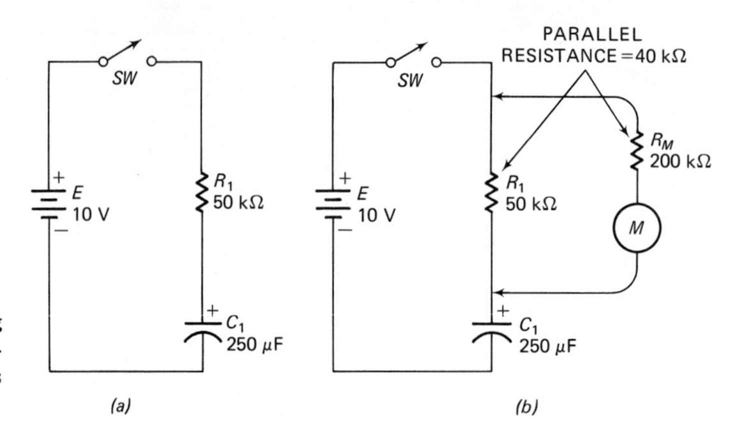

Fig. 14-17 Problem in observing time-constant effect. (*a*) The time-constant circuit. (*b*) Meter reduces circuit resistance.

voltage drop across the resistor will decrease to 37 percent of its maximum value. Since the applied voltage is 10 volts, 63 percent of the applied voltage is 6.3 volts, and 37 percent of the applied voltage is 3.7 volts.

You could watch the result of the time constant by connecting a voltmeter across R_1 in Fig. 14-17*a*. But many voltmeters are made by simply placing a meter multiplier resistor R_M in series with a sensitive meter movement M. This is shown in Fig. 14-17*b*. A typical multiplier value for such a meter on a 10-volt scale would be 200,000. If you place this 200,000 ohms in parallel with the 50,000 ohms, then the parallel resistance will be only 40,000 ohms. This will reduce the time constant, and it will take less time for the voltage across R_1 to drop to 37 percent of the applied voltage.

We will now look at some practical applications of a time-constant circuit. You will remember in the television system that the television signal consisted of both the vertical and horizontal sync pulses. It is necessary to separate these pulses in the sync separator. Only the vertical pulses should affect the vertical sweep, and only the horizontal pulses should affect the horizontal sweep.

In most TV sets the sync separator consists of simple *RC* circuits like the ones shown in Fig. 14-18. You can think of these circuits as being simple filters.

The vertical sync pulses occur at a rate of 60 per second. This is a low frequency. The *integrating circuit* (see Fig. 14-18) permits these low-frequency pulses to pass, but the high-frequency horizontal pulses are short-circuited to ground by capacitor C_1. This is a simplified explanation of how the integrating circuit works. Its design, however, is based on the fact that the time constant of R_1 and C_1 is long compared to the amount of time represented by the horizontal sync pulses.

For the *differentiating circuit* a sharp pulse will be produced at the front edge of each pulse. Thus the output will be a series of sharp pulses for the serrated vertical sync pulse and the horizontal sync pulses. These are all used to control the frequency of the horizontal scan oscillator at 15,750 hertz.

The output of the integrating circuit goes to the vertical oscillator and controls its frequency so that there are 60 fields per second. The output of the horizontal differentiating circuit goes to the horizontal oscillator and controls its frequency so that there are 15,750 lines of scan generated every second.

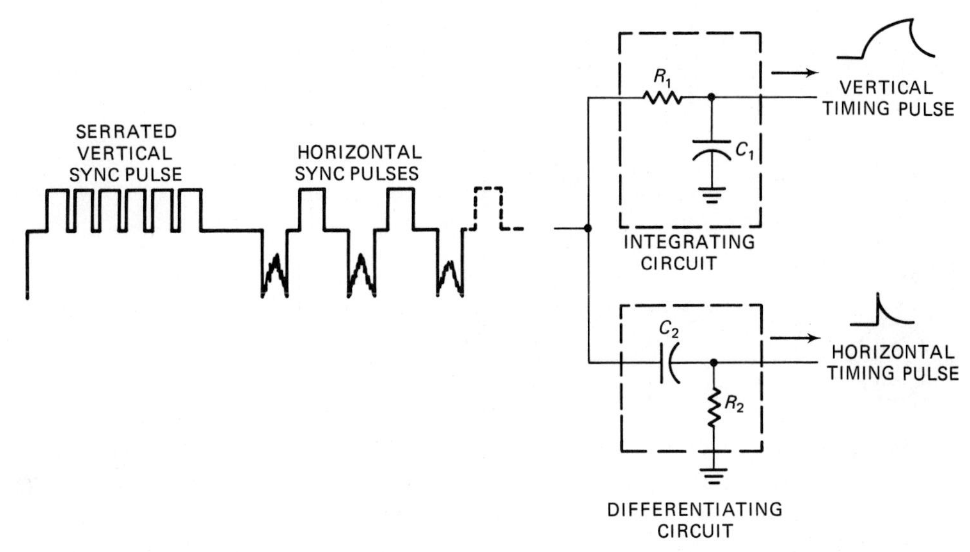

Fig. 14-18 Application of *RC* circuit.

Figure 14-19 shows another application of an *RC* circuit. This is the circuit for a *neon oscillator.* It uses an *RC* circuit to determine its frequency.

When the frequency of an oscillator is controlled by an *RC* circuit or by an *RL* circuit, it is called a *relaxation oscillator.*

When the switch *SW* is closed, capacitor C_1 begins to charge through R_1 (solid arrows). The neon lamp cannot conduct until it reaches a certain minimum value, called the *firing potential.* Since the voltage across the capacitor is very low at the instant the switch is closed, the neon lamp is OFF and no current can flow through it.

When the voltage across the capacitor reaches the firing potential of the neon lamp, it begins to conduct. This is indicated by a glow of the lamp. Its resistance value is low when it is glowing, and the capacitor is discharged. The path of discharge current is shown by the dotted arrows.

After the capacitor has discharged to a lower voltage value, the lamp goes OFF and the capacitor again begins to charge. The overall result is that the neon lamp flashes ON and OFF at a rate depending upon the time constant of R_1 and C_1. The simple relaxation oscillator shown in Fig. 14-19 is sometimes used as a highway flasher. The neon lamp, of course, is very large in that case.

The *multivibrator* shown in Fig. 14-20 is a very important relaxation oscil-

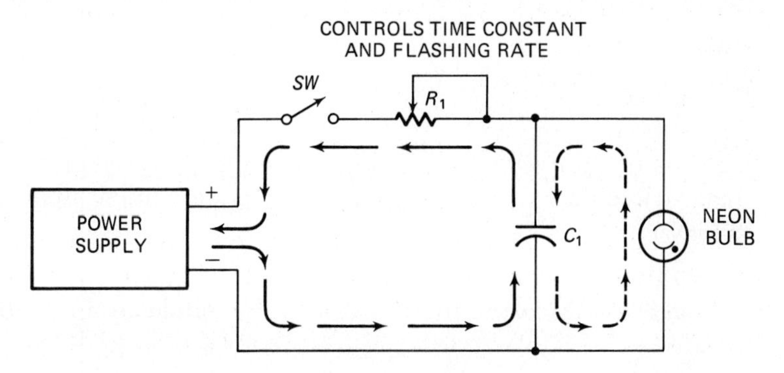

Fig. 14-19 A neon oscillator.

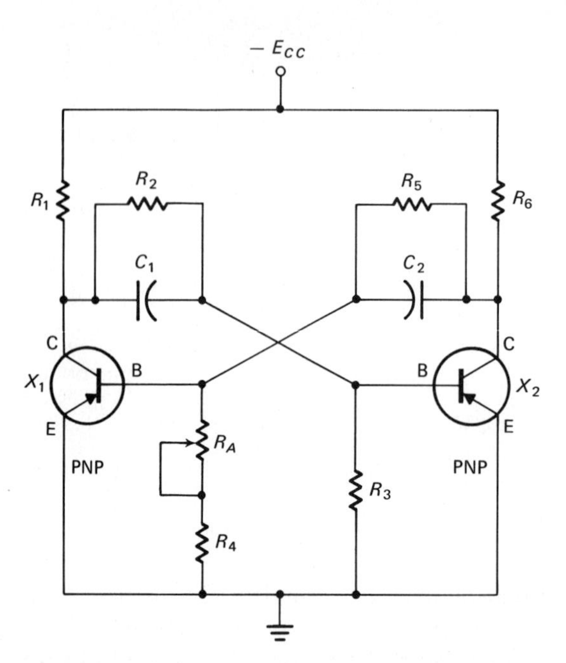

Fig. 14-20 A multivibrator circuit.

lator. The principle of its operation is based on the fact that there are two transistors and that each one controls the conduction of the other. In other words, when X_1 is conducting it shuts X_2 OFF, and when X_2 conducts it shuts X_1 OFF. Each conducts for a short period of time, and oscillation is thus produced.

The amount of time that each transistor conducts is dependent upon the time constant of the two *RC* circuits. One *RC* circuit is R_1, R_2, R_3, and C_1. The other time constant circuit is R_4, R_A, R_5, R_6, and C_2. Since the time constant of these two circuits determines how long each transistor conducts, they control the frequency of oscillation. Changing the time constant of either (or both) of the circuits will change the frequency at which the circuit oscillates. Multivibrators are often used for generating vertical and horizontal scan signals in television receivers.

In this experiment you are going to hook up a simple time-constant circuit and draw a time-constant curve. You are also going to hook up a multivibrator and show that the frequency of oscillation depends upon the time constant of the *RC* circuit.

PART I

Test Setup Wire the circuit shown in Fig. 14-21. Figure 14-21*a* shows the schematic drawing and Fig. 14-21*b* shows the pictorial diagram. Since the meter is in parallel with resistor R_1, the circuit resistance is going to be less than 47 kilohms. The actual resistance is dependent upon the resistance of the voltmeter.

Procedure

step one Measure the power-supply voltage and record the value.

E_A = ______________

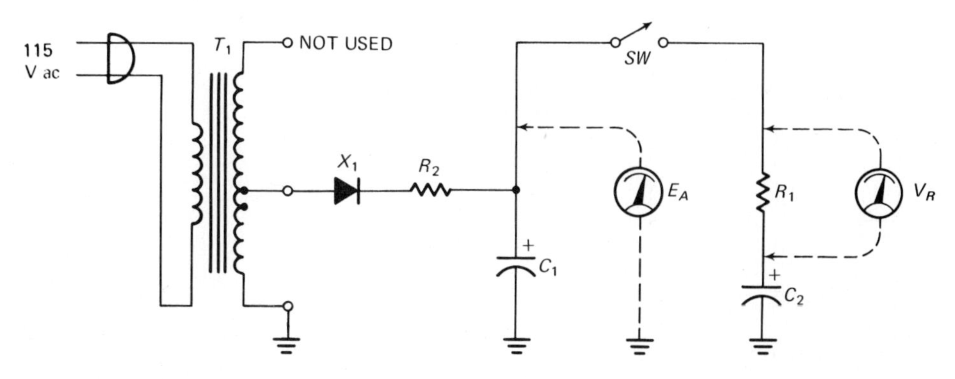

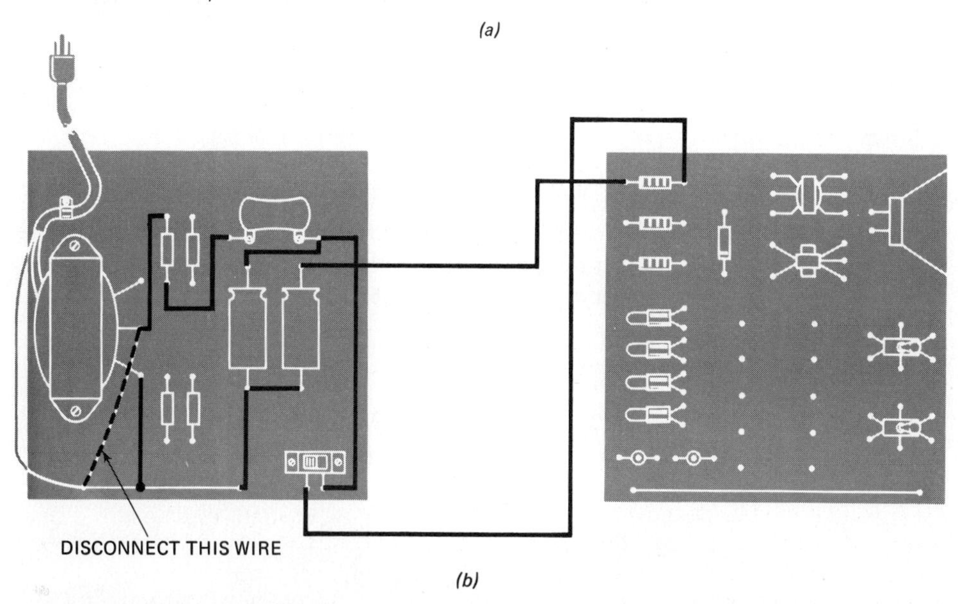

Fig. 14-21 Time-constant setup. (*a*) Schematic drawing. (*b*) Pictorial diagram.

step two The voltage across the resistor will drop to 0.37 times the power-supply voltage when the switch is closed. What is 0.37 times the value of voltage you measured in Step 1?

$$0.37 \times E_A = \underline{\hspace{5cm}} \text{ volts}$$

step three Make sure the switch for the circuit is open. Momentarily connect a jumper across the capacitor to make sure that the capacitance voltage is 0 volts.

step four Close the switch and record the voltage across the resistor V_R at 10-second intervals. Record your voltage readings in Table 14-2.

step five Using the values from Table 14-2, draw a graph showing the voltage across R_1 versus the amount of time. Your graph should look something like the one shown in Fig. 14-22.

Table 14-2

Time, in Seconds	Voltage, in Volts
0	
10	
20	
30	
40	
50	
60	
70	
80	
90	
100	

step six In the graph you have drawn, determine how long it takes the voltage to drop to 37 percent of the applied voltage. Record the value here. (The voltage you found in Step 2 is 37 percent of the applied voltage.)

____

step seven Calculate the time constant for the circuit.

____

$$\begin{aligned} T &= RC \\ &= 47{,}000 \times 0.000\,250 \\ &= 11.75 \text{ seconds} \end{aligned}$$

step eight Determine the time constant of the circuit graphically from the graph you drew in Step 5. To do this, draw a horizontal line from the point that is 0.37 times the maximum voltage. From the point where this voltage strikes the curve, draw a vertical line to the time value. The procedure is shown in Fig. 14-23. Use the voltage and time values of Table 14-2.

step nine Does the graphical time-constant value equal the time-constant value that you calculated in Step 7? ____________
Yes or No

Your graphical value of time constant may be less than the calculated value. Remember that if two resistors are connected in parallel, their resistance is

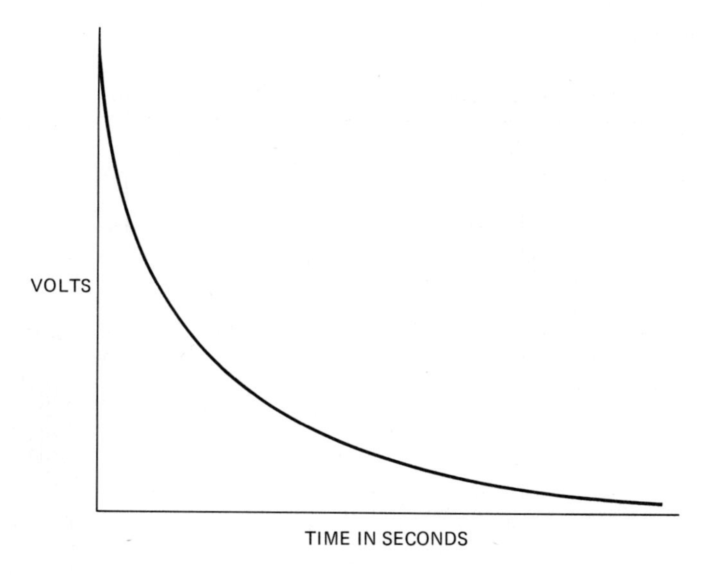

Fig. 14-22 Your time-constant curve should look like this.

always less than the resistance of the smaller one. Therefore you must have less than 47,000 ohms when the voltmeter is connected across the 47-kilohm resistor. But remember that both the resistor R_1 and the capacitor C_2 have tolerances and will be different from the stated values. So you cannot predict the graphical value of the time constant, but can only obtain it from your graph (Fig. 14-23).

PART II

Test Setup The circuit is shown in Fig. 14-24. The schematic is shown in Fig. 14-24*a*, and the pictorial diagram is shown in Fig. 14-24*b*. Note that this circuit is different from the multivibrator circuit of Fig. 14-20 because of the

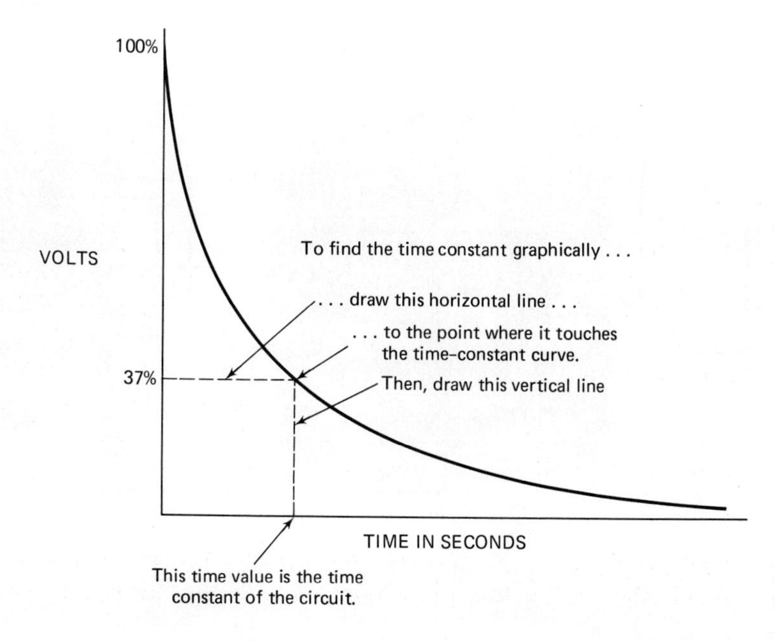

Fig. 14-23 Graphical method of finding the time constant.

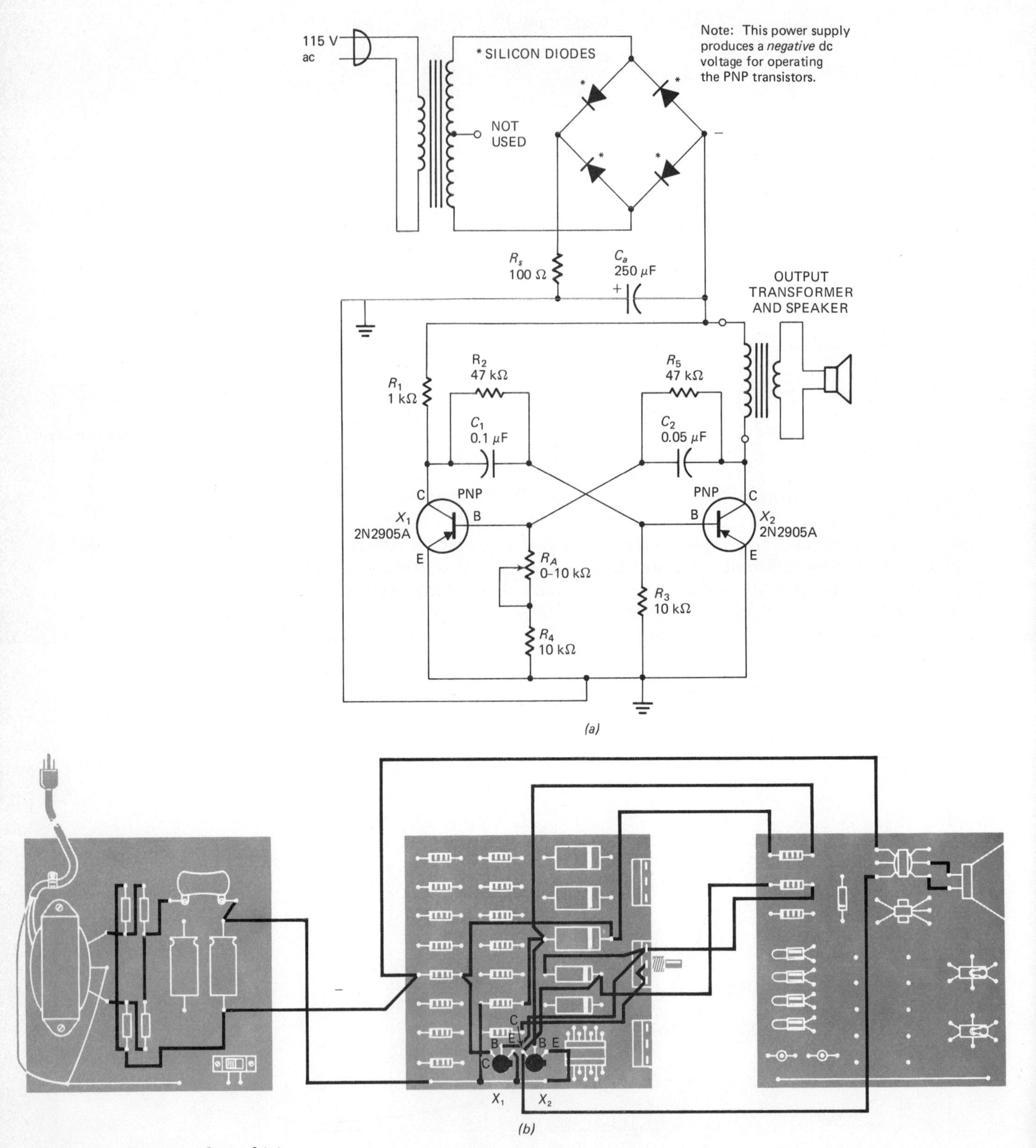

Fig. 14-24 Test setup for multivibrator experiment. (*a*) Schematic drawing. (*b*) Pictorial diagram.

speaker in place of resistor R_6. This relaxation oscillator will actually operate the speaker, but you may not get a loud volume.

Procedure

step one Wire the circuit of Fig. 14-24.

step two Apply the voltage.

step three Listen carefully to the tone in the speaker. In your opinion, is this a high frequency, medium frequency, or low frequency?

____

step four Adjust resistor R_A while listening to the frequency in the speaker. Does the frequency change? ______________
Yes or No

When you change the resistance value, the frequency of the tone should change. The value of R_A affects the time constant. This experiment shows that the RC time constant affects the frequency of oscillation.

Conclusion In Part I of this experiment you showed that the time constant of an RC circuit can be determined graphically. However, in this case the time constant of the circuit must be very long so that you have time to make accurate readings.

You learned in Part II that RC time-constant circuits can be used in an oscillator and that the frequency of the oscillation is dependent directly upon the RC time constant.

These are not the only applications in electronics where you will encounter time-constant circuits, but they are two very popular examples.

SELF-TEST WITH ANSWERS

(Answers with discussions are given at the end of the chapter.)

1. Here are two equations for time constant:

$$\text{(I)} \quad T = RC$$

$$\text{(II)} \quad T = \frac{L}{R}$$

   Which of the following statements is correct regarding these equations? (*a*) Both equations are wrong; (*b*) Only Equation I is correct; (*c*) Only Equation II is correct; (*d*) Both equations are correct.
2. Which of the following is the correct name for an oscillator that depends upon a time constant for its operation? (*a*) Pierce; (*b*) Armstrong; (*c*) Crystal; (*d*) Relaxation.

3. Which of the following is the higher frequency? (*a*) Field frequency; (*b*) Frame frequency.
4. The lighted rectangular area on a receiver picture tube is called the (*a*) AGC; (*b*) time out; (*c*) reflex; (*d*) raster.
5. The difference between the picture carrier frequency and the sound carrier frequency is (*a*) 4.5 megahertz; (*b*) 3.58 megahertz; (*c*) 10.9 megahertz; (*d*) 455 kilohertz.
6. A certain AGC circuit has a stage that permits only the sync pulses to be used for generating the dc AGC voltage. This stage is called (*a*) a ringing AGC circuit; (*b*) a tryon AGC circuit; (*c*) a keyed AGC circuit; (*d*) a delayed AGC circuit.
7. The section of the receiver that permits you to select only one station is the (*a*) tuner; (*b*) detector; (*c*) video amplifier; (*d*) retractor.
8. The type of TV voltage supply that depends on the operation of the horizontal output stage is called the (*a*) bridge rectifier; (*b*) doubler; (*c*) high-voltage power supply; (*d*) rectodyne.
9. The number of horizontal lines generated in a TV receiver each second is (*a*) 60; (*b*) 30; (*c*) 15,750; (*d*) $262\frac{1}{2}$.
10. The part of the TV signal that causes the vertical retrace on the receiver CRT is (are) the (*a*) blanking pedestals; (*b*) vertical sync pulses; (*c*) subcarrier; (*d*) video signal.

ANSWERS TO SELF-TEST

1. (*d*)
2. (*d*)—Pierce, Armstrong, and crystal oscillators are all forms of sine-wave generators. A relaxation oscillator generates a nonsinusoidal waveform. As explained in the chapter, a relaxation oscillator uses either an *RC* or an *RL* circuit.
3. (*a*)—Since there are two fields per frame, the field frequency is twice the frame frequency.
4. (*d*)
5. (*a*)—It is this 4.5-megahertz frequency difference that is used for the sound intermediate frequency.
6. (*c*)—A delayed AGC system is one in which there is no AGC control voltage for very weak signals. Any type of AGC circuit can produce this.
7. (*a*)
8. (*c*)—Bridge rectifiers and voltage doublers are used in low-voltage power supplies. There is no such thing as a rectodyne power supply.
9. (*c*)
10. (*b*)—The vertical sync pulse is located on top of the vertical blanking pedestal.

what are the basic logic circuits used in computers and industrial electronics? 15

INTRODUCTION

Since the golden age of Greece, philosophers have been trying to devise systems of logic that would enable them to reason in a step-by-step procedure. Early Greek philosophers had established the basic rules related to *syllogisms*—a form of logical argument in three steps. A syllogism consists of a main statement called the *major premise,* a statement about the major premise, called a *minor premise,* and a *conclusion.* In its simplest form a syllogism looks something like this:

All people are mortals	(Major premise)
Socrates is a person	(Minor premise)
Therefore Socrates is a mortal	(Conclusion)

Through the years philosophers developed this system of logic to a very high degree of refinement. By the early 1800s the rules for formal logic (of which a syllogism is an example) were fairly well established.

There has always been a language problem in trying to arrive at a conclusion through formal logic. Language makes it hard to express major premises and minor premises which are related and which are true. Much of the problem lies in the fact that many words have double meanings. Another problem is that there are words for which there are no clear meanings. When these words are used in formal logic, the conclusions may be valid according to the rules, but they may at the same time have no relation to the real physical world.

In 1854 George Boole printed a paper which tried to get around the language problem in formal logic. His idea was to eliminate the pitfalls of word meanings and emotional words which make it difficult to follow arguments. Boole devised a method of reducing the logic statements to symbols which are very much like mathematical symbols. When statements are reduced to symbols, it is a simple mathematical procedure to arrive at the correctness (or incorrectness) of a conclusion.

The laws of boolean algebra have been greatly enlarged upon and perfected since Boole introduced them. In the study of electronics boolean algebra is used for designing switching circuits and logic circuits. A complete study of boolean algebra is a lifetime work, and it cannot be completely discussed in a single chapter. However, we will present some basic logic circuitry and show how these circuits can be used in combinations. The output signals of these circuits will be identified with boolean algebra symbols.

You will be able to answer these questions after studying this chapter.

- ☐ **What are some examples of numbering systems?**
- ☐ **What are logic circuits?**
- ☐ **What are NOT circuits?**
- ☐ **What are OR circuits?**
- ☐ **What are NOR circuits?**
- ☐ **What are NAND circuits?**
- ☐ **What are flip flops?**

INSTRUCTION

What Are Some Examples of Numbering Systems?

When you were much younger you learned to count by tens. In other words, your system of counting has 10 digits. They are 0 through 9. This system of counting probably developed because you have 10 fingers and 10 toes.

It is possible to count in other numbering systems as well as by 10s. You are already familiar with one system called the *duodecimal* system. In that system you count by 12s rather than by 10s. This system may have developed as a result of the fact that there are about 12 full moons in a year. We still have traces of the duodecimal system in our civilization. For our method of keeping time the clock is divided into two 12-hour periods for one day. There are 60 minutes in an hour (five 12s) and 60 seconds in a minute (again, five 12s).

If you had learned to count by 12s, the duodecimal system would seem to be quite convenient. However, since you are trained in the decimal system, the duodecimal system seems hard to work with. For example, if you want to know the number of minutes and seconds between 3:51 P.M. and 7:23 A.M., it is not a simple matter of putting one number over the other and subtracting.

You can use any number as a base (called the *radix*) for a numbering system. In electronics a logical numbering system is one that has only two digits. The reason for this is that so many of our transducers and electronic components are capable of producing only two distinct outputs. A switch is ON or OFF. A transistor is *conducting* or *cut off*. A capacitor is *charged* or *discharged*. A diode is *forward-biased* or *reverse-biased*. These are a few examples of components with two distinct outputs, or *states*. No doubt you could name many others. A component with two distinct states can be used for counting. It is much more difficult to obtain components which have 10 distinct outputs for use with the decimal system.

The numbering system based on only two symbols is called the *binary system*. Since much of the logic and control systems are based on binaries, we will take a brief look at this system of counting.

How Do You Count with Binary Numbers?

In the decimal system—that is, the system of counting by 10s—you use 10 digits to represent the numbers. In other words, the decimal system has a radix of 10. These digits are 0 through 9. When you reach the number 9, you have used all the digits. To go one number higher you must begin using the digits in combinations. Therefore, 10 is represented by a 1 and a 0, 11 by a 1 and 1, and so forth. When you reach the number 99, you have used all the two-digit combinations possible, and it is necessary to use three digits in combinations for numbers above 99.

To count in the binary system you have only two symbols to represent the numbers. These symbols are 0 and 1. After you have reached the number 1 in counting with binaries, it is necessary to begin using the 0 and 1 in combinations. To see how this is done, refer to Table 15-1. In this table the decimal numbers from 0 to 16 are given. You are already familiar with this numbering system. In the binary numbering system you can count 0 and 1, but in order to go above 1 you must begin to use the two symbols in combinations. Likewise, when you reach the binary number for 4 (100) and the binary number for 8 (1000) you must again add a digit.

For comparison the octal number system, which has eight symbols, is also shown in Table 15-1. In the octal system, after you have reached the number 7, you must begin to use the symbols in combination.

From Table 15-1 it can be clearly seen that it is going to take a lot more digits to represent large binary numbers than are needed for large numbers in the decimal system. In spite of this the binary numbering system is much more convenient for use in control systems and computers. As mentioned before, this is because of the large number of components which are readily available with two distinct levels of operation.

Figure 15-1 illustrates some components used for producing binary signals.

Table 15-1 Comparison of Number Systems

Decimal Number	Binary Number	Octal Number
0	0	0
1	1	1
2	10	2
3	11	3
4	100	4
5	101	5
6	110	6
7	111	7
8	1000	10
9	1001	11
10	1010	12
11	1011	13
12	1100	14
13	1101	15
14	1110	16
15	1111	17
16	10000	20

As shown in Fig. 15-1*a*, a switch can be used to represent binary numbers. When the switch is open, it represents the binary number 0, and when the switch is closed, it represents the binary number 1. This is an arbitrary choice, because you could just as easily use the switch to represent the number 1 when it is open and the number 0 when it is closed. Either way,

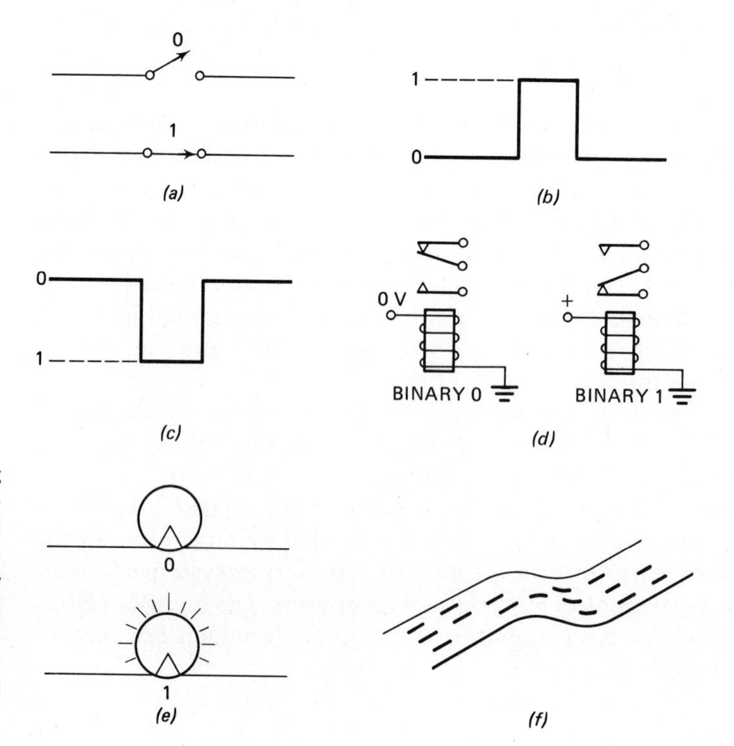

Fig. 15-1 Methods of representing binary numbers. (*a*) Switch positions can represent binary numbers. (*b*) A pulse used for positive logic. (*c*) A pulse used for negative logic. (*d*) A relay can be used to represent binary numbers. (*e*) Light bulbs can represent binary numbers. (*f*) Punched tape is coded for 0 and 1.

the switch has actually only two conditions, *open* and *closed*. Therefore it is ideal for representing binary numbers.

Figure 15-1*b* shows how a pulse can be used to represent binary numbers. In this illustration, when the voltage is at the 0-volt value, it represents the number 0. When the voltage is at its maximum amplitude, it represents the number 1. When pulses are used in this manner to represent binaries, they are said to be *positive logic*.

Figure 15-1*c* shows a pulse representation of *negative logic*. The 0-volt level represents the binary 0, but a negative-going pulse is used to represent the binary 1.

Relays are ideal for representing binary numbers. An example is shown in Fig. 15-1*d*. A relay only has two conditions, *energized* and *not energized*. The energized condition is usually used for representing digital number 1, and the deenergized condition represents the number 0.

Figure 15-1*e* shows how light bulbs can be used to represent binaries. In the illustration, when the lamp is ON, it represents the binary 1, and when the lamp is OFF, it represents the binary 0. This is the standard way to represent binary numbers with lamps.

Figure 15-1*f* shows a very important way of representing binaries. It is a punched tape. A hole in the tape represents the binary 1, and a place where there is no hole represents the binary 0. The punched tape is useful because it can be used to store a large number of binary numbers. The stored binaries can be converted to electrical commands for operating machinery. Magnetic tapes are also used instead of punched tapes for the same purpose. A magnetized region is 1, and a nonmagnetized region is 0.

Figure 15-1 does not show all the possible ways to represent binaries, but rather typical methods which you are likely to see used in electronics equipment. Diodes, triodes, transistors, and FETs also can be used to represent binary numbers.

Figure 15-2 shows an example of how a transistor can be used to represent the binaries. In Fig. 15-2*a* the transistor is cut off, and the voltage drop across the emitter resistor is 0 volts. This is the output of the transistor circuit and represents the binary 0. The reason the transistor is cut off is because there is no forward bias between the emitter and base. You will remember that there must be an emitter-to-base current flow in a transistor before current can flow in the emitter and collector circuit.

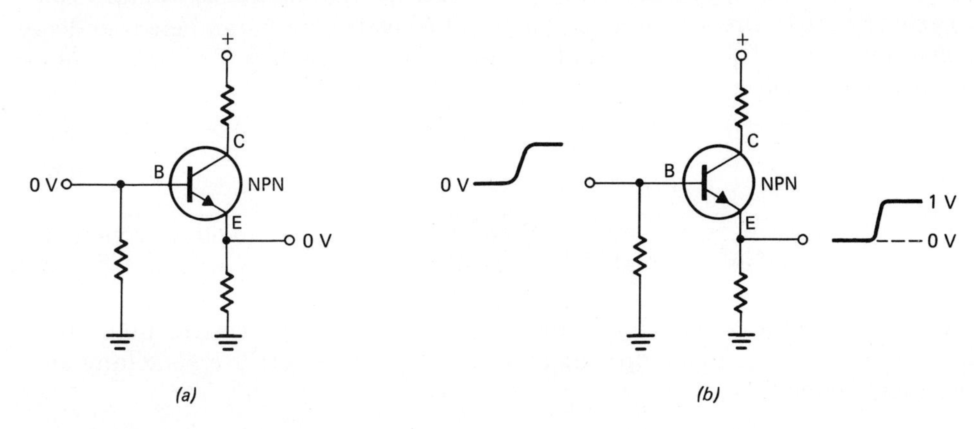

Fig. 15-2 A transistor can be used to represent binaries 0 and 1. (*a*) This transistor is in the 0, or *low*, state. (*b*) A positive-going pulse switches it to the 1, or *high*, state.

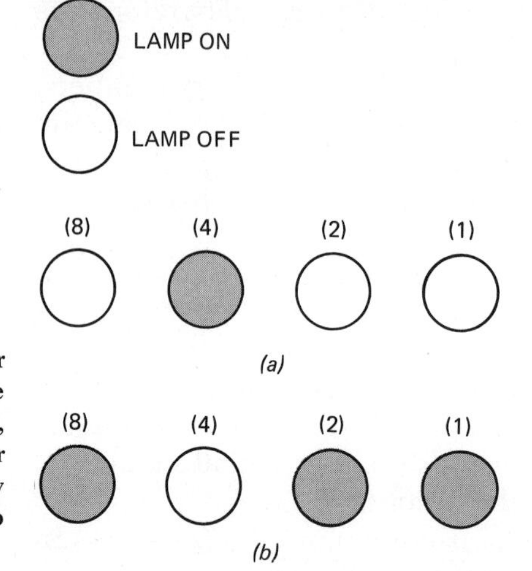

Fig. 15-3 A lamp display for showing binary numbers. (*a*) The lamps show binary number 0100, which is equal to decimal number 4. (*b*) The lamps show binary number 1011, which is equal to decimal number 11.

In Fig. 15-2*b* a positive pulse is applied to the base of the transistor. This causes a flow of emitter-to-base current, which in turn causes a relatively large collector current. As a result the emitter goes positive and represents the binary 1. Since the transistor circuit of Fig. 15-2 has only two possible conditions (conducting and nonconducting), it is very useful for representing binary numbers.

Figure 15-3 shows how four lamps can be used to represent binary numbers up to and including the number 15. There are four lamps, marked 1, 2, 4, and 8. These positions are sometimes marked 2^0, 2^1, 2^2, and 2^3. In either case each lamp represents a binary digit. The binary numbers are represented by turning lamps ON and OFF. In Fig. 15-3*a* only the lamp representing four (2^2) is ON. You can read the lamps directly as a binary number. In this case it is 0100. Table 15-1 shows that this is equal to decimal number 4.

Figure 15-3*b* shows the number 1011 represented with binary lamps. As shown in Table 15-1 this represents the decimal number 11.

While it is true that computers and control systems can operate more easily with binary numbers, it is not true that people can readily adapt to these binary systems. Therefore there are electronic circuits which convert the decimal system into binary numbers. They are called *decimal-binary converters.* Also there are circuits which can convert binary numbers into decimal numbers. They are called *binary-decimal converters.* Normally these converters come in small integrated-circuit packages.

Summary

1. George Boole developed a method of writing formal logic arguments in symbols. This eliminated the pitfalls of emotional words and double meanings for words.
2. Boolean algebra has been brought to a high degree of perfection. It is used in designing computers and control circuits that use switching and logic circuits.

3. There is a wide variety of electrical and electronic components that have two distinct levels of operation, but very few that have ten distinct levels. For this reason computer and control systems operate with binary signals.
4. The number of different symbols used in a numbering system is equal to its base, or *radix*. For example, the radix in the decimal system is 10.
5. The radix of the binary numbering system is 2. The symbols used are 0 and 1.
6. To convert from the decimal system to the binary system, an electronic circuit called a *decimal-binary converter* can be used.
7. A *binary-decimal converter* is an electronic circuit used for converting from the binary to the decimal system.

What Are Logic Circuits?

The logic circuits within a computer or control system operate with binary-input and binary-output signals. The basic logic circuits will be discussed here by using switches and relays. However, remember that in every case a diode, tube, transistor, or FET can be used in place of the relay because these electronic components can be operated as ON/OFF devices. Use of relays here is only for the purpose of simplifying the discussion of basic logic circuits.

What Are NOT Circuits?

A NOT *circuit* is one in which the output is the inverse of the input. For this reason NOT circuits are also known as *inverters*. Figure 15-4 shows the principle involved.

In the relay circuit of Fig. 15-4*a* the lamp is connected through the normally closed contacts of the relay. Operating switch *A* will cause current to flow through the relay coil, energizing it and switching the relay terminals. This will open the circuit for the lamp, and the lamp will be OFF. We can

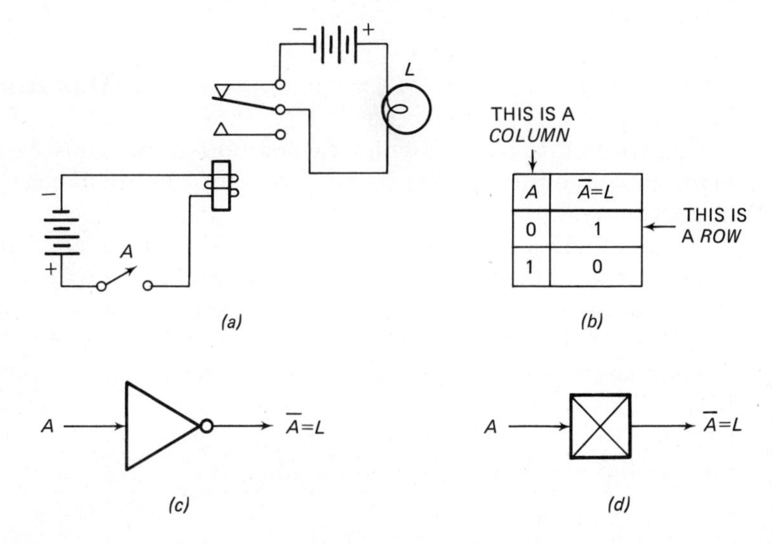

A	Ā=L
0	1
1	0

Fig. 15-4 The NOT circuit. (*a*) A relay NOT circuit. (*b*) Truth table for the NOT circuit. (*c*) MIL symbol for a NOT circuit. (*d*) NEMA symbol for a NOT circuit.

describe the operation of this circuit by saying the lamp is ON whenever switch A is OFF (or open), and the lamp is OFF whenever switch A is ON (or closed).

A simpler way to describe the operation of the NOT circuit is to use the *truth table.* This is shown in Fig. 15-4*b*. A truth table simply tells all the conditions under which the circuit operates. The horizontal direction in the truth table is called a *row,* and the vertical direction is called a *column.* The first row of the truth table shows that when the condition of switch A is open (symbolized by 0), the lamp is ON (symbolized by 1). The second row shows that when the switch is closed (symbolized by 1), the lamp is OFF (symbolized by 0). The truth table tells all the possible conditions of operation for the simple NOT circuit of Fig. 15-4*a*.

The symbol $\overline{A}$ in the right column of the truth table is a logic symbol meaning NOT A. You should read $\overline{A} = L$ as "NOT A equals L." This means that when the switch is open (not closed), the lamp is ON.

There are a number of different ways to show the schematic symbols for a logic circuit. The two most common ones are the MIL symbol (Fig. 15-4*c*) and the NEMA symbol (Fig. 15-4*d*). For each of the logic circuits discussed in this section, we will show both symbols. As a technician you should become familiar with both types of symbols and should learn to work with either type on an illustration. Notice that on the MIL symbol there is a small circle at the output of a standard amplifier symbol. That small circle on a MIL symbol always indicates that the output (or input) is inverted.

Summary

1. A truth table can be used to show all the possible operating conditions for a logic circuit.
2. A NOT circuit is also known as an inverter.
3. An overbar is used to indicate NOT. Thus, $\overline{A}$ means NOT A.
4. To represent logic circuits on schematic diagrams there are two types of symbols, MIL and NEMA. You should know both types.

What Are AND Circuits?

An *AND circuit* is one in which two or more inputs must be present in order to produce an output. Figure 15-5 shows the important characteristics of AND circuits.

Figure 15-5*a* shows two switches (A and B) in series with a lamp L. In order for the lamp to be ON it is necessary that both switches must be closed. If an open switch represents binary 0 and a closed switch represents binary 1, then A and B must be in the 1 state in order for the lamp to be in the 1 state.

There are two inputs (A and B) required for the simple AND circuit of Fig. 15-5*a*. This is an example of a *two-input* AND. Some electronic AND circuits have five or ten inputs. The total number of inputs is called the *fan-in,* and the number of outputs is called the *fan-out.*

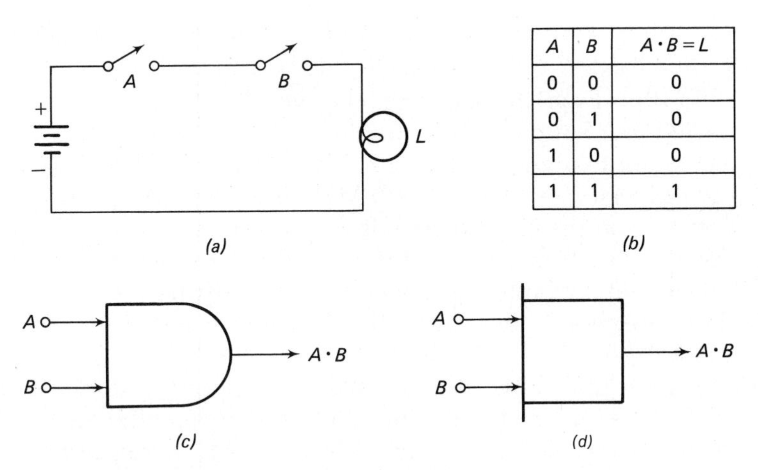

A	B	A·B=L
0	0	0
0	1	0
1	0	0
1	1	1

Fig. 15-5 An AND circuit. (*a*) A simple AND circuit. (*b*) Truth table for the AND circuit. (*c*) MIL symbol for an AND circuit. (*d*) NEMA symbol for an AND circuit.

In logic circuitry the symbol for AND is a times sign like the one used in ordinary arthimetic. All the following statements mean A AND B:

$$AB$$
$$A \cdot B$$
$$A \times B$$

The truth table of Fig. 15-5*b* tells the complete AND story. In this truth table columns A and B represent the conditions of the switches, and the last column represents the output, or condition of the lamp. The first row shows that when both switches are open (0) the lamp is OFF (0). In the second row switch A is open, but switch B is closed. With only one switch closed, the lamp is OFF. The third row shows that when switch A is closed and switch B is open, the lamp is OFF. The fourth row shows that when both switches are closed or in the 1 condition, the lamp is ON, or in the 1 condition. You should study this truth table carefully because it is an important way of representing AND circuits.

An AND circuit is also known as an *AND gate*. The MIL symbol for a two-input AND circuit is shown in Fig. 15-5*c*. The two inputs A and B must both be present to produce the output which is shown in logic form as $A \cdot B$ (read "A and B"). Figure 15-5*d* shows the NEMA symbol for the two-input AND circuit. The input and output for the NEMA symbol are the same as for the MIL symbol. If there were more inputs, there would be simply more terminals on the symbol.

Summary

1. With an AND circuit all the inputs must be present in order to get an output.
2. An AND circuit is also called an *AND gate*.
3. All the following mean A AND B: AB, $A \cdot B$, and $A \times B$.
4. In symbols the expression "the lamp is ON when switches A and B are both closed" can be written $AB = L$.

What Are OR Circuits?

In the English language the word *or* has two completely different meanings. Suppose, for example, you say "Mary or John can go to the store." There are two ways to interpret this sentence. One way to interpret it is to say that Mary, or John, or both Mary and John, can go to the store. In this interpretation the *or* is known as *INCLUSIVE OR* because it is possible for both of them to go. Another way to interpret this sentence is that either Mary or John can go, but not both. In that case it is called an *EXCLUSIVE OR*. In logic circuits there are two kinds of OR circuits, one for *INCLUSIVE OR* and one for *EXCLUSIVE OR*.

Figure 15-6 shows the INCLUSIVE OR circuit, truth table, and symbols. In Fig. 15-6*a* two switches are connected in parallel in the lamp circuit. If switch *A* is closed and switch *B* is open, the circuit for the lamp will be completed and the lamp will be ON. Likewise, if switch *B* is closed and *A* is open, the lamp will be ON. In this particular circuit, if both *A* and *B* are closed the lamp will be ON. Therefore this is an INCLUSIVE OR circuit. A plus sign is used to represent INCLUSIVE OR. $A + B = L$ means *A* or *B* equals *L*.

The truth table for the INCLUSIVE OR circuit is shown in Fig. 15-6*b*. The first row shows that if both *A* and *B* are in the 0 condition, or open, the lamp is OFF, or in the 0 condition. The second row shows that if *B* is closed and *A* is open, the lamp will be ON. Row three shows that if *A* is closed and *B* is open, the lamp will be ON. Row four shows that if both *A* and *B* are closed, the lamp is ON.

The symbols for INCLUSIVE OR circuits are shown in Fig. 15-6*c* and *d*. Figure 15-6*c* shows the MIL symbol, and Fig. 15-6*d* shows the NEMA symbol.

The circuitry for an EXCLUSIVE OR is slightly more complicated. Figure 15-7 shows an EXCLUSIVE OR relay circuit. Figure 15-7*a* shows the relay circuit when both relays are in the deenergized state. The lamp circuit is not completed. However, if either relay 1 or relay 2 is energized, the lamp circuit will be completed. The relays are energized by closing switches *A* and *B*. If you close both *A* and *B* at the same time, both relays will become energized, and the lamp circuit will be open. This is an EXCLUSIVE OR circuit because you can turn the lamp ON by switching either *A* or *B*, but not both.

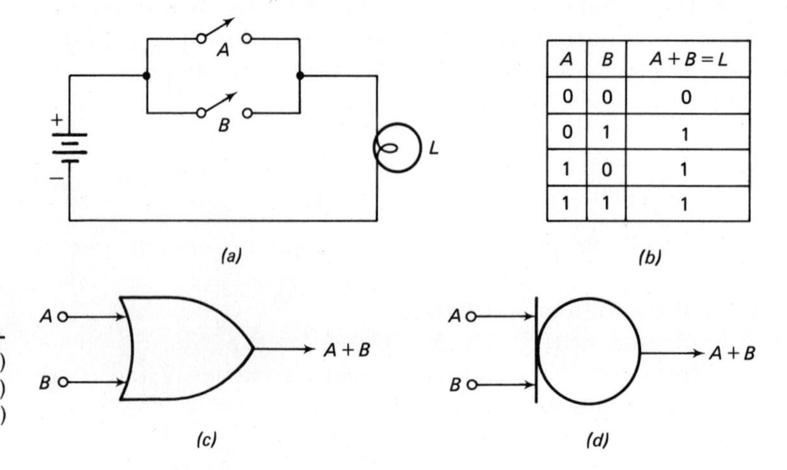

A	B	A + B = L
0	0	0
0	1	1
1	0	1
1	1	1

Fig. 15-6 The INCLUSIVE OR circuit. (*a*) A simple OR circuit. (*b*) Truth table for the OR circuit. (*c*) MIL symbol for an OR circuit. (*d*) NEMA symbol for an OR circuit.

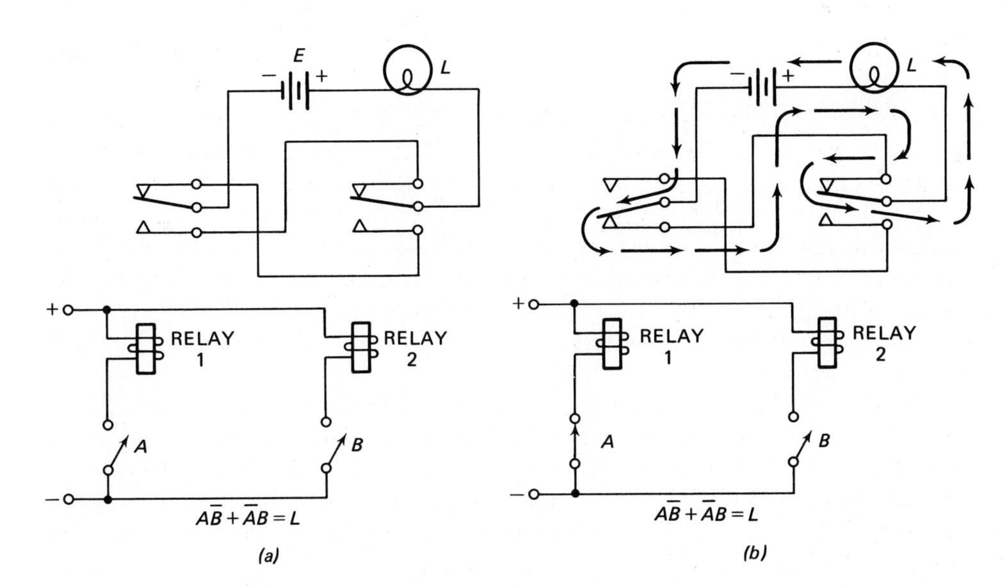

Fig. 15-7 An EXCLUSIVE OR circuit. (*a*) Both switches open, both relays are deenergized. (*b*) Switch A closed, relay 1 energized.

You should trace the lamp circuit carefully in the circuit of Fig. 15-7*a*. With both switches *A* and *B* open there is no complete path for battery *E* and lamp *L*. Figure 15-7*b* shows what happens when switch *A* is closed. Relay 1 is shown in the energized condition, but relay 2 is not energized. Follow the arrows that represent current in the circuit. Starting at the negative terminal of *E*, trace the current path through the closed contacts of relay number 1, through the closed contacts of relay number 2, through *L*, and back to the positive terminal of *E*.

The plus sign with a circle around it always means EXCLUSIVE OR. In symbols:

$$\overline{A}B + A\overline{B} = A \oplus B$$

Summary

1. There are two possible interpretations when the word *or* is used in a sentence.
2. "Switch *A* or *B* can be used to turn the lamp ON": If this means switch *A*, or switch *B*, or both, then it is an INCLUSIVE OR.
3. If the expression means switch *A* or switch *B*, but not both, then it is an EXCLUSIVE OR.
4. The expression "*A* or *B*" is written $A + B$. This is an INCLUSIVE OR.
5. $A \oplus B$ means *A* or *B*. This is one way of writing EXCLUSIVE OR in symbols.
6. $\overline{A}B + A\overline{B}$ is another way of writing EXCLUSIVE OR in symbols.

What Are NOR Circuits?

A NOR *circuit* is one in which there will be an output when the two or more inputs are not present. The NOR circuit is different from the NOT circuit in

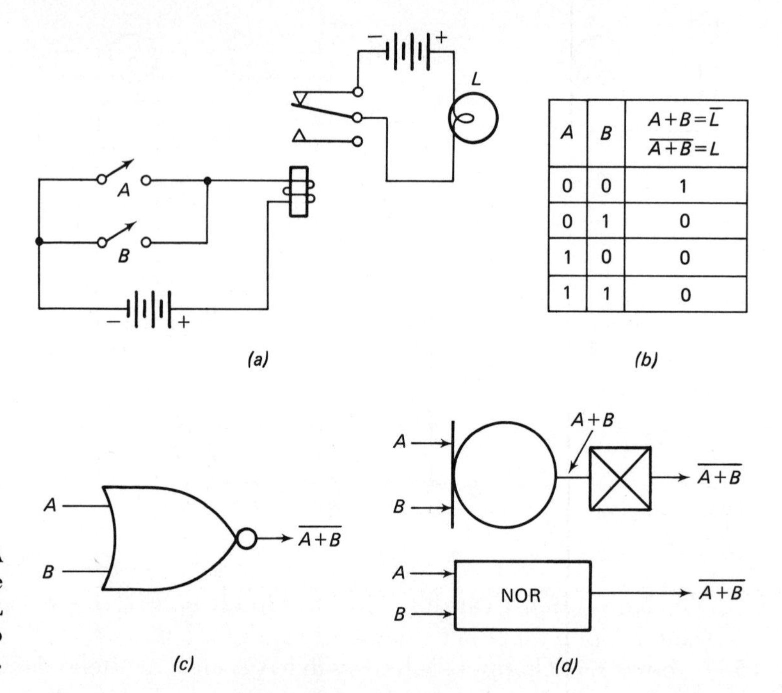

Fig. 15-8 **The NOR circuit.** **(*a*) A relay NOR circuit.** **(*b*) Truth table for the NOR circuit.** **(*c*) MIL symbol for a NOR circuit.** **(*d*) Two examples of NEMA NOR symbols.**

that the NOT circuit has only one input, while the NOR circuit has two or more inputs. Figure 15-8 shows the NOR circuit story.

Figure 15-8*a* shows a simple relay NOR circuit. When the relay is deenergized, the lamp circuit is closed through the normally closed contacts of the relay. If either *A* or *B* is in the 1 condition (or if both are in the 1 condition), the relay will be energized and the lamp circuit will be open. Therefore the only condition under which the lamp can be ON is if there is no input at either *A* or *B*.

Figure 15-8*b* shows the NOR truth table. The first row shows that if there is no input, there is an output. The second, third, and fourth rows show that if either or both of the inputs are in the 1 condition, there is no output.

Figure 15-8*c* and *d* shows the MIL and NEMA symbols for the NOR circuit. There are two NEMA symbols for a NOR circuit. One shows an OR circuit with the output negated, or inverted. The output of the OR in this symbol would be *A* or *B*, and after it is inverted it is NOT *A* or *B*. Another way to show the symbol is simply to show a box with NOR written in it. This is also shown in Fig. 15-8*d*.

What Are NAND Circuits?

A NAND *circuit* is one in which both inputs must be in the 1 condition in order for the output to be in the 0 condition. Figure 15-9 shows the NAND story.

Figure 15-9*a* shows a simple relay NAND circuit. Note that the lamp is ON only when *A* and *B* are both in the 0 condition. If either *A* or *B* is in the 1 condition, but not both, the lamp will still be ON. The only way for the lamp to be turned OFF is if *A* and *B* are both in the 1 condition.

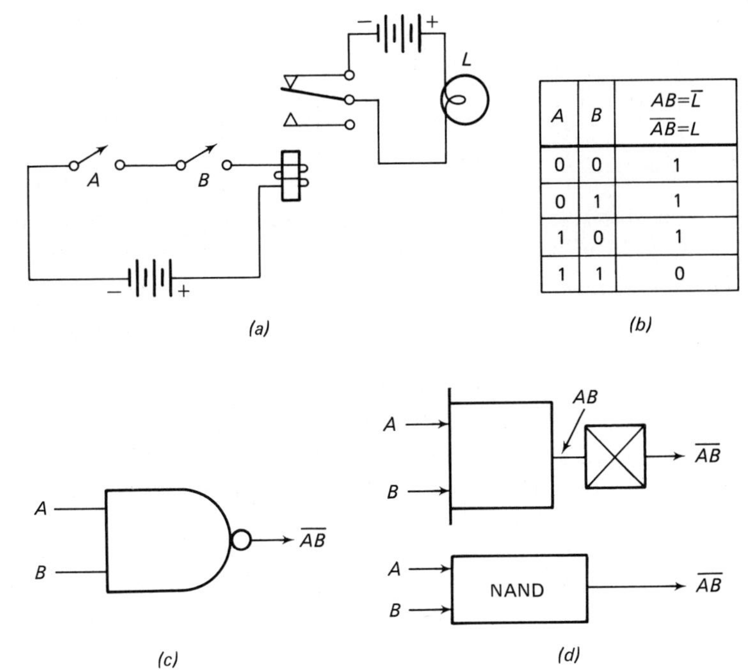

Fig. 15-9 **The NAND circuit.** ***(a)*** **A relay NAND circuit.** ***(b)*** **Truth table for the NAND.** ***(c)*** **MIL symbol for a NAND circuit.** ***(d)*** **Two examples of NEMA NAND symbols.**

The NAND truth table is shown in Fig. 15-9*b*. The first three rows in this truth table show that if either or both of the switches are in the OFF condition, the light is ON. The fourth row shows that if both of the inputs are in the 1 condition, the lamp is OFF.

The MIL symbol for the NAND circuit is shown in Fig. 15-9*c*, and Fig. 15-9*d* shows two ways to draw a NEMA symbol. One way is to use an AND circuit with an inverter at the output. The output of the AND is *A* and *B*. After it is inverted, it becomes NOT *A* and *B*. Another way to show a NAND circuit with a NEMA symbol is just simply to show a box with the word NAND on it. This is also shown in Fig. 15-9*d*.

Summary

1. In a NOR circuit there will be no output if one or both of the inputs is present.
2. A NOR circuit can be made by using an OR gate and a NOT circuit in series.
3. The symbolic way of saying that a lamp *L* is not ON in a circuit, provided an input *A* or *B* is present, is: $A + B = \overline{L}$. From the truth table of Fig. 15-8*b* you can see that it is also true that $\overline{A + B} = L$. In other words, the light is in the 1 condition when neither *A* nor *B* is ON.
4. In a NAND circuit there will be an output only if all the inputs are not present.
5. A NAND circuit can be made by using an AND gate and a NOT circuit in series.
6. The symbolic way of saying that a lamp *L* is OFF in a circuit when either or both the inputs *A* and *B* are present is $AB = \overline{L}$. From the truth table you

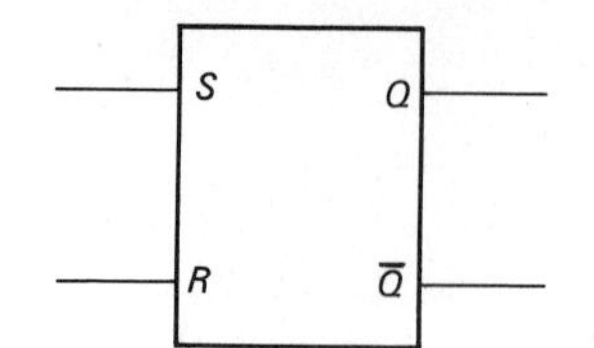

Fig. 15-10 Symbol for a basic flip flop.

can also see that $\overline{AB} = L$. In other words, the lamp is in the 1 condition when A and B are both OFF.

What Are Flip Flops?

In Chapter 11 you studied the operation of multivibrators, which were described as being astable circuits. They are made with two amplifying devices, and one or the other device—but not both—can be conducting at any instant. In operation, conduction continually shifts back and forth between the two conducting devices.

A *flip flop* is closely related to the multivibrator because it also has two amplifying devices that have two stable states. However, unlike the multivibrator, it is *bistable*—that is, it can remain indefinitely with one or the other (but not both) of the amplifying devices conducting.

Figure 15-10 shows a simple representation of a flip flop. There are two input signals, marked S and R, which stand for *set* and *reset*, respectively. This type of flip flop is called R-S. The output terminals are marked Q and $\overline{Q}$ on this illustration, but other identifying characters are sometimes used for the output.

There are two possible conditions for the flip flop. One is when the output of the Q terminal is high and the output of the $\overline{Q}$ terminal is low, and the other condition is when the output of the Q terminal is low and the output of the $\overline{Q}$ terminal is high. When a terminal output is *high,* it is in the binary 1 condition; and when it is said to be *low*, it is in the binary 0 condition.

The only two possible conditions for operating the R-S flip flop are shown in Fig. 15-11. The condition shown in Fig. 15-11*a* is known as the ON condition or *high* condition of the flip flop. Thus when the Q terminal is high the flip flop is in a *high* condition. When the Q terminal is low, as shown in Fig. 15-11*b*, the flip flop is in the *low* condition. It is very important that you understand that either output of the flip flop can be high or low, but the two outputs must always be in an opposite condition—that is, it is not possible for both the Q and $\overline{Q}$ to be in the high condition or for both the Q and $\overline{Q}$ terminals to be in the low condition.

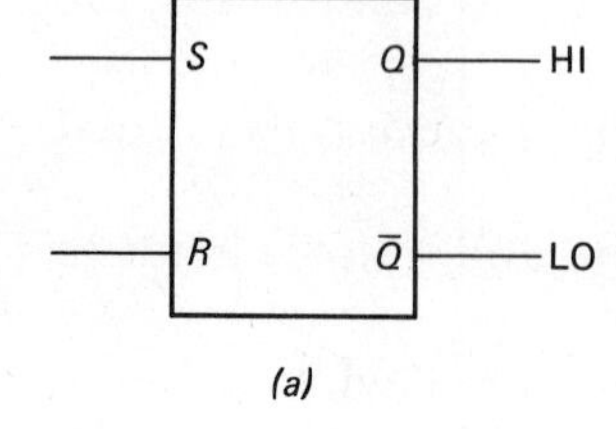

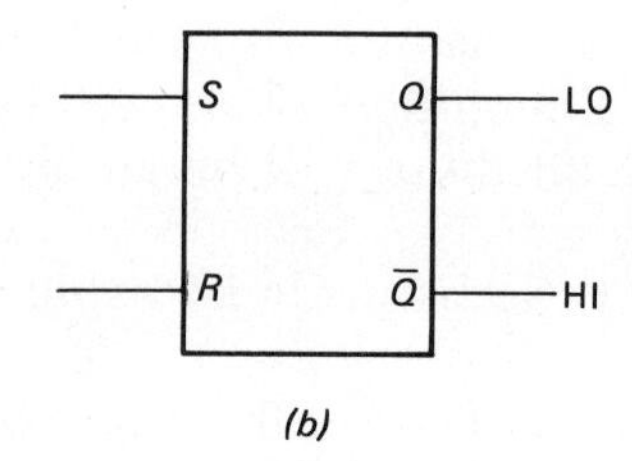

Fig. 15-11 The two possible states of a flip flop. (*a*) Flip flop in the high state. (*b*) Flip flop in the low state.

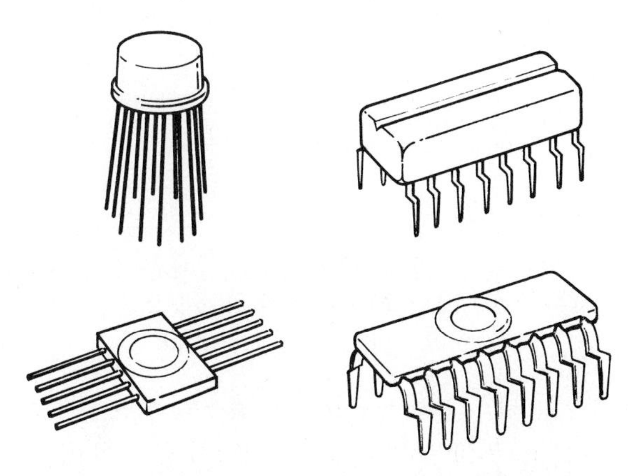

Fig. 15-12 In modern control circuits, flip flops are in IC packages like this.

Flip flops can be made from discrete components such as tubes, transistors, and relays, but in modern control circuitry you are more likely to see flip flops made from combinations of basic logic gates. These logic gates are integrated circuits which are used in combinations.

The flip flop itself is usually in a single enclosed package and is treated as a single component. (See Fig. 15-12.) For this reason it is important for you to know how the flip flop behaves with different input signals and what its output should be for various input-signal combinations.

In the experiment section of this chapter you will make a flip flop circuit and study its output for various input signals.

Summary

1. A flip flop is a bistable circuit. It remains in either of two states of operation until it is switched.
2. An *R-S* flip flop is ON (or high) when the *Q* output terminal is in the high condition.
3. When the *Q* output terminal of an *R-S* flip flop is in the low condition, the flip flop is OFF (or low).

PROGRAMMED REVIEW QUESTIONS

(Instructions for using this programmed section are given in Chapter 1.)

We will review the important concepts of this chapter. If you have understood the material, you will progress easily through this section. Do not skip this material because some additional theory is presented.

1. The lamps in Fig. 15-13 represent binary numbers. A dark circle represents a lamp that is ON, and a light circle represents a lamp that is

  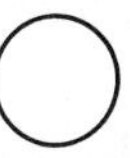 

Fig. 15-13 The circuit condition for the question in block 1.

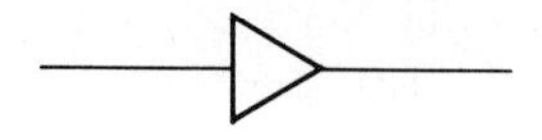

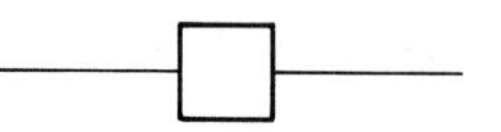

Fig. 15-14 The circuit for the question in block 2.

OFF. The decimal number represented by the lamps in Fig. 15-13 is
A. 9. (Proceed to block 9.)
B. 11. (Proceed to block 17.)

2. *The correct answer to the question in block 9 is* **A.** *When output terminal Q is in the low condition and output terminal $\overline{Q}$ is in the high condition, the flip flop will operate. Conversely, when output terminal Q is high, terminal $\overline{Q}$ is low.* Here is your next question.
Regarding the schematic symbols shown in Fig. 15-14,
A. they are both representations of NOT circuits. (Proceed to block 27.)
B. neither is a representation of a NOT circuit. (Proceed to block 4.)

3. *The correct answer to the question in block 10 is* **B.** *AND circuits, OR circuits, and other logic circuits are also referred to as gates.* Here is your next question.
The inputs to a logic circuit may be referred to as the
A. fan-in. (Proceed to block 13.)
B. starters. (Proceed to block 26.)

4. *The correct answer to the question in block 2 is* **B.** *The MIL circuit for the NOT symbol has a small circle at the output of the amplifier symbol. This indicates that the output is negated. The NEMA symbol has an X in the box. Therefore neither of the illustrations is correct for representing a NOT circuit.* Here is your next question.
In Fig. 15-15 the switches are arranged so that the light will be ON when switch *A* and switch *B* or switch *C* are closed. Which of the following boolean expressions describes this condition?
A. *ABC*. (Proceed to block 15.)
B. $A \times (B + C)$. (Proceed to block 12.)

5. *Your answer to the question in block 16 is* **B.** *This answer is wrong. The expression* AB + $\overline{AB}$ *means that you can have A and B or you can*

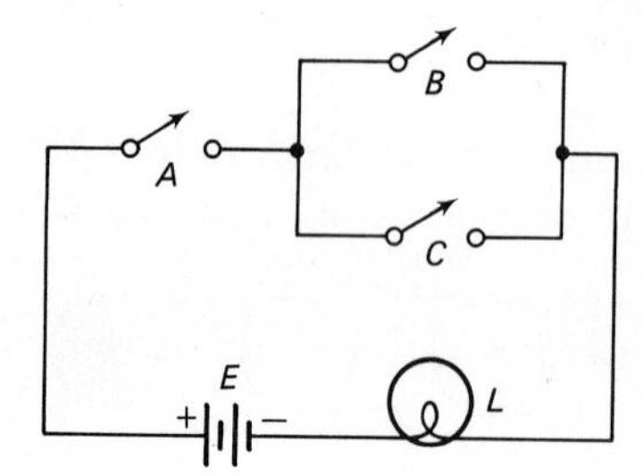

Fig. 15-15 The circuit for the question in block 4.

have NOT *A and* NOT *B in order to achieve an output. This is not a characteristic of an* EXCLUSIVE OR *circuit.* Proceed to block 11.

6. *The correct answer to the question in block 25 is* **B.** *With an* EXCLUSIVE OR *you do not have an output when both inputs are in the 1 state.* Here is your next question.
In the octal numbering system the radix is
A. 7. (Proceed to block 24.)
B. 8. (Proceed to block 10.)

7. *Your answer to the question in block 9 is* **B.** *This answer is wrong. It is never possible for an operating flip flop to have two outputs that are high, or two outputs that are low, at the same time.* Proceed to block 2.

8. *The correct answer to the question in block 12 is* **B.** *The* EXCLUSIVE OR *system is shown in Fig. 15-7.* Here is your next question.
Which of the following is a truth table for a NOT circuit?

A. (Proceed to block 16.)

A	*L*
0	1
1	0

B. (Proceed to block 20.)

A	*L*
0	1
0	1

9. *The correct answer to the question in block 1 is* **A.** *The lamps show 1001. This binary number equals decimal number 9. (Verify this in Table 15-1.)* Here is your next question.
Which of the following is a possible operating condition for a flip flop?
A. Output terminal Q is in the low condition, and output terminal $\overline{Q}$ is in the high condition. (Proceed to block 2.)
B. Output terminals Q and $\overline{Q}$ are both in the high condition. (Proceed to block 7.)

10. *The correct answer to the question in block 6 is* **B.** *Counting 0, there are eight digits in the octal system, and its radix is 8.* Here is your next question.
Another name for an AND circuit is
A. AND bridge. (Proceed to block 23.)
B. AND gate. (Proceed to block 3.)

11. *The correct answer to the question in block 16 is* **A.** *The expression* $A \oplus B$ *means* EXCLUSIVE *A* OR *B. Another way to write this is* $A\overline{B} + \overline{A}B$. Here is your next question.

Which number is missing in this two-input AND-gate truth table?

A	B	L
0	0	
0	1	0
1	0	0
1	1	1

A. Cannot be determined. (Proceed to block 14.)
B. 1. (Proceed to block 22.)
C. 0. (Proceed to block 25.)

12. *The correct answer to the question in block 4 is* **B.** *The expression* $A \times (B + C)$ *means "A and either B or C, or both." Note that + is the sign for* INCLUSIVE OR. Here is your next question.
Figure 15-16 shows the truth table for
A. A two-input NOR gate. (Proceed to block 18.)
B. An EXCLUSIVE OR gate. (Proceed to block 8.)

13. *The correct answer to the question in block 3 is* **A.** *The inputs to a logic circuit are referred to as the* ***fan-in,*** *and the outputs are referred to as the* ***fan-out.*** *In general it is considered to be an advantage if a logic circuit can handle a large fan-in and a large fan-out. Despite the relative disadvantages of using relays in electronic circuitry, there is one definite advantage: They can have a relatively large fan-in and fan-out.* Here is your next question.
An *R-S* flip flop that is made with NOR logic is in the low condition. A positive pulse applied to the *R* terminal will
A. have no effect on the output. (Proceed to block 28.)
B. switch the flip flop to the high condition. (Proceed to block 19.)

14. *Your answer to the question in block 11 is* **A.** *This answer is wrong. For a two-input* AND *gate there is no output when both inputs are 0.* Proceed to block 25.

15. *Your answer to the question in block 4 is* **A.** *This answer is wrong. The expression ABC means "A and B and C."* Proceed to block 12.

A	B	L
0	0	0
0	1	1
1	0	1
1	1	0

Fig. 15-16 The truth table for the question in block 12.

16. *The correct answer to the question in block 8 is* **A.** *As shown by the truth table, when the input A is 0 the output is 1, and when A is 1 the output is 0. This is a characteristic of the* NOT *circuit.* Here is your next question.
Which of the following is the same as $A \oplus B$?
A. $\overline{A}B + A\overline{B}$. (Proceed to block 11.)
B. $AB + \overline{AB}$. (Proceed to block 5.)

17. *Your answer to the question in block 1 is* **B.** *This answer is wrong. Study the binary numbers in Table 15-1, then* proceed to block 9.

18. *Your answer to the question in block 12 is* **A.** *This answer is wrong. Review the truth table for* NOR *and for* EXCLUSIVE OR *gates, then* proceed to block 8.

19. *The correct answer to the question in block 13 is* **B.** *The flip flop is in the low condition and will be switched by a positive pulse to the reset terminal.* Here is your next question.
When a flip flop is in the *high* condition, which output terminal (Q or $\overline{Q}$) is high? (Proceed to block 29.)

20. *Your answer to the question in block 8 is* **B.** *This answer is wrong. It is very important that you learn to identify and understand truth tables. In a* NOT *circuit, when the input is 1, the output is 0. When the output is 1, the input is 0.* Proceed to block 16.

21. *Your answer to the question in block 25 is* **A.** *This answer is wrong. The symbol used in the equation is for an* EXCLUSIVE OR. Proceed to block 6.

22. *Your answer to the question in block 11 is* **B.** *This answer is wrong. The two inputs are 0; therefore the output must be 0.* Proceed to block 25.

23. *Your answer to the question in block 10 is* **A.** *This answer is wrong.* AND *circuits are not normally called* AND *bridges.* Proceed to block 3.

24. *Your answer to the question in block 6 is* **A.** *This answer is wrong. Although the highest digit in the octal system is 7, there are eight total digits because you have to also count 0.* Proceed to block 10.

25. *The correct answer to the question in block 11 is* **C.** *Figure 15-5 shows the* AND *truth table.* Here is your next question.
The expression $A \oplus B = L$ means that
A. L is 1 if either or both inputs (A and B) are 1. (Proceed to block 21.)
B. L is 1 if A or B is 1, but not if both A and B are 1. (Proceed to block 6.)

26. *Your answer to the question in block 3 is* **B**. *This answer is wrong. The inputs to logic circuits are never called starters.* Proceed to block 13.

27. *Your answer to the question in block 2 is* **A**. *This answer is wrong. Neither of the two blocks shown represents* NOT *circuits. You should study the schematic representation for* NOT *circuits very carefully, then* proceed to block 4.

28. *Your answer to the question in block 13 is* **A**. *This answer is wrong. R-S flip flops made with* NOR *logic are switched with positive pulses. If the flip flop is in the low condition, the positive pulse to the reset (R) terminal will switch it back to the high condition.* Proceed to block 19.

29. *The Q output is high when the flip flop is high.*
You have now completed the programmed questions. The next step is to put some of these ideas to work in laboratory experiments. Proceed to the Experiment section of this chapter.

EXPERIMENT

(The experiment described in this section may be performed on the circuit board described in Appendix C or on a similar laboratory setup.)

Purpose In this experiment you will study examples of how a light-sensitive resistor (LSR) performs in some transducer circuits. These types of circuits have many applications in industrial electronics and in measuring instruments. Note that while this experiment does not directly relate to logic circuits, it incorporates vital principles you must learn in your study of basic electronics.

Theory In electronics a transducer is sometimes also known as a *sensor*. It is a device that permits one type of energy to control another type of energy. Two good examples of transducers are microphones and loudspeakers. In a microphone, sound energy controls the amount of current or voltage output. In a loudspeaker, electrical energy controls the amount of sound energy coming from the speaker. In both cases the transducers have one kind of input energy and another kind of output energy.

Usually a transducer which generates an output signal is connected into the system through some special kind of circuitry or component. This is necessary because the output of this type of transducer may be too small in voltage or current to operate the system directly.

One popular circuit in which transducers are connected is the *bridge*. Figure 15-17 shows a special bridge circuit, called a *Wheatstone bridge* (this circuit does not yet show a transducer), which is used for measuring resistance values. The unknown resistance R_x is connected into the circuit. Variable

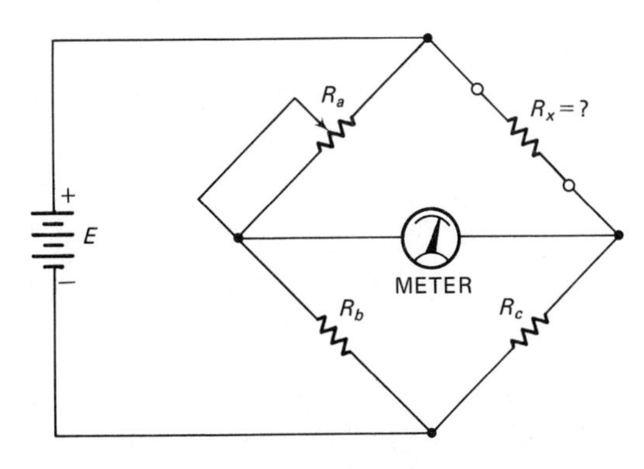

Fig. 15-17 Wheatstone bridge circuit.

resistor R_a is adjusted until there is no current indicated by the meter. Usually the dial for R_a is calibrated so its resistance can be read directly. When there is no current flowing in the meter the bridge is said to be *balanced,* and the value of R_x is found from the equation

$$R_x = \frac{R_a R_c}{R_b}$$

An important feature of this Wheatstone bridge and one of the reasons it is popular for measuring resistance is the fact that changes in applied voltage E will not affect its operation. The reason for this is that the operation is based on the ratios of voltage drops across the resistors. Even though the applied voltage is changed, the ratios will not change, and therefore the bridge can operate correctly with a wide range of voltage values. This is a very important feature of the bridge that makes it popular for transducer circuits. In some cases it is important to connect a transducer into a circuit which will not be affected by changes of the applied voltage. The voltage or frequency output of the transducer should be proportional only to what is being measured. It should *not* be affected by power-supply voltage changes. That is why you will find certain transducers connected into bridge circuits. In fact, in this experiment you will demonstrate this type of circuit.

As mentioned before, a transducer output may be quite low and therefore may have to be amplified in some applications. A transducer amplifier will also be demonstrated in this experiment.

PART I

Test Setup Figure 15-18 shows the test setup for the first part of this experiment. It consists of a halfwave rectifier circuit (X_1 and C_1). The output voltage of this rectifier circuit is about 9 volts dc. It is applied across the bridge in place of E of Fig. 15-17.

The transducer (LSR) is a light-sensitive resistor. Its resistance decreases with an increase in the amount of light. It should be possible to adjust resistor R_1 so that the voltmeter reads 0 volts with normal room lighting.

You should understand that in some practical circuits the voltmeter may be replaced by an amplifier or in some cases by a very sensitive current-indicating device such as a microammeter.

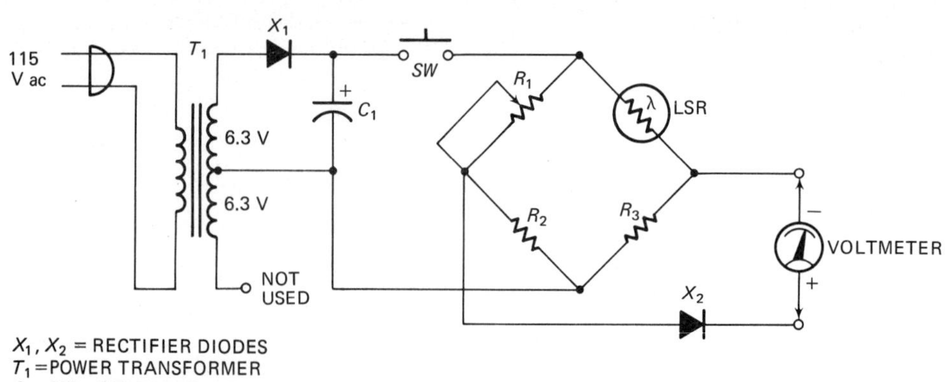

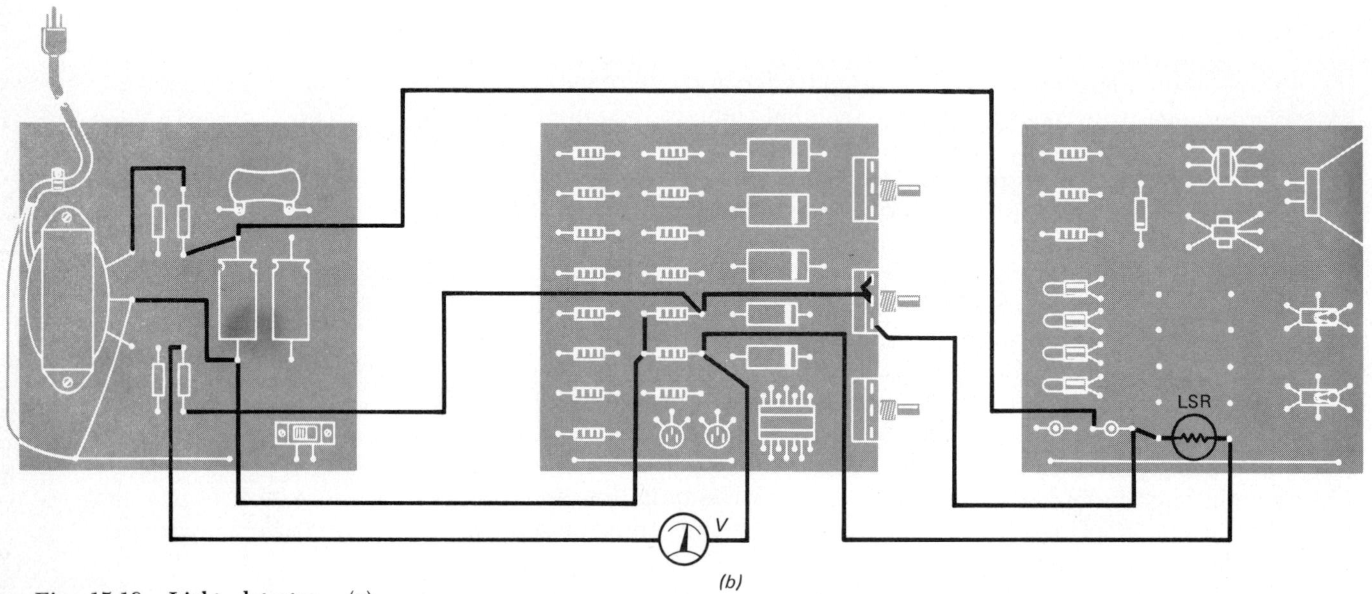

Fig. 15-18 Light detector. *(a)* **Schematic diagram.** *(b)* **Wiring diagram.**

Procedure

step one Connect the circuit as shown in Fig. 15-18. Figure 15-18*a* shows the schematic, and Fig. 15-18*b* shows the wiring diagram. Observe the polarity of the voltmeter carefully in making this connection. It may be necessary to reverse the connections of the voltmeter and the diode X_2 in order to get the meter to read upscale. The diode X_2 is used to prevent the voltmeter from reading downscale when the bridge is out of balance.

step two Set the scale selector of the dc voltmeter to some value greater than 9 volts. Operate the switch and observe the voltmeter reading. It should read upscale. If not, adjust resistor R_1 and see if it will read upscale. If you get no voltage reading, then it will be necessary to reverse both the diode X_2 and the voltmeter connections and repeat the procedure.

step three Adjust R_1 carefully so that the voltmeter reads 0 volts. The LSR should be exposed to the room light at this time. If you have a voltmeter with a zero-center scale, it will be ideal for this application. In such cases you can remove the diode in series with the voltmeter connection.

step four Once you have zeroed the voltmeter, cover the LSR with a piece of cardboard. Does the voltmeter read upscale?

____
Yes or No

If you have properly connected the circuit, the voltmeter will read upscale when the LSR is covered. The reason for this is as follows: First, you balance the voltmeter by adjusting its voltage to zero with variable resistor R_1. This means that the voltmeter is balanced with the LSR resistance exposed to light. When you cover the LSR, its resistance changes, and that unbalances the bridge, causing the voltmeter to read upscale. This type of circuit is used extensively in certain measuring instruments and in industrial electronics systems.

PART II

Test Setup Figure 15-19 shows the test setup for this part of the experiment. Figure 15-19*a* shows the test setup, and Fig. 15-19*b* shows the wiring diagram. Transistor Q_1 has a 3.3-kilohm load resistor and a 10-kilohm resistor in series with the LSR for the bias circuitry. The power supply consists of a half-wave rectifier diode X_1 and filter capacitor C_1. The theory is quite simple. When the amount of light falling on the LSR changes, the bias resistance for the transistor changes. This will cause the amount of conduction to change and cause a change in the output reading on the voltmeter.

Procedure

step one Wire the circuit shown in Fig. 15-19.

step two Measure the output voltage of the transistor at the collector and record the value (normal room light). ________________ volts

You should get a low dc voltage. This indicates that the transistor is conducting a high current. Much of the voltage drop in the circuit is across the load resistor. There is little voltage drop across the transistor.

step three Cover the LSR and observe the change in the voltmeter reading. Does the voltmeter reading go up? ________________
Yes or No

The voltmeter reading should go up when the LSR is covered. The reasoning is as follows: With the LSR exposed to light, its resistance is low, and there is a

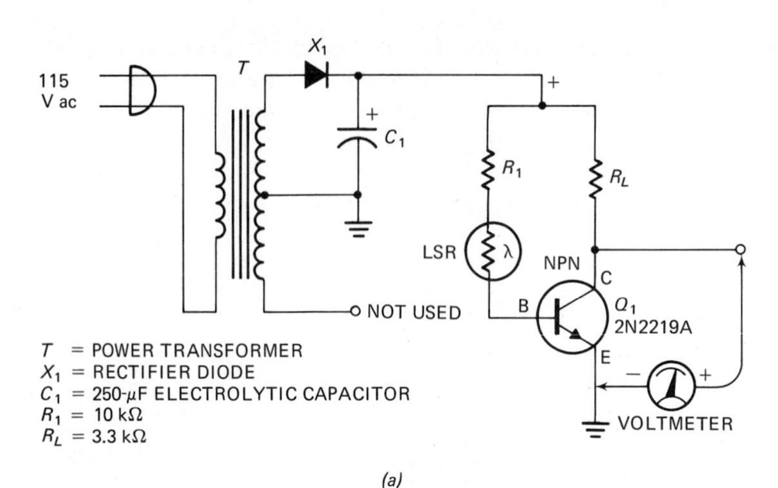

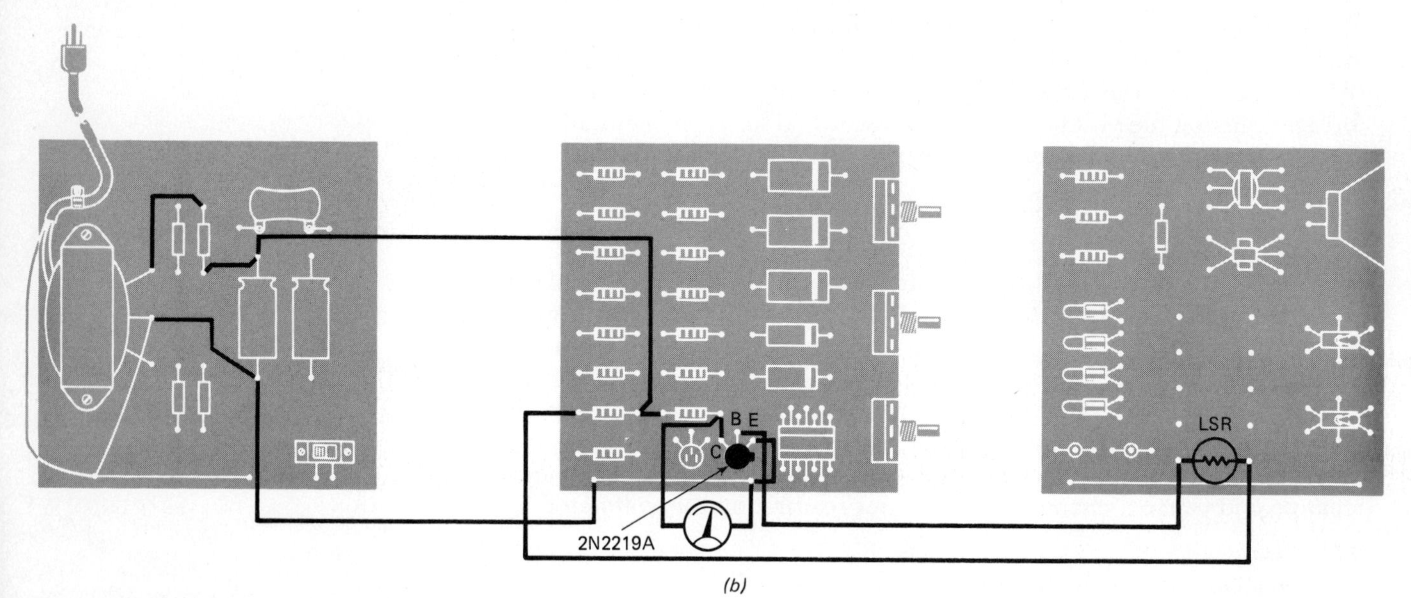

Fig. 15-19 Transducer with amplifier. (*a*) Schematic diagram. (*b*) Wiring diagram.

large base current flowing. This base current is sufficient to cause the transistor to be almost saturated. Another way of saying this is that the base current is large enough so that the transistor is conducting almost its maximum possible current. Under this condition the voltage across the transistor is low, and a large voltage drop occurs across the 3.3-kilohm resistor. Observe that the voltmeter is really measuring the voltage across the transistor, so the voltmeter indicates a low reading.

When the LSR is covered, its resistance increases. This decreases the base current and decreases the collector current. When you decrease the collector current, you decrease the voltage drop across the load resistor R_L, and you increase the voltage across the transistor. This causes the voltage across the transistor to go up. Notice that the voltage across the transistor plus the voltage across R_L must always equal the applied voltage. (This is basic Kirchhoff's voltage law.) Therefore, if you increase the voltage drop across R_L, you decrease the voltage across the transistor and vice versa.

You can also consider the action in terms of phase reversal. When the LSR is covered, its resistance increases. The greater drop across the LSR causes the base voltage to become less positive. When the base of a transistor amplifier becomes less positive, its collector becomes more positive.

Conclusion The LSR is a light-sensitive resistor that acts as a transducer. It has a resistance value that varies with the amount of light. The greater the light, the lower the resistance.

When the LSR is placed in a bridge circuit (Fig. 15-18), changes in light will cause the bridge to become unbalanced and produce an output.

In the amplifier circuit of Fig. 15-19, changes in LSR resistance cause a change in the base current, which also changes the amount of collector current. In modern applications a high-gain operational amplifier is used as a transducer amplifier to make the change in output greater for a given change in light. In other words, the high-gain op amp increases the sensitivity of the circuit.

SELF-TEST WITH ANSWERS

(Answers with discussions are given at the end of the chapter.)

1. The decimal number 17 is equal to binary number (*a*) 10101; (*b*) 10010; (*c*) 100001; (*d*) 10001.
2. The radix in the duodecimal system is (*a*) 12; (*b*) 2; (*c*) 10; (*d*) 8.
3. Which of the following more accurately describes the action of a flip flop? (*a*) It has no stable state. It continually switches back and forth between two states. (*b*) It has two stable states. It will remain in one state until it is switched to the other.
4. For an *R-S* flip flop the output cannot be determined when (*a*) *S* and *R* are both 0; (*b*) *S* is 1 and *R* is 0.
5. An *R-S* flip flop is in the high condition (*a*) when *S* is 1 and *R* is 0; (*b*) when *S* is 0 and *R* is 1.
6. In the binary system of counting the radix is (*a*) 0; (*b*) 1; (*c*) 2; (*d*) none of these.
7. In logic circuitry which of the following means *x* and *y*? (*a*) $x \div y$; (*b*) $x \oplus y$; (*c*) $x + y$; (*d*) xy.
8. Which of the following truth tables is for $\overline{A + B}$?

(*a*)

A	B	L
0	0	1
0	1	0
1	0	0
1	1	0

(*b*)

A	B	L
0	0	1
0	1	0
1	0	0
1	1	1

9. A machine is wired with a manual ON/OFF switch and a thermal cutout switch. If the temperature of the machine becomes too high, the thermal cutout switch turns it OFF. Also, it can be shut OFF with the manual switch. Which of the following is true? (*a*) The switches are in parallel or in an OR connection; (*b*) The switches are in series or in an AND connection.
10. Would it be possible to have a numbering system with a base of 7—that is, with a radix of 7? (*a*) No; (*b*) Yes.

ANSWERS TO SELF-TEST

1. (*d*)—Table 15-1 does not go to decimal number 17. However, if you study the sequence of binary numbers, you will see that the next number after 10000 will be 10001.
2. (*a*)—In the true duodecimal system you would not write numbers 10 and 11. These are combinations of lower digits, and in the duodecimal system you would not start to combine digits until you got to number 12. Therefore you would have separate symbols to represent 10 and 11.
3. (*b*)—The description in (*a*) is for a multivibrator. The description in (*b*) defines a flip flop.
4. (*a*)—The output of the *R-S* cannot be determined if the inputs are both 0, and it cannot be determined if the inputs are both 1. This is a disadvantage of the *R-S* flip flop. More elaborate circuits—called *J-K* flip flops—can be used if this undetermined output causes a problem in a circuit.
5. (*a*)—The high, or ON, condition of a flip flop occurs when the 1 output is high and the 0 output is low. This occurs when *S* is 1 and *R* is 0.
6. (*c*)—The radix is 2 and the binary digits are 0 and 1.
7. (*d*)
8. (*a*)—The expression $\overline{A + B}$ means "not *A* or *B*." It is not EXCLUSIVE OR, so the output is zero if *A* or *B* (or both) equal 1.
9. (*b*)—If the switches were in parallel, you could not shut the machine OFF unless you operated the manual switch at a time when the machine overheats. In other words, both switches would have to be open to shut the machine OFF. When they are in series (in the AND connection) either switch can turn the machine OFF. Compare the OR and AND circuits of Figs. 15-5*a* and 15-6*a*.
10. (*b*)—You can have a numbering system with any radix.

how do you locate troubles in electronic equipment? 16

INTRODUCTION

Technicians who perform troubleshooting and circuit analysis on electronic equipment must develop fast and efficient methods of locating *faults* (that is, troubles in the system). One of the most important aids that a technician has is a good understanding of how electronic circuits work. Without this understanding the troubleshooting procedure may become hit or miss. A technician who clearly understands how a system is supposed to work has a better chance of quickly locating a fault when the system is not working.

A knowledge of electronics, by itself, does not mean that a technician can locate a fault quickly. The second thing that a technician must have is a knowledge of special troubleshooting methods. Some of these methods will be discussed in this chapter.

Armed with a good basic knowledge of circuitry and a knowledge of troubleshooting methods, the next thing technicians must do is to learn as much as possible about each electronic system that they are going to work with. Sometimes a manufacturer will provide charts or tables for locating faults. When these are not available, good technicians will make their own.

Much of the subject matter in this book has dealt with methods of locating troubles in circuits and components. For one example, you know that an emitter-to-base short circuit will turn a transistor OFF. This technique is used to determine if the base current controls the collector current.

The troubleshooting material that you have already studied will not be reviewed here. Instead you will learn the methods used by technicians to find troubles in electronic systems. Block diagrams are used to explain these methods. These diagrams are similar to the ones provided by companies that make electronic systems.

You will be able to answer these questions after studying this chapter.

- ☐ **What is the best way to locate a fault in a system?**
- ☐ **How do you analyze a system?**
- ☐ **What are the procedures for signal tracing and signal injection?**
- ☐ **How is signal injection used in troubleshooting?**
- ☐ **How is signal tracing used in troubleshooting?**
- ☐ **How do you find a defective circuit and a defective component?**

INSTRUCTION

What Is the Best Way to Locate a Fault in a System?

There is one method of locating faults in a system that can be used for most occasions: *Go from the general to the specific.* In other words, the first step is to look at the operation of the complete system. Then determine which is the faulty unit. Next, locate the faulty circuit within the unit. Finally, locate the component that is defective. In this discussion we define a *system* as being the complete assembly, whereas a *unit* is a part of that assembly which performs a special function. A *circuit* exists within a unit, and circuits are made up of *components.*

An automobile can be used as an analogy for making these meanings clear. The complete automobile is a system. The engine is a unit within the system. The ignition assembly is a circuit within the unit, and a spark plug is a component in the ignition circuit.

As another example, consider a numerical control system that uses punched tape for operating a drill press. The complete assembly from the punched-tape input to the drill-press output is a system. The tape reader, which changes the information provided by punched holes into electrical signals, is a unit within the system. Logic circuits are part of the tape reader, and a resistor in a logic circuit is a component.

Figure 16-1 shows a logical procedure for locating a fault and making a repair. As shown in the block diagram, the first step is to analyze the complete system. This means to check the overall function. In this step it is important for technicians to listen to complaints made by the operators of the equipment, but it is also important not to allow this information to distract them. Many valuable minutes can be wasted on troubleshooting only from information given by the equipment operator. You must learn to listen carefully to the complaint, then see for yourself if the complaint is accurate.

When the complete system has been analyzed, the next step is to isolate the defective unit that is causing the problem. It is useful to learn the symptoms of troubles that occur most often. Here again we can make an analogy. If a patient goes to the doctor with a bullet wound, it is not necessary for the doctor to give a complete physical. In this case the symptoms are obvious. The doctor can go immediately to the area that needs attention.

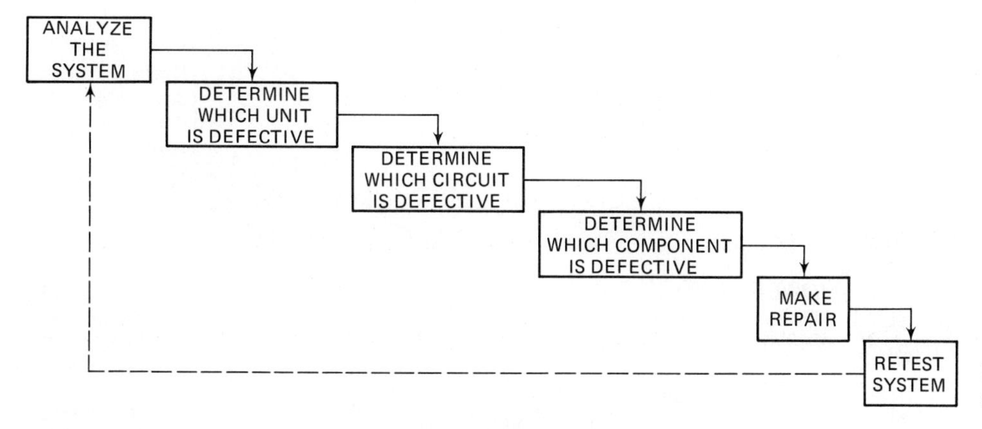

Fig. 16-1 Procedure for troubleshooting a system.

Once the defective unit has been located, the technician must determine what is the quickest way to get the system working. Sometimes there are complete units that can be substituted for a defective one.

A good example of this is in television repair. The complete audio section may be defective due to a bad resistor. Rather than take time in the customer's home to locate the defective component, the technician may be able to plug in a complete audio section. In our definition the audio section would be called a *unit*, but the manufacturer calls it a *module*. Figure 16-2 shows how a module can be used to replace a defective unit in a television receiver. The technician can troubleshoot the defective unit after replacing it and making sure that the system is working correctly.

Figure 16-1 shows that finding the defective circuit is the next step. Once this is done, the procedure for finding the defective component usually requires the use of component test equipment. Examples are volt-ohm-milliammeters, transistor testers, tube testers, and capacitance testers. You are familiar with these instruments from your study in previous chapters.

As shown in Fig. 16-1, finding the defective component is not the end of the job. The component must be repaired or replaced. When the equipment is made with printed circuit boards and integrated circuitry, special methods must be used for making repairs. These methods have been discussed in earlier chapters.

The final step, as shown in Fig. 16-1, is to retest the complete system. Always keep in mind that a unit may operate properly on the bench, but when it is reinserted into the system, it may not be able to perform its job. An example of this is a power-supply failure. After the power supply has been repaired, it may be necessary to reset the voltages for proper operation. In some systems these voltages can be reset only when the power supply is connected to the system. This is necessary in order to assure that the voltages are correct when the power supply is working under its normal load.

Summary

1. A very important aid to finding faults in a system is a good knowledge of electronics.
2. Besides knowing electronics, it is also important to know the special methods used for finding faults.

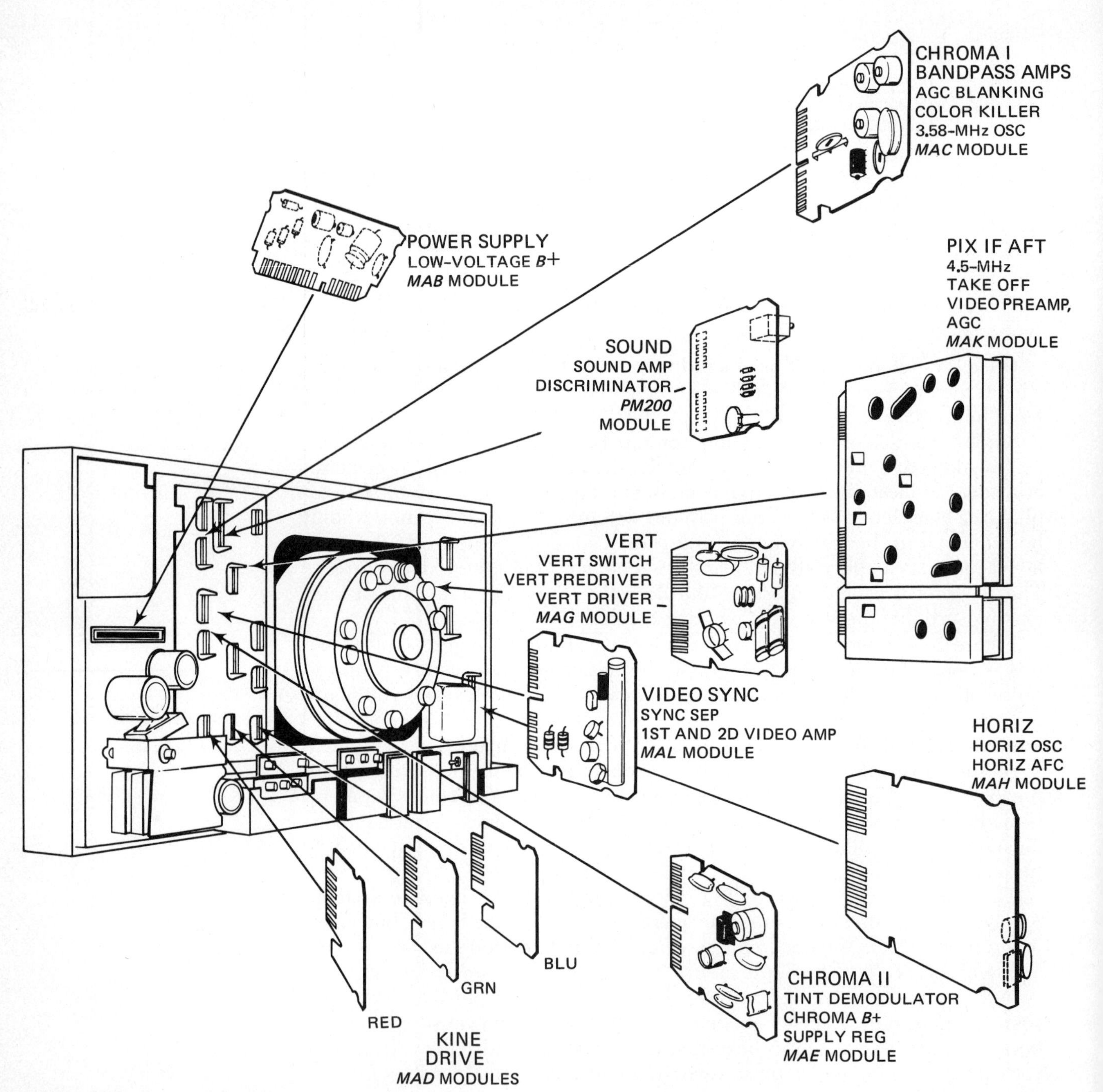

Fig. 16-2 It is possible in some systems to replace a complete unit with a plug-in module. (*Courtesy of RCA.*)

3. A third important aid for finding faults is to learn as much as possible about the system. This includes a study of the manufacturer's literature.
4. A method that is used to find a fault is to go from the general to the specific.
5. Complaints made by the operators of the equipment may be helpful in locating faults.

6. By replacing a module, or unit, it may be possible to get the system into operation quickly.
7. The module, or unit, can be repaired after it has been replaced and the system is operating correctly.
8. A technician must be able to find the trouble and also be able to correct it.
9. The final step in troubleshooting is to retest the system to make sure it is working the way it should.

How Do You Analyze a System?

We have completed an overall look at the procedure for troubleshooting and repair. We will now discuss special methods for each of the blocks in the diagram of Fig. 16-1. We will begin by discussing the method used for analyzing the system which is shown in Step 1 of Fig. 16-1. Figure 16-3 shows a block diagram of the step-by-step procedure for analyzing a system. The first step calls for an overall inspection. In this step look for obvious faults such as broken wires, burned spots, and circuit breakers that have been tripped. Inspect all plugs and cables.

Your sense of smell can often tell you if a transformer has been overheated, if a resistor has been burned out, or if a part has been destroyed by too much current. If the system is in operation, listen for arcing or rubbing sounds and any other unnatural sounds in the system. To summarize, the first step in looking for trouble in a system is to look for obvious faults. Use your senses to locate the part of the system that has a fault.

The block diagram of Fig. 16-3 shows that the second step in locating a system fault is to check the output of the system. There are two possibilities:

1. The system has no output.
2. The output of the system is not correct.

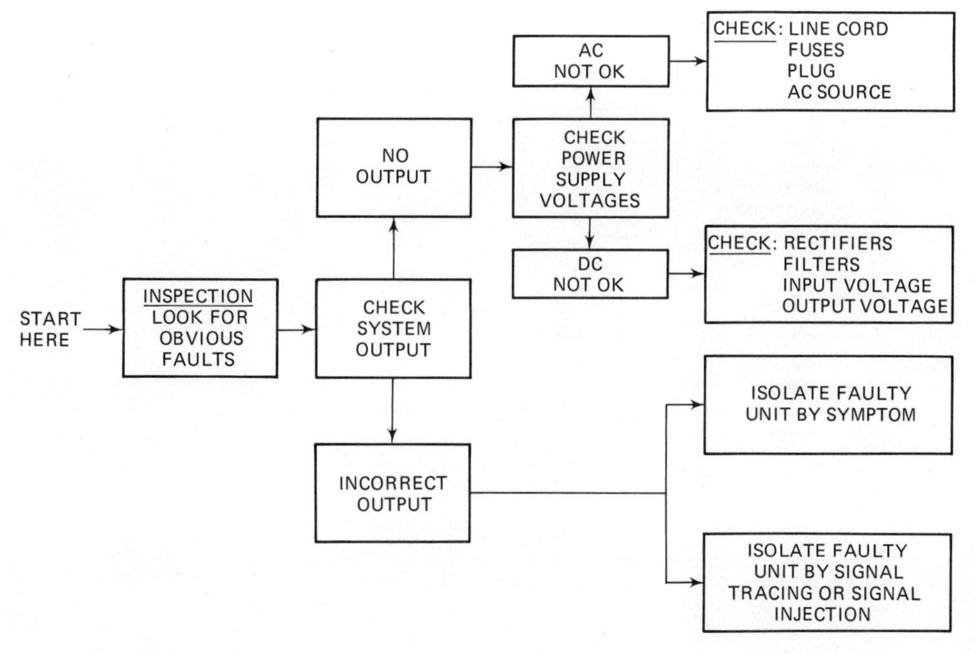

Fig. 16-3 Procedure for system analysis.

Suppose that a technician is asked to fix a television set. First, the technician will look for obvious faults, as described for the first block in Fig. 16-3. This will be done *before* power is applied to the TV set. If the technician sees burned spots or smells the effects of an overheated transformer, he or she will *not* plug in the set. The fault will be located by using the ohmmeter and component testers.

If inspection does not show any obvious faults, the technician will go to Step 2. Power will be applied to the TV set, and the system output will be checked.

If there is no output of any kind (no sound, no picture, no raster), then the technician will have to look for a trouble in a unit that affects all parts of the TV system.

However, if the output is not correct, the technician may be able to determine which unit is defective from the type of output the system has.

Figure 16-4 shows two examples of how a fault in an electronic system can be located. A faulty TV set is used in this example, but any system can be analyzed with the method described here. In Fig. 16-4*a* it is assumed that there is no output when the TV set is energized. In other words, there is no sound, no picture, and no raster. The faulty unit is most likely to be the low-voltage power supply, since that is the only part of the system that is common to all the units mentioned.

Figure 16-4*b* shows the procedure if there is an output, but it is faulty. In this example, there is a picture but no sound. The obvious place to start is the audio section, since the video section is working. Note also that all sections that amplify the video signal must be working.

When a technician has experience with a system, he or she will learn to recognize the symptoms that go with a fault. For example, no sound, no picture, and no raster means "trouble in the power supply" (or its power input) to a TV technician.

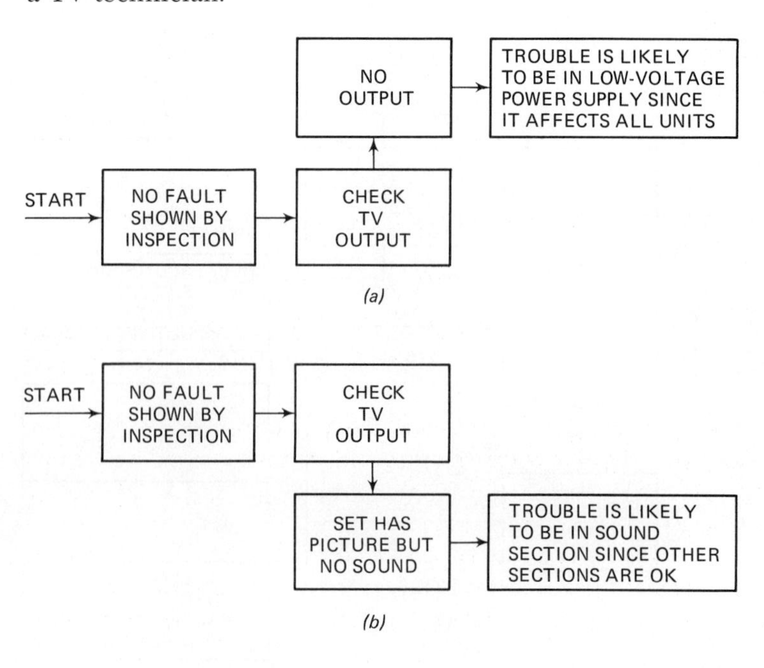

Fig. 16-4 Locating a fault in a TV system. (*a*) When there is no output. (*b*) When the output is incorrect.

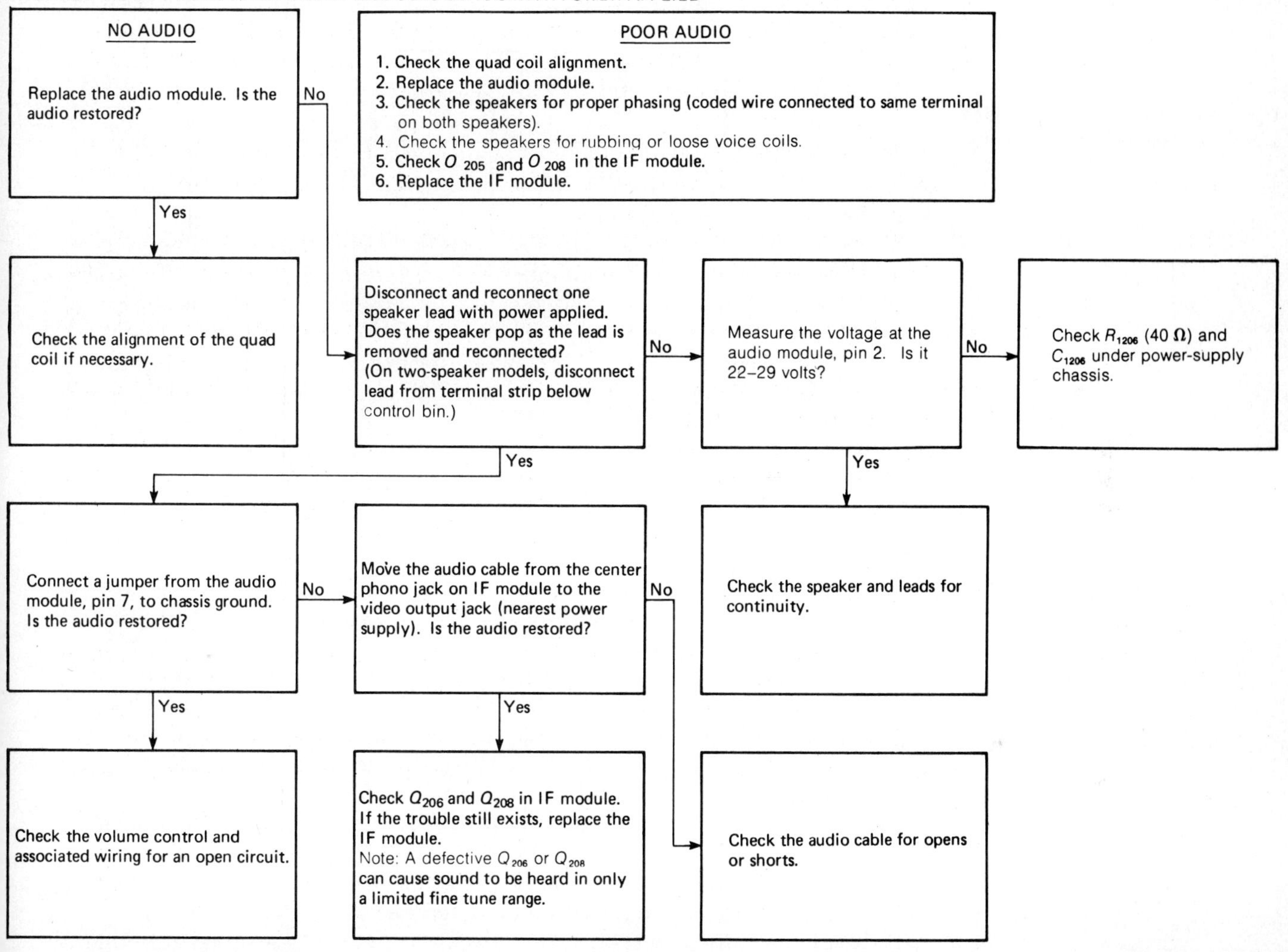

Fig. 16-5 Typical troubleshooting chart. (*Courtesy of General Electric.*)

Manufacturers often supply a troubleshooting table or list of symptoms. Figure 16-5 shows a typical troubleshooting chart for finding a fault in a TV system that has no audio.

What Are the Procedures for Signal Tracing and Signal Injection?

When you inspect the complete system, you will very often determine which unit is defective. This is done by studying the symptoms and using some basic logic. You say to yourself, "No power—so check the power supply," or "No sound—so check the sound section."

In a few cases the faulty unit cannot be determined this way. An example will be given with reference to a television receiver. Again, it should be noted that you are studying methods of locating a trouble in electronic

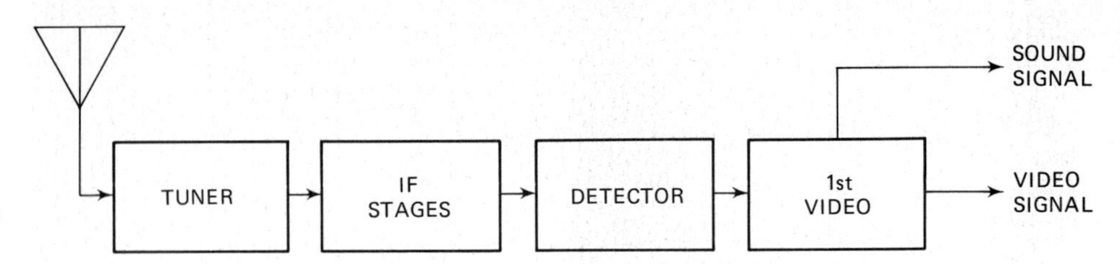

Fig. 16-6 The sound and video signals are amplified in these units.

systems. We are using the TV set as an example, but the procedure is the same for *any* electronic system.

Suppose your system analysis shows that there is no sound and no picture in a TV receiver. You will remember that an intercarrier receiver amplifies both the sound signal and the video signal in some stages. This is shown in Fig. 16-6.

If there is no sound and no video, the fault can be in any unit that has these signals. This includes the antenna, tuner, IF stages, detector, and in some sets the first video amplifier.

The next step is to determine which of these units is faulty. Either of two methods can be used. They are called *signal tracing* and *signal injection.* These procedures are used when power is delivered but does not produce the correct output from the system.

How Is Signal Injection Used in Troubleshooting?

For the signal-injection method you start at the output of the system and work toward the input. You inject the proper operating signals into each unit. This procedure is shown in block diagram form in Fig. 16-7. This is part of a TV system that is being tested. It consists of four units and a picture tube. The output of the picture tube is also a major output of the system.

The first step is to inject a signal at the input of the picture tube—point *a* of Fig. 16-7. If the picture tube is working, the signal should produce a pattern on the screen. If the picture tube is not working, then injecting a signal at point *a* will not produce a pattern on the screen.

If the picture tube is working, the next step is to inject a signal at point *b*. If this does not produce a pattern on the screen, then the trouble must be the video amplifier.

If there is a pattern, the next step is to inject a signal at point *c*. You keep doing this, step by step, until you come to a point where there is no output when you inject a signal. This tells you which unit is defective.

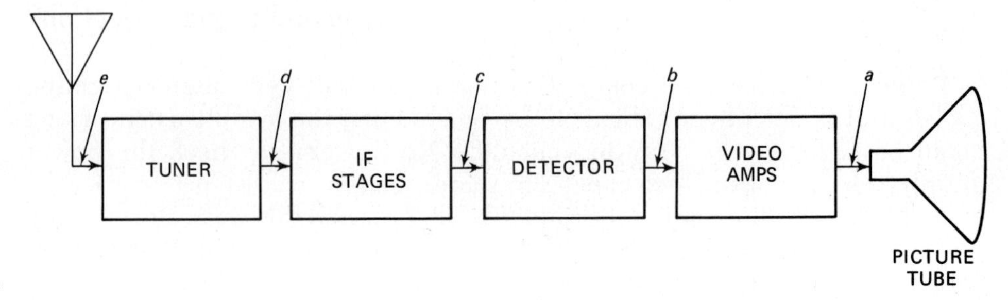

Fig. 16-7 Steps in signal injection.

Fig. 16-8 An audio generator for signal injection. (*Courtesy of the Heath Company.*)

Suppose, for example, that you get a pattern on the screen when you inject a signal at point *c*, but you do not get a pattern when you inject a signal at point *d*. This means that the trouble must be between points *c* and *d*.

In order to perform the signal-injection test just described, you must inject the proper signal at each point. Technicians use a signal generator for this purpose. Figure 16-8 shows one type of signal generator.

How Is Signal Tracing Used in Troubleshooting?

The signal-tracing method is shown in Fig. 16-9. In this case you start at point *a*, the tuner output. Using an oscilloscope (or signal tracer) you determine if there is an output. If the output of the tuner is correct, then the next step is to go to the output of the IF stage at point *b*. This procedure continues, tracing the signal from its input to the output of the system.

If the system has a bad unit, the signal will be lost at some point between the input and the output. Suppose, for example, that there is an output at

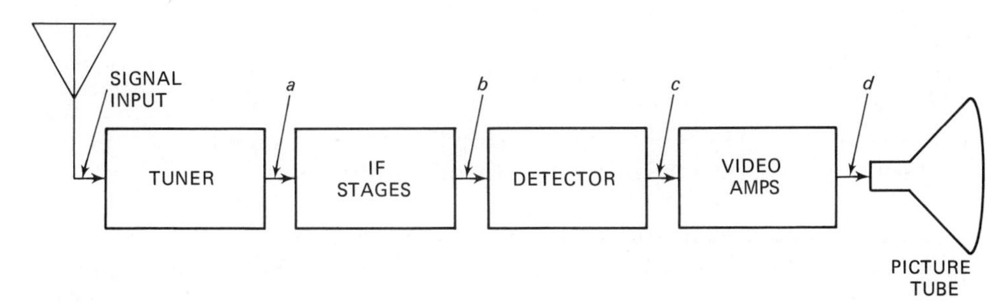

Fig. 16-9 Steps in signal tracing.

point b, but there is no output at point c. Obviously, then, the trouble is between points b and c, and the detector circuit should be suspected.

How Do You Find a Defective Circuit and a Defective Component?

When you have located the defective unit in the system, the next step is to locate the defective circuit and then the defective component. This is usually done by making voltage measurements.

Manufacturers usually provide information needed for analyzing circuits by making voltage and resistance measurements. An incorrect circuit-voltage or resistance value can be a clue to a faulty component.

In earlier chapters you learned how voltage measurements are used in troubleshooting. You will review these in the programmed section and the experiment section of this chapter.

Summary

1. A good place to start analyzing a system is to look for obvious faults such as burned places and broken wires.
2. If the system is in operation but the output is not correct, it may be possible to locate the faulty unit by studying the type of output the system has.
3. Manufacturers often supply troubleshooting material for technicians. This makes it possible to locate faults quickly.
4. When a system is not working properly, the faulty unit can be located by signal injection or signal tracing.
5. After the faulty unit has been located, the defective circuit and component can be found.

PROGRAMMED REVIEW QUESTIONS

Some of the material in this programmed review is taken from earlier chapters. All the questions are related to troubleshooting and circuit analysis.

(Instructions for using this programmed section are given in Chapter 1.)

We will review the important concepts of this chapter. If you have understood the material, you will progress easily through this section. Do not skip this material because some additional theory is presented.

1. In the circuit of Fig. 16-10 the voltage at point a is measured and found to be + 10 volts with respect to ground. Which of the following is correct?

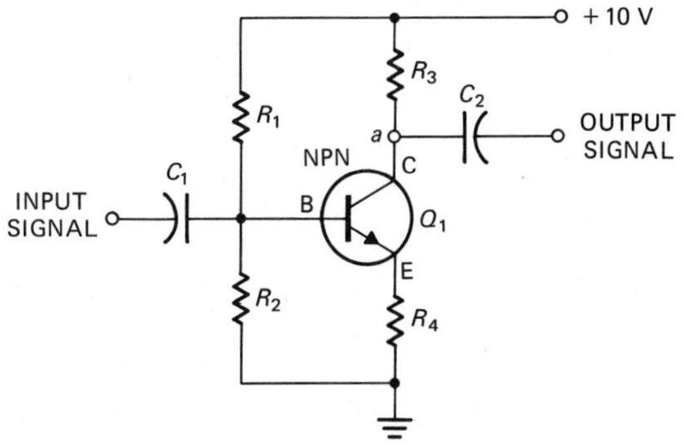

Fig. 16-10 A transistor amplifier circuit.

A. Transistor Q_1 is not conducting. (Proceed to block 17.)
B. Resistor R_3 is open. (Proceed to block 9.)
C. This is a proper voltage for point a. (Proceed to block 19.)

2. *Your answer to the question in block 17 is* **A**. *This answer is wrong. It would be a correct answer if the system were an audio amplifier, but the question does not say what kind of system it is. Many systems do not have audio amplifiers.* Proceed to block 3.

3. *The correct answer to the question in block 17 is* **B**. *The power supply is common to all parts of a system. If there is no output from a system, the power supply is a likely cause.* Here is your next question.
   In the circuit of Fig. 16-11, resistor R_2 is open. The voltmeter should read approximately
   A. 0 volts. (Proceed to block 21.)
   B. 10 volts. (Proceed to block 23.)

4. *The correct answer to the question in block 23 is* **A**. *A system is a complete electronic package. This includes an input stage, a number of units, and an output stage.*
   The input signal is usually in a form that cannot be used directly by humans. The RF input to a radio receiver is an example of such an input signal.

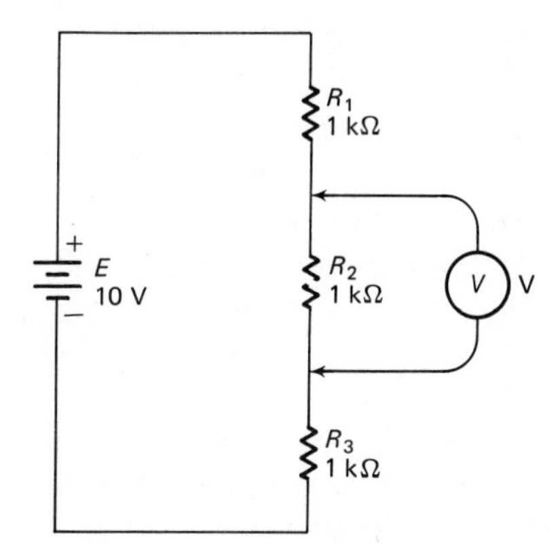

Fig. 16-11 What should the voltmeter reading be?

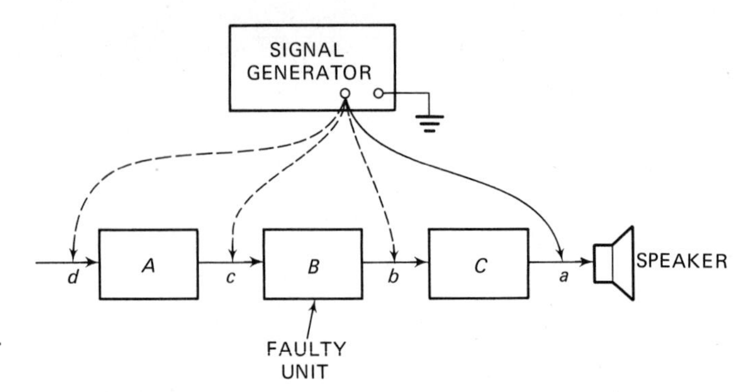

Fig. 16-12 Test setup for signal injection.

The output of the system is in a form that can be used for operating machinery or can be used directly by humans. Using a radio as an example, the output is sound.

The units are often called stages, which modify the input signal and amplify it. In a radio, examples are the mixer, IF stage, detector, and audio amplifier. Here is your next question.

You are going to test the audio system of Fig. 16-12 by signal injection. The unit marked *B* is actually at fault. You will not get an output sound from the speaker when the signal is injected at

A. point *b*. (Proceed to block 24.)
B. point *c*. (Proceed to block 22.)

5. *The correct answer to the question in block 22 is* **B**. *The system should be completely retested after major adjustments or repairs have been made.* Here is your next question.

To inject a signal into the IF stage of a receiver, you should use

A. an RF generator. (Proceed to block 10.)
B. an IF amplifier. (Proceed to block 13.)

6. *Your answer to the question in block 23 is* **B**. *This answer is wrong. A television tuner is part of a TV receiver, which is a system.* Proceed to block 4.

7. *The correct answer to the question in block 25 is* **B**. *Two cases where the emitter is not 0 volts are shown in Fig. 16-13. When there is an*

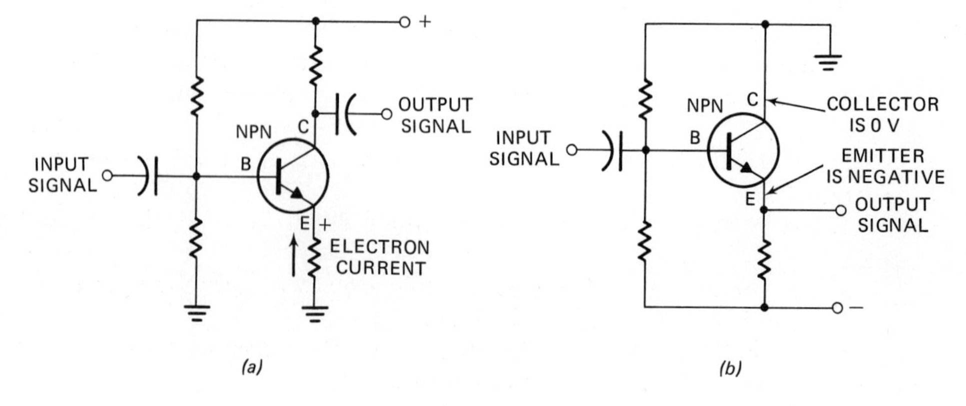

Fig. 16-13 Two circuits in which the emitter dc voltage is not 0 volts. (*a*) The emitter voltage is positive with respect to ground. (*b*) The emitter voltage is negative with respect to ground.

emitter resistor connected between the emitter and ground, the emitter current will cause a voltage drop across that resistor. This is shown in Fig. 16-13a.

Another example is when the collector is at dc ground potential and the dc operating voltage is applied to the emitter. This is shown in Fig. 16-13b.

It is important for you to know what voltages are to be expected at different points in a circuit. If you do not know what a voltage should be, you will not be able to tell if the voltage that you are reading is correct or incorrect. Here is your next question.

Which of the following would almost surely destroy a transistor in an amplifying circuit?

A. Short from the base to the collector. (Proceed to block 11.)

B. Disconnect the base lead. (Proceed to block 12.)

8. *Your answer to the question in block 25 is* **A**. *This answer is wrong. You have connected circuits in your experiments in which the emitter is not at 0 volts dc ground potential.* Proceed to block 7.

9. *Your answer to the question in block 1 is* **B**. *This answer is wrong. If R_3 is open, the voltage at point a will be 0 volts with respect to ground.* Proceed to block 17.

10. *The correct answer to the question in block 5 is* **A**. *An RF signal generator can produce signals in the intermediate-frequency range. The IF signal must be modulated with an audio signal. Remember that an IF signal does not pass through a detector stage. The detector passes only the audio, or modulated, portion of the signal.* Here is your next question.

Is the antenna part of a TV system?

A. Yes. (Proceed to block 25.)

B. No. (Proceed to block 27.)

11. *The correct answer to the question in block 7 is* **A**. *If the base is shorted to the collector, a large amount of base current will flow. This in turn will cause a high emitter and collector current. These high currents flowing in a transistor will usually destroy it. This is a good reason for not soldering in a transistor circuit with power applied. The soldering iron could accidentally short the base of a transistor to its collector and destroy the transistor.* Here is your next question.

When looking for a fault in an electronic system, a good place to start is

A. to replace all resistors, since they are most likely the source of a trouble. (Proceed to block 20.)

B. to look for obvious faults, then check the output of the system for symptoms. (Proceed to block 26.)

12. *Your answer to the question in block 7 is* **B**. *This answer is wrong. If you disconnect the base lead of a transistor, there will be no emitter-to-collector current. In other words, the transistor will be cut off. It will not be destroyed.* Proceed to block 11.

13. *Your answer to the question in block 5 is* **B**. *This answer is wrong. An IF amplifier does not produce a signal. Its job is to amplify an intermediate-frequency signal. Therefore it cannot be used to inject a signal.* Proceed to block 10.

14. *Your answer to the question in block 26 is* **C**. *This answer is wrong. A logic probe is used to determine if there is a 1 condition or a 0 condition. A logic probe is not used to measure a voltage drop.* Proceed to block 15.

15. *The correct answer to the question in block 26 is* **A**. *A low-resistance meter across 1 megohm will reduce the resistance and therefore reduce the voltage measurement. You must know the limits of your test equipment so that you can use the equipment for troubleshooting.* Here is your next question.
When a medium-power silicon transistor is conducting current from emitter to collector, the emitter-to-base voltage should be _______.
(Proceed to block 28.)

16. *Your answer to the question in block 26 is* **B**. *This answer is wrong. A low-resistance voltmeter cannot be used for measuring voltage in a high-resistance circuit.* Proceed to block 15.

17. *The correct answer to the question in block 1 is* **A**. *When Q_1 conducts through load resistor R_3, there is a voltage drop across that resistor. The voltage at point a should be less than 10 volts. Since there is no voltage drop across R_3, Q_3 must not be conducting.* Here is your next question.
If there is no output of any kind from a system, a likely cause is the
A. audio amplifier. (Proceed to block 2.)
B. power supply. (Proceed to block 3.)

18. *Your answer to the question in block 22 is* **A**. *This answer is wrong. After the equipment has been repaired, it is very important to retest the system.* Proceed to block 5.

19. *Your answer to the question in block 1 is* **C**. *This answer is wrong. The voltage at point a should be less than the supply voltage.* Proceed to block 17.

20. *Your answer to the question in block 11 is* **A**. *This answer is wrong. In the first place, you should not start troubleshooting by replacing components. Also, it is not true that resistors are the most likely source of trouble in electronic systems.* Proceed to block 26.

21. *Your answer to the question in block 3 is* **A**. *This answer is wrong. The voltmeter will read approximately 10 volts.* Proceed to block 23.

22. *The correct answer to the question in block 4 is* **B**. *When a signal generator is used to inject an audio signal at point b, a sound will be heard in the speaker. When the audio signal is injected at point c, there is no sound heard from the speaker. This tells you that the trouble is between points b and c. In other words, the trouble is in unit B.* Here is your next question.
Which of the following is the final step in troubleshooting?
A. Repair the circuit by replacing the component. (Proceed to block 18.)
B. Retest the system. (Proceed to block 5.)

23. *The correct answer to the question in block 3 is* **B**. *The voltmeter resistance is generally very much greater than 1 kilohm. Thus for practical purposes almost the entire 10 volts will appear across the voltmeter, and it will read about 10 volts.* Here is your next question.
Using the term system as defined in this chapter, which of the following is a system?
A. A radio receiver. (Proceed to block 4.)
B. A television tuner. (Proceed to block 6.)

24. *Your answer to the question in block 4 is* **A**. *This answer is wrong. When you inject a signal at point c, you will not hear an output from the speaker. The signal cannot get through faulty unit B.* Proceed to block 22.

25. *The correct answer to the question in block 10 is* **A**. *The transmitting antenna converts electrical signals into radio waves. (They are often called radio waves even though they are actually television signals.) The receiving antenna converts the waves into electrical signals for operating the receiver circuits.* Here is your next question.
Does the emitter of a transistor have to be at dc ground potential?
A. Yes. (Proceed to block 8.)
B. No. (Proceed to block 7.)

26. *The correct answer to the question in block 11 is* **B**. *Before you energize an electronic system (that is, before you turn the power* ON*), look for obvious faults. Use your senses.*

If there are no obvious faults, then you should energize the system. Sometimes you can tell which unit is faulty by observing the system output. Here is your next question.
To measure the dc voltage drop across a 1-megohm resistor you should use a
A. high-resistance voltmeter. (Proceed to block 15.)
B. low-resistance voltmeter. (Proceed to block 16.)
C. logic probe. (Proceed to block 14.)

27. *Your answer to the question in block 10 is* ***B****. This answer is wrong. The television transmitting antenna is absolutely necessary to radiate the signal. The receiving antenna is needed for changing the transmitted waves into electrical signals.* Proceed to block 25.

28. *The answer to the question in block 15 is about 0.7 volts. Technicians measure this voltage when troubleshooting in a transistor circuit. This is one way of telling if the transistor is conducting.* You have now completed the programmed questions. The next step is to put some of these ideas to work in laboratory experiments. Proceed to the Experiment section of this chapter.

EXPERIMENT

(The experiment described in this section may be performed on the circuit board described in Appendix C or on a similar laboratory setup.)

Purpose The 555 timer can be used as an audio oscillator. It produces an output signal in the audio range, and this signal can be used for signal injection. It is not an ideal signal generator because the output is in the form of pulses, but it can be used for signal-injection purposes.

A capacitor is an important part of a signal-injection probe. During the signal-injection procedure the probe may be touched to a point where there is a dc voltage, such as the plate of a vacuum-tube circuit, or at the collector of a transistor. This will place a high value of dc voltage on the probe. The capacitor prevents the dc voltage from destroying the output circuit of the signal generator. You will find that most signal generators have an *isolation capacitor* in the probe or output circuit.

You will remember from your study of power supplies and power amplifiers that a Darlington circuit is used when a high-gain and a high-output power is needed. Now you are going to connect a Darlington amplifier and use it as an audio power amplifier.

In this experiment you are going to connect a 555 timer and show that it is generating an audio signal by connecting its output to a speaker through the audio amplifier. Then you will use the 555 timer as a signal-injection test signal fed through the audio amplifier.

Test Setup All the circuits in this experiment will obtain their dc operating voltage from the dc power supply shown in Fig. 16-14. Figure 16-14*a* shows the schematic diagram, and Fig. 16-14*b* shows the pictorial diagram. You will recognize this as being a bridge rectifier circuit.

Figure 16-15 shows the test setup for connecting the 555 timer as an audio oscillator. Figure 16-15*a* shows the schematic diagram, and Fig. 16-15*b* shows the pictorial diagram.

Figure 16-16 shows the Darlington amplifier. Figure 16-16*a* shows the schematic diagram, and Fig. 16-16*b* shows the pictorial diagram.

Figure 16-17 shows the test setup for using the 555 timer to inject an audio

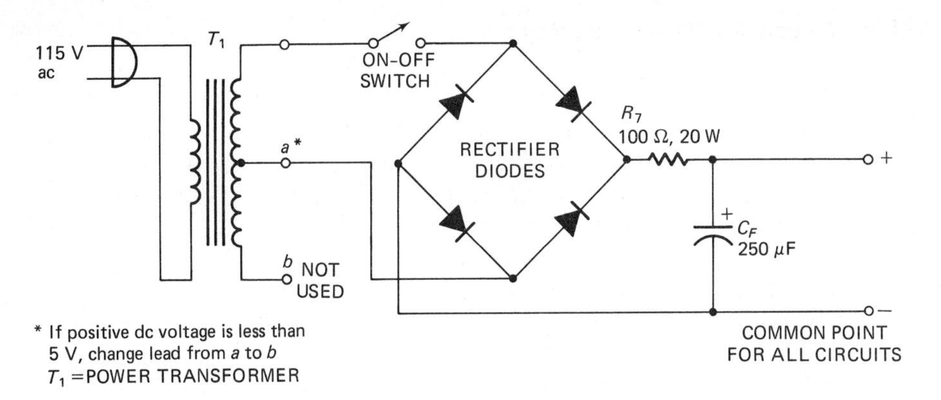

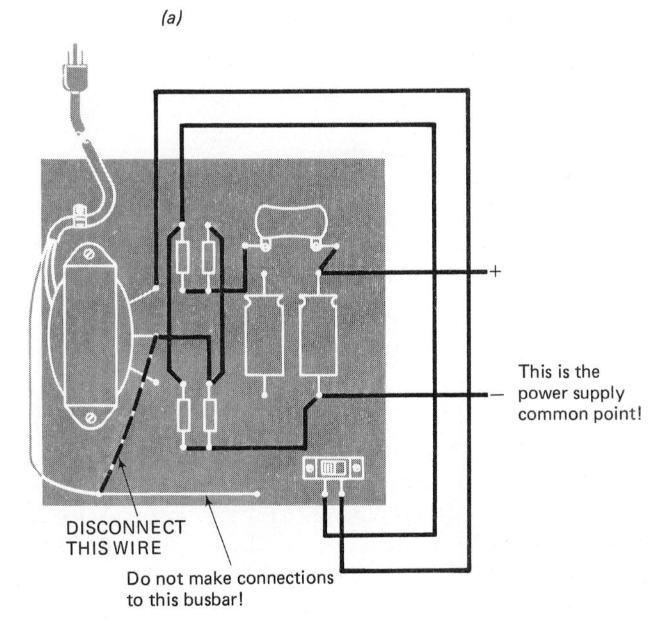

Fig. 16-14 Dc power supply for the experiment. (*a*) The schematic diagram. (*b*) The pictorial drawing.

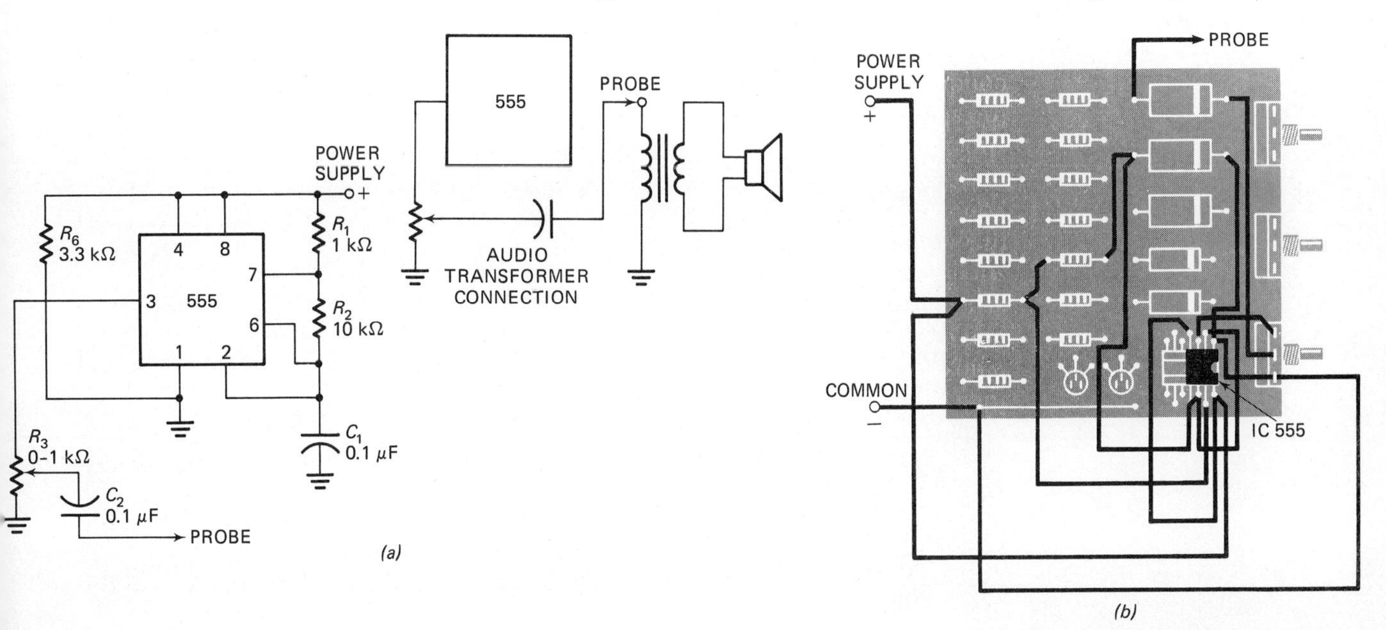

Fig. 16-15 Generator for audio test signal. (*a*) The schematic diagram. (*b*) The pictorial diagram.

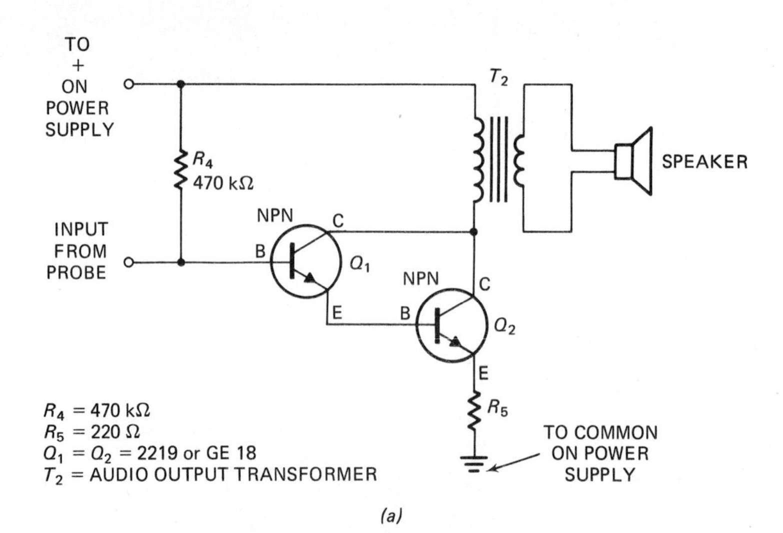

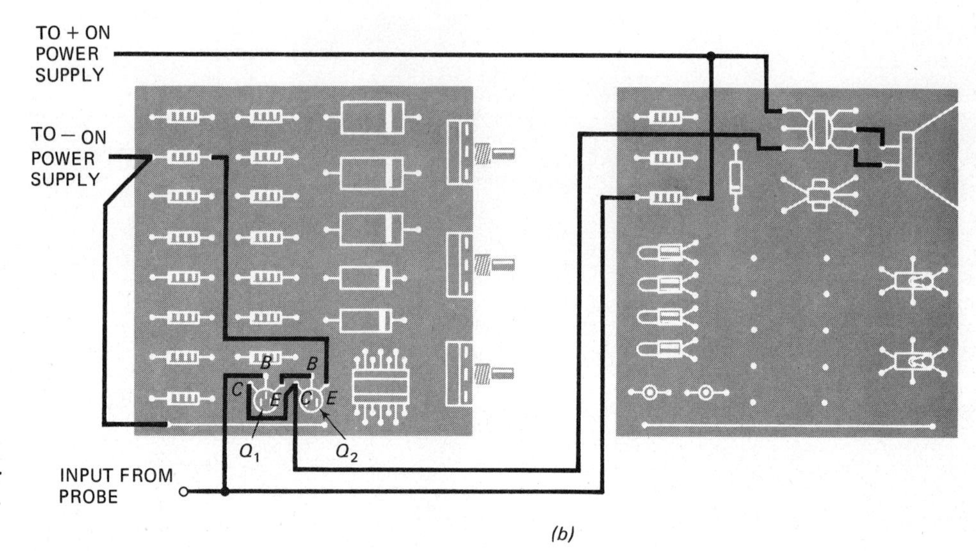

Fig. 16-16 Darlington audio power amplifier. (*a*) **The schematic diagram.** (*b*) **The pictorial diagram.**

signal into the Darlington amplifier. Note that Fig. 16-17*a* shows the schematic diagram and that Fig. 16-17*b* shows the pictorial drawing.

Procedure

step one Turn on the power-supply circuit shown in Fig. 16-14.

step two Measure the dc voltage across the power-supply output terminals. Record the value here. ______________ volts

This voltage should be between 5 and 10 volts.

step three Turn OFF the power supply. Connect the timer circuit of Fig. 16-15 to the power supply at point *a* and to the common point (Fig. 16-17). Be sure the power-supply switch is OFF while you are making the connection to the 555 timer. This means there will be no dc voltage on the terminals when you are connecting solid-state circuits to the supply.

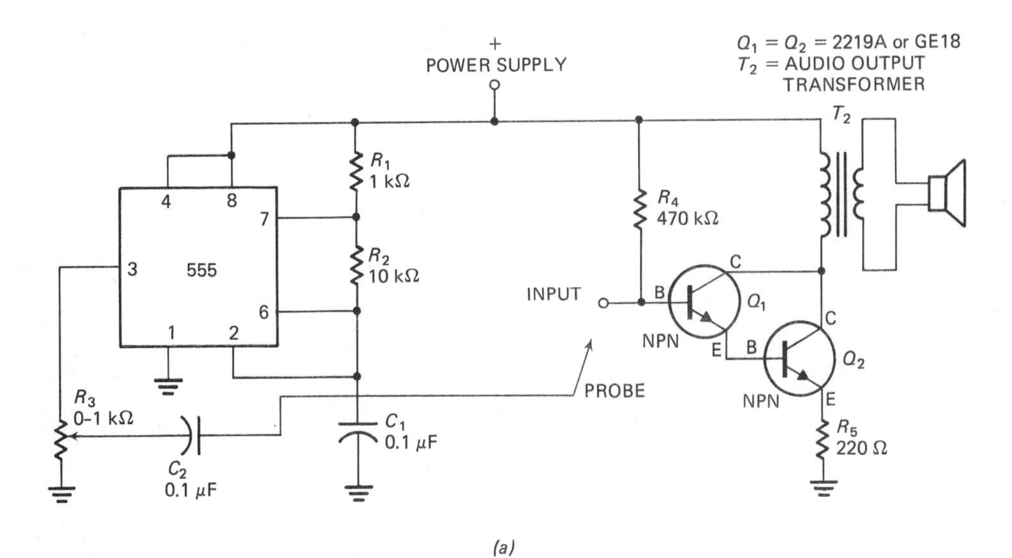

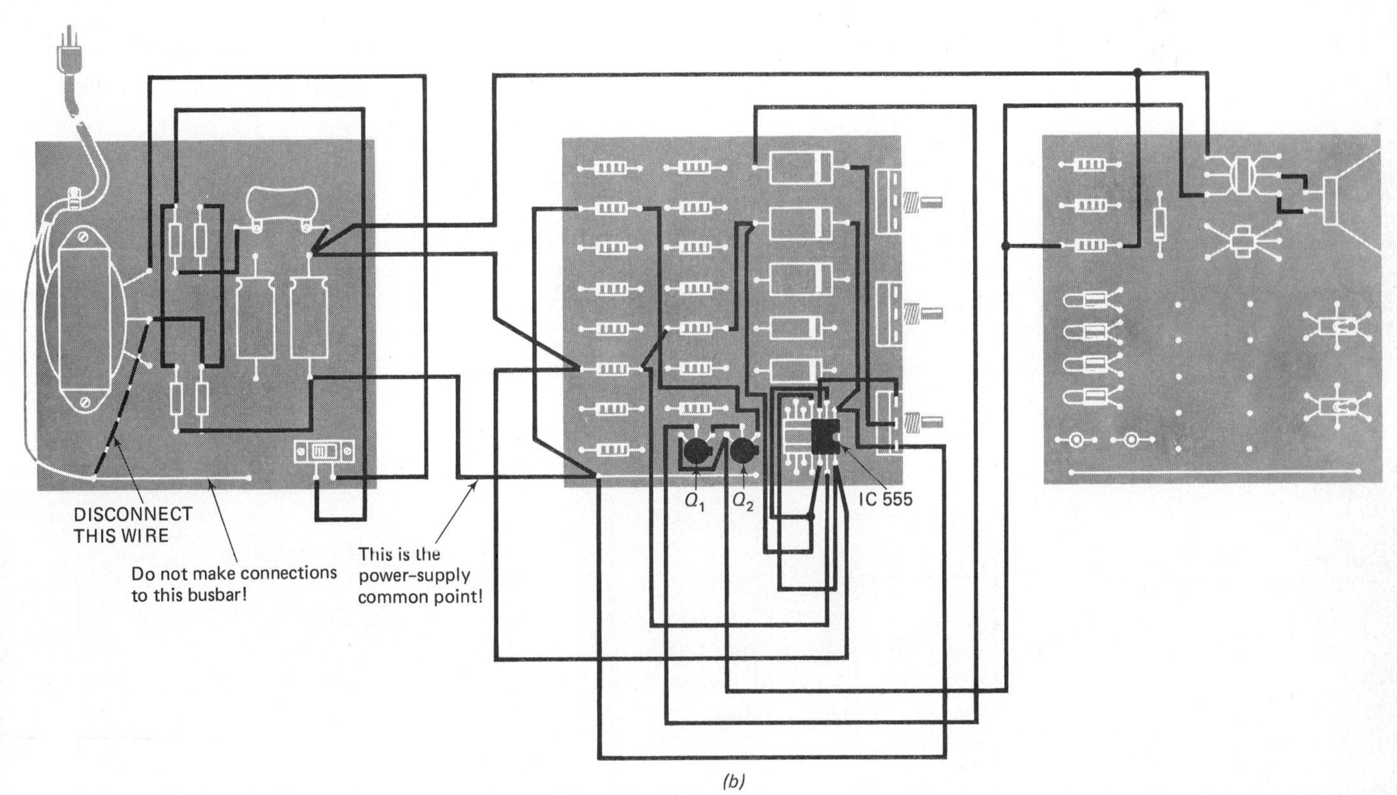

Fig. 16-17 The test setup for signal injection. (*a*) The schematic diagram. (*b*) The pictorial diagram.

step four Connect the audio probe directly to the output transformer. The connection is shown in the inset of Fig. 16-15*a*.

step five Turn the power-supply switch ON. Do you hear a sound from the speaker? ________________
Yes or No

You should hear an audio output from the speaker.

step six Adjust variable resistor R_6 and note that the sound volume varies. The variable resistor serves as a volume control in this circuit.

step seven Set variable resistor R_6 to the OFF position. This is an important step. When you inject the signal at the input of the Darlington amplifier, it will be amplified. It is very important to start with the minimum amount of signal so that you do not overdrive the amplifier. Remember that signal injection involves injecting the signal into points where there is more and more gain. So you must keep reducing the amount of injection signal as you go from point to point. Turn off the power supply.

step eight Connect the Darlington amplifier to the power supply and speaker as shown in Fig. 16-16. Turn on the power supply.

step nine Touch a screwdriver blade to the input terminal of the Darlington amplifier. Listen carefully to see if there is a noise in the speaker.

Do you hear a sound? ________________
Yes or No

In many circuits a screwdriver blade can be used for injecting a signal. The metal blade picks up signals from electrical equipment. These signals are injected into the circuit.

Also the blade may have a static charge—that is, it may be slightly positive or negative with respect to the input terminal. This causes a clicking sound when the blade is touched to the input.

You do not always get a sound with the screwdriver blade. As a matter of fact, there are so many things it depends upon that you cannot consider it to be a reliable test. However, technicians often use it as a quick test method. Turn off the power supply.

step ten Using the complete system of Fig. 16-17, connect the 555 timer audio-injection probe to the input terminal of the Darlington amplifier. Turn on the power supply. Adjust the output with R_3 to obtain a sound in the speaker. Does it require less signal input from R_6 (by its adjustment) to the

Darlington to get a sound in the speaker? ________________
Yes or No

You may expect it to take less signal to operate the Darlington power amplifier. However, in this particular circuit the Darlington circuit is designed to deliver only a small amount of power. For this reason you may not observe much increase in output with the power amplifier. Turn off the power supply.

Conclusion In this experiment you learned that a 555 timer can be used to generate an audio signal for testing. This signal is actually powerful enough to operate a speaker directly or to drive a power amplifier.

If there were more stages, the audio signal could be injected into each stage separately, as described in this chapter.

There are two important features to remember in the use of a signal generator: (1) The probe should be connected through a dc isolation capacitor, such as C_2 of Fig. 16-17, to protect the signal generator and the circuit being tested.

(2) The generator output-signal strength should be adjustable. This makes it possible to readjust the signal amplitude as you go from point to point when performing the signal-injection test.

SELF-TEST WITH ANSWERS

(Answers with discussions are given at the end of the chapter.)

1. Which of the following is the usual way of troubleshooting in a system? (*a*) Check components first, then the complete system; (*b*) Check components first, then the units; (*c*) Replace all tubes and transistors; (*d*) Use the senses (sight, hearing, and smell) to determine if there are any obvious faults.
2. Is this statement true or false? The most reliable information you can get about a faulty system is that given by the operator of the system. (*a*) True; (*b*) False.
3. Some companies package one or more units in a plug-in container. This is referred to as (*a*) mix and match; (*b*) a test compartment; (*c*) an after-product; (*d*) a module.
4. The last step in troubleshooting an electronic system is (*a*) soldering; (*b*) component testing; (*c*) capacitor discharge; (*d*) system testing.
5. Which of the following is sometimes used by technicians for signal injection? (*a*) A plastic straw; (*b*) The eraser end of a pencil; (*c*) A screwdriver blade; (*d*) A wooden stick.
6. Which of the following is the least useful (of the ones listed) for servicing electronic systems? (*a*) A cordless soldering iron; (*b*) A knowledge of electronics; (*c*) Special methods for locating faults; (*d*) A knowledge of the system.
7. Suppose you are asked to repair a faulty system. Which of the following is the wrong move as a first step? (*a*) Plug it in and turn it ON; (*b*) Check fuses and look for other signs of an overloaded circuit or component.
8. An audio-signal generator is going to be used for signal injection in audio systems. The probe should have (*a*) a pointed copper tip; (*b*) a cloth handle; (*c*) an isolation capacitor; (*d*) a dc voltage of at least 10 volts at the tip.
9. In addition to the item mentioned in Question 8, it is also important for the signal generator to have (*a*) an adjustable output-signal amplitude; (*b*) a method of inverting the polarity of the output signal.
10. Which of the following units produces an output that is usually used by more than one unit in a system? (*a*) The power supply; (*b*) The output transducer; (*c*) The power amplifier; (*d*) The output transformer.

ANSWERS TO SELF-TEST

1. (*d*)—Do not start to replace components before you have analyzed the system. The best place to start is to use your senses to see if there is an obvious fault.

2. (*b*)—Always remember that the operator may not be trained to recognize clues for faults. An important clue could easily be overlooked. You should listen carefully to what the operator has to say, then you should check the system yourself.
3. (*d*)—A disadvantage of modular construction is that a complete unit (or units) must be replaced even though the trouble is actually an inexpensive resistor. This could be costly if the customer is required to pay the cost of the module.
4. (*d*)—You have not completed the job until you make sure that the complete system is working.
5. (*c*)—This was described in the Experiment section.
6. (*a*)—In some cases repair may be made without the need for a soldering iron. All the others listed are useful.
7. (*a*)—Never do this! Look for signs of an overload before you energize the system.
8. (*c*)—The purpose of the isolation capacitor was described in the Experiment section.
9. (*a*)—This is needed for signal injection so you can reduce the amplitude of the test signal as you move from amplifier to amplifier.
10. (*a*)—The power supply provides the dc operating voltage for all units that have amplifiers.

appendix a
safety

The experiments described in this book are done with low voltages (usually 6 or 12 volts) applied to the circuitry. This minimizes the chance of electrical shock. If you are reading this book, you may be considering electronics as a career. This safety appendix contains information which you should know for doing even the most basic jobs in electronics. Safety should be one of your most important concerns on a job. This is true not only if you are working with electricity and electronics, but also if you are working on any job where accidents can be fatal.

Much work is being done in industry to reduce the job hazards. The first accurate statistics available on industrial accidents date to 1912, when it was reported that 35,000 people lost their lives in industrial accidents! These are not people that lost their lives working with electricity, but rather people working in all the industrial trades. Today the number of deaths by accidents in industry has been reduced to about one-third that number, but this is still considered to be too many by safety experts.

Of course, the most tragic part of an accidental death is the grief that it causes. Accidents that do not result in death can result in a serious impairment of physical ability and unnecessary cost and waste. Do not be surprised when you go to work in industry if a great amount of emphasis is placed on your safety and the safety of your fellow employees.

The Importance of Training

One of the first things to avoid is trying to do jobs for which you are not trained or which you are not familiar with. You might find yourself doing this even when working on simple laboratory experiments. If you catch yourself saying, "I wonder what would happen if I connected this wire to that point," you are probably about to make a serious mistake.

If you understand the theory of a circuit (which means that you have been trained for that particular work), you will *know* what will happen if you "connect this wire to that point." In other words, *an important first step in safety is to learn as much as you can about the equipment that you are working with.* Do not experiment in electronic circuits by touching wires just to see what would happen. Here is a good motto:

STUDY FIRST
WORK LATER

Electricity Is a Serious Business

Horseplay and practical jokes just are not funny in a laboratory. They *can* be fatal. What may seem like a practical joke can turn into a terrible disaster. If someone were to make a survey of the dumbest remarks in history, here is one that would certainly have to be included: "I didn't know it would hurt him."

Do Not Pay Attention to Myths

To a person who has not studied electricity, it is a mysterious and very dangerous thing. This is probably one of the reasons that there are so many myths about electricity. One of the most dangerous of these is that "you can build up a resistance to electricity by taking repeated shocks of increasing intensity." Frankly, this just is not so. Not only is it *untrue,* it is an extremely dangerous thing to try. Your health can be ruined—especially your heart—by repeated shocks with electricity.

Another myth, which is just as dangerous as the one just mentioned, is the idea that a "little bit of electrical shock is good for you," or that it is fun to see who can take the biggest electrical shock. The amount of electrical shock that you can take is not in any way related to how strong you are or how big you are. There are many, many factors involved which do determine whether a shock will be fatal, but the final determining factor is *how much current flows through your body.* Note this important fact: It is *not* the *voltage* that kills you, but rather, it is the amount of *current* that passes through your body.

A current of only 1/1000 of an ampere (or 1 milliampere) can produce a shock which you can feel. At about 1/100 of an ampere (or 10 milliamperes) the shock is so severe that it paralyzes the muscles, and you cannot release the conductor. If you see someone being shocked by a conductor that he cannot let go of, do not be foolish enough to grab him and try to pull him away. If you do, you will receive a shock too. Instead, deenergize the circuit when possible. If you cannot do this, then use a dry board, or rope, or other insulating material to force him away.

At about a 1/10 of an ampere (100 milliamperes) shock may be fatal if it lasts for a second or more. These are extremely small values of current when you stop to think that the fuses in your house are rated for 15 amperes of current. In other words, *a shock can be fatal long before the fuse blows.*

If someone tries to tease you by saying "You are not afraid of a little electricity, are you?", remember that the answer to that question should be *yes*. Airplane pilots have a favorite saying: "There are old pilots and there are bold pilots but there are no old, bold pilots." The same could be said of people working with electricity.

People that play in electronics labs are eliminated from the profession in two ways. First, they are likely to make a serious mistake. Second, they will find it hard to get jobs because nobody wants to work around a clown.

Another myth that should be disregarded is the idea that you cannot be injured when working on a circuit if you keep one hand in your pocket. This idea is based on the fact that you are much more likely to be hurt with both hands across a hot circuit. However, it is not true that putting one hand in your pocket will keep you from getting hurt.

Safety Features

There are safety features installed on equipment, and the purpose of these features should not be defeated. *Fuses* and *interlocks* are two examples.

You already know that a fuse will *not* protect you from a serious injury. However, it will open the circuit to the defective equipment. When a fuse burns out, always determine the reason why. Do not just simply replace the fuse or defeat the fuse circuit.

An interlock is a device that shuts the equipment OFF when its cabinet is open or when it is partially disassembled. *Interlocks are for the protection of people.* In this way they are different from fuses, which are for the protection of equipment. In some cases it is necessary to work on the equipment when it is energized. To do this an experienced technician can *defeat* the interlock—that is, bypass it so that the equipment will work even though the cabinet is open. This procedure is acceptable if you know exactly what you are doing in the circuit, but if you are not familiar with the equipment, you should *never* defeat the interlock.

Tools Are Important in a Good Safety Program

There are a number of operations related to electronics which require special safety precautions. Regardless of what type of work you are doing, you should keep your tools and equipment in top-notch order. Be sure that screwdrivers are sharp and tools are clean. Always wear safety glasses and safety gloves on jobs that require them, and before you work on any equipment be sure that it is deenergized (whenever possible).

It is especially important to know where the main power switch is in the area where you are working. You might find it necessary to turn the power OFF quickly. If you find that it is necessary to work around high-voltage equipment, never work alone.

Fire Hazards

You no doubt know that electricity can start fires. For example, it is not uncommon for a house with poor wiring to go up in flames. You should always keep in mind the fact that electricity can create fire hazards, and you should have a good understanding of how to deal with these hazards.

You may find it necessary to work with soldering irons or soldering guns, and you should remember that these guns or irons get hot enough to set fires. You should not get into the habit of flicking solder from an iron or gun to clean it. Flecks of solder can hit you or someone near you. This could cause a skin burn, or worse yet, the solder may go into an eye or ear.

When you are soldering, be especially careful not to breathe in smoke that is created by the soldering and the flux. As a matter of fact, you can make a general rule about smoke:

It is never good to inhale smoke regardless of where it comes from.

If electronic equipment catches on fire, you must be very careful *not* to throw water on the fire, and you must not use a foam-type fire extinguisher. Electricity can actually follow the water or foam and cause you a serious injury. Even if it were not for the possibility of injury, it would not be a good idea to use water or foam extinguishers on electrical fires because both could destroy the equipment. What might have been a small burnout can end up to be a complete destruction of the equipment if water or foam is applied while the equipment is energized.

Some types of equipment use selenium rectifiers, which produce a considerable amount of smoke when they burn out. This smoke can actually be injurious to your health, and again, the above general statement can be repeated: *Never breathe in smoke of* ***any kind****, regardless of where you think it is coming from.*

Take a few seconds to investigate the fire extinguishers in your working area. There should be an identification plate on the extinguisher that tells whether or not it can be used for an electrical fire. The kind usually recommended for electrical fires is the CO_2 type.

The following steps should be taken if a fire breaks out in an electronics lab.

1. Warn everyone in the area.
2. Deenergize the circuit immediately. You may have to do this by throwing the main switch.
3. Call the fire department or turn in a fire alarm. Remember that firefighters are experts in all kinds of fires, and you need their help as quickly as you can get it.
4. Using the correct fire extinguisher (a CO_2 type), direct the extinguisher at the *base* of the flame. Be careful not to let the fire get between you and the exit of the building.

Know Your First Aid

Persons who receive a serious shock may have their breathing impaired, and it is important to get their breathing started again *as quickly as possible.* Do not stop to loosen their clothes—let someone else do that. Instead, start artificial resuscitation immediately.

Two methods of artificial resuscitation are described here. The mouth-to-mouth method is preferred, but in some cases (such as a severe injury to the

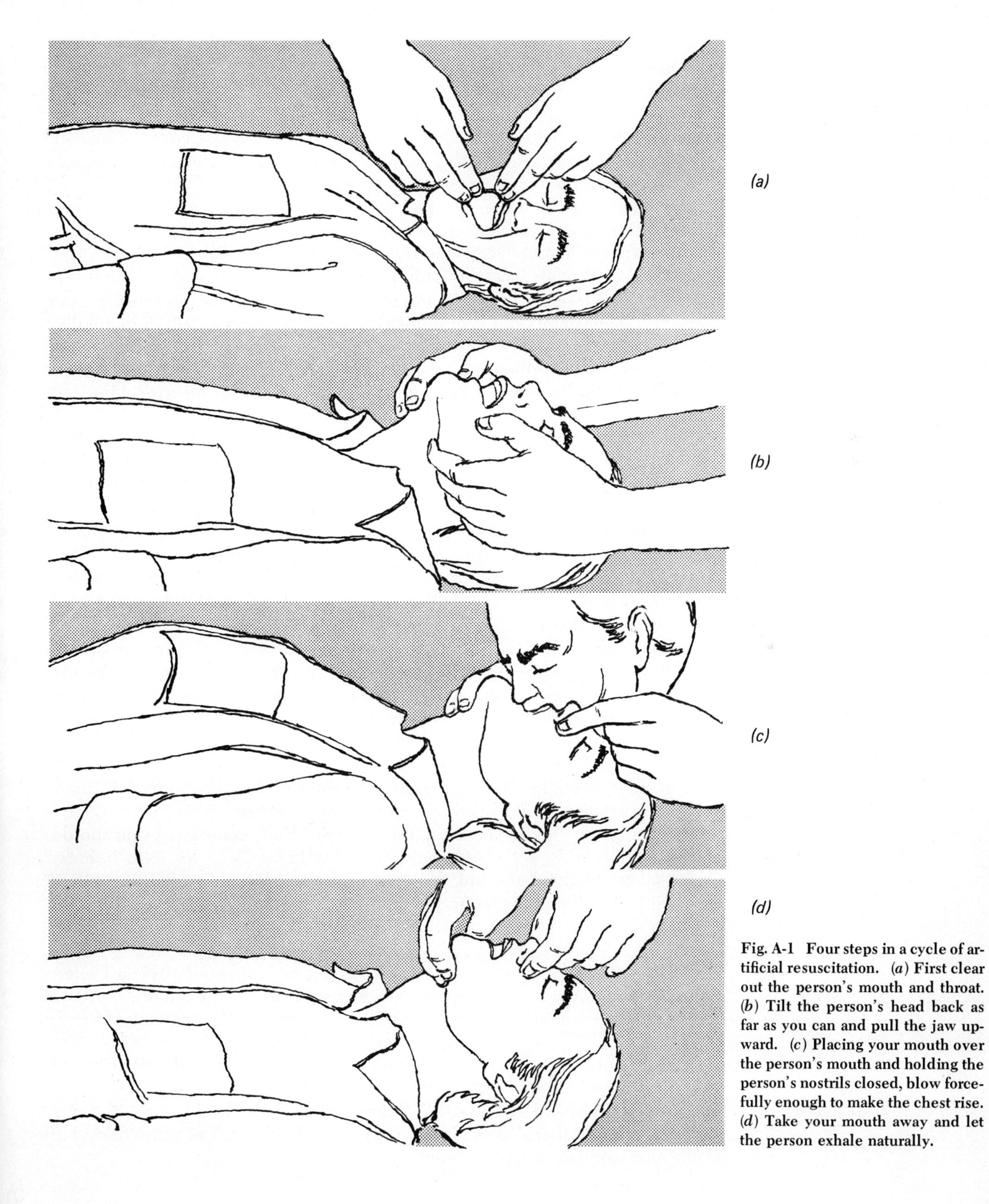

Fig. A-1 **Four steps in a cycle of artificial resuscitation. (*a*) First clear out the person's mouth and throat. (*b*) Tilt the person's head back as far as you can and pull the jaw upward. (*c*) Placing your mouth over the person's mouth and holding the person's nostrils closed, blow forcefully enough to make the chest rise. (*d*) Take your mouth away and let the person exhale naturally.**

face) the other method should be used. You should learn and practice both methods.

Regardless of which method you use, start by making sure that the victim's mouth and nose are free from restrictions, and that the victim does not have something in the mouth, like gum or candy, that he or she could choke on.

Figure A-1 shows the mouth-to-mouth method. Cover your fingers with a handkerchief or cloth and wipe out the person's mouth and throat.

Tilt the victim's head back as far as you can and pull the jaw upward until the lower teeth are in front of the upper teeth. Then take a deep breath. Place your mouth over the victim's mouth, holding the victim's nostrils. Make sure that no air leaks out around your mouth.

Blow into the person's mouth forcefully enough to make the chest expand and rise. Then take your mouth away and let the person exhale naturally.

Repeat this cycle 12 to 15 times a minute. Each complete cycle should take 4 or 5 seconds. Once you establish the rhythm, do not break it regardless of what else is being done to the victim. Keep it up until professional help arrives.

Figure A-2 shows the Shaeffer method of artificial resuscitation. Your hands should be placed in the middle of the victim's back—just below the shoulder blades—so that the fingers are spread downward and outward and the thumb tips are almost touching. Move forward until your arms are approximately vertical and use the weight of your body to press down. Exert a slow, steady pressure on your hands until a firm resistance is met. The purpose of this action is to force the air out of the victim's lungs.

The next step is to release the pressure by backing away, but be careful not to make sudden moves. It is a good idea to release the pressure of your hands from the victim's ribs with a gradual outward motion. As you move back, draw the victim's arms upward and toward you, keeping your arms nearly straight. The purpose of this motion is to relieve the pressure on the victim's chest and to allow the air to move into the lungs freely. After you get to this position, repeat the process over and over again, being sure to use a steady rhythm.

In order to keep the rhythm smooth, you can repeat over and over to yourself the expression "Out goes the bad air" (when you are pushing down), "and in goes the good" (when you are pulling back). Each complete cycle should take 4 or 5 seconds, and you should do about 12 or 15 cycles each minute. Once you establish the rhythm, do not break it regardless of what else is being done to the victim. Keep it up until professional help arrives.

Regardless of which method is used, it is important to keep the victim warm.

Have someone wrap him or her in clothing or blankets between cycles. You should not try to do this yourself, or it will break the rhythm. Avoid moving the person unless it is *absolutely* necessary (such as the case of fire or other hazard).

It is very important to remember that you should not give up too soon. It may take hours to actually bring victims around to the point where they can breathe by themselves. Once they start to breathe, do not have them sit up. Be sure to watch them carefully. If they stop breathing, you must start your artificial resuscitation again immediately.

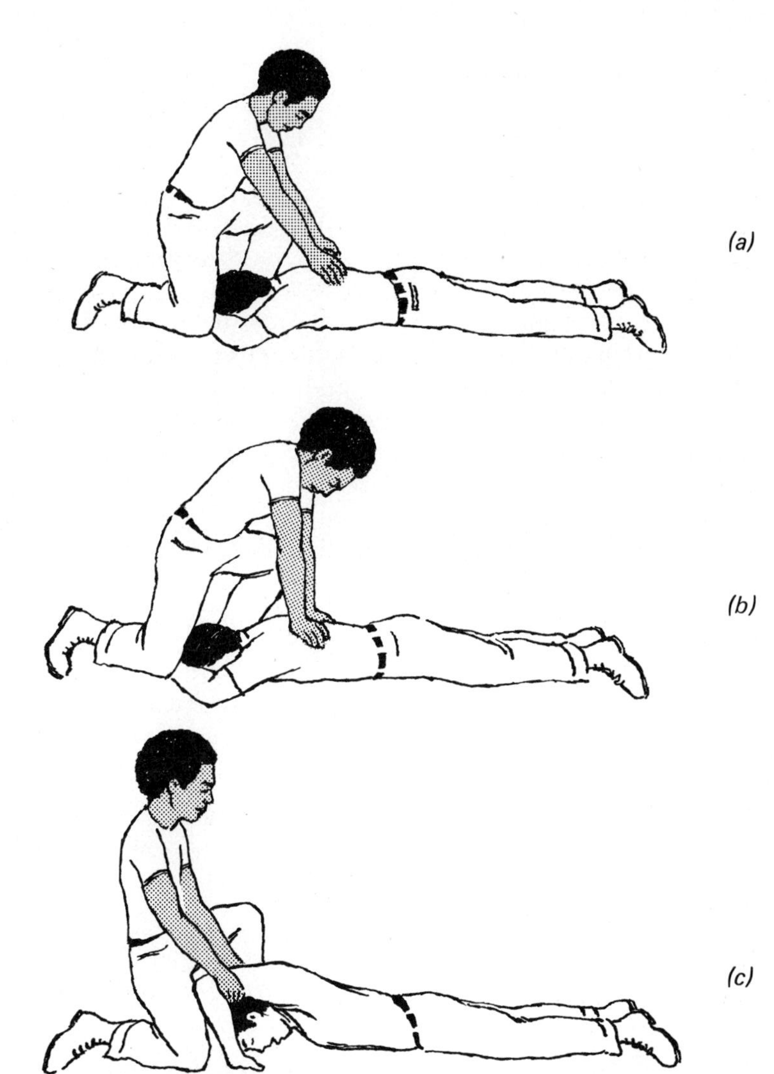

Fig. A-2 Three steps in a cycle of artificial resuscitation. (*a*) The first step is to get the victim into this position. Place your hands on each side of the victim's back as shown. (*b*) In this step pressure is exerted on the back to force the air out of the victim's lungs. Your arms should be straight and nearly vertical. (*c*) As you move back, lift the victim's arms and pull them forward as shown here. The victim's arms should be raised enough to arch the back, and the chest should be barely lifted from the ground in order to ease the intake of fresh air into the victim's lungs.

If a person becomes burned, cut, or wounded in some way, you must seek medical help immediately. If you have been trained in first aid, you can apply pressure at the recommended pressure points to reduce the problem of bleeding, but at the same time you should make sure someone has gone for medical help.

The important thing to remember in the case of burns is to *never* apply iodine, antiseptic, or powder to the burned area. Also you should avoid using cotton directly on the burn because this can cause additional injury.

You can often make a burn less serious by covering it with cold water or ice to cool the burn. Also, most of the pain from a burn comes from air moving across the area. Therefore, if the skin is merely reddened (but not broken), you can cover the area with a coating of water-soluble ointment (do not use grease). Avoid breaking any blisters, and do not pull clothing away that is stuck to a burned area of skin. Instead, cut around it.

The information given here is not intended to substitute for first aid training.

Summary of Safety Precautions

Here is a summary of some of the important safety precautions to be taken when working around electrical and electronic equipment.

- ☐ Avoid loose clothing and jewelry that can be caught in machinery.
- ☐ Wear shoes with thick insulating soles. Avoid the kinds of shoes that have cleats and metal hobnails.
- ☐ Keep your tools clean and sharp, and make sure that they are properly insulated.
- ☐ Avoid inhaling vapors and smoke of any kind.
- ☐ Never operate defective electrical equipment, such as a power tool in which the three-way plug has been defeated so that it can fit into a two-way receptacle. Also equipment in which insulating mounts or knobs have been removed should never be operated.
- ☐ Never use alcohol or carbon tetrachloride ("carbon tet") as a cleaning agent around electrical systems. Alcohol will burn if it gets too hot, and carbon tet will emit a poisonous gas when in contact with hot metals.
- ☐ Do not rely on interlocks, fuses, and high-voltage safety relays to protect you in an electrical circuit. Your best safety precaution is knowledge of what you are doing.
- ☐ When possible, completely deenergize a circuit when working on it.
- ☐ If you find it necessary to replace electric fuses, use an approved type of fuse puller. Do not reach into the box with your bare hands to remove cartridge fuses even though the circuit is OFF.
- ☐ Never take *anything* for granted.
- ☐ Do not work on electrical equipment that you are unfamiliar with. Take a little time to learn how a circuit works and what it is supposed to do. Your knowledge of electricity is one of your best safety devices.
- ☐ Always know where the main switches are and where the fire extinguishers are in the area where you are working. Determine which types of fires the fire extinguishers are designed for.
- ☐ Never put water, carbon tetrachloride, or foam extinguishers on electrical fires.

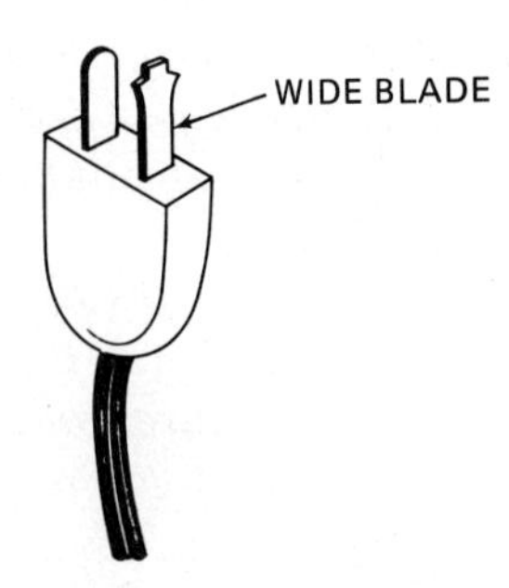

Fig. A-3 **A polarized plug.**

- ☐ Avoid horseplay and make it clear that you do not appreciate practical jokes and horseplay.
- ☐ Take some time to learn first aid procedures because they can help you (not only in the electronics lab, but also in many other life situations).
- ☐ If the equipment has a polarized plug like the one in Fig. A-3, it is a safety feature. The wide blade goes to the low side of the ac line, and the plug can be inserted into the socket in only one way. Do not defeat this safety feature!

PROGRAMMED REVIEW QUESTIONS

(Instructions for using this programmed section are given in Chapter 1.)

We will now review some of the important ideas of safety in this programmed section. If you have understood the material, you will progress easily through this section. Do not skip this section because some additional material is presented.

1. Which of the following is true?
   A. If there is a fire extinguisher in an electronics lab, you can presume that it is safe for use on electrical fires. (Proceed to block 9.)
   B. Always take time to learn which fire extinguishers are safe to use *before* a fire starts. (Proceed to block 17.)

2. *Your answer to the question in block 19 is* ***B****. This answer is wrong. In some cases a voltage of 70 volts can be fatal, but in other cases a voltage of 20,000 volts will not be fatal. Obviously it is not the voltage that is the determining factor.* Proceed to block 22.

3. *The correct answer to the question in block 18 is* ***B****. You will find that those marked "Safe for Electrical Fires" are CO_2 extinguishers.*
   Here is your next question.
   In case of an electrical fire you should first
   A. turn in a fire alarm. (Proceed to block 7.)
   B. deenergize the circuit. (Proceed to block 16.)

4. *Your answer to the question in block 21 is* **A.** *This answer is wrong. It can be dangerous and costly to "try a few things" in a circuit when you do not know anything about the equipment.* Proceed to block 13.

5. *Your answer to the question in block 13 is* **A.** *This answer is wrong. Although it is a good practice to keep one hand in your pocket, this is no sure guarantee against injury. Your best protection is to know what you are doing.* Proceed to block 19.

6. *Your answer to the question in block 18 is* **A.** *This answer is wrong. If you spray water on an electrical fire, two things will happen.* ***First,*** *the water will probably ruin the circuit. (What might have been a small fire can turn into an expensive loss.)* ***Second,*** *electricity can follow the spray to your body, and this could be fatal.* Proceed to block 3.

7. *Your answer to the question in block 3 is* **A.** *This answer is wrong. It is very important to get help, so you* **should** *turn in an alarm. However, you should* ***first*** *deenergize the circuit. The best way to do this is to turn* OFF *the main power switch or circuit breaker. Again, you should know where this switch is* **before** *a fire starts so that you do not have to look for it in an emergency.* Proceed to block 16.

8. *The correct answer to the question in block 16 is* **B.** *Of course, it is important to loosen the clothing and get obstructions away from the person's nose and mouth, but let someone else do this. If there is no one else around, do it between cycles. The breathing cycle must be started as soon as possible.* Here is your next question.
   Is this statement true or false? If you plan to work around electronic equipment, you should build up your resistance to electricity by taking repeated shocks.
   A. This statement is true. (Proceed to block 14.)
   B. This statement is false. (Proceed to block 21.)

9. *Your answer to the question in block 1 is* **A.** *This answer is wrong. There may be fire extinguishers placed in the lab for solvent or grease fires or for paper or wood fires. You should* **not** *use these for electrical fires.* Proceed to block 17.

10. *The correct answer to the question in block 25 is* **A.** *Besides not putting iodine on burns, you should avoid the use of cotton swabs and avoid the use of medicated powders.* Here is your next question.
    The complete cycle of artificial resuscitation should take about
    A. 20 seconds for each complete cycle, with a total of 12 to 15 complete cycles per minute. (Proceed to block 12.)
    B. 4 or 5 seconds for each complete cycle, with a total of 12 to 15 complete cycles per minute. (Proceed to block 18.)

11. *Your answer to the question in block 18 is* **C.** *This answer is wrong. Most foam-type extinguishers should never be used on electrical fires for the same reasons that water should not be used.* Proceed to block 3.

12. *Your answer to the question in block 10 is* **A.** *This answer is wrong. A human could not live with such a small amount of air. Besides, if each cycle takes 20 seconds, you cannot have 12 to 15 cycles per minute. Review the procedure described, then* proceed to block 18.

13. *The correct answer to the question in block 21 is* **B.** *It is not safe, and it is not good practice, to experiment with electronic equipment. The correct procedure is to learn about the equipment,* **then** *work with it.* Here is your next question.
Is this statement true or false? When working in high-voltage circuits, you cannot be injured when working on electrical equipment as long as you remember to keep one hand in your pocket.
A. This statement is true. (Proceed to block 5.)
B. This statement is false. (Proceed to block 19.)

14. *Your answer to the question in block 8 is* **A.** *This answer is wrong. You cannot build up a resistance to electricity, and it is dangerous to your health to try it.* Proceed to block 21.

15. *Your answer to the question in block 17 is* **A.** *This answer is wrong. A fuse is used to protect the* **equipment,** *not to protect the user or the technician working on the equipment.* Proceed to block 25.

16. *The correct answer to the question in block 3 is* **B.** *If you take a moment to deenergize the circuit, you may prevent additional fire hazards. In some cases deenergizing the circuit will also put the fire out.* Here is your next question.
Which of the following steps should be taken first if you are going to apply artificial resuscitation?
A. Loosen the person's shoes and belt. (Proceed to block 20.)
B. Get the person into position as quickly as possible and start artificial resuscitation. (Proceed to block 8.)

17. *The correct answer to the question in block 1 is* **B.** *It is possible that more than one type of extinguisher may be located in the lab. You should make it your own job to learn which are for electrical fires. A CO_2 extinguisher is usually used on electrical fires.* Here is your next question.
Which of the following statements is true?
A. If a circuit is properly fused, you cannot receive a fatal shock by touching it. (Proceed to block 15.)
B. Even though there is a fuse in the circuit, you could receive a fatal shock by touching it. (Proceed to block 25.)

18. *The correct answer to the question in block 10 is* **B.** *You should practice this rate until you can do it accurately.* Here is your next question.
For an electrical fire you should use
A. water. (Proceed to block 6.)
B. CO_2 extinguishers. (Proceed to block 3.)
C. foam-type extinguishers. (Proceed to block 11.)

19. *The correct answer to the question in block 13 is* **B.** *Keep one hand in your pocket as a safety precaution whenever possible, but always be careful. Be sure that you are insulated from the electrical conductors, even though you believe the circuit to be deenergized.* Here is your next question.
Which of the following is the thing that actually determines if a shock is fatal?
A. The amount of current that flows through the body. (Proceed to block 22.)
B. The amount of voltage in the circuit. (Proceed to block 2.)

20. *Your answer to the question in block 16 is* **A.** *This answer is wrong. You should start* **immediately** *to get the breathing cycle started. Let someone else loosen the person's clothing.* Proceed to block 8.

21. *The correct answer to the question in block 8 is* **B.** *The rumor that a person can build up a resistance to electricity is a dangerous one.* Here is your next question.
Is this statement true or false? If you are not familiar with a piece of electronic equipment, try a few things. After all, that's how you learn new things.
A. This statement is true. (Proceed to block 4.)
B. This statement is false. (Proceed to block 13.)

22. *The correct answer to the question in block 19 is* **A.** *A current of 1/10 of an ampere could be fatal.* Here is your next question.
When you are working near high-voltage circuits you should
A. work alone. (Proceed to block 24.)
B. work with someone. (Proceed to block 26.)

23. *Your answer to the question in block 25 is* **B.** *This answer is wrong. You should* **never** *put iodine on a burn.* Proceed to block 10.

24. *Your answer to the question in block 22 is* **A.** *This answer is wrong. It is never a good idea to work alone in any system where accidents may occur.* Proceed to block 26.

25. *The correct answer to the question in block 17 is* **B.** *Defective equipment* **may** *blow a fuse, and in this way you know that the equipment has a fault which may be dangerous. However, the fuse is mostly for* *protecting against fires and against destruction of electrical equipment. Remember that a 0.1-ampere shock can be fatal, and a 15-ampere fuse would be no protection at all.* Here is your next question.

Which of these statements is true?

A. Never put iodine on a burn. (Proceed to block 10.)
B. The best thing that you can put on a burn is iodine. (Proceed to block 23.)

26. *The correct answer to the question in block 22 is **B**. At the very least, a partner could go for help if an accident should occur.*
You have now completed the programmed questions. The next step is to take the true-false Self-Test.

SELF-TEST WITH ANSWERS

Answer each of the following questions true or false.
(Answers are given at the end of the chapter.)

1. Safety should be one of your most important concerns.
2. You cannot build up a resistance to electricity.
3. Keeping one hand in your pocket when working around high voltages is a safe practice, but this is not an absolute guarantee of safety.
4. You should not work on equipment that you are not familiar with.
5. A current of only 1/10 of an ampere for about 1 second could be fatal.
6. There should be about 12 to 15 cycles per minute when applying artificial resuscitation.
7. It is not true that a little bit of electricity is good for you.
8. A fuse in a circuit will not protect you from a serious shock.
9. You should not use cotton to swab a burn.
10. You should not experiment with expensive equipment.
11. With a current of 1/100 of an ampere you cannot let go of the circuit.
12. You should never inhale smoke or fumes of any kind.
13. You should not engage in horseplay in the lab.
14. When there is an electrical fire, you should deenergize the circuit.
15. A current of only 1/1000 of an ampere can produce a shock that you can feel.
16. You should never put iodine on a burn.
17. It is a good safety practice to keep tools sharp and clean.
18. You should find out where the main power switch and the fire extinguishers are when working in an area.
19. Each cycle in artificial resuscitation should take about 4 or 5 seconds.
20. Never use water or foam-type fire extinguishers on electrical fires.

ANSWERS TO SELF-TEST

All answers are true.

appendix b vocabulary words and symbols

VOCABULARY WORDS

A BATTERY The battery used to power the filaments in portable vacuum-tube equipment.

ACCELEROMETER A transducer that is designed to sense acceleration.

ACTIVE COMPONENT A component used for generating or amplifying signals.

ACTIVE TRANSDUCER A transducer that generates a voltage which is related to the amount of input energy.

AMPLIFICATION A small change in input voltage (or current) causing a larger change in output voltage (or current).

ANALOG A system that produces an output which is proportional to some input quantity.

ANODE A positive terminal, or the electrode connected to a positive terminal.

B BATTERY The battery used to operate the plate circuit in portable vacuum-tube equipment.

BANDWIDTH The range of frequencies the amplifier can handle.

BASE A section of a bipolar transistor that is equivalent to the control grid of a tube and gate of an FET. The input signal is usually delivered to this electrode. (For illustration, see Symbols section.)

BLACKER-THAN-BLACK REGION That portion of the television signal that has a greater amplitude than the blanking pedestal. The sync pulses are in the blacker-than-black region.

BINARY SYSTEM A numbering system based on two symbols such as 0 and 1.

BIPOLAR COMPONENT A component that depends upon both holes and electrons for its operation.

BISTABLE COMPONENT A component (or circuit) that has only two conditions for operation. It can remain in one condition or the other, but not both.

C BATTERY The battery used to obtain the grid bias voltage in portable vacuum-tube equipment.
CAPACITIVE REACTANCE The opposition that a capacitor offers to the flow of alternating current.
CAPACITOR A component that stores energy in the form of an electrostatic field and opposes any change in voltage across its terminals. (For symbols, see Symbols section.)
CAPSTAN A small metal cylinder used in tape recorders to pull the tape across the tape head at a constant speed.
CATHODE A negative terminal, or the electrode connected to a negative terminal.
COERCIVE FORCE The amount of magnetizing force needed to remove magnetism.
COIL A component that stores energy in the form of an electromagnetic field. Also called an *inductor* or a *choke.*
COLLECTOR The electrode of a bipolar transistor that is equivalent to the plate in a tube and the drain of an FET. The output signal is normally taken from the collector. (For illustration, see Symbols section.)
COMMON The place where the signal voltage is considered to be 0 volts.
CONVENTIONAL CURRENT FLOW Current through the circuit from the positive terminal of the voltage source to the negative terminal.
CONVERTER A type of power supply used to change dc to another value of dc voltage or current.
CROSSOVER DISTORTION A type of distortion that occurs when the output waveform is distorted near the zero-amplitude points.
CRYSTAL OVEN A container that holds the temperature of a vibrating crystal constant. Its purpose is to make the frequency more constant.
DAMPED WAVE A transmitted signal produced by drawing a spark across a gap, with the spark producing electromagnetic waves over a wide range of frequencies. It is identified by an amplitude that decreases with time. (For illustration, see Symbols section.)
DARLINGTON AMPLIFIER A type of amplifier in which two transistors are connected so that the base current of one transistor is determined by the amount of current flowing through the other. (For illustration, see Symbols section.)
DEFLECTION COIL A coil with current that is used to deflect the beam in a cathode-ray tube. It is the magnetic field around the coil that deflects the electron beam.
DEFLECTION PLATE A metal plate with voltage that is used to deflect the beam in a cathode-ray tube.
DEGENERATIVE FEEDBACK Feedback produced when the output signal is returned to the input in such a way that it subtracts from the output signal. Also called *negative* feedback.
DETECTOR A radio stage that separates the RF and audio signals and delivers the audio signal to the amplifier.
DIELECTRIC A material that does not readily conduct electron-current flow.
DIGITAL A system in which the output is a series of digits or numbers.
DIODE A component that conducts electron current in one direction

(cathode to anode) but not in the reverse direction (anode to cathode). (For symbols, see Symbols section.)

DIRECTLY HEATED CATHODE A cathode on which the emitted electrons come directly from the surface of the filament.

DISCRETE QUADRAPHONICS A method of quadraphonics in which the outputs of four microphones are recorded on four separate tracks. At playback each track is treated as a separate recording to a four-speaker system.

DRAIN The electrode of a JFET, equivalent to the plate of a tube or the collector of a bipolar transistor.

ELECTROMAGNETIC COMPONENT A component whose operation depends upon both electric currents and magnetic fields.

ELECTRON One of the particles in an atom. It is a negatively charged particle.

ELECTRON CURRENT FLOW The current which is traced through the circuit from the negative terminal of the voltage source to the positive terminal.

ELECTRONICS The science of putting electrons to work.

ELEMENTS The 92 basic ingredients which are combined in various ways to produce all materials that are known.

EMITTER FOLLOWER Common collector circuit; the input signal goes to the base and the output signal comes from the emitter. The comparable tube circuit is a cathode follower, and the equivalent FET circuit is a source follower.

EMITTER STABILIZATION RESISTOR A resistor in the emitter circuit that protects the transistor against thermal runaway. It also stabilizes the amplifier against temperature changes.

ENCLOSURE The box in which the speaker is located.

ENCODED QUADRAPHONICS A method of quad sound that takes the electrical signals from four microphones and combines them electrically into two signals. Also called *matrix* quadraphonics.

ENHANCEMENT MOSFET A MOSFET that will not conduct unless there is a forward bias applied.

EQUIVALENT CIRCUIT A circuit that imitates another circuit.

FAN-IN The total number of inputs in a logic circuit.

FAN-OUT The total number of outputs in a logic circuit.

FARADAY'S LAW OF ELECTROMAGNETIC INDUCTION Law stating that whenever there is motion between a magnetic field and a conductor, a voltage is produced.

FERRITE BEADS Small beads of magnetic material that act like RF chokes. (For symbols, see Symbols section.)

FIELD EMISSION Emission produced when electrons are torn away from the surface of the cathode by the positive voltage applied to the anode.

FILTER A circuit that will pass only one frequency or range of frequencies.

FIRING POTENTIAL A certain minimum value of voltage required to make a neon lamp conduct.

FLIP FLOP A bistable circuit. It has only two possible levels of operation, ON and OFF. The output states are often called *high* and *low*.

FLUX Magnetic field or field lines.

GAS-REGULATING DIODE A diode with a gas-filled envelope. The voltage across this type of diode is constant.

GATE The control electrode in an FET, an SCR, and a triac.

GATED AGC An AGC circuit in a TV receiver that is turned ON and OFF by the sync pulse. This prevents noise spikes that occur at any time, except during the sync pulse, from affecting the picture.

GRID AUDION The first triode. It was invented by DeForest.

GRID-LEAK BIAS A type of bias that is obtained from the signal.

GROUND WAVE A radio wave that moves along the earth's surface.

HIGH In logic circuitry, describing a binary 1 state, or ON condition.

HIGH-PASS FILTER A filter that will pass all frequencies above the cutoff point.

HIGH-VOLTAGE SECTION A circuit that produces the high voltage needed to operate the cathode-ray tube in a TV receiver.

HOLE-IN-THE-MIDDLE EFFECT A problem in stereo when there is sound coming from the right and left but no sound in the middle. This is also called Ping-Pong effect.

HOLES Positive charge carriers.

HYBRID IC An integrated circuit with tiny discrete components attached to it.

HYSTERESIS CURVE A curve that shows the amount of magnetism produced by a magnetizing force. The area within the curve is directly related to the amount of hysteresis loss. (For illustration, see Symbols section.)

HYSTERESIS LOSS Loss occurring in transformers because the iron core becomes magnetized during each half cycle of current.

IMITATION SYSTEM A form of quad sound in which there is a very short delay of the sound signal before it reaches the rear speakers. Also called a *simulated quad* system.

INCANDESCENT Made so hot that it gives off light.

IN-CIRCUIT CHECKER A transistor checker that determines if the transistor can amplify. It is not necessary to remove the transistor to check it with this instrument.

IN PHASE When two signals both go positive at the same instant and both go negative at the same instant. (For illustration, see Symbols section.)

INTEGRATED CIRCUIT (IC) A circuit that has many transistors, resistors, and capacitors all engraved on a single chip of silicon.

INTERCARRIER TV Television receivers that amplify both the sound and video signals in the IF stage. The sound and video signals are separated at or after the detector.

INTERCOMS Intercommunications systems.

IONOSPHERE Atmospheric layer that reflects radio waves back to earth so they can be used for long-distance reception. Also called the Kennelly-Heaviside layer.

JUNCTION FIELD-EFFECT TRANSISTOR (JFET) A unipolar transistor whose majority charge carriers are either electrons or holes.

KENNELLY-HEAVISIDE LAYER Another name for ionosphere. (See Ionosphere.)

KEYED AGC The same thing as gated AGC. (See Gated AGC.)

KIRCHHOFF'S CURRENT LAW Law which states that, if you call the currents that enter a junction positive and the currents that leave the junction negative, then the sum of the currents at the junction must be zero.

KIRCHHOFF'S VOLTAGE LAW Law which states that, if you call the voltage drops around a circuit negative and if you call the voltage rises (generated voltages) positive, then the sum of the voltages around any closed circuit path must be zero.

LARGE-SCALE INTEGRATION (LSI) An integrated circuit with more than 50 active components.

LEFT-HAND RULE Rule which states that, if you grasp a wire (mentally) with your left hand so that your thumb points in the direction of electron-current flow, then your fingers will circle the wire in the direction of the magnetic field.

LIGHT-EMITTING DIODE (LED) A diode that emits visible light at its junction. (For symbol, see Symbols section.)

LINEAR INTEGRATED CIRCUIT Type of IC that is used where it is desired to amplify signals. The output is always directly related to the input at all times in a linear IC.

LOW In logic circuitry, meaning in the binary 0 or OFF condition.

LOW-PASS FILTER A circuit that rejects all frequencies above the cutoff point.

MAJORITY CHARGE CARRIERS The holes in P-type material and the electrons in N-type material.

MATRIX METHOD A method of quad sound that takes the electrical signals from four microphones and combines them into two signals. Also called *encoded* quadraphonics.

MEDIUM-SCALE INTEGRATION (MSI) An IC with less than 50 active components.

MINORITY CHARGE CARRIERS The electrons in P-type material. In N-type material these are the holes.

MODULE A unit within an assembly which performs a special function.

MONAURAL SOUND Sound coming from only one location.

MONOLITHIC Describing a complete circuit on a single chip of silicon.

N-CHANNEL JFET An FET that uses only electrons for majority charge carriers. (For symbols, see Symbols section.)

NEGATIVE CHARGE CARRIERS Electrons which may be considered to be the carriers of electric current.

NEGATIVE FEEDBACK Feedback which is produced when the output signal is returned to the input in such a way that it subtracts from the output signal. Also called *degenerative* feedback.

NEGATIVE-RESISTANCE REGION Region in the tube-characteristic curve where the voltage across the tube is increased but the plate current decreases.

NPN TRANSISTOR A transistor consisting of three layers of material—two N types and one P type. It is a three-terminal amplifying component. (For symbol, see Symbols section.)

OHM'S LAW A law that gives the relationship between voltage, current, and resistance in a circuit. It states that the amount of circuit current I in amperes equals the voltage E divided by the resistance R. $I = E/R$

OPEN CIRCUIT A circuit that does not have a complete path for current to flow from the generator and back to the generator.

OPERATIONAL AMPLIFIER (Op Amp) A basic amplifier used at one time for math operations in analog computers. It is now being used in many applications. (For symbol, see Symbols section.)

OPTOELECTRONIC COMPONENT A component that (1) gives off light when it is provided with the correct voltages, or (2) the operation of which changes in some way when exposed to light.

OVERLOAD Excessive current flow.

PADDER CAPACITOR A capacitor connected in series with a variable capacitor used to change the capacitance range.

PASSIVE CIRCUIT ELEMENT A component that does not generate a voltage.

PASSIVE TRANSDUCER A transducer in which the energy being sensed produces changes in resistance, capacitance, or inductance. Must always be connected into a circuit that has a source of electrical energy.

P-CHANNEL FET An FET that employs only holes for majority charge carriers. (For symbols, see Symbols section.)

PEAK INVERSE VOLTAGE RATING (PIV) The maximum reverse voltage that can be applied across a diode before a breakdown occurs.

PENTODE A tube with a suppressor grid.

PHANOTRON A gas-filled diode.

PHASE INVERSION Condition in which the output signal is upsidedown, or 180° out of phase, from the input signal. (For illustration, see Symbols section.)

PHOTOCELL An active photoelectric transducer that produces an output voltage which depends upon the amount of light striking it.

PHOTOCONDUCTIVE DIODE A diode that conducts when exposed to light, provided it has a positive voltage on its anode (with respect to its cathode).

PHOTORESISTIVE TRANSDUCER A component in which changes in light energy produce changes in resistance.

PIEZOELECTRIC METHOD Description of certain materials generating a voltage across their surface when they are under pressure or under force.

PING-PONG EFFECT See Hole-in-the-middle effect.

PNP TRANSISTOR A transistor made of three layers of material—two P types and one N type. It is a three-terminal amplifying component. (For symbols, see Symbols section.)

POSITIVE CHARGE CARRIERS Holes.

POSITIVE FEEDBACK A type of amplifier feedback in which part of the output signal is fed to the input in phase. This increases the amplitude of the input signal and increases the amplifier gain. Also called *regenerative* feedback.

POTENTIOMETER A variable resistor connected to control a voltage. Also called a *pot*.

POWER AMPLIFIER An amplifier used for changing a signal voltage into a signal current for operating a transducer.

POWER GAIN The output-signal power divided by the input-signal power.

QUADRAPHONIC SOUND An audio playback system that uses four speakers to provide surrounding sound.
QUIESCENT CURRENT Idling current. The direct current that flows when there is no signal input to the tube, transistor, or FET.
RASTER The lighted rectangular area on a TV screen.
RECTIFIER A diode circuit used for converting the alternating current of the ac power line to a direct current.
REGENERATIVE FEEDBACK See Positive feedback.
REGULATION A measure of how well the power-supply output voltage remains constant under varying load-current values.
RESISTANCE WIRE A wire with a high resistance per centimeter.
RESPONSE CURVE A curve showing gain versus frequency.
RF CARRIER Radio-frequency signal radiated from a transmitting antenna.
RF CHOKE A component used for opposing changes in radio-frequency current.
RHEOSTAT A variable resistor connected into a circuit in such a way that it varies the current.
SECONDARY ELECTRONS Electrons jarred loose from a metal plate as a result of the plate being struck by other high-speed electrons.
SEMICONDUCTOR DIODE A diode which does the same job as a vacuum-tube diode—that is, it permits current to flow in only one direction. Also called a *solid-state* diode.
SERIES-FED OSCILLATOR An oscillator which has some part of its amplifier current flowing through the tuned circuit.
SERIES REGULATOR A power-supply circuit in which the load current flows through the regulator.
SHUNT-FED OSCILLATOR An oscillator circuit designed so that the amplifier current does not flow through the tuned circuit.
SHUNT REGULATOR A power-supply circuit in which the load current does not flow through the regulator.
SIGNAL INJECTION A method of troubleshooting in which the proper signal is injected into the system starting at the output and going toward the input.
SIGNAL INPUT The point where the signal enters the amplifier.
SIGNAL OUTPUT The point where the signal leaves the amplifier.
SIGNAL TRACING A method of troubleshooting by starting at the input and tracing the signal to the output.
SILICON CONTROLLED RECTIFIER (SCR) A fast-acting switch. (For symbol, see Symbols section.)
SOLID-STATE DIODE See Semiconductor diode.
SPACE CHARGE A cloud of electrons emitted by the filament that exists within the tube.
SQUAWKER A speaker designed for reproducing sound in the center range of frequencies.
STEREOPHONIC SOUND An audio system in which the sound is coming from two locations, left and right of the listener.
SYNC PULSE The part of the signal that tells the receiver to start a new line or start a new field.
SYSTEM The complete electronic assembly.

THYRISTOR A semiconductor switching component. Examples are triacs and SCRs.

TRIAC A semiconductor switching component. It performs the same job as an SCR, except that the triac can conduct equally well in two directions. (For symbols, see Symbols section.)

TRIODE The grid audion, which makes it possible for radio receivers to tune in weak signals from stations a great distance away.

TROUBLESHOOTING Locating a fault in a system.

TRUTH TABLE Table that tells all the conditions under which a logic circuit operates.

TUBE TESTER A piece of equipment to test vacuum tubes.

TWEETER A speaker designed for reproducing high-frequency sounds.

TWO-INPUT AND A logic circuit that requires two inputs in order to produce an output. Some AND circuits have five or ten inputs.

VARISTOR A resistor that changes by a great amount when the voltage across it is changed. Also called a *voltage-dependent* resistor, or VDR. (For symbol, see Symbols section.)

VESTIGIAL SIDEBAND Method of television transmission in which one complete sideband and a portion of the other are transmitted. It saves space in the electromagnetic spectrum.

VIDEO Picture.

VIDEO SIGNAL VOLTAGE A voltage that is dependent upon the number of secondary electrons at any instant.

VOLTAGE AMPLIFIER An amplifier used for making a small signal voltage into a large signal voltage.

VOLTAGE GAIN The output-signal voltage divided by the input-signal voltage.

VOLTAGE-DEPENDENT RESISTOR See Varistor.

VOLTAGE-REGULATED SUPPLY A supply that has the same output voltage regardless of the amount of load resistance.

ZENER DIODE A diode that is normally operated with a reverse voltage (anode negative and cathode positive). The voltage across the zener diode is constant. (For symbol, see Symbols section.)

SYMBOLS

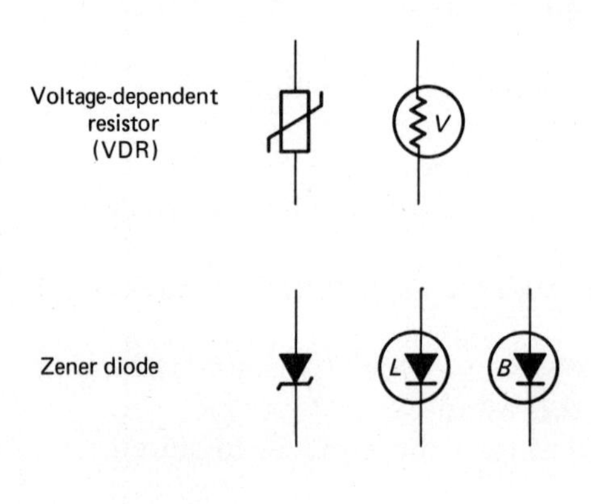

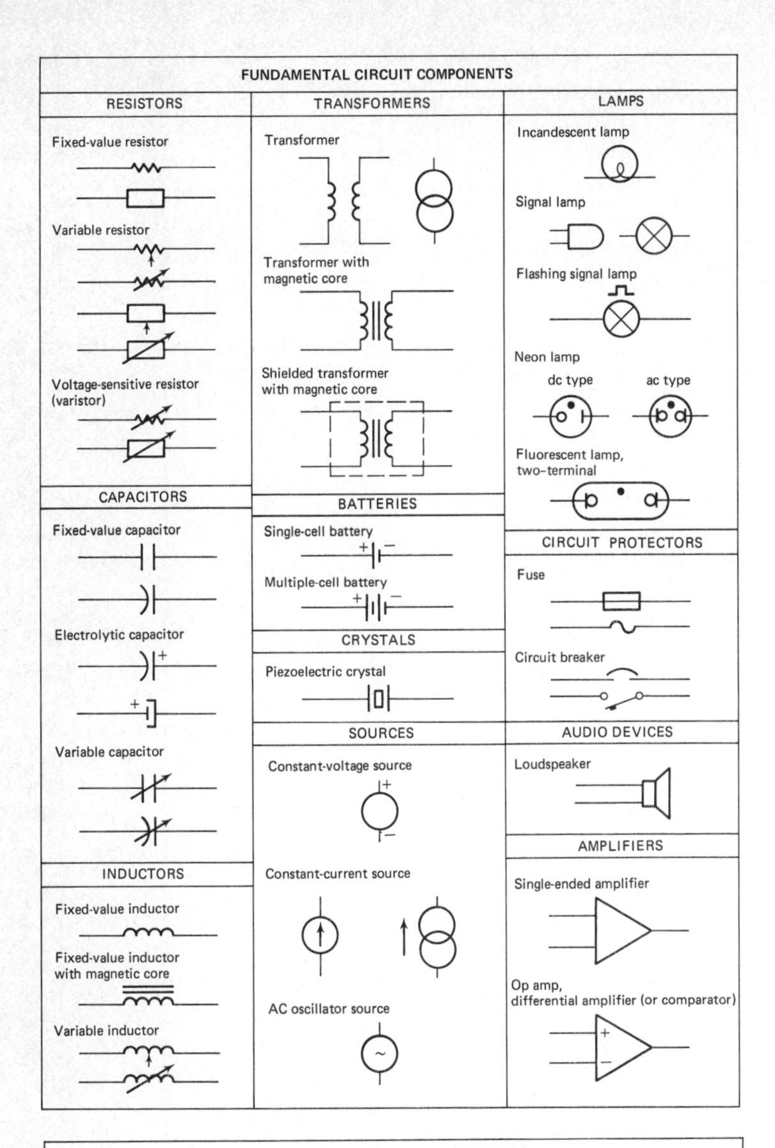
FUNDAMENTAL CIRCUIT COMPONENTS
RESISTORS
Fixed-value resistor
Variable resistor
Voltage-sensitive resistor (varistor)
TRANSFORMERS
Transformer
Transformer with magnetic core
Shielded transformer with magnetic core
LAMPS
Incandescent lamp
Signal lamp
Flashing signal lamp
Neon lamp
dc type
ac type
Fluorescent lamp, two-terminal
CAPACITORS
Fixed-value capacitor
Electrolytic capacitor
Variable capacitor
BATTERIES
Single-cell battery
Multiple-cell battery
CRYSTALS
Piezoelectric crystal
SOURCES
Constant-voltage source
Constant-current source
AC oscillator source
CIRCUIT PROTECTORS
Fuse
Circuit breaker
AUDIO DEVICES
Loudspeaker
AMPLIFIERS
Single-ended amplifier
Op amp, differential amplifier (or comparator)
INDUCTORS
Fixed-value inductor
Fixed-value inductor with magnetic core
Variable inductor

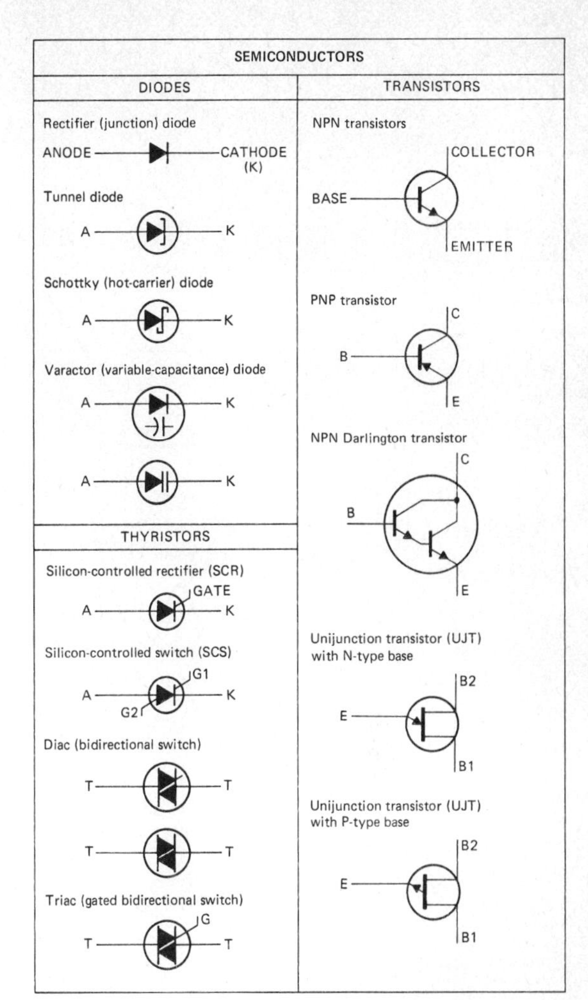
SEMICONDUCTORS
DIODES
Rectifier (junction) diode
ANODE
CATHODE (K)
Tunnel diode
Schottky (hot-carrier) diode
Varactor (variable-capacitance) diode
THYRISTORS
Silicon-controlled rectifier (SCR)
GATE
Silicon-controlled switch (SCS)
G1
G2
Diac (bidirectional switch)
Triac (gated bidirectional switch)
TRANSISTORS
NPN transistors
COLLECTOR
BASE
EMITTER
PNP transistor
NPN Darlington transistor
Unijunction transistor (UJT) with N-type base
B2
B1
Unijunction transistor (UJT) with P-type base

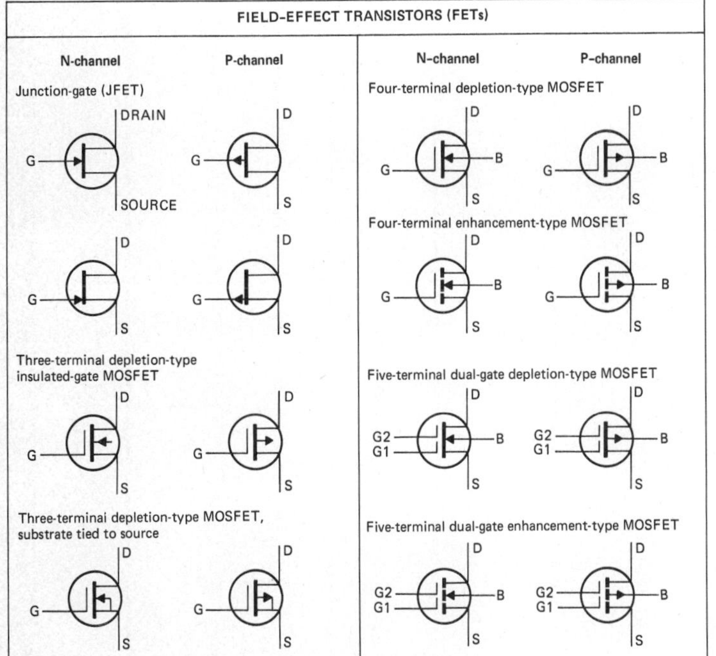
FIELD-EFFECT TRANSISTORS (FETs)
N-channel
P-channel
Junction-gate (JFET)
DRAIN
SOURCE
Three-terminal depletion-type insulated-gate MOSFET
Three-terminal depletion-type MOSFET, substrate tied to source
Four-terminal depletion-type MOSFET
Four-terminal enhancement-type MOSFET
Five-terminal dual-gate depletion-type MOSFET
Five-terminal dual-gate enhancement-type MOSFET

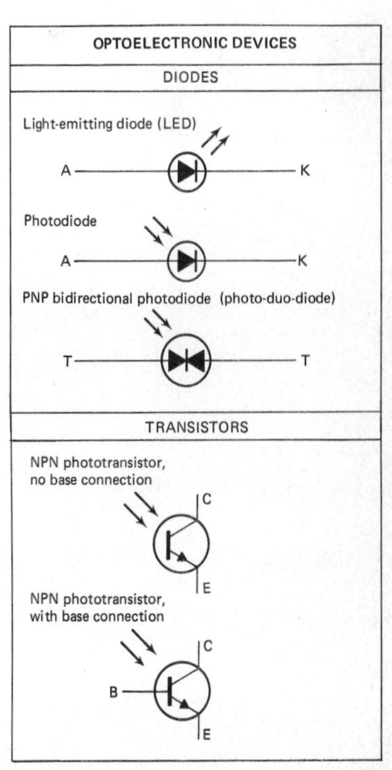
OPTOELECTRONIC DEVICES
DIODES
Light-emitting diode (LED)
Photodiode
PNP bidirectional photodiode (photo-duo-diode)
TRANSISTORS
NPN phototransistor, no base connection
NPN phototransistor, with base connection

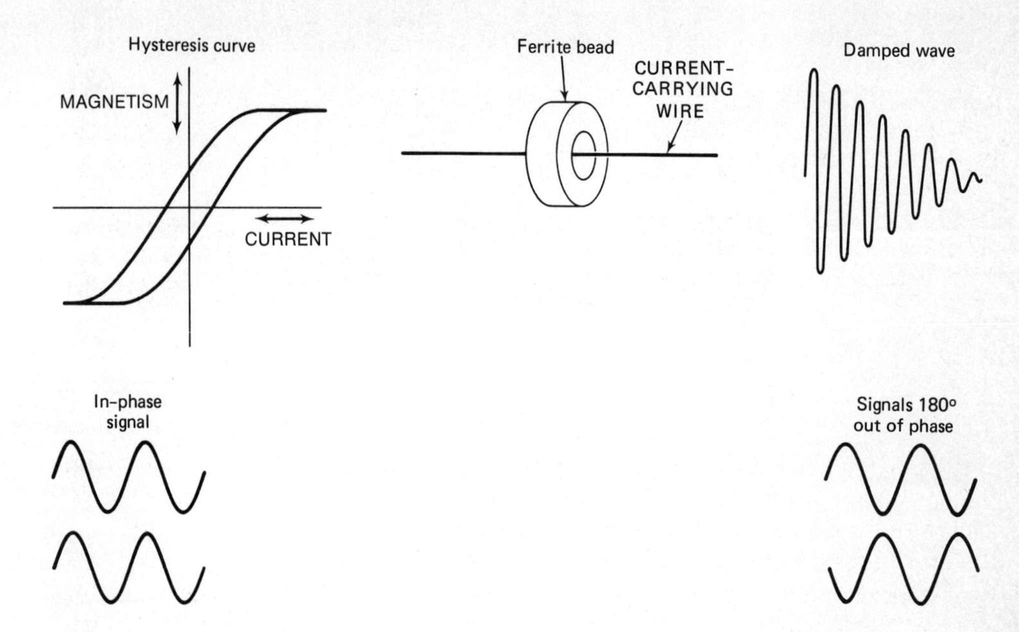
Hysteresis curve
MAGNETISM
CURRENT
Ferrite bead
CURRENT-
CARRYING
WIRE
Damped wave
In-phase
signal
Signals 180°
out of phase

appendix c construction of circuit boards

The circuit board is designed to permit experiments to be added to meet the specific requirements of an instructor. Some of the components, such as transistors and integrated circuits, are provided with sockets to permit the use of newer types and universal replacement types.

The components on the boards are identified with capital letters. This avoids any possible confusion that would be caused by giving them letters and subscripts. For example, electrolytic capacitor G may be called C_1 or C_2 (or C with any subscript) in the experiment schematics. Therefore it cannot be called C_1 on the circuit board.

PARTS LIST

Power-Supply Board

A—Power Transformer

Primary: 117 volts
Secondary: 12.6 volts, 2 amperes, center-tapped

B, C, D, E—Rectifier Diodes

Silicon, 2 amperes, 50 volts PIV

F—Resistor

Wirewound, 5 watts, 100 ohms

G, H—Capacitors

Electrolytic, 250 microfarads, 50 volts

I—Switch

SPST, 2 amperes

J—Busbar

#12 solid copper wire

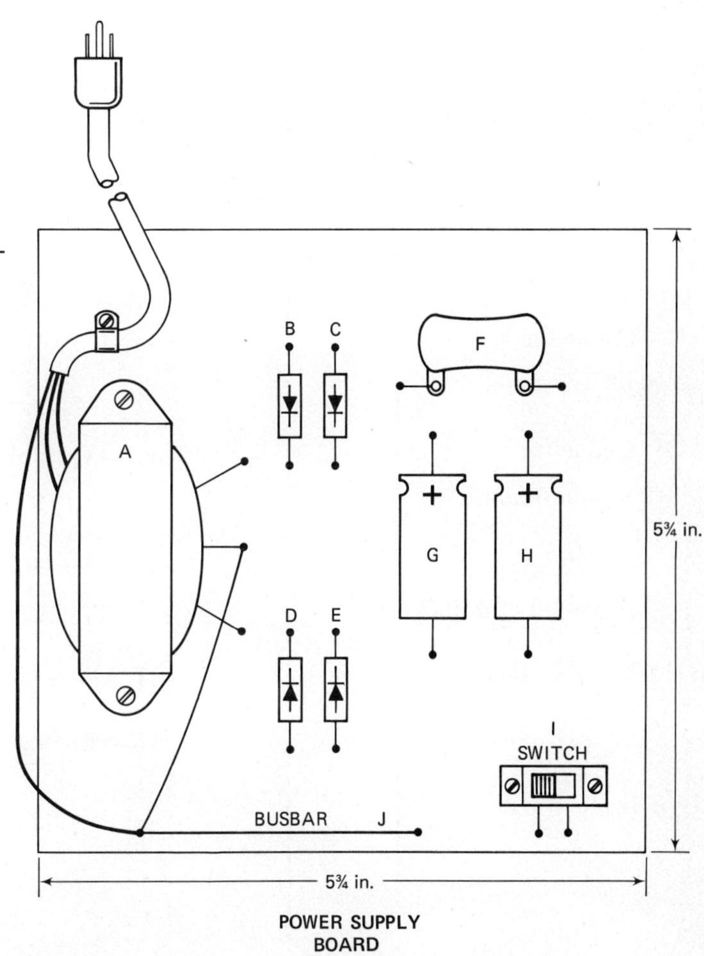

POWER SUPPLY BOARD

Fig. C-1

Component Board #1

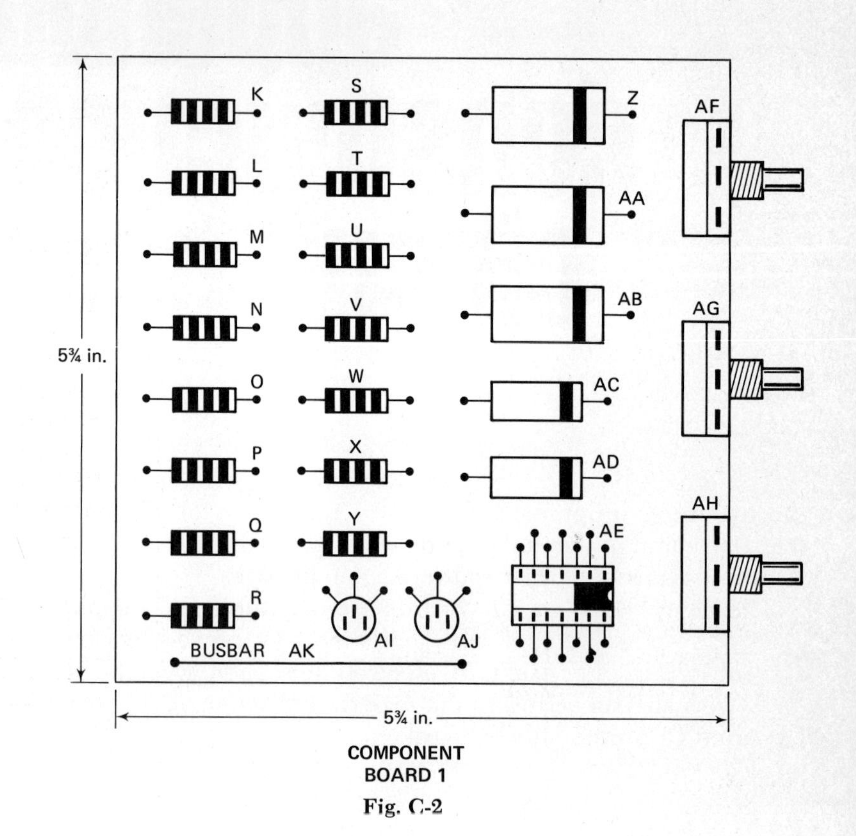

Fig. C-2

K—Resistor

1 watt, 47 ohms

L—Resistor

1 watt, 220 ohms

M—Resistor

1 watt, 330 ohms

N—Resistor

1 watt, 470 ohms

O, P—Resistors

1 watt, 1000 ohms

Q, R—Resistors

1 watt, 3.3 kilohms

S—Resistor

1 watt, 4.7 kilohms

T, U, V, W, X, Y—Resistors

1 watt, 10 kilohms

Z, AA, AB—Capacitors

0.1 microfarad

AC—Capacitor

0.05 microfarad

AD—Capacitor

0.01 microfarad

AE—Socket

Integrated circuit, 14 pin

AF—Variable Resistor

0 to 1 kilohm

AG—Variable Resistor

0 to 10 kilohms

AH—Variable Resistor

0 to 50 kilohms

AI, AJ—Transistor Sockets

AK—Busbar

Component Board #2

AL, AM—Resistors

1 watt, 47 kilohms

AN—Resistor

1 watt, 470 kilohms

AO, AP—Lamps

12.6 volts

AQ, AR—Lamps

6.3 volts

AS—Audio Output Transformer

0.4 watts
Primary impedance: 200 ohms*
Secondary impedance: 3.2 ohms*
Secondary impedance: 8 ohms*

AT—Relay

6-volt dc SPDT

AU—Capacitor

Electrolytic, 1000 microfarads, 50 volts

AV—Speaker

3-inch PM type

AW—Terminals for unmounted components

AX—Push-button switch, normally closed

AY—Push-button switch, normally open

AZ—Busbar

BA, BB—Switches

DPDT

* Depending upon connection made.

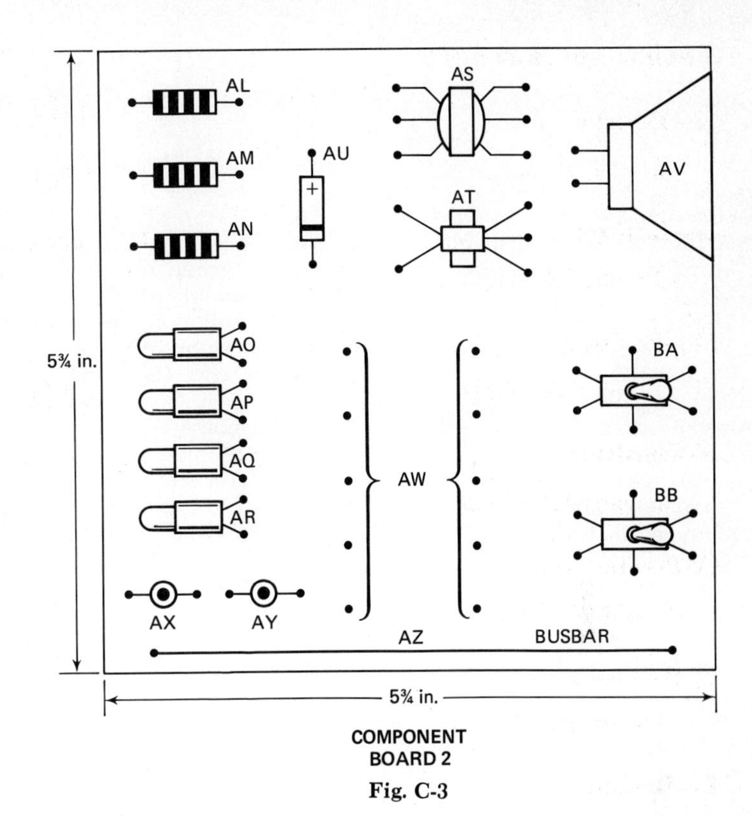

COMPONENT
BOARD 2
Fig. C-3

Additional Parts:

Integrated Circuits

555 timer
741 operational amplifier

Transistors:
2N2219A, or GE 18 (2 required)
2N2905A (2 required)
2N5296 or D44C8 (Power transistors)
Field-effect Transistor (N-channel JFET)
2N4868 or GE-FET-1
Unijunction Transistor 2N2646
Line cord
SCR—C15F or equivalent
LSR—Light-sensitive resistor, cadmium sulfide photo cell, Archer number 276-116 or equivalent

index